Perfect Horoscope (Predictions)

Author: Ram Babu Sao

Shri Ram Babu Sao is meritorious, genius and very intellectual person since childhood and is a Mechanical Engineer (1970) from Patna University. He secured the "National Merit Scholarship" of India during Secondary Board Examination. He has deep interest in Astrology since childhood. He is deeply committed and involved in Astrological research and development work. Sincerity, deep study and his hobby has put him into the use of many new techniques and methods for horoscope prediction. He has studied many books and magazines on astrology. He thought of the necessity of a consolidated Book covering all the aspects, topics on subject matters pertaining to astrology at one place. This Book **"Perfect Horoscope (Predictions)"** is the jest of most of the topics on astrology. This will result in improving the general quality levels of the reader to a greater satisfaction. He has taken a lead in upgrading the process of awareness of various matters benefiting the fresh learner in astrology. Any body can avail of practical knowledge on various topics related to astrology and can make predictions in detail of himself or his family member with the help of this Book.

Contact: Mobile: 9819506068
Email: rbsao8844@yahoo.co.in

@ Copyright: Author-2004

CAUTION

All rights reserved. No part of this book may be reproduced or transmitted in any form or by any means, electronic or mechanical including photocopy without permission in writing from the publishers.

Disclaimer

The book **"Perfect Horoscope (Predictions)"** is not a writer's whole & sole product. It is a combination of the knowledge and expertise of the author and the data collected from different Books, specially researched to meet the objective and to enhance the knowledge on Astrology. Wherever necessary, the reference of the above books has been given. Because the prediction is not based on the contents of the book but on the skills, purity and perfection of the astrologer, any prediction based on this book shall be the responsibility of the person making predictions.

ISBN-13: 978-1537197197

ISBN-10: 1537197193

First Edition: August 2016

Preface

The book, **"Perfect Horoscope (Predictions)"**, is a unique book, which is very informative and also easy to understand. One book is truly the equivalent of several books on astrology. You can make perfect predictions of Horoscope of any member of your family with the help of this single book. This provides some of the elementary and in depth essential elements on complete horoscope predictions, especially wth the planetary combinations (Yoga). Many of the basics on astrology are explained in detail.

Astrology is not knowing your future, but planning your future by averting the misshapenness by action in the right time, wearing Gems, wearing Yantras, chanting Mantras and Prayers. It is important to realize that success comes only with the right actions at the right timing. The whole secret of Astrology is "Right Timing". This Book provides the best times for successful ventures such as starting a business, buying a home, or investing in the Stock Market. Mr. Gate made a fortune using astrology for "Right Timing". This knowledge is made available to you though this book. By using the book, your life will be more prosperous than ever before. It is important to work "Smartly" but not hard. This Book gives you the followings:

1. The prospective tools to make your life more rewarding.
2. Pictures of your career and love life at its ultimate zenith.
3. Guidelines to ever dream of becoming a Star.
4. Discovering your financial fortune in life.
5. Points out the secrets of looking the "Best you can be every day".
6. An opportunity to start out a professional practice and setting your fees.
7. Making the Horoscope Predictions of yourself or any member of your family.

Many people need to know about their future and financial status, most important events in their future and timing when it is going to improve. Money controls the way we live our lives, and this amazing readings will give you the insight, you need to take control of your financial future. It could be one of the most important readings that you will get through this Book that can truly change your life. This Book will isolate time to capture the situation and reveal its significance. Our life is speedy. It is ever active and is changing every moment. Each one of us is facing difficulty at every step. This book will facilitate to reach your destination by moving ahead with ease even in the storming situation. This is so much strife and struggle in the present time as it was never before. This is a time of ready-made food and fast food. Nobody has time to cook the food and then eat. Only this feeling motivated me and necessitated me making this book. It is full of all information on various aspects of predictions including formation of "Yoga" at one place to be referred easily and quickly by anybody whether busy in any profession.

Vedic Astrology is the best in the world and covers almost every aspect of the individual life along with the "planetary combinations" in the horoscope chart and timings of the events to happen in individual life. This contains all the aspect of astrology with illustrations so that complete information is conveyed in a simple language. I am confident that this book will help you in achieving your object and success in every field of astrology.

I have tried to make clear about what is the correct astrology work. These are all correct and true facts & figures collected from various books and incorporated here in a single book for the first time for use by the common men. Behind all this, there is our exhaustive study and collections. More than the study is the presentation of the subject matter and even much more than the presentation of the subject matter is long years of experience and association with the astrology work all over India and abroad. This gives the authenticity to the book. This book is a tool for the Jyotish Students, for the Beginners, for the somewhat Advanced Students and for the Professionals too.

The recognition and importance, which this book has received within a short time, is a positive proof of its efforts and impartiality. Hundreds and thousands of persons from our country and from foreign countries are responding and referring this book. There is a great demand created for this book. This is no credit to us but is the genuineness and absolutely authenticity and clarity of this book. We are confident that the readers and experts will consider my effort.

Table of Content

1.	**Introduction**	**1-16**
1.1	Astrology History	1-2
1.2	Astrological Terms	2-10
1.3	Panchanga (Hindu calendar)	10-12
1.4	Kundali (Chart)	12-15
1.5	Reference Books	15-16
2	**Janma Phala**	**17-30**
2.1	Predictions by Janma Phala	17-30
3.	**Muhurata**	**31-32**
3.1	Muhurata Predictions	31-32
3.1	Muhurata Predictions	31-32
4.	**Nakshatra**	**33-76**
4.1	Nakshatra Predictions	35-76
5.	**House (Bhava)**	**77-84**
5.1	Concept of House of House working	77-78
5.2	House-to-House Relationship	78-80
5.3	Classification of House/Bhava	80-81
5.4	House Signification (Karaktva)	81-84
6.	**Zodiac Sign**	**85-124**
6.1	Sign Characteristics	85-91
6.2	Predictions by Sign Element	91-93
6.3	Predictions by Sign in House	93-124
7.	**Moon Sign (Rashi)**	**125-130**
7.1	Predictions by Moon Sign	125-130
8.	**Sun Sign**	**131-146**
8.1	Predictions by Sun Sign	131-146
9.	**Planet**	**147-256**
9.1	Planets in Transit, Motion, & Retrograde	148-151
9.2	Planet's Characteristics	151-161
9.3	Benefic and Malefic Planet	161-163
9.4	Shani and Divorce	163-170
9.5	Predictions by Cluster of Planets in Houses	179-171
9.6	Predictions by Exalted & Debilitated Planets	171-252
9.7	Predictions by Combust Lords & planets	252-256
10.	**Lords of House**	**257-284**
10.1	Predictions by Lords in 12 House	257-267
10.2	Predictions by Ascendant Lord-Yoga	267-284
11	**Lagna wise Horoscope (Predictions)**	**285-488**
11.1	Aries (Mesha) Lagna	285-299

11.2	Taurus (Vrishabh) Lagna	299-317
11.3	Gemini (Mithuna) Lagna	317-333
1.4	Cancer (Karka) Lagna	333-348
11.5	Leo (Simha) Lagna	348-365
11.6	Virgo (Kanya) Lagna	365-381
11.7	Libra (Tula) Lagna	381-402
11.8	Scorpio (Vrischika) Lagna	402-418
11.9	Sagittarius (Dhanu) Lagna	418-435
11.10	Capricorn (Makara) Lagna	435-452
11.11	Aquarius (Kumbha) Lagna	452-471
11.12	Pisces (Meena) Lagna	471-488
12	**Divisional Charts**	**489-502**
12.1	Predictions by Divisional Charts	489-502
13	**Dasha Period**	**503-566**
13.1	Vimshotari Dasa Systems	503-13.2
13.2	Predictions by Maha Dasha	507-517
13.3	Predictions by Antar Dasha	517-562
13.4	Predictions by Pratyantar Dasha	562-566
14	**Planetary Combinations (Yoga)**	**567-608**
15.1	Planetary Combinations (Yoga)	570-608
16	**Jaimini Astrology**	**609-620**
16.1	Predictions by Jaimini Astrology	608-621
17	**Previous Birth Curse**	**621-622**

1

Introduction

1.1 Astrology History

The growth and achievement in life of a Mankind (Individual) depends on so many factors, such as, 1) Astrological Effects, 2) Genetic Effects, 3) Environmental Effects, and last but not the least, 4) Society & Contacts. Astrological Effects can be found out by "Astrology" or "Jyotish" which means the 'science of light' and is related with the Light and Magnetic Field emitted by the planets (Graha). Hora Astrology deals with the Phalitha Jyotish, which means predictions about individual's life. 'Hindu Astrology' is founded by the Maharashi Aryabhatta, Parasara, Varaha Mihira, Jaimini, Garga, Kalidasa and Kalyan Varma. The astrology fully knows the individual's future as indicated by horoscope but can't certify the same as it is not 100 percent Mathematics or Science".

In classical Jyotish the Moon has equal or greater power to the lagna (Ascendant). Accurate readings cannot occur from the Rashi-Lagna only. Accurate readings require at least two preliminary scans (predictions): first from the Rashi-Lagna, and second from Chandra-Lagna. Then more detailed scans such as from the Mahadasha-Lord and the Karaka are also required. More accurate Jyotish predictions are often produced by reading significations from the Chandra Lagna first, Radical Lagna second, Navamsha Lagna third, and Varga Lagna additionally. Thus it is easier to identify the behaviours and environments that provide the best psycho-emotional support while predictions of a Horoscope. Horoscope is a mirror in which an astrologer can see one's past, present and future. Horoscope is like a snapshot of a particular place in time and space. For casting the natal horoscope of an individual the time of birth, date of birth and place of birth is needed. There are 12 houses and 12 Signs in a horoscope from which an astrologer can predict about various areas of the life of an individual. This **"Perfect Horoscope (Predictions)"** Book enables the astrologer to know that what the future has in store for the native. For the calculation of the timing of various events indicated in the horoscope the knowledge of impact of major period/sub period and transit is used.

1.2 Astrological Terms

Affliction: Affliction of a planet is formed by, (1) its placement in the 6th, 8th or 12th houses or (2) association with the ruler of the 6th, 8th or 12th lords; or (3) association with natural malefic; or (4) association of Badhaka planet; or (5) its Combustion or placement between two natural malefic. Affliction of a planet is such a bad condition in which his energies or powers are considered to be zero and give an adverse effect.

Angle (Kendra): The Ascendant (first house), Descendant (fourth house), Mid-heaven Seventh house) and I Mum Collie (tenth house) are called Kendra (Angle).

Antar Dasa Periods: Antar Dasa is the sub period or shorter period of the Maha Dasa division proportionately with the period of that planet. Total period of the nine Antar Dasa periods is equal to the period of Maha Dasa of planet.

Ascendant (Lagna): The Ascendant or Lagna or Rising Sign is the Sign in which an individual is born.

Aspect (Drishti): When one planet "sees", i.e. influences, another planet (or house) by being in a specific angular relationship with the other planet (or house). Vedic astrologers use Whole Sign aspects, where planets affect every planet in the aspect house, regardless of orb.

Aspects to Houses: All planets aspect the house opposite to the one that they occupy, whether there are any planets in that house or not.

Association: It is a relationship between two or more planets by their position in a house.

Atma Karaka: The Sun is called the natural Atma Karaka. In Jaimini astrology, the planet having the highest longitudinal progression in the Horoscope; irrespective of the sign in which it is placed, is called Atma Karaka planet.

Auspicious Planet: Planet, which gives positive effects and good result to the individual is called auspicious planet.

Barren Signs: Gemini, Leo and Virgo are associated with infertility and are considered Barren Signs. Aries, Sagittarius and Aquarius are considered semi-barren Signs.

Benefic Planet: The benefics in Vedic astrology are Venus, Jupiter, and the Moon, plus Mercury when it is either not influenced by a malefic, or is only influenced by a benefic. However, in the creation of yogas, Mercury is always treated as a benefic.

Malefic Planets: The malefic planets in Vedic astrology are Mars, Saturn, Rahu (the North Node), and Ketu (the South Node), plus Mercury if it is only influenced by other malefic (not counting the Sun), and, for some assessments, the Sun.

Bhava (House): The complete Zodiac is divided into twelve parts for the purpose of complete study of astrology. Each division is called a Bhava (House).

Bhukti: Bhukti is the period, which indicates how many years of Maha Dasa of the planet have passed before the time of birth of the individual.

Birth Time: This is the moment of first breath of a new born individual.

Bright Moon: If the Moon is in the sign opposite to the Sun, or in either of the two adjacent signs, then it is considered bright. E.g. if the Sun is in Gemini, the Moon is bright in Scorpio, Sagittarius, the opposite sign to the Sun, and Capricorn.

Chart: It is a figure or sketch consisting of 12 houses, in which the position of the planets and the Signs are given.

Combustion (Astangatha/Vikala): The planet associated with an identical longitude or equal longitude or in Conjunction to the Sun within a certain orb or reaching nearer to the Sun by the distance mentioned in the degree, such as, Moon-12^0, Mars-17^0, Mercury-13^0, Jupiter-11^0, Venus-9^0, and Saturn-15^0 is called Combust is called Combustion. The energy of that planet is considered burnt by Sun and the planet does not operate independently any more. It is true that a combust planet is week. When a true planet is within 6 degrees of the Sun, it is said to be combust. If within 3 degrees, it is seriously combust. A combust planet is weaker and has difficulty expressing its gifts or power, and the themes of the houses it rules can become challenging or not much effective to the native.

Conjunction: Two planets situated together in a house or occupying position close to each other within a certain orb or reaching nearer are called in Conjunction.

Constellation: It is a set of stars or a group of fixed stars and is called Nakshatra.

Dasa: Every human being is scheduled to experience each of the nine planet's major specified period against the total of 120 years in his life time. It is called Dasa or Maha Dasa. Dasa are the most important tool for finding out timing of events in individual life.

Debilitated: The sign a planet is weakest in. This is called Fall in Western astrology, i.e. Sun in Libra, Moon in Scorpio, Mercury in Pisces, Venus in Virgo, Mars in Cancer, Jupiter in Capricorn, and

Saturn in Aries get debilitated. The degree of deepest debilitation, after which strength begins to return to the planet, is, respectively, 10, 3, 15, 27, 28, 5, 20, e.g. Jupiter is weakest at 5 degrees Capricorn. Thus the debilitated planet in loose his power and gets weakened

Dark Moon: When the Moon is within 72 degrees of the Sun, it grows dark and loses strength. A waning Moon is significantly weaker than a waxing Moon (given their distances from the Sun being equal.

Debilitation (Khala): When a planet occupies its Sign of Fall for Neecha Bhanga, the planet is in Debilitation and is considered weak or Neecha.

Decanate/Drekana: When a sign is divided into three parts, each of 10 degrees, each part is called 1st Drekana, 2nd Drekana and 3rd Drekana of that sign respectively.

Defeated Planet: When a planet is in association with an enemy planet within a certain degrees or less than specify, then the weakest planet is said to be a defeated planet.

Degrees of Maximum Exaltation & Debilitation: Maximum degree of Exaltation or Debilitation is the defined degree of planet position in a Sign.

Derivative House: This is a house related to another house in the Chart, which signify the events of individual. Example: The 3rd house is the house of the brothers and sisters and the fourth house is the mother's house. Accordingly, the third house from the fourth house, i. e. the 7th house in a natal chart will describe the signification of the brothers and sisters of the native mother.

Dig Bala: Aka directional strength. Each planet has its favorite house (akin to the Traditional Western notion of Joy). When in that house, the planet gains both strength and stability. The houses in which planets gain dig bala in are: Jupiter and Mercury in the 1st, the Moon and Venus in the 4th, Saturn in the 7th, and the Sun and Mars in the 10th.

Dignity: A planet is dignified when it occupies its Own Sign, its Moolatrikona or its Exaltation Sign, aspect by a benefic planet without any aspect or affliction by a malefic, when it is not retrograde or when it is increasing in its light. Such planet gives good results.

Direct Motion: It is a motion of the planet that follows the natural order of revolution cycle. The letter "D" marked against a planet indicates the direct motion of planet.

Dispositor: The lord of the sign is called Dispositor of a planet positioned in that Sign. The Dispositor of a planet is the soul essence of that planet, like prime minister. It dictates the planet to reacts in the way he likes. Example: Venus is the Dispositor of Saturn as because of ruler ship of Libra and Saturn has to act as per choice of Venus.

Dusthanas: They are also called Trikasthanas or Trik houses, these are the "bad" houses. Houses 6, 8, and 12 govern many of the unpleasant themes of life (e.g. debt, obstacles, loss, death, disease, anxiety, enemies, etc.), and therefore planets ruling these houses embody challenging themes, as do planets occupying these houses.

Element: Twelve Signs have been divided into four groups of each three Signs having same nature, called Elements. These Elements are called Fire, Earth, Air or Water.

Exaltation: The planet in a particular Sign is called the planet in Exaltation Sign. Planet is dignified during his Exaltation.

Face: The "faces" arise from a subdivision of zodiac Sign into six equal parts. This is an obsolete term meaning the division of each sign into six equal parts of 5° each. The faces derive from the ancient Egyptian decants.

Friendly Planet: The planets of one group are called friendly planets.

Functional benefic Planets: As per Lagna Sign of the Natal Chart, some planets are defined as Functional benefic Planets even though they are Natural Malefic. Their influence is thought to be positive or constructive for the native. It makes the house strong in which he is sitting. The planet gives good result during his Main Dasa and the Antar Dasa fouling together in the same period.

Functional Malefic Planets: As per Lagna Sign of the Natal Chart, some planets are defined as Functional Malefic Planets whose influence is thought to be negative or destructive for the native. The lord of the 6th, 8th, and 12th are called Functional Malefic (Inauspicious) Planets and will make the house weak in which he is sitting. The planet does not give the good result during his Main Dasa and the Antar Dasa fouling together in the same period.

Ghati: The Ghati is a measure of the Time. One day = 24 hours = 60 Ghati. 2 ½ Ghati =1 hour; one Vighati = 24 seconds; and 60 Vighati = 1 Ghati.

Gochara (Transit): The revolution or usual movement of planets in the Zodiac Sign during the course of revolution around the Sun is known as Gochara (Transit).

Graha Yudha (planet war): Whenever one planet comes within 5^0 orb of another one among the planets Mars, Mercury, Jupiter, Venus and Saturn, as viewed from the Earth, it causes planetary war, or Graha Yudha. One of the two planets involved in this war is said to be vanquished and another is a victor. The victorious planet produces powerful auspicious effects, while the vanquished or defeated one becomes inauspicious. The house in which this phenomenon occurs is destroyed and the individual suffers throughout his life with respect to that house events.

Hora: Each zodiacal sign is divided into two parts of 15 degree each and is called Hora.

Horizon: It is the visible juncture of Earth and the sky and represented in a horoscope.

Horoscope: It is the Janma Kundali, which depicts the positions of different signs and planets at the time of birth. It also represents the Rising Sign at the place of birth and the location of planets in various signs.

Increasing light (waxing): When Moon moves from the position of New Moon to full Moon, the shining portion of Moon grows larger and is called Increasing or Waxing Moon.

Inimical: It means unfriendly and enemy planet or house.

Ketu (Dragons Tail/South Node of Moon): Ketu is the dead body of the lusty demon, killed by Vishnu. So, Ketu is the tail of the dragon and called the South Node of moon, which losses and symbolizes the death along with Saturn.

Latitude: This is the celestial angular distance measured north or south of the plane of the ecliptic. This is the distance of a planet from the Equator.

Longitude: It is the distance in degrees or in arc on Earth from 0° Aries eastward to any given point that intersects the ecliptic, such as, 10° Taurus is expressed as longitude 40°.

Moolatrikona: The position of the planets in the particular Sign is considered the Moolatrikona of planet. Planet is dignified during his Moolatrikona. The Moolatrikona sign is usually a part of its own house with the exception of the Moon, where it behaves almost as favourably and gets more strength.

Moon sign: The Moon sign is the zone of the zodiac in which the Moon is positioned when a person is born. This is also called Rashi too.

Muhurata: This is an auspicious moment for starting any enterprise, marriage or journey. One Muhurat is equals 48 minutes.

Mutual Aspects: In Vedic astrology, when planets occupy the same house or are in opposite houses, they are said to aspect each other, I.e. to see and influence and impact each other, mutually. In classical Jyotish texts, planets in the same house are said to be in association with each other.

Nakshatra: The Nakshatra ruled by the Moon at the birth time of a native is called Janma Nakshatra.

Natal Chart: The horoscope cast at the birth time of the individual, showing the position of Signs and Planet with respect to house, is called a Natal Chart or Nativity.
Native: It refers to a person (male or female) for whom a horoscope is cast and studied.
Natural Benefic: Moon (waxing), Mercury, Venus, Ketu and Jupiter are called Natural Benefic Planets in order of increasing Benefic.
Natural Malefic: Sun, Mars, Saturn and Rahu are called Natural Malefic Planets in order of increasing malefic. Moon remains a malefic from 9th day of Krishna Paksha to 7th lunar day of Shukla Paksha.
Navamsa Chart: It is a nine Divisional Chart of a Sign with 3° 20' segments each.
New Moon: It is the beginning of a lunation cycle and of the waxing phase.
Opposite: It is point at a distance of 180° in the Zodiac or the 7th house apart.
Orb: This is the degrees of Longitude in the Zodiac.
Own house: A house ruled by a planet is known as own house.
Pad: Each Nakshatra is divided into four divisions or Parts. Each Part is called Pad of Nakshatra. Accordingly, there are 27 x 4 = 108 Pad comprising the whole of the zodiac.
Paksha: The 15 days period during which the Moon goes on in its orbit from the Full Moon to Zero or vice versa is called Paksha. The first fortnight of the month is the period during which Moon is waxing and is called Shukla Paksha. The second fortnight during which the Moon is waning is called Krishna Paksha.
Planet: Sun, Moon, Mars, Mercury, Jupiter, Venus, Saturn, Rahu, Ketu, Uranus, Neptune and Pluto are the twelve heavenly bodies which appear to move in the Zodiac and influence the human body and are called Planets.
Planetary War (Graha Yudha): Except Sun, if any planet is in association
Planetary war: When a true planet is within 1 degree of another true planet, a planetary war takes place. It is as if they are fighting for territory, and even the victor of the war gets bloodied. The victor can be the planet having higher latitude and/or brightness. Thus the weaker planet in war gets weakness. Planetary wars never involve the Sun, Moon, Rahu, or Ketu.
Prediction: Knowing about natives past, present and future with the help of horoscope is prediction. But, in one accident many lives are taken, does that indicate similarity in everyone's horoscope? No, because the horoscope of the place or a vehicle in which passengers

are travelling, supersedes the horoscopes of the natives. Hence, in a calamity everyone's horoscope does not necessarily indicate death. However, this point needs further research and views from the readers

Rahu (The Dragon Head/ North Node of Moon): Rahu is the dead body of the lusty demon, killed by Vishnu. So, Rahu is the head of the dragon and symbolizes the trouble. He is also called the North Node of Moon.

Rashi: Janma Rashi is the Sign occupied by Moon at the time of birth of the native.

Retrograde motion (Vakra/Saktha): When observed from Earth, it appears the apparent backward motion of a planet or moving in reverse direction than its natural direction of travel and is called Retrograde Motion.

Retrograde: Vedic astrology recognize planet in retrograde as the movement of a planet in the opposite direction across the sky, Vedic astrologers utilize the fact that retrograde planets are at their brightest and largest (closest to the Earth), and that therefore, retrograde planets are strong.

Rising Planet: The planets which are positioned in the Rising Sign or Ascendant in the natal chart are Rising Planets.

Rising Sign (Lagna): The earth is rotating once a day around its axis. One of the 12 zodiacal signs is entering the 1st house every two hours. The Sign entering into 1st house at the birth time is known as Rising Sign or Lagna.

Ruler Ship: Vedic astrology employs sign ruler ships. Mars rules Aries and Scorpio, Venus rules Taurus and Libra, Mercury rules Gemini and Virgo, the Moon rules Cancer, the Sun rules Leo, Jupiter rules Sagittarius and Pisces, and Saturn rules Capricorn and Aquarius. Planets are strong and stable in the signs they rule.

Exalted: The sign opposite to the debilitation sign is the sign of Exaltation and a planet is strongest in this sign. Sun in Aries, Moon in Taurus, Mercury in Virgo, Venus in Pisces, Mars in Capricorn, Jupiter in Cancer, and Saturn in Libra. The degree of deepest exaltation is the same as the degree of deepest debilitation, just in the opposite sign. E.g. Saturn is strongest at 20 degrees of Libra.

Ruling Planet: The planet which rules the Ascendant or Lagna Sign is called the Ruling Planet.

Signs: The zodiac is divided into twelve parts like Aries, Taurus, Gemini, Cancer, Leo, Virgo, Libra, Scorpio, Sagittarius, Capricorn, Aquarius, Pisces and are called Sign.

Sun Sign: In astrology, Sun Sign is defined a period of one month with respect to the person Birth Date and is called Sun Sign. Generally, Sun Sign is 2nd house from the position of Sun in Hindu Astrology.

Temporal Benefics and Malefics: They are also called Functional Benefics or Malefics. When a planet only rules bad houses it becomes a temporal malefic, bringing challenge into the person's life. If a planet only rules good houses (especially Kendra or trikonas), it brings positive experiences into the person's life. (Note: for assessing a planet's temporal nature, houses 2 and 11 are considered mildly well, and house 3 as mildly bad. So Jupiter ruling the 8th and 11th would still be considered a temporal malefic, and Mars ruling the 3rd and 10th a temporal benefic.)

Trikona (Trine): The Houses 1, 5 and 9 are called Trikona or Trine.

Trikonas: Also called Trines, houses 1, 5, and 9 are the most fortunate houses, in that they represent blessings, opportunities, good fortune, wisdom, and spirituality, and planets in them and ruling them bring positive themes to the life.

Kendras: The angles or angular houses in the West, such as, houses 1, 4, 7, and 10 are Kendras and boost the power, action, and initiative of planets, and therefore can support greater success.

Lagna: The Ascendant is the Rising sign or the First House itself, as in the phrase "a planet in the Lagna". The owner of the Ascendant and the planets in the First House both represent the person more than any other planets. Planets aspecting either the ruler of the First House or the First House itself also are quite influential, and their influence is integrated into the person's makeup.

Chandra Lagna: A chart constructed by rotating the horoscope until the Moon occupies the First House. This is also used to calculate yogas, and a yoga that occurs in both the Lagna chart and the Chandra Lagna chart will be much more powerful.

Surya Lagna: A chart constructed by rotating the horoscope until the Sun occupies the First House. This is also used to calculate yogas, and a yoga that occurs in both the Lagna chart and the Surya Lagna chart will be more powerful.

True Planet: Mercury, Venus, Mars, Jupiter, and Saturn are classified as true planets for determining planetary war and combustion and in Yogas.

Ugra (Fierce): The nature of some of Nakshatra is fierce as per astrology and is called Ugra Nakshatra. Hitler Nakshatra was Ugra (Fierce).

Unfriendly house (Deena): The house with the lord ship of enemy group of planet with respect to the planet position in that house is defined as the Unfriendly House.

Upachaya: The 3rd, 6th, 10th and 11th houses from the Lagna are called the Upachaya.

Vishnu House: The houses 1st, 5th & 9th stand for Vishnu and hence called Vishnu House.

Waning: The phase of the Moon during which the visible portion of the Moon decreases, is called Waning.

Waxing: The phase of the Moon during which the visible portion of Moon grows larger, is called Waxing.

Whole Sign Houses: Unlike modern Western astrology, Vedic astrology uses whole sign houses, i.e. where each house is one whole sign, i.e. begins at zero degrees and ends at thirty degrees of the sign, and every planet in one sign occupies the same house as the others in the same sign.

Winning Planet: When a planet is in association with an enemy planet and its degrees are more than that of enemy planet, then that planet is said to be a winning planet.

Yoga: A specific configuration defined by one or more planets in specific signs, houses, or relationship with other planets. While there are many challenging yogas, Vedic astrologers usually focus upon the yogas that uplift ones life. Therefore, in this article and most others, unless otherwise specified, the word yoga implies a positive yoga.

Trikas (Badhakasthana): The Houses 6, 8 and 12 are called Trikas or Badhakasthana. These are the Badhaka houses and considered the evil houses of suffering.

Zodiac: It is literally the circle of stars. Zodiac is defined a band of the heaven approximately 14° wide, centred on the Ecliptic, against which the Sun and other planets are seen to move, as seen from the Earth.

1.3 Panchanga (Hindu calendar)

The Panchanga is the "Lunar Calendar" based on Moon's monthly movement. One month is divided into two Paksha, such as, Krishna Paksha and Shukla Paksha. On the average year as 29.53 x 12 = 354.36 days, which is less by 11 days compared to earth's cycle days (365 days)? Thus, in every three years there is an Adhika Maah to cover the gap of these 11 days. Panchanga is used to determine the most ideal or auspicious time for carrying out various activities like getting married, stepping into a new home, starting education, laying the foundations of new homes and buildings work for the first time. It gives the exact time when a particular task can be undertaken to reap maximum benefits. It provides a list of festivals. Panchanga means five organs to understand the Phalita. These five things are (1) Vara/Din (Day); (2) Tithi (Date), (3) Nakshatra (Group of Stars), (4) Yoga and (5) Karana.

i) Vara (Day): The day starts from the sunrise and continue till next sunrise. It is the solar day of the week. There are 7 days in a weak.

ii) Tithi (Date): There are 15 Tithis, commencing from next day of Amavasya. The 1st Tithi is the 1st day of bright half of a lunar month to Purnima (Full moon), the 15th Tithi of the bright half. They are called the Tithi of the Shukla-Paksha (brighter phase). Similarly, there are 15 Tithi commencing from next day of Purnima to Amavasya called Krishna Paksha (darker phase). The 15 Tithi are called Pratipada (Prathama), Dwitiya, Tritiya, Chaturthi, Panchami, Shasthi, Saptami, Ashtami, Navami, Dasami, Ekadasi, Dwadashi, Trayodashi, Chaturdasi, and Purnima or Amavasya.

iii) Nakshatra (Naal): There are 27 Nakshatra (group of stars) that Moon occupies. It is also called the constellation.

iv) Yoga: Yoga is an auspicious moment. There are 27 Yoga in a month. They are called 1) Vishakumbha, 2) Preeti, 3) Aayushman, 4) Saubhagya, 5) Shobhana, 6) Atiganda, 7) Sukarma, 8) Dhriti, 9) Shoola, 10) Gand, 11) Vriddhi, 12) Dhruva, 13) Vyaghaata, 14) Harshana, 15) Vajra, 16) Siddhi, 17) Vyatipaata, 18) Variyaana, 19) Parigha, 20) Shiva, 21) Siddha, 22) Saddhya, 23) Shubha, 24) Shukla, 25) Brahma, 26) Indra, and 27) Vaidhriti.

v) Karana: There are one Karana in a one and a half of the Tithi a (Day). There are 11 Karana altogether, which repeats. They are: 1) Bala, 2) Baalava, 3) Kaulava, 4) Taitil, 5) Gara, 6) Vanija, 7) Vishti, 8) Shakuni, 9) Chatushpada, 10) Naaga, 11) Kinstughna.

Hindu calendar (Maah): The Hindu calendar has twelve months (Maah) in a lunar year corresponded to the following English Calendar Months. The solar month is given against Maah in which they begin and ends. Each Maah starting and the end date are as follows:

Chaitra 30 days (March 22 – April 20); Vaisakha 31 days (April 21 – May 21); Jyaistha 31 days (May 22 - June 21); Asadha 31 days (June 22 – July 22); Sravana 31 days (July 23 – August 22); Bhadra 31 days (August 23 – September 22); Asvina 30 days (September 23 – October 22); Kartika 30 days (October 23 – November 21); Agrahayana 30 days (November 22 – December 21); Pausa 30 days (December 22 – January 20); Magha 30 days (January 21- February 19) and Phalgun 30 days (February 20 – March 21).

1.4 Kundali (Chart)

A) South Indian style Kundali (Chart): In South Indian Style Chart, the position of the signs is always fixed and the position of the Ascendant is always changing. The houses are counted in a clockwise direction. The upper top left but one Rectangular box, being denoted by the digit 1 is always Aries. The next Rectangular box right to it, being denoted by the digit 2, is always Taurus and so on as written in the Chart. The digit 1 through 12 indicates the position of the Signs fixed in clockwise direction. It is always fixed in the South Indian style of Chart. The Ascendant (Lagna) falls in one of the Sign depending on rising Sign and is called Lagna or Ascendant of the Chart. The counting of the Houses is always done in clockwise direction from the Ascendant (Lagna) as first house through twelfth house. The planets occupy the House according to their longitudinal position.

12 (Pieces) Venus	1 (Aries)	2 (Taurus)	3 (Gemini) Ascendant Mars
11 (Aquarius) Mercury Ketu			4 (Cancer) Moon
10 (Capricorn) Sun	Lagna Chart-1		5 (Leo) Rahu
9 (Sagittarius) Jupiter Saturn	8 (Scorpio)	7 (Libra)	6 (Virgo)

FIG 1: South Indian Style Lagna Chart

C) East Indian (Bengali) style Kundali (Chart): In the East Indian (Bengali) style Chart, the Houses are always fixed and the rising Sign falls in the 1st House, which is called Lagna or Ascendant, which is at the top in the centre. The Lagna (Ascendant) as well as the other houses are always fixed and are counted in anti-clockwise direction from the 1st House or the Lagna or the Rising Sign. The planets occupy the House according to their longitudinal position. The numbers shown in this format tell us which sign is in the Lagna and other Houses as shown in the Chart. In Bengali style zodiac is again fixed and ascendant & planets move anti clock wise along the zodiac unlike South Indian System. In western style the ascendant is fixed again and placed on the left hand side whereas

D) East Indian (Bengali) & Western style Kundali (Chart): It is a circular Chart divided in twelve parts in which Sign and Planets position are given as shown below.

Classification Charts: The different type of Horoscope Charts are prepared to study the different aspects of life, such as Natal Chart; Chalit Chart; Transit Chart; Moon Chart; Tithi Parivesha Chart and Divisisional Chart

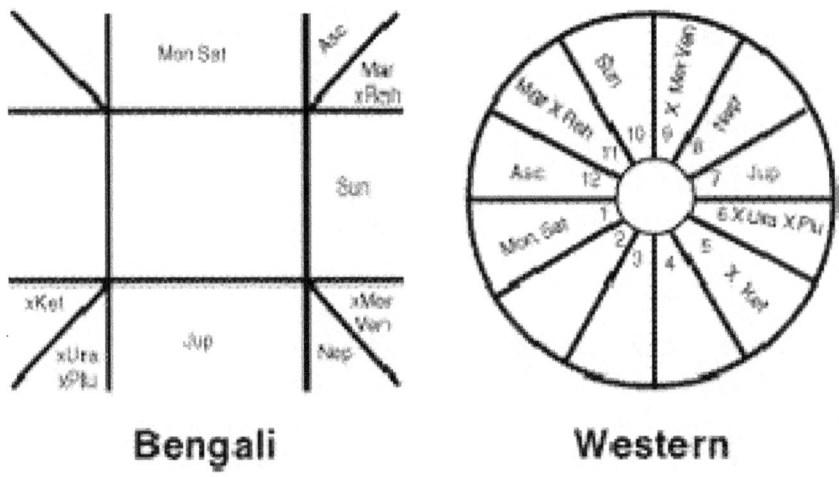

FIG 3: East Indian (Bengali) & Western Country Style Chart

B) North Indian style Kundali (Chart): In the North Indian style Chart, the Houses are always fixed and the rising Sign falls in the 1st House, which is called Lagna or Ascendant, which is at the top in the centre. The Lagna (Ascendant) as well as the other houses are always fixed and are counted in anti-clockwise direction from the 1st House or the Lagna or the Rising Sign. The planets occupy the House according to their longitudinal position. The numbers shown in this format tell us which sign is in the Lagna and other Houses as shown in the Chart.

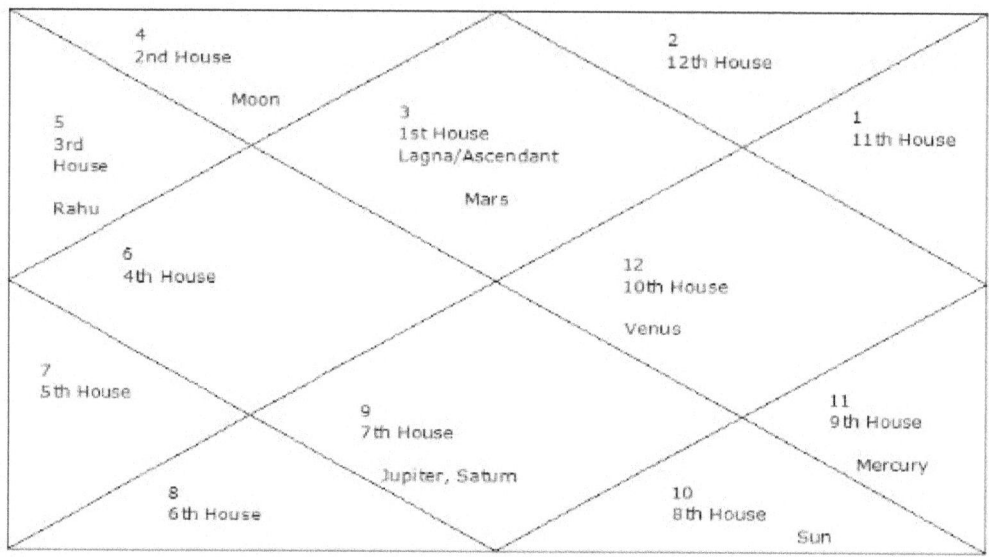

FIG 2: North Indian Style Chart

Natal Chart: This is the Main Lagna-Chart.

Chalit Chart: Lagna Chart shows the house only approximately with a full Zodiac Sign in it. Chalit Chart is shows the actual Zodiac Sign in a house. Hence Chalit chart is very important to indicate the actual planet position in a particular house and cannot be ignored in astrology. But Chalit Chart shall not be used for determining the aspects or knowing the sign in which a planet is posited or knowing the strength of a planet. For example, if a Sun moves from lagna to 12th house in Chalit Chart while it was in Aries in Lagna chart, then it does not mean that Sun has changed the Sign and shifted to Pisces. It remains in Aries only, but is placed in 12th house because of the actual division of house calculation. Chalit Chart is the actual calculation of the 12 houses of the Lagna Chart and therefore, actual position of planet by making the house division for corrects predictions, because the planets show their behavior according to the house they occupy.

Transit Chart: The Transit Chart or Progression Chart is prepared by positioning the rising Sign in 1st House and planets in the particular fouling Sign at that particular time. Then, this Chat is compared to find the aspect of transiting planet on Natal chart planet to find daily, weakly or monthly forecast of the horoscope when the native wish to find the best time for important event happening such as a change of job, the best time for marriage or having children etc. In short, the Transit Charts are a study aid for planetary influences in the individual life during a particular period.

Moon-Sign (Rashi) Chart: In Moon-Sign (Rashi) Chart, the Sign with Moon placed in the first House as Ascendant and rests other Signs with their Planets follow the Ascendant.

Sun-Sign Chart: In Sun-Sign Chart, the Sun Sign according to Month ofSun-Sign is placed in the first House as Ascendant and rests other Signs with their Planets follow the Ascendant.

Tithi Parivesha Chart: The Tithi Parivesha Chart manifests concretely the Vimshotari Maha Dasa and Antar Dasa sequence. In this chart, the lord of the Vimshotari Maha Dasa is placed in the first House as Ascendant. In relation to this, the inner workings of the Antar Dasa are revealed with the help of the Antar Dasa lord planet positioned in different Houses. The Antar Dasa Lord brings the timing of the events that come about as a result in life. Tithi Parivesha Chart distils all this along with the transit of the Moon in the Rashi Chakra (Rashi-Chart).

Divisional Charts: In Divisional Charts, each Sign is divided into various divisions, as per the requirements of the type of the Chart. The Divisional Charts are used for study of (1) strength of the planets, (2) important aspects of life. Each Chart gives a clear history of one of aspects of life of the native, such as (1) The physique of the native is known by Lagna Chart, (2) wealth by Hora Chart, (3) happiness through co-born/sibling by Drekana Chart, (4) fortunes from Chaturthamsa Chart, (5) sons and grandsons from Saptamsa Chart, (6) spouse and planet strength from Navamsa Chart, (7) power and position from Dasamsa Chart, (8) parents from Dvadasamsa Chart, (9) pleasure and adversities through conveyances from Shodasamsa, (10) worship from Vimsamsa, (11) learning from Chaturvimshamsa, (12) strength and weakness from Saptavimshamsa, (13) evil effects from Trimsamsa, (14) auspicious and inauspicious effects from Khavedamsa, (15) all indications from Akshavedamsa and (16) Shashtiamsa charts.

1.5 Reference Books

1. Hora, Ganita and Samhita: The great sage Parasara narrated the science of astrology as heard through Lord Brahma in three divisions, viz. Hora, Ganita and Samhita.

2. The Brihat Parasara Hora Shashtra: The great sage Parasara lived at the time of the Mahabharata war, about 3000 BC. The Brihat

Parasara Hora Shashtra (a compendium on astrology) is the primary textbook of Vedic astrology.

3. Saravali Translated by R. Santhanam: This focuses on planets in houses and Decanate.

4. Jaimini Sutras: The Jaimini sutras by Rishi Jaimini are a unique classic, and considered as next only to the Brihat Parasara Hora Shashtra.

5. Bhrigu Samhita: Rishi Bhrigu was the first compiler of predictive Astrology. His famous compilation, Bhrigu Samhita, which contains the predictions for thousands of combinations, is popular even today.

6. Bhrigu Sutras: Rishi Bhrigu was the first compiler of predictive Astrology. His famous compilation, Bhrigu Samhita, which contains the predictions for thousands of combinations, is popular even today.

7. Prasana Tantra: By Neelakanta Daivagyna around 1550 AD, is a great classic dealing with the Prasana or Horary Astrology and a must for any astrologer.

8. Brihat Parasara Hora Shashtra Translated by R. Santhanam: The "bible" of Vedic astrology.

2

Janma Phala

There are the Panchanga - Janma Phala and is given as below:

1. Samvatsara Janma Phala	9. Ayan Janma Phala
2. Ritu Janma Phala	10. Maasa Janma Phala
3. Shukla Paksha Janma Phala	11. Krishna Paksha Janma Phala
4. Ratri Janma Phala	12. Tithi Janma Phala
5. Din Janma Phala	13. Karan Janma Phala
6. Yoga Janma Phala	14. Yoni Janma Phala
7. Gana Janma Phala	
8. Lagna Janma Phala	

2.1 Predictions by Samvatsara Janma Phala

Samvatsara: It is a lunar year which starts from the first day of the bright half of Chaitra month (i. e. the Chaitra Shukla Pratipada to Phalgun Krishna Amavasya). Samvatsara is a measure of time taken by Jupiter in traversing on its average speed in zodiacal Sign. In one Samvatsara, there are 361.02672 days. The most prevalent Samvatsara is "Saka Samvat". Samvatsara are grouped in a cycle of 60 years and is known as the Jovian Cycle. The names of different Samvatsara in the cycle are given in the Table below.

Method of calculation of Samvatsara: The methods for identifying the name of the Samvatsara in which a person is born are given below:

Multiply the "Saka Samvat" by 22 and then add 4291 to the product and divide the result by 1875. Add the quotient to the given Samvatsara, neglecting the remainder. Again divide the total by 60. Add one to the remainder, neglecting the quotient; we get the Samvatsara position in the cyclic order.

Example: A native is born on 20.12.1902. So, the given "Saka Samvat" is 1902.

Calculation: So, 1902 x 22 = 41844; Then, 41844 + 4291 = 46135; Then, divide the product by 1875 = 46135/1875 = 24.6053. We get 24 as quotient and 1135 as remainder. Disregard the remainder. Now, add the Samvatsara to the quotient = 24 + 1902 = 1926; Then, divide the sum by 60 = 1926/60 = 32.1. Quotient is disregarded, and the remainder is 6. Add 1 to 6 = 6 + 1 = 7. So, the 7th Samvatsara in the cyclic order as per the table shows "Shrimukh".

Table 1: List of the 60-year Cycle of Samvatsara

1. Prabhava	16. Chirabhanu	31. Hemlambi	46. Paridhavi
2. Vibhava	17. Subhanu	32. Vilambi	47. Pramadi
3. Shukla	18. Tarana	33. Vikari	48. Ananda
4. Pramoda	19. Parthiva	34. Sarvari	49. Rakshas
5. Prajapati	20. Vyaya	35. Plava	50. Nala
6. Angira	21. Sarvajit	36. Shubhkrita	51. Pingala
7. Shrimukh	22. Sarvadhari	37. Shobhakrita	52. Kalayukta
8. Bhava	23. Virodhi	38. Krodhi	53. Siddharti
9. Yuva	24. Vikriti	39. Vishwavasu	54. Raudra
10. Dhata	25. Khara	40. Paraabhava	55. Durmati
11. Ishwara	26. Nandana	41. Plavang	56. Dundubhi
12. Bahudhanya	27. Vijay	42. Kilak	57. Rudhirodgari
13. Pramathi	28. Jaya	43. Saumya	58. Raktakshi
14. Vikram	29. Manmatha	44. Sadharana	59. Krodhan
15. Vrisha	30. Durmukha	45. Virodhakrit	60. Kshaya

Predictions by sixty Samvatsara: Samvatsara is arranged alphabetically so that it can be located easily.
Ananda: He/She has more than one wife, clever, efficient in his occupation, has obedient sons, a sense of indebtedness and is charitable. **Angeera:** He/She is endowed with beauty and happiness. He/She is fortunate, friendly, has many sons, conceals his thoughts and enjoys a long life. **Bahudhaanya:** He/She is clever in business and trade, receives favour from the powerful people, and has knowledge of scriptures, charitable and rich. **Vibhava:** He/She is generous at heart, famous, has tremendous qualities, and is charitable, jolly and lovable. **Chitrabhanu:** He/She loves a variety of clothes and fragrances, fulfils his desires and is kind. **Dhatri:** He/She has many good qualities, respects his teachers, expert in arts and crafts, has knowledge of the scriptures and is kind in nature. **Dundubhi:** He/She gets honour and respect from authorities, has all

the comforts in life, and owns house, vehicle and loves fine arts. **Durmati:** He/She believes that whatever he does is right; others are wrong. Hence, he remains most of the time unhappy, engaged in irreligious activities and unintelligent. **Durmukha:** He/She is cruel, mean-minded, and greedy and has bad habits and practices. **Hemalamba:** He/She possesses great looks and vehicles, is rich, has a happy and contented family life, and is an accumulator of things. **Ishvara:** He/She is efficient, gentle, and happy but loses his temper quickly. **Jaya:** He/She gets respect and honour from the common people, is destroyer of his enemies, has expertise in the knowledge of scriptures and is attracted to sensual pleasures. **Kaalayukta:** He/She is a useless teacher, is wickedly intelligent, unfortunate and quarrelsome. **Keelak Sanvatsar:** He/She has mediocre appearance, talks sweetly, is kind and has well developed forehead. **Khara:** He/She is lustful, has a filthy body, has a rough nature, talks unnecessarily and is shameless. **Krodhana:** He/She obstructs other's work and causes confusion. **Krodhi:** He/She is of cruel nature, loved by women and obstructs other's work. **Kshaya:** He/She is a squanderer of accumulated wealth, believes in serving others, and has a strong heart and little inclination towards virtuous actions. **Manmaatha:** He/She is lover of ornaments, sensualist, and speaker of truth, loves songs and dance and is in pursuit of worldly pleasures. **Nala:** He/She is intelligent, strikes riches in the business of aquatic things, clever in farming work and commerce, is a supporter, patron of people and has meagre wealth. **Nandana:** He/She does social services like digging ponds and wells, is charitable, pious and has a happy family. **Paarthiva:** He/She is religious minded, has expertise in the scriptures and arts and runs after worldly pleasures. **Paraabhava:** He/She is an accumulator of little to moderate wealth, talks bitterly about others and has no manners. **Paridhavee:** He/She is a learned man with a kind heart, has a command over the arts, is intelligent and achieves success in business. **Pingala:** He/She has yellowish eyes, commits mean deeds, has unsteady wealth, charitable but talks bitterly. **Plava:** He/She is lusty, conceals his thoughts and restless. **Plavanga:** He/She is restless, cunning, inclined to bad actions and has a weak body. **Prabhava:** He/She accumulates everything, enjoys every worldly pleasure, and is gifted with children and intelligence. A person born in this Sanvatsara possesses many things. He enjoys worldly pleasures, has children and is intelligent. **Prajapati:** He/She is kind, a carrier of his family's traditions, humble, and worshipper of deities and Brahman. **Pramaathee:** He/She possesses royal signs, vehicles, has knowledge of scriptures, a defeater of his enemies, and is a politician.

Pramadi: He/She is wicked, proud, quarrels a lot, greedy, poor, possesses little intelligence and performs mean deeds. **Pramoda:** He/She is generous and charitable, endowed with beauty, believer of truth, has tremendous qualities, has a helpful nature and efficient in all faculties and is proud. **Rakshasa:** He/She is cruel, commits bad and destructive deeds, quarrelsome, greedy, is poor, has little intelligence and is mean. **Raktakshi:** He/She has a natural tendency towards virtuous actions and religious activities, lusty, is jealous of other's success and suffers from diseases. **Raudra:** He/She has a horrible appearance, an animal rarer, criticizes others. **Rudhirodgaari:** He/She possesses red eyes, a weak body, frequently suffering from jaundice, excessive anger, disfigured nails and faces the danger of an attack by weapons. **Sadharana:** He/She likes travelling, is efficient in writing, discriminative, has excess of anger, is pious and is free from sensual pleasures. **Sarvadhaari:** He/She has many servants in his service, is always in the pursuit of pleasures, possesses beauty, loves sweets, and is patient and honourable. **Sarvajeeta:** He/She gets respect from the powerful people, enthusiastic, pious, has a huge body and is a defeater of his enemy. **Saumya:** He/She is learned, rich, pious, has reverence towards the deities and is hospitable to guests, but possesses a weak body. **Shaarvari:** He/She is expert in buying and selling, is repelled by those friends who are in pursuit of pleasures and is studious. **Shobhakrita:** He/She is well advanced in every faculty, possesses an auspicious quality, is kind, endowed with beauty and is efficient. **Shreemukha:** He/She has knowledge of the sacred books and scriptures, intelligent, powerful and famous. **Shubhakrita:** He/She is very fortunate, humble, possesses knowledge, enjoys a long life on account of his pure and pious activities and is rich. **Shukla:** He/She is happy, simpleton, blessed with the pleasure of having a nice wife and sons, rich, fortunate, a learned man and humble. **Siddharthee:** He/She is generous, jolly, a winner, beautiful, respectable, rich and capable. **Subhanu:** He/She has curly hair, beauty, is kind-hearted, defeats his enemies and is rich. **Taaran:** He/She is cunning, brave but restless, has knowledge of arts, cruel, does deplorable acts, enjoys pleasures but is poor. **Vibhava:** He/She enjoys every worldly pleasure, is beautiful or handsome, an artist, the superior person in his family and clan, a learned man and humble. **Vijaya:** He/She is brave, gentle, an orator, kind hearted, charitable and defeater of his enemies. **Vikaari:** He/She is full of prejudice but he is also a master of arts, restless, cunning, and very talkative and does not trust his friends. **Vikrama:** He/She is of violent nature and accomplisher of difficult tasks is brave. **Vikriti:**

He/She is poor, horrible physical appearance, lacks intelligence and cunning. **Vilamba Naam:** He/She is cunning, greedy, lazy phlegmatic, weak and a believer in destiny. **Virodhakrita:** He/She is short tempered and disobeys his father. **Virodhee:** He/She is a lecturer, lives in foreign countries far away from his family, quarrels with his friends. **Vishvavasu:** He/She is blessed with a beautiful spouse and obedient sons, is generous, patient, loves sweetmeats and possesses many qualities. **Vrisha:** He/She blows his own trumpet, possesses deplorable traits, keeps company of bad and mean people, and is dirty and lazy. **Vyaha:** He/She has happiness, comfort and is an addict, a debtor, restless spends recklessly. **Yuva:** He/She is tall, of serious nature, believes in charity, intelligent, contented and lives a long life.

2.2 Predictions by Janma Ritu (Season) Phala

Spring Season (September, October): He/She is endowed with beauty and intelligence, is famous, expert in music and arithmetic, and follows the teachings of scriptures. **Summer Season (May, June):** He/She is a lecturer, is sensual and also interested in aquatics and is blessed with opulent wealth and fortune and has long hair. **Rainy Season (July, August):** He/She is a winner, intelligent, is famous and powerful, endowed with beauty and loves to possess vehicles. **Autumn Season (January, February):** He/She has an excess of wind element in his body, is rich, ready to perform his duties all the time. Pious and possesses vehicles. **Hemant Season (March, April):** He/She is clever, generous, and full of qualities, religious minded and a philosopher. **Winter Season (November, December):** He/She is a lover of sweetmeats, has excess of appetite, has a happy family life, performs virtuous actions and is handsome and beautiful but short tempered, anger and powerful.

2.3 Predictions by Janma in Shukla Paksha Phala

He/She is always happy and joyous, enjoys long life, humble and has a happy family life, a tender body and passes his time happily and without any tension.

2.4 Predictions by Janma in Krishna Paksha

He/She is brilliant, resembles his father, possesses wealth and honour, has beautiful eyes, receives love and affection from the king, is respected by friends and is rich.

2.5 Predictions by Ratri Samaya Janma Phala

He/She has poor eyesight, lusty, suffers due to diseases, is cruel and commits sinful actions without letting anybody know about them.

2.6 Predictions by Day Janma Phala

Sunday: He/She is brave, with little hair, a winner, enthusiastic, brilliant, has excessive element of bile in his body and is dark in colour. **Monday:** He/She is a scholar, possesses tranquillity, talks sweetly, well-behaved and has discriminative intelligence. **Tuesday:** He/She loves battles and wars, earns his livelihood with the help of land, is inclined towards virtuous actions and has a sharp nature. **Wednesday:** He is blessed with beauty, talks sweetly, possesses wealth, and is expert in art forms, well-behaved in his conducts. **Thursday:** He/She is a scholar, possesses numerous qualities, wealth, is endowed with beauty, acquires powers (Siddhis) and is loved by his elders. **Friday:** He/She has curly hair, is jolly, intelligent, loves clothes of white and light colours and inclined towards virtuous path. **Saturday:** He/She gets affected by premature aging, possesses weak and a feeble body.

2.7 Predictions by Yoga Phala

Atiganda Yoga Phala: The Yoga is danger or obstacles and provides difficult life due to numerous obstacles and accidents; revengeful and angry. He/She will be very arrogant, is afflicted by pain in his throat, cunning, a quarrelsome and a hypocrite. This Yoga provides danger or obstacles and provides difficult life due to numerous obstacles and accidents; revengeful and angry. **Ayushman Yoga Phala:** The Yoga is long-lived and gives good health and longevity, energy. He/She will like to travel, endeavour to search for the different possibilities of earning his livelihood, and enjoy a long life. This Yoga provides long-lived and gives good health and longevity, energy. **Brahma Yoga Phala:** The Yoga is priest, God, trustworthy and confidential, ambitious, good discernment and judgement. He/She will be very studious, a follower of virtuous path, respectable, has tranquillity and peace of mind, is generous.

Brahma: This Yoga provides priest, God, trustworthy and confidential, ambitious, good discernment and judgement. **Dhriti Yoga Phala:** The Yoga is determination and enjoys the wealth, goods and spouses of others; indulges in the hospitality of others. He/She has knowledge, loves to live happily, humble with patience and follows routine life. This Yoga provides determination and enjoys the wealth, goods and spouses of others; indulges in the hospitality of others. **Dhruva Yoga Phala:** The Yoga is constant and steady character, able to concentrate and persist, wealthy. He/She will be blessed by goddess Lakshmi as well as by Saraswati, has knowledge and unblemished fame. This Yoga provides constant and steady character, able to concentrate and persist, wealthy. **Ganda Yoga Phala:** The Yoga is danger or obstacles, flawed morals or ethics, troublesome personality. He/She will be not helping, has rough and hard nature and has excessive anger. This Yoga provides danger or obstacles, flawed morals or ethics, troublesome personality. **Harshana Yoga Phala:** The Yoga is thrilling, intelligent, delights in merriment and humour. He/She has a delicate and tender body, interested in the study of sacred books, loves ornaments and an annihilator of his enemies. This Yoga provides thrilling, intelligent, delights in merriment and humour. **Indra Yoga Phala:** The Yoga is chief and has interest in education and knowledge; helpful, well-off. He/She will be powerful, radiant, has excess of phlegm in his body and the best person in his family. This Yoga provides chief and has interest in education and knowledge; helpful, well-off. **Parigha Yoga Phala:** The Yoga is obstruction and encounters many obstacles to progress in life; irritable and meddlesome. He/She gives false witness, bails out many people from the prison, has little hunger and defeats his enemy. This Yoga provides obstruction and encounters many obstacles to progress in life; irritable and meddlesome. **Preeti Yoga Phala:** The Yoga has fondness and is well-liked, attracted to the opposite sex, enjoys life with contentment. He/She will be an excellent orator, is blessed with beauty, charitable, is of jolly nature and is a virtuous man feels happy seeing others happy. This Yoga provides fondness and is well-liked, attracted to the opposite sex, enjoys life with contentment. **Sadhya Yoga Phala:** The Yoga is amenable, well behaved, accomplished manners and etiquette. He/She will be humble, clever, jolly, efficient in his occupation, always a winner, and attains the blessings of his deity with the powers of mantras. This Yoga provides amenable, well behaved, accomplished manners and etiquette. **Saubhagya Yoga Phala:** The Yoga gives good fortune and enjoys a comfortable life full of opportunities, happy.

He/She will be rich, has tremendous knowledge, follows virtuous paths and is fortunate but proud. The Yoga gives good fortune and enjoys a comfortable life full of opportunities, happy. **Siddha Yoga Phala: Siddha:** The Yoga is accomplished, accommodating personality, pleasant nature, interest in ritual and spirituality. He/She speaks truth, has his sense organs under control, efficient in every faculty and an accomplisher of many tasks. This Yoga provides accomplished, accommodating personality, pleasant nature, interest in ritual and spirituality. **Siddhi Yoga Phala:** The Yoga gives success and is skilful and accomplished in several areas; protector and supporter of others. He/She will be of generous heart, clever, has a sweet and soft behaviour, believes in sacred texts, is fortunate and has the knowledge of the self. The Yoga gives success and is skilful and accomplished in several areas; protector and supporter of others. **Shiva Yoga Phala:** The Yoga is auspicious and is honoured by superiors and government, placid, learned and religious, wealthy. He/She has knowledge of the sacred hymns, has his sense organs under control, blessed with beauty, is fortunate and is blessed by the Lord Shiva. This Yoga provides auspicious and is honoured by superiors and government, placid, learned and religious, wealthy. **Shobhana Yoga Phala:** The Yoga is splendour, lustrous, demeanour, and sensualist and obsessed with sex. He/She will be endowed with beauty, has presence of mind, is intelligent and always on the path of virtuosity. This Yoga provides splendour, lustrous, demeanour, and sensualist and obsessed with sex. **Shoola Yoga Phala:** The Yoga is spear and Painful, confrontational and contrary, quarrelsome, angry. He/She will be poor, suffers form diseases, does evil deeds, an idiot and suffers from colic in his stomach or rheumatic pain. This Yoga provides spear and painful, confrontational and contrary, quarrelsome, angry. **Shubha Yoga Phala:** The Yoga is auspicious, lustrous, personality, wealthy, irritable, but gives problems with health. He/She will be virtuous, soft-spoken and a preacher of virtuosity. Shukla Yoga Phala: He/She speaks truth, has full control over his sense organs, has power and strength, and is an efficient debater and a winner in the battle. This Yoga provides auspicious, lustrous, personality, wealthy, irritable, but gives problems with health. **Sukarma Yoga Phala:** The Yoga is Virtuous and performs noble deeds, magnanimous and charitable, wealthy. He/She will be of jolly nature, Master of Arts, brave, an enthusiast, a benefactor and follows the path of virtuosity. This Yoga provides virtuous and performs noble deeds, magnanimous and charitable, wealthy. **Shukla Yoga Phala:** The Yoga is bright White, garrulous and flighty, impatient and impulsive; unsteady and changeable mind. This Yoga provides bright White, garrulous and flighty, impatient and impulsive; unsteady and changeable mind. **Vaidriti Yoga Phala:** The Yoga has poor support and is critical, scheming nature; powerful and overwhelming mentally or physically. He/She will be of restless nature, condemns order, is cruel, a believer in the slaying of the sacred books yet possesses dirty thoughts and heart. The Yoga has poor support and is critical, scheming nature; powerful and overwhelming mentally or physically. **Vajra Yoga Phala:** The Yoga is diamond, Thunderbolt, well-off, lecherous, unpredictable, and forceful. He/She will be intelligent, powerful speaks the truth, has the ability to judge the authenticity of the jewels, and loves ornaments. This Yoga provides diamond, thunderbolt, well-off, lecherous, unpredictable, and forceful. **Vridhi Yoga Phala:** The Yoga has growth and is intelligent, opportunistic and discerning, life constantly improves with age. He/She will be an accumulator of things, makes profit in his business, and is clever and fortunate. This Yoga provides growth and is intelligent, opportunistic and discerning, life constantly improves with age. **Variyana Yoga Phala:** The Yoga is gives comfort, loves, ease, and luxury, lazy, lascivious. **Vishakumbha:** The Yoga is supported and prevails over others, victorious over enemies, obtains property, wealthy. He/She will be a sensualist, humble, spends free-handily in spite of possessing little wealth and is inclined towards virtuosity. The Yoga is gives comfort, loves, ease, and luxury, lazy, lascivious. **Vishakumbha Yoga Phala:** He/She has a happy and satisfied family life, is of independent nature and cares for his physical appearance. This Yoga provides supported and prevails over others, victorious over enemies, obtains property, wealthy. **Vyagata Yoga Phala:** The Yoga is beating, cruel, intent on harming others. He/She will be cruel, a liar is abusive and calumniator and is of violent nature. This Yoga provides beating, cruel, intent on harming others. **Vyatipata Yoga Phala:** The Yoga gives calamity and is prone to sudden mishaps and reversals, fickle and unreliable.

He/She obeys his parents, is inflicted with venereal diseases, rough and hard natured and tries to obstruct others' efforts. The Yoga gives calamity and is prone to sudden mishaps and reversals, fickle and unreliable.

2.8 Predictions by Gana Phala

Deva Gana Phala: He/She has a pleasant voice, a kind heart, eats little, appreciates others' qualities and is wealthy. 2) **Manushya Gana Phala:** He/She is religious, proud, wealthy, kind-hearted, proficient in arts, radiant and disciplined. 3) **Rakshasa Gana Phala:** He/She is very talkative, patient and quarrelsome.

2.9 Predictions by Janma Lagna Phala

Mesh Lagna (Aries Ascendant): He/She is proud, wealthy, virtuous, having excessive anger, valiant but is still dependent on others. **Vrishabh Lagna (Taurus Ascendant):** He/She is a pleasant talker, is a scholar and loves all. **Mithuna Lagna (Gemini Ascendant):** He/She is proud, loves his friends, charitable, seeks sensual pleasures, and is wealthy. He is lusty, annihilator of his enemies, and progresses slowly in life. **Karka Lagna (Cancer Ascendant):** He/She is religious but also seeks sensual pleasures, loves sweets, handsome and admires virtuous people. **Simha Lagna (Leo Ascendant):** He/She is a sensualist, an annihilator of his enemies, eats little, has few children, is an enthusiast and shows bravery and valance in the battlefield. **Kanya Lagna (Virgo Ascendant):** He/She is endowed with beauty, has knowledge of the scriptures, has a good fortune, but is lusty. **Tula Lagna (Libra Ascendant):** He/She is a scholar, earns his livelihood by virtuous means, proficient in arts, wealthy and is respected by everybody. **Vrischika Lagna (Scorpio Ascendant):** He/She is wealthy and a scholar. **Dhanu Lagna (Sagittarius Ascendant):** He/She is an expert in policy matters, religious, intelligent and an important person in his family. **Makara Lagna (Capricorn Ascendant):** He/She is inclined towards evil deeds, has many children, and is greedy, lazy but hard working. **Kumbha Lagna (Aquarius Ascendant):** He/She has a steady mind, is promiscuous, slow at work but leads a happy and contented life. **Meena Lagna (Pisces Ascendant):** He/She is wealthy and has a frail body.

2.10 Predictions by Ayan Janma Phala

Uttarayana (northward movement): The Sun passes through the signs of Capricorn, Aquarius, Pisces, Aries, Taurus and Gemini, it is called Uttarayana. Uttarayana Janma Phala: He/She is of jolly nature, has a happy family life, lives long, has inclination towards virtuous actions, is generous and has tremendous patience.
Dhakshinayana (southward movement): The Sun passes through the signs of Cancer, Leo, Virgo, Libra, Scorpio and Sagittarius, it is called Dhakshinayana. Dhakshinayana Janma Phala: He/She is a

man of glory and distinction, is engaged in farming works, rears animals and has an unbearable nature.

2.11 Predictions by Maasa (Month) Janma Phala

Chaitra Maasa (later part of March till 20th of April): He/She performs virtuous actions, is a scholar, humble, interested in sensual pleasures, loves sweet things, loves friends, is a state minister and discriminative. **Vaishakha Maasa (later part of April till 20th of May):** He/She possesses the best characteristics, is virtuous, humble, powerful, religious, but lusty and enjoys a long life. **Jyestha Maasa (later part of May till 20th of June):** He/She is of 'Forget and forgive type of person, restless, lives abroad, are brilliant, intelligent and a procrastinator. **Ashada Maasa (later part of June till 20th of July):** He/She wastes his time in useless talks, is careless and negligent, respects his teacher, and has less appetite, engaged in virtuous actions but very proud. **Shravana (later part of July till 20th of August):** He/She obeys his father's command, enjoys happy and satisfied family life, and is famous and full of qualities. **Bhadra Maasa (later part of August till 20th of September):** He/She is rich, possesses a weak body, enjoys a happy family life, and is unaffected by joy and sorrow. **Ashwini Maasa (later part of September till 20th of October):** He/She is a scholar, possesses wealth, performs virtuous actions, accepts the good qualities of others, and possesses wealth and vehicles. **Kartika Maasa (later part of October till 20th of November):** He/She performs virtuous deeds, very talkative, rich, brave, lusty, and carries out the business of buying and selling and is very famous. **Magha Maasa (later part of November till 20th of December):** He/She loves pilgrimage, is humble, has expertise in art forms, and is a sensualist, benevolent, always on the path of virtuousness and rich. **Pushya Maasa (later part of December till 20th of January):** He/She is benevolent but not so fortunate to possess ancestral wealth, spends wealth acquired after great hardship, has the ability to conceal his thoughts, is a practitioner of the scriptures and has a weak body. **Magha Maasa (later part of January till 20th of February):** He/She is a practitioner of Yoga, has the knowledge of sacred hymns (mantras), loves to meet religious people and defeats his enemies using his wit. **Phalgun Maasa (later part of February till 20th of March):** He/She is benevolent, efficient in his occupation, kind hearted, powerful, possesses a delicate and tender body, a flirt and talks illogically. **Adhika Maasa (20th of March**

till end of March): He/She is born in this month, he is uninterested in the sensual pleasures, has inclination towards virtuosity, loves performing pilgrimages, and is free of diseases and loved by all.

2.12 Predictions by Tithi (Date) Janma Phala

Pratipada/Prathama (1st day of the lunar month): He/She has a large family, is a scholar, full of wisdom and discrimination. Has fair complexion, possesses wealth in the form of gems and jeweller, is humble, handsome and receives wealth and favour from the king. **Dwitiya (2nd day of the lunar month):** He/She is charitable, kind hearted, discriminative, jolly and virtuous. **Tritiya (3rd day of the lunar month):** He/She is lusty, scholar, powerful, receives benefits from the king or an authority, and resides in foreign countries, clever, sensualist and proudly. **Chaturthi: (4th day of the lunar month):** He/She is a debtor, brave, efficient in the battles, narrow hearted, a gambler and restless. **Panchami (5th day the lunar month):** He/She has healthy body, has a happy family life, kind towards animals, receives benefits from the king and is charitable. **Shasthi (6th day the lunar month):** He/She is a follower of truth, possesses wealth and a happy family, very brave and clever. **Saptami (7th day of the lunar month):** He/She lives on others' wealth, is a destroyer of his enemies, has tremendous knowledge, appreciates others' good qualities, has large eyes, respects sages and is religious, begets a daughter. **Ashtami (8th day the lunar month):** He/She is fortunate in having tremendous wealth and sons, fortunate in receiving education with the help of the king and is in restless nature. **Navami (9th day the lunar month):** He/She is disinterested in the affairs of friends, talks bitterly, opposes learned people and is without any manners. **Dasami (10th day the lunar month):** He/She is religious, possesses wealth, and has knowledge of many scriptures, generous, humble, handsome and lusty. **Ekadasi (11th day of the lunar month):** He/She is a devotee of the deities, and respects the Brahmins, is charitable, pious and virtuous soul and is always jolly. **Dwadashi (12th day the lunar month):** He/She loves water, well behaved in his conducts, loves to live in the house constructed by him, gives alms to the poor and receives benefits from the king. **Trayodashi (13th day the lunar month):** He/She is bestowed with beauty, has inclination towards virtuosity, has a long neck, many children, and is brave and clever. **Chaturdasi (14th day the lunar month):** He/She is cruel, brave, and clever, loves humour, lusty, intolerant, and angry and opposes other people. **Purnima: (Full Moon day):** He/She is blessed with beauty, acquires wealth by just means, sensualist, kind hearted and possesses many qualities. Amavasya (Dark Moon day): He/She is tranquil, devotee of his father, thinker, suffers severe hardships but acquires wealth ultimately, sensualist and with great qualities.

2.13 Predictions by Karana Janma Phala

Balava Karana: He/She is brave, sensualist, poet, charitable and intelligent. **Bala Karana:** He/She is lusty, kind hearted, endowed with beauty and has a pleasant nature, is a scholar, fortunate and is wealthy. **Chatushpada Karana:** He/She lacks virtuosity, incapable of accumulating wealth, has a frail and weak body, and is engaged in rearing animals. **Gara Karana:** He/She is benevolent, respectable, discriminative, and clever, defeats his enemies, generous and is blessed with beauty. **Kaulava Karana:** He/She is loved by everybody, independent and a soft talker. **Kinstughna Karana:** He/She treats his friends and enemies alike and is full of lust and weak. He person born in this Karana is of evil nature, restless and unsteady, wicked, has excessive anger, and spoils his family's reputation. **Shakuni Karana:** He/She is efficient in knowledge of scriptures, ever alert, fortunate and has knowledge of the auspicious moments. **Taitil Karana:** He/She is tender and delicate, blessed with beauty, proficient in various arts, an excellent orator, intelligent but has shifty eyes. **Vanijya Karana:** He/She is proficient in art, ever smiling, has knowledge and makes profit in business. **Vishti Karana:** He/She is blessed with beauty, restless, powerful, a winner and sleeps too much.

2.14 Predictions by Yoni Janma Phala

Ashva Yoni (Horse class): He/She is independent, full of qualities, has expertise in playing musical instruments and is a devotee. **Chaaga Yoni (Goat class):** He/She is a flirt, an enthusiast, expert in conversational skill but has a short life. **Gaja Yoni (Elephant class):** He/She is honoured by the authority, is powerful, a sensualist and an enthusiast. **Maarjaat/Manjar Yoni (Cat class):** He/She is efficient in his occupation, loves sweetmeats and is unkind. **Maheesha Yoni (Buffalo class):** He/She is a fighter and warrior, lusty, has many children, religious and has excessive wind element in his body. **Mesha Yoni (Ram class):** He is valiant, loves battle, wealthy and benevolent. **Mushaka Yoni (Rat class):** He is intelligent, wealthy, always ready to do his work, and a sceptic. **Mriga Yoni (Deer class):** He/She is of independent nature, earns his livelihood with excellent means, always speaks truth, has love and affection towards his dependent and is brave. **Nakula Yoni (Mongoose class):** He/She is benevolent, wealthy of high order, obeys his parents. **Sarpa Yoni (Snake class):** He/She is angry, cruel, and unfaithful and misappropriate of others' wealth. **Shwana Yoni (Dog's class):** He/She is enthusiastic, a rebel of his caste, a devotee of his parents. He person born in this class is religious, virtuous, practical, and full of qualities and the best person in his family. **Vaanar Yoni (Monkey class):** He/She is of restless nature, loves sweetmeats, quarrelsome, lusty and has good, obedient children. **Vyaghra Yoni (Tiger class):** He/She is of independent nature, efficient in accumulating wealth, receptor of virtuous preaching and full of self praise.

3

Muhurata

3.1 Muhurata Predictions:

Panchanga is used to determine the most ideal or auspicious time for carrying out the undertakings based on the Tithi (Lunar day), Vaar (Day), Karana, Yoga and Nakshatra which is known as Muhurata. Tithi (Date) brings prosperity; Vara (Day) prolongs life, Nakshatra removes misdeeds, Yoga gives immunity from diseases, and Karana and Hora leads to success. Sun and Mars rules forceful punishment (danda), Mercury and Saturn rules diplomacy (bheda), Moon rules temptation (Dana) and Jupiter and Venus rules good counsel (Sama).

a) Muhurata by Tithi (Waightage value = 1 Point):

Tithi: The Tithi (Hindu Date) is indicated in Panchang by **T.** The time next to the T is the ending time of Tithi. If the Tithi ends after midnight but before next sunrise, the end time will have a value greater than 24 hours. To get actual time, subtract 24 from that time. Remember that Indian Tithi does not change at midnight 0:00 hours like English date. Example: If the Panchang for October 23rd is (T: Dwadashi 27:10:11). It means it ends after midnight of that day at 27 hours, 10 minutes and 11 seconds. So it means Dwadashi will end at 3:10:11 AM of the next day (October 24th).

Good Tithi for General purpose: 2 (Dwitiya), 3 (Tritiya), 5 (Panchami), 7 (Saptami), 10 (Dasami), 11 (Ekadasi), 13 (Trayodashi -- Shukla Paksha only), and 1 (Prathama - Krishna Paksha Only). Please avoid Krishna Paksha 13 (Trayodashi), 14 (Chaturdasi), and Amavasya in all good work. Please avoid Shukla Paksha 1 (Prathama) and Rikta Tithi 4-9-14 (from both Paksha) in all good work.

Good Tithi for Wedding: All Tithi are good for wedding. Please avoid Rikta 4-9-14, Krishna Trayodashi, Amavasya, and Shukla Paksha Prathama 1 Tithi.

Good Tithi for Griha Pravesha: All Tithi are good for Griha Pravesha Avoid Shukla Paksha 1-4-9-14, and Krishna Paksha 4-9-13-

14-Amavasya.

Good Tithi for Buying a New Vehicle: Best Tithis are 7-11-15, but avoid Amavasya.

Good Tithi for Starting Business: Tithis 2, 3, 5, 7, 10, 11th are good for new Business.

b) **Muhurata by Vaar (Day) (Waightage value = 8 Point):**

Good Day (Vara): Monday, Wednesday, Thursday and Friday are good days. Avoid Sunday, Tuesday and day of eclipse.

c) **Muhurata by Karana (Waightage value = 16 Point):**

It is half of the part of Tithi, which is called Karana. Karana name is indicated by K: There are two Karana per Tithi. Hence there are two lines for Karana, where the time next to the Karana indicates when it is going to end. Example: (Karan: 08:32:33). There are 11 Karana altogether. They are: 1) Bala, 2) Balava, 3) Kaulava, 4) Taitil, 5) Gara, 6) Vanijya, 7) Vishti, 8) Shakuni, 9) Chatushpada, 10) Naaga, and 11) Kinstughna.

Good Karana: All Karan are good for Muhurat except Vishti (Bhadra). Try to avoid Vishti (Bhadra) Karana in all auspicious ceremonies.

d) **Muhurata by Yoga (Waightage value =32 Point):**

Yoga name is followed by **Y**: Ganda 08:31:32 indicates it is going to end at 08:31:32.

Good Yoga: All Yoga is good for Muhurat. Avoid Vyatipata, Vaidriti, Parigha, Vishkumbha, Vajra, Shoola, Atiganda, and Vyagata.

e) **Muhurata by Nakshatra (Waightage value = 60 Point):**

Muhurata prohibited by Nakshatra: The 16th Nakshatra (Star) from his/her birth star and is prohibited for any kind of muhurata, such as, travel/yatra, elebration, puja or undertakings any work and if performed, it will cause grief

Nakshatra Bala (Tara Bala): Tara Bala is to find a day suitable for Muhurata. To calculate Tara Bala, count the position of muhurata's day Nakshatra from the Janma (birth) Nakshatra (both Nakshatra inclusive). Divide that by 9. The reminder is Tara Bala. If the remainder (Tara Bala) is 2, 4, 6, 8, 9 or 0 then it's very good otherwise, if the remainder (Tara Bala) is 1, 3, 5, 7, then they are not auspicious, but it is bad for muhurata.

f) **Muhurata by Rahu Kalam:** The RK means the Rahu Kalam. The rising period of Rahu is considered inauspicious in the South as he is considered a malefic for auspicious functions. The time frame indicates Rahu Kalam duration. **Example:** (RK: 09:03-10:06). Avoid this time for your Muhurata.

4

Nakshatra

The Nakshatra are derived from the constellations. The belt of fixed stars is divided into 27 segments, the boundary of each being marked by a particular fixed star located near the zodiacal belt. These segments are known as the Nakshatra or Asterism. Each Nakshatra is 13 degree 20 minutes, superimposing the 12 Rashi. Each Nakshatra is further sub-divided into four equal arcs, which are called Pada. This way, we have total 108 divisions of Nakshatra called Pada, each 3 degree 20 minutes. The Nakshatra of the day is determined by the position of the Moon in the Zodiac. $2\,^1/_4$ Nakshatra make one Rashi or Zodiac Sign. The Nakshatra has a significant meaning. It is Na + Kshatra, which means that it, can never be destroyed. It is permanent. Nakshatra are permanent. The Nakshatra are far away but yet they exercise their influence on our lives in an unmistakable manner. The Moon enters a new Nakshatra, approximately each day of the month. The 27 fixed stars are used heavily in Vedic Astrology, which appear to us as stars within the 12 signs of the zodiac. Actually, some of them are stars, and some are clusters of stars. They are also called demigods.

Table 1: Nakshatra Characteristics

Sr. No	Nakshatra Name	Ruler/Lord	Deity/Demigod	Caste/Varan
1	Ashwini	Ketu	Ashwini Kumar	Merchant
2	Bharani	Venus	Yama (Death)	Lower cast
3	Krittika	Sun	Agni (Fire)	Brahmin
4	Rohini	Moon	Brahma	Shudra
5	Mrigashira	Mars	Moon	Servant
6	Aridra	Rahu	Rudra	Butcher
7	Punarvasu	Jupiter	Aditi	Merchant
8	Pushya	Saturn	Brihaspati	Warrior
9	Ashlesha	Mercury	Sarpas (Snake)	Outcast
10	Magha / Makha	Ketu	The Pitris	Shudra

11	Purva Phalguni	Venus	Bhaga	Brahmin
12	Uttara Phalguni	Sun	Aryaman	Warrior
13	Hasta	Moon	Savitri	Merchant
14	Chitra	Mars	Tvashtri	Farmer
15	Swati	Rahu	Vayu	Butcher
16	Visakha	Jupiter	Indragni	Outcast
17	Anuradha	Saturn	Mitra	Shudra
18	Jyeshtha	Mercury	Indra	Farmer
19	Moola	Ketu	Nirriti	Butcher
20	Purva Ashada	Venus	Apah	Brahmin
21	Uttara Ashada	Sun	Vishva	Warrior
22	Shravana	Moon	Vishnu	Outcast
23	Dhanishta	Mars	Vasus	Farmer
24	Shathabisha	Rahu	Varuna	Butcher
25	P. Bhadrapada	Jupiter	Ajaikapada	Brahmin
26	U. Bhadrapada	Saturn	Ahirbudhnya	Warrior
27	Revati	Mercury	Pushan	Shudra

Table 4: Nakshatra Characteristics

Name	Gender	Nature (Tatva).	Yoni	Gana
Ashwini	Male	Vata	Horse	Deva
Bharani	Male	Pitta	Elephant	Manushya
Krittika	Female	Kapha	Sheep	Rakshasa
Rohini	Male	Kapha	Serpent	Manushya
Mrigashira	Female	Pitta	Serpent	Deva
Aridra	Female	Vata	Dog	Manushya
Punarvasu	Female	Vata	Cat	Deva
Pushya	Male	Pitta	Sheep	Deva
Ashlesha	Male	Kapha	Cat	Rakshasa
Magha / Makha	Male	Kapha	Rat	Rakshasa
Purva Phalguni	Female	Pitta	Rat	Manushya
Uttara Phalguni	Male	Vata	Cow	Manushya
Hasta	Female	Vata	Buffalo	Deva
Chitra	Female	Pitta	Tiger	Rakshasa
Swati	Male	Kapha	Buffalo	Deva
Visakha	Male	Kapha	Tiger	Rakshasa
Anuradha	Female	Pitta	Rabbit	Deva
Jyeshtha	Male	Vata	Rabbit	Rakshasa

Moola	Male	Vata	Dog	Rakshasa
Purva Ashada	Male	Pitta	Monkey	Manushya
Uttara Ashada	Male	Kapha	Mongoose	Manushya
Shravana	Female	Kapha	Monkey	Deva
Dhanishta	Female	Pitta	Lion	Rakshasa
Shathabisha	Female	Vata	Horse	Rakshasa
P. Bhadrapada	Male	Vata	Lion	Manushya
U. Bhadrapada	Female	Pitta	Cow	Manushya
Revati	Female	Kapha	Elephant	Deva

4.1 Nakshatra Predictions:

Marriage compatibility is an important contribution of Jyotish to the world as only it can foresee any problems in the married life of a couple and can precisely determine which area the problem shall root from.

1. Ashwini

General: He/She is daring, handsome, truthful, prosperous, self-sufficient, intelligent, popular, knowledgeable, rich, well mannered, expert, outspoken, explorer, sportsmanship, and has vitality, courage, dynamism, initiative, action and a contented family life.

1. Physical features: Male has a beautiful countenance, bright and large eyes, broad forehead and big nose. Female eyes will be bright, but small like a fish with a magnetic look.

2. Character: He will remain faithful and will not hesitate to sacrifice anything for beloved persons. He keeps his patience even at the time of greatest perils. He is the best advisor to the persons in agony. He is a firm believer of God. He keeps the entire surroundings neat and clean. Female attracts people with her sweet speech. She maintains utmost patience. She indulges in too many sexual operations. She respects elders.

3. Education and sources of earnings / profession: Male will be full of struggle up to his 30th years of age. There will be steady and continuous progress after 30 years and will continue up to 55 years of age. He meets his desires and needs at any cost. Female will do administrative job in office. She may quit the job or seek voluntary retirement after 50 years of age.

4. Family life: Male loves his family with sincerity. His own family members will hate him due to his adamant behaviour. He cannot derive affection and care from his father. In other words, his father will neglect him as also his co-born. He may derive affection and care only from his maternal uncle. Maximum possible help will come from outside his family circle. Normally marriage takes place between 26 to 30 years of age. There will be more sons than daughters to him. Female marriage takes place at a young age i.e., before the age of 23 years and marriage ends either with divorce or separation or even death of the husband. There will be more female children than male children. She responds to the needs and desires of her children. She will be busy in the welfare of the family and society.

5. Health: Male health will generally be good. He will have mental worry and anxiety and may face brain disorder of mild type at a later stage. Female is healthy but brain disorder of mild type may take place at a later stage.

Positive traits: They have courage, daringness and swiftness of action, action oriented nature, intellectual ability and acute sharpness.

Negative traits: They give the rash impulsiveness with unnecessary risks and blunders creating danger and troubles and have revengefulness and stubbornness in nature.

Career: They excel as physicians, entrepreneurs and in areas of adventure sports, military or armed forces, law enforcements and athletic. He/She does business related to machinery, iron, steel, mineral, writing, publishing and engineering. Some has officer grade posting in architecture, stock broking, police, army, interior designer and flying, driving, riding and sports field. He/She is predisposed to muscular injuries and should be careful, as accidents are common.

2. Bharani

General: He/She is disciplined, a ruthless, truthful, law-abider, cruel, wicked, intelligent, dark, healthy, courageous, long lived, very creative and does painting. He/She dislikes being controlled and manipulated. He/She may have problems in romance and prone to putting on weight and never shies away from a fight. He/She is excessively indulged sexually. They are long lived. He/She desires to attract opposite sex and is marked by large expressive eyes and mesmerizing smile.

1. Physical features: Male is of medium size, and has less hair, large forehead, bright eyes and beautiful teeth. Female will have a very beautiful figure with white teeth.

2. Character: Male does not like to harm or buttering anybody. He has to face a lot of resistance and failures. He will spoil the relationship even with near and dear ones on small matters. When the opponent comes forward with folded hands, he will completely forget the enmity. His arrogance quite often leads him to miseries. Subordination is equal to death for him. He will lead into trouble. He is fond of spreading rumours and wasting time in humours. Female possesses a clean, admirable and modest character. She respects her parents and elderly persons. She is bold, impulsive and over optimistic.

3. Education, sources of earnings / profession: Male will have a positive change in his life and livelihood after his 33 years of age. He can be successful in the business of tobacco items or cultivation of tobacco. Female earns her own livelihood and will be successful as a receptionist, guide or sales woman or in sports activities.

4. Family life: Male will get married around 27 years of age. He is fortunate in conjugal bliss. His spouse will be expert in household administration and well behaved. His father will die, if his birth is in the 1st or 2nd quarter of Bharani. He dislikes being away from his family members even for a day. Female gets married around her 23rd years of age. She will have an upper hand in all matters and she will behave like a commander. She will enjoy full confidence, love and satisfaction from her husband but her in-laws will trouble her. She will over power her husband in all walks of life, but is not aggressive.

5. Health: There won't be any serious health problems. Normally, male is a very poor eater. There is chance of injury in the forehead and just around the eyes. He is a chain smoker. Female health will be good. She will have frequent menstrual problems, uterus disorders, anaemia and in some cases tuberculosis has also been noticed.

Positive traits: They have poise, calmness, multiplicity of interests, sincerity and dutifulness towards their liabilities.

Negative trait: They are indulge in material and physical pleasures, vulnerable to luxurious ness and extravagance, unnable to fit up with any form of domination and source of authority and has disregard for others' sentiments.

Career: His/Her careers will be in military, chemical industry, medicine or agriculture. He/She may be a social reformer, activists and philosophers. They excel in arts and curve a niche in careers based on singing, dancing, acting and painting. They also make excellent administrators, businessman and ingenuous surgeons and adjudicators.

3. Krittika

General: He/She is brilliant, very prosperous, learned, powerfully built, dark and possesses wealth. He/She has a magnetic & commanding personality and popularity in the opposite sex, leadership skill, penetrating ability to detect any imperfection that stands in the way of success. He/She will be good at carpentry, sculpture, metal work, and interior decoration. He/She runs after other sex and is heavy eaters and fiery. He/She also gets into bad habits like gambling, drinking and arguing. He/She is stubborn, aggressive, and very angry. He/She is thin and good cook. He/She is endowed with son and enjoys life.

1. Physical features: Male is of middle stature with prominent nose, sympathetic eyes, thick and stout neck, solidly built body with big shoulders and developed muscles. He has a commanding appearance. Female has extremely beautiful body, medium height.

2. Character: Male does not like to achieve name, fame and wealth through unfair means or at the mercy of others. His need is less and does not believe in age-old blind belief and customs. He does hard work. He does public work but looses and fails. Frustrations lead him to outburst. Once he got heated up, subsequent steps will be dangerous. Female will not subdue to the pressure of others. Therefore, she has to suffer mentally. She is fond of quarrels. She expresses arrogance not only in her appearance but also in the domestic environment.

3. Education, source of earning/profession: Male will earn his livelihood in the foreign land or away from the place of birth. Partnership business is not suitable to him. In case he is interested in business, he can derive maximum benefits from the yarn export, medicines and decorative industries. He is very slow in work. Female is not highly educated. She earns by working in the paddy field, agriculture field, fisheries and other small-scale industries. She also earns as a musician, artist, tailor or leather products manufacturer.

4. Family life: Male is lucky in his Love marriage with a faithful and virtuous wife, expert in household administration. The health of the spouse will be a concern for him and he may have to quite often live separately due to work or due to ill health of the spouse. He is more attached to mother and will enjoy more favour and love from the mother. His father will be a pious man and well-known person; he cannot enjoy comforts and benefits from father. The period between 25 to 35 years and 50 to 56 years will be very good. Female cannot enjoy full comforts of her husband. In some cases childlessness or separation from husband is also indicated or no marriage takes place. She will not keep cordial relation with relatives. She lives in an illusion world and she shows discourtesy to the people who actually are her well wishes.

5. Health: Male is prone to dental problem, weak eyesight, tuberculosis, wind and piles, brain fever, accident, wounds, malaria or cerebral meningitis. Female health will be severely afflicted either due to excess work or lack of mental peace. Glandular tuberculosis has also been noticed.

Positive traits: They have honesty, frankness and respect of people around, protective ability, and are extremely hard working.

Negative traits: They have tendency of fault finding, anger, lack of diplomacy and are centre of negative criticism on account of their bluntness.

Career: He prospers as administrators, leaders, lawyers and excels in career away from their home land. The partnership will never be his cup of tea; he/she can excel in business related to yarn, artistic goods and medicines, engineering and draftsman ship.

4. Rohini

General: He/She is long-lived, religious, experts, well-behaved, master in conversation, genius, popular, wealthy and has excellent persuasive skills, and earns his livelihood through agricultural occupation. He/She gains others' confidence and motivates others to work hard and achieves goals. His/Her marital life is blissful. He/She is devotee of parents and kind hearted. His/Her work areas are business. He/She can be over sexual. He/She can rise to the top and achieve his/her desires and has children. He/She respects gods and Brahmins. He/She is able servants.

1. Physical Features: Male is slim in physique; normally dark, short structured and has very attractive eyes with a special magnetic touch. Appearance is very beautiful and attractive, big shoulder and well developed muscles. Female is beautiful. Her eyes are very attractive. She is of medium height and fair complexion.

2. Character: Male is short tempered but courageous. He has special knack to find faults of others and his brain rules him. There is no tomorrow for him. He spends everything for today's comfort. He lacks patience and forgiveness. He fails to pass through the path of requisite system and has the downfall. He blindly believes others. Female is well behaved and well dressed with a lot of pomp and show. She has a very weak heart, but is short tempered and invites troubles.

3. Education, sources of earning / profession: Male may earn from business of milk products, sugar cane or engineering goods. He is best adapted for mechanical or laborious work. He faces a lot of problems economically, socially and on health grounds during the age of 18 to 36 years. He will enjoy life best between the age of 38 to 50 years and 65 to 75 years of age. Female earns from oils, milk, hotels, and paddy fields and as a dressmaker. Middle level education is indicated for her.

4. Family life: Male cannot enjoy full benefit from his father and is more attached to his mother and maternal uncle. His married life will be marred with disturbances. Female family life is good with a full comfort from her husband and children. She must avoid doubting her own husband as otherwise marriage may end in divorce.

5. Health: Male is prone to diseases connected with blood cancer, jaundice, urinary disorders, blood sugar, tuberculosis, respiratory problems, paralysis and throat trouble. Female health is good. She will have pain in the legs, in the breast, and may have breast cancer, irregular menses and sore throat, pimples and swelling above the neck.

Positive traits: He has temperamental sweetness and gentleness and they are so fine tuned that they seldom get into fits of temper. Their softly affectionate nature does not allow offence or revenge and a blend of perseverance.

Negative traits: They have lack of purpose, stability and imagination; lack of endurance and patienc; struggle, indecision and fickle mindedness.

Career: Natives are efficient sculptors, artistes, musicians, danseuse and creative directors. They can make mileage out of career options including photography, editing, agriculture, environment, advertisement, writing, marketing and jewellery designing.

5.   Mrigashira

General: He/She is truthful, handsome; enjoy wealth and prosperity, pure in heart, knowledgeable in architecture, well at administration, sharp shooter, loved by the king, full of energy, very popular and very creative. He/She travels a lot. He/She is successful research fellows and highly successful in conducting investigations, highly intelligent, and has beautiful children and many sexual relationships. He/She is restless, gentle, peaceful, and always travelling.

1. Physical feature: Male will have a beautiful and stout body, tall, moderate complexion, thin legs and long arms. Female is tall, sharp look, leaning body. Her countenance and body are very beautiful.

2. Character: Male has a doubting nature and is always suspicious. He is very sincere in his dealings. He may get cheated in partnership business. He puts blind trust on others and gets frustration and repentance. He may be ditched. He is an inborn coward. He will not have peace of mind and get irritated even on small matters. His life will be painful up to 32 years of age. His life starts settling down to the maximum satisfaction after that. She has a poisonous tongue, but is educated. She acquires considerable wealth and enjoys good food. She will be lucky and have ornaments and fine clothing. She is greedy for wealth.

3. Education, sources of earning / profession: Male will have good education. He is a very good financial adviser. He achieves success in business after the age of 32 years. He achieves the benefits unexpectedly during the good period at 33 to 50 years of age, but it will be wasted later on due to his own fault. Female will attain good knowledge in the mechanical or electrical engineering, telephones, electronics. She is more interested in the jobs that are normally done by males.

4. Family life: Male will not derived benefit from the co-born. Co-born will cause troubles and problems to him and will maintain extreme enmity and harm him. His sincere love and affection will not have any effect on his relatives. His spouse is attracting romantic, may not keep good health. Moreover, his married life is not cordial due to his adamant nature. Female keeps her husband under her control. She may have one or two love affairs in her early life, which will not culminate into marriage. However, after marriage she is very much attached to her husband, as if, nothing has taken place in the past. She will have children and is devoted to her husband.

5. Health: Male may face ill health during his childhood, like constipation ultimately leading to stomach disorder, cuts and injuries, pains in the shoulders near collarbones. Female is prone to goitre, pimples, venereal diseases, menstrual troubles and shoulder pains.

Positive traits: He/She is truthful, clean at heart, intelligent and has good administration powers and are obedient, respecting his teachers and always keen to learn and observe.

Negative trait: He/She is restless and nervous, impatient, takes wrong decisions and commits mistakes.

Career: He/She is musician, tailor, engineer, communicator or traveller. Careers are sports, advertising, communication, and environment campaigns, travel related industry. He/She will fare well in career when given freedom.

6. Aridra

General: He/She is hard working, proud, cruel, lower levels person, bereft of money, expert in trade and commerce, and does jobs which are forbidden. He/She is very much clam and has helpful nature.

He/She life undergoes many ups and downs and U turns, which means He/She is to face many challenges in his life. He/She is little educated, long lived, and little interested in doing things and more inclined towards sex. He/She causes pain to her loved one and suffers due to hunger and hates all. She likes to renovate old houses or cars. He/She has a lust for power and material things and is calculating. His/Her violent temperament causes many tears and depression to him and leads to his/her early destruction and death.

1. Physical features: Male has different shape and structure and is slim. Female has handsome body with charming eyes, prominent nose.

2. Character: Male is good psychologist. His dealings with his friends and relatives will be of very much cordial type. He does not have a constant type of behaviour Female is well behaved and peace minded. She is intelligent, helpful to others and clever in finding fault in others. Some of them may have two mothers or two fathers.

3. Education, sources of earning/profession: Male is over sincere in the work or his business. He is selfless social workers and earns his livelihood away from his home and family. He is settled in foreign places. He has golden period between 32 to 42 years of age. Career best suited is lower level in fire fighting, farming, gardening, writing, postal services, transport, education, police and defence. Female will attain distinction in the educational or scientific field. She may specialize in electronic or pharmaceutical. She also has consultation works.

4. Family life: Male marriage will be delayed. In case marriage takes place early he will be compelled to live separately from the family due to difference of opinion. When the marriage takes place at a late stage his married life will be good. His spouse will exercise full control over him. Female marries at a late stage. She enjoys love and affection from her husband and husband's family. Her married life will be full of thorns. Even children cannot give her the required happiness. In some cases, there will be either death of the husband or divorce takes place.

5. Health: Male may have some incurable diseases like paralysis, heart trouble and dental problems. He is also prone to Asthma, esonophilia, and dry cough, ear trouble. Female is prone to menstrual troubles, asthma, spoiled blood, lack of blood or uterus trouble, ear trouble, and mumps, bilious and phlegmatic.

Positive traits: He/She is hard working, intelligence. He/She enovates old things and produce new things with their fertile thinking and so they are strong and mentally stable and at the same time empathetic towards others.

Negative traits: He/She is always sad and a kind of impulsive mentality, cleaver and calculating, stubborn and may exhibit violent temperament, which can not help them to make very close social bonds.

Career: His career lies in fields like transport industry, communication industry, financial broking firms, shipping industries.

7. Punarvasu

General: He/She is charitable, endowed with children, long lived, loved by wife, and possesses wealth. He/She has harmony, utmost happiness and satisfaction, pleasant nature and is friendly. He/She is marked by a strong power of intuition, and is blessed with divine power and grace. He/She will share his material possessions with others and will choose a religious or spiritual career. His/Her deep-rooted faith in God and high morals will give his/her family a deep sense of security. He/She is beautiful, well-spoken, artistic, rich, and satisfied. He/She loves to travel and is content with little.

1. Physical features: Male is handsome and has long thighs and long face. There will be some identification mark on the face or on the backside of the head. Female eyes are red and have curly hair, sweet speech and high nose.

2. Character: Male does not like to cause trouble to others. He will lead a simple life. Female has argumentative tongue, which leads her into frequent friction with her relatives and her neighbours. She will have many servants and a comfortable life.

3. Education, sources of earning/profession: Male is in business, journalism, publishing, auditing and writing and is contended with little and sticks to ancient tradition and belief. He attains much name and fame as a teacher or as an actor and physician. He will not do so well up to 32 years of age. He will not accumulate wealth but attain public honour. He lacks business trick and is straightforward. He is an innocent and frustrated looking. Female gets mastery over dances.

4. Family life: Male will have very happy married life. Children feel loved and secured. He is the most obedient child of his parents. He respects his father and mother as also his teacher. His married life may not be good. He may either divorce his wife or get involved in another marriage. In case he does not go for the second marriage the health of the spouse will give a lot of problems and mental agony. However, his spouse will have all the qualities of a good housewife. Female's husband will be most handsome man.

5. Health: Male has diseases related to lungs, blood, waist, and ear and dental. There is not any serious health problem. He drinks lot of water. He has strong digestion. Female cannot enjoy good health and has Jaundice, tuberculosis, goitre and pneumonia, stomach upset and ear trouble.

Positive traits: He/She is spiritual, sobre, pleasant, social, and has tendency to discern a spiritual meaning of life, essential domestic or home spun nature.

Negative traits: He/She has lack of forethought, carefulness, susceptibility, and disregard for people, fickle mind and instability.

Career: He/She excels as spiritual teachers, psychologists and mystic philosophers, architecture, innovations, civil engineering and maintenance of buildings. Acting, dramatics and writing are also appealing for them.

8. Pushya

General: He/She is religious, humble, handsome, fortunate, intelligent, self-sufficient, helping hand, important person in the world, knowledgeable, respected by kings, fair, wealthy, and obeys his parents and possesses wealth and vehicles of novelty. He/She is caring, nurturing, kind, and helpful. He/She is honest, devotee of parents. He/She believes in the rules, and forces others to follow the laws. He/She wants to share his/her inner wealth and so he/she is frequently teachers, preachers and professors. He/She could be predisposed to ailments like gall stones, gastric ulcers, skin diseases.

1. Physical Features: Male will have a scar or a black mole on his face. Female has a short stature, moderate complexion, well-proportioned. Generally, he will be beautiful.

2. Character: Male is very weak by heart and doesn't reach at conclusion on any matter. His outward expression is hypocritical against his consciousness i.e. the inner expression is negative and the outward expression is positive. He is fond of good dresses. Female is peace-minded, very submissive to the elders but oppressed by all. She is sincere, affectionate but moody. She is religious and god fearing. She respects her elders. She is conservative. She does systematically and methodical action.

3. Education, sources of earning/profession: Male is employed in state government jobs like medicine, social work, health care. He will have utter failure. He has several obstacles because of no basic education. He does work with utmost sincerity and certainty. His success is as probable. He faces the grip of poverty up to the age of 15-16 years of age and thereafter he will enjoy a mixture of good and bad up to the age of 32 years. There will be remarkable all round improvements i.e. economically, socially and in health from the age of 33 years. He will stay away from his native place and family for his bread earning. He undertakes the work where maximum travelling is involved. Female may have income from Agriculture, land and buildings. She may be employed in a job where maximum trust and secrecy is required e.g. private secretaries, secret departments of a country.

4. Family life: Male will have lot of problems in the family circle and he will have to depend on others for even day-to-day requirements. He will be forced to stay away most of the time from his family. He is very much attached to his parents. Female is devotee and attached to her husband only, but her husband quite often mistakes her for her moral character. She wants to say a lot of things but she is prevented by her inherent quality of shyness and others form a negative opinion about her. She will have good duty bound children.

5. Health: Male will not keep good health during his childhood up to the age of 15 years. Gallstones, gastric ulcer, jaundice, cough, eczema and cancer are the possible diseases. Female will have respiratory problem. She is prone to tuberculosis, ulcers, breast cancer, and jaundice, eczema, bruises in the breast or gastric ulcer.

Positive traits: He/She is benevolent, philanthropy and humanitarian, cool and has good manners, protection against destruction and violence, inherent qualities of creation and expansion.

Negative traits: He/She is zealously protective of his family, society or community to which they belong, orthodox, narrow-minded and possessive, conservative and prejudice.

Career: He/She excels in careers related to counselling, public administration, planning, research, good priests or clergies. He/She is geologist, developer of land, aquatic biologist, land or agricultural merchants, philosophers, religious leaders, teachers and professors.

9.   Ashlesha

General: He/She is born leaders, cruel to all beings, daring, angry, wicked, short-tempered, born wanderer i.e. travels unnecessarily, does undesirable jobs, harmful to people; and suffers on account of sex starvation, pay fines by doing sinful acts and causes anguish to others. He/She spends his wealth for evil purposes, and losses on investments. His/Her work areas will be business, contracts or teaching. He/She can hypnotize any one with glare of eyes and uses it for black magic. He/She is cold blooded dangerous man. He/She is always sad, vary bad person, highly stingy, and develops enmity with his own people and others.

(i) If born in the first pad, he/she is impotent. (ii) If born in the second pad, he/she is a servant always serving others. (iii) If born in the third pad, he/she suffers from diseases. (iv) if born in the fourth pad, him/her, though fortunate, has medium life.

1. Physical Features: Male thinks and walks fast. He is normally dark, possesses rude appearance and rude features but inside built up is void. Female is not good looking. However, she will have beautiful figure.

2. Character: Male is grateful to anybody, very talkative; shines in the political field and can give suitable leadership to a country. He does not believe others blindly. He mostly associates himself with black marketers, thieves and murderers. He does not keep any distinction between rich and poor, good and bad people. He always supports the persons who are not in need and reject the request of the persons who are in need. He is luckiest, popular but hot-tempered person. Female is shy. Her moral is of a very high order. She will enjoy good respect and recognition from her relatives. She conquers her enemies.

3. Education, source of income/profession: Male has education in Arts or Commerce and changes careers many times. There will be a heavy loss of money at the age of 35-36th years and unexpected and unearned income at 40 years of age. Female does official work. She may be employed in an administrative capacity. Illiterate females may be engaged in selling fishes or work in the agricultural field.

4. Family life: Male shoulders family responsibilities but his wife does not want to share her wealth with other relatives such as sister-in-law or cousin brother Female is very efficient in the household administration. Her in-laws can make some plot against her so as to create friction with her husband.

5. Health: Male will have flatulence, jaundice, and pains in legs and knees, stomach problem. He will be addicted to drugs. Female will have frequent nervous breakdown, Joint pains, hysteria and jaundice.

Positive traits: He/She has intuition, overcoming a situation steeped in danger, planning, cool calculation, and courage and leadership disposition.

Negative traits: He/She displays cold ruthlessness, deceit and an aura of suspicion and is endowed with unattractive features and has suspicion and cunningness and merciless guile, insidious harmfulness, dependence on others and shamelessly plotting nature, deception, dishonesty and falsehood, miserliness, selfishness, ungratefulness, acute suspicion, depression, anxiety, schizophrenia and total absence of benevolence.

Career: He/She excels in careers related to cunning politicians, filing law suits, lawyers, advocates, and business men. He/She also does well in entertainment industry, occult mysticism and astrology. With the planetary lordship of Mercury, he/she succeed more in business than in profession.

10. Magha

General: He/She is reserved, highly sexy, endowed with comforts, learned, honest, helpful, great souls, devotee of parents, courageous, and destroy enemies. He/She has royal and respectable positions in his life. It is the symbol of authoritative status, domination and high social respect. He/She is leader and has material pleasures of the world. He/She is enterprising nature, thoughtful interaction with other peoples of the society and righteousness. He/She respects his father and others and has leadership skills and gets to the top. He/She will enjoy a harmonious marital life. He/She may do well in business and can amass a lot of wealth. He/She is a brilliant planner and organiser and is always in command of a situation. He/She may take top most charge and has a drive for power and wealth like a C E O. He/She is long lived, served by many servants, helps their relatives.

1. Physical features: Male has a prominent neck and a hairy body, a mole in the hands and also beneath the shoulder and is medium size height and innocent looking countenance. Female has a most beautiful and attractive feature. If Saturn aspects Moon in this Nakshatra, she will have long bunch of hair.

2. Character: Male is God fearing, soft spoken and leads a noiseless life. He receives honour and recognition from the learned persons. He is hot-tempered and cannot tolerate any action or activity, which is not within the purview of truth.

2. Character: She is charitable and God fearing. She will enjoy royal comforts. She can attain mastery over both household and official activities. She helps without selfish motive.

3. Education, sources of earning/profession: Male attains much success due to his straightforwardness. His dealings with his superiors and subordinates are very cordial and most technical. Hence, he is a link between his bosses and subordinates. Female is employed in managerial position like a queen if Jupiter is placed in this Nakshatra. She is married to a wealthy person.

4. Family life: Male will enjoy a happy and harmonious married life. He shoulders several responsibilities and burden of his co-born.

Female invites friction in the family; makes conflict between her husband and in-laws, resulting in mental torture for everybody in the family. She will have good children; preferably the first will be a son and next two daughters.

5. Health: Male may suffer from night blindness, cancer, asthma or epilepsy if Saturn and Mars jointly aspects Moon in this Nakshatra or conjoin in this Nakshatra. Female's eyes may be affected, and has blood disorders, uterine trouble and hysteria.

Positive traits: He/She is religious and traditional in nature, god fearing and respects his elders, deeply follows his forefather's teachings and abides by that strictly. His empathetic behaviour, care towards other's feelings and possessing a careful approach towards life is another positive traits.

Negative trait: He/She is often attracted to material pleasures and is proudly due to enjoyment of high positions in the society. He/She is often over confident being in powerful position, and imposes his decisions on others without much consideration, which causes resentment in some people about him.

Career: He/She works hard and enjoys a very good position. His career interest is to work independently as bosses, leaders and switch his jobs frequently to achieve the position hedesires.

11. Purva Phalguni

General: He/She is has joy and creation, prosperity and enjoyment and brings good fortune and luck. He develops tkeenness towards arts from childhood and pursues the interest to have a respectable and profitable livelihood through his chosen interests. He/She is an artist. He/She is experts in love making, handsome, clever but cunning, wealthy, givers, long lived and loves his/her brothers, gets few children. He/She may incur a lot of debt. He/She lapses into a depression. His/Her marital life will be happy. He/She is devotee of parents, religious, good speaker and generous. He/She is very sexual and passionate. He is naturally attractive, very polite, loving; very compassionate and well spoken which makes him the focus of attention by the opposite sex. He has an empathetic and genial attitude towards mankind, a large number of people surrounding him. He is peace loving and keep distance from disputes, and tries to bring in an amicable solution to the problems without fighting. He is greatly liked and loved by people around them.

Male natives

1. Physical features: He has attractive personality with a stout body and a snubbed nose.

2. Character and general events: He has intuitive powers to know others. He extends his helping hand to the needy. He has sweet speech. He is fond of travelling.

3. Education, sources of earning/profession: He cannot take up a job of subordination. He cannot be a 'yes master' and so can not derive much benefit from superiors. He does not like to get any benefit at the cost of others pocket. He is able to crash his enemies and attain much success in all the work he undertakes. He gives preference to the position rather than money. He frequently changes job. The borrowers will not return the money to him. He will reach a good position where power and authority vests in him after his 45th years of age. In the business field he can shine well.

4. Family life: His married life will be happy. He will have good wife, children and derives much happiness from them. He leads life away from native place and family members.

5. Health: His health is good and no permanent nature of disease is noticed in the native.

Female natives

1. Physical features: She will be medium height and has overall an attractive personality.

2. Character and general events: She is polite, artist and does charitable deeds. She has self-imposed showy image and believes that there is none above her.

3. Education, sources of earning/profession: She attains a good degree of education of scientific subjects. She may be a lecturer. She will have reasonable wealth.

4. Family life: She has a loving husband and children. She is duty bound wife. Her daredevil attitude creates friction between the family members. She cannot have a cordial relation with her neighbours due to her arrogance nature and position of her husband.

5. Health: She suffers from frequent menstrual, asthma, jaundice or breathing trouble.

Positive traits: He maintains a cordial and warm relationship with others and relatives. He is very loving and affectionate and he cares for the people close to him and helps them to the utmost when it is needed. He keeps homes and surrounding clean and decorates them with beautiful things.

Negative traits: He is over confident and arrogant about his appearance and deeds. He is restless and loses patience easily.

Career: He studies in areas like criminal psychologies, research and administrative studies. The suitable career areas for him are government services, transport management, automobile industries and hotel industries. He/She will do well in the legal or law enforcement field.

12. Uttara Phalguni

General: He has kind heart, helpfulness, kind, honest, truthfulness, and knowledgable. He has overwhelming wisdom in the fields of science and arts and due to essential goodness he wins over friends and foes, keeping away from disputes and quarrels, peacefulness and geniality. He/She is wealthy, famous, kingly, flexible, hard working, impotent, and achieves fame. He/She earns well out of commissions and royalties. Marital life is pleasant and contented. He/She is business minded and has an association with bad people. He/She is courageous, good with people, leading him/her to powerful positions.

Male Natives

1. Physical features: Normally, he is a tall and fat figure, large countenance and long nose and a black mole in the right side of his neck.

2. Character: He is extreme sincere and good in social work. He does not have patience or tolerance. However, he will not admit his fault at any cost.

3. Education, sources of earnings / profession: He shines in the profession of public contact and earns a good amount as commission out of the public dealings. He is hard-working and reaches a good position. He is suited to the profession as a teacher, writer or research. He earns extra money out of tuitions. There will be complete darkness up to 32 years of age. Thereafter, his progress is much faster and achieves much of his desires from his 38th years of age up to his 62nd years of age. He will earn fame and wealth during his fifties. He has self-acquired assets. He is good in mathematics or engineering.

4. Family life: His married life is good. He is contended and wife will be most efficient.

5. Health: His health will be generally good. He is prone to dental problems, gastric trouble, liver and intestine problems.

Female Natives

1. Physical features: She has medium height and a black mole on the face.

2. Character and general events: She will have very calm and simple nature. She does not keep enmity for long. She is principled, always joyful.

3. Education, sources of earnings / profession: She has mathematical aptitude or scientific background. She is a teacher or lecturer or in administrator in sanitary department or hospitals. She may also earn money as a model or as an actress.

4. Family life: She has happy family life. She is wealthy and very clever in managing domestic work. Neighbours may create problems in family life in alliance with her in-laws.

5. Health: She enjoys good health with diseases like Asthma, menstrual and headache.

Positive traits: He has intrinsic greatness of head and heart, determination, conviction, esteem, knowledge and passion for the delicate subtleties of life.

Negative traits: He has extra marital affairs, obsession with cleanliness and arrogance and he can be inconsiderate and disdainful.

Career: He does well in careers related to scientific research, astronomy, media, sales, philanthropy, acting, writing, armed forces and military.

13. Hasta

General: He/She is warrior, religious minded, learned, wealthy, daring, helpful to others, well built, givers, God fearing and pious and gets the properties of his/her father and respects Brahmins and possesses wealth. He/She will get name and recognition. He/She has a very competitive nature that makes a success. He/She is honest, calm and good at managing things and excelling at jobs with lot of travelling. His/Her marital life will be happy and contented. He/She has good sense of humour and good speakers. His/Her early life may be plagued by hardships restraints and possible impediments.

Male Natives

1. Physical features: He is tall, stout with short hands comparing to the body structure. There will be a scar mark on his upper right hand or beneath the shoulder.

2. Character: He has the magnetic power to attract others and gets respect and honour from the public. He helps the needy and does not deceive others. He is simple. His life is full of frequent ups and down due to some hidden curse on him.

3. Education, sources of earning/profession: He keeps strict discipline and works at managerial level. He/She is engaged in business or, in a high position in the service. He even with a preliminary academic background will possess excellent all-round knowledge. It will bring in unexpected circumstantial changes both in the family front as well as in his academic and professional or business field up to 30th years of age. He/She will settle down in his life between 30 to 42 years of his age. There will be remarkable accumulation of wealth and all-round success in the business beyond 64 years of age.

4. Family life: He is able to enjoy an ideal married life. His wife is a good housewife. His wife may indulge in homosexual activities in her youth period.

5. Health: He is prone to cough, cold, asthma or sinuous.

Female Natives

1. Physical features: She is extremely beautiful and entirely different body.

2. Character: She has shyness of a female sex. She respects the elders and will not hesitate to express her views openly, even; she is subjected to enmity from her relatives. **3. Education, sources of earning/profession:** She is generally not employed or employed in the agricultural field and in the construction labour activities.

4. Family life: She can enjoy a happy married life. Her husband will be wealthy and loving. She enjoys good benefit from her children. There is one son and two daughters.

5. Health: Her health is good except high blood pressure and asthma in her old age.

Positive traits: He has ability to win people and situations over counts, exceptional sharpness, self control, faithfulness, and strength of intellect, innate ability to inspire and motivate people.

Negative traits: He is dominating and cruel sometimes and is prone to rude callousness and has lack of sensitivity.

Career: He shines exceptionally in careers related to entrepreneurship, counselling and consultancy and business, skill, craftsmanship and wisdom, and makes his way up the ladders of competition to reach top notch ranks and designation in technical lines and art. With his ability towards reconciliation he does better in mediating and settlement of disputes.

14. Chitra

General: He has prosperity, intuition, dynamism, craftsmanship of the highest order, and visual palatability. He/She is well dressed, courageous, intuitive, popular, expert in politics, rich, endowed with wife and children, God fearing and pious and defeats his enemies. He/She may pursue higher education. He/She is extremely knowledgeable and very concerned with political and social issues. He/She will do research, teaching and intellectual pursuits. He/She has very beautiful eyes with well-proportioned bodies. He/She is wonderful conversationalist and says the right thing at the right time. Many artists are born in this Asterism.

Male Natives

1. Physical features: He is lean and is identified even in a crowd of hundreds of people through his magnificent dealings and expressions.

2. Character: He is peace loving. He goes to any extreme for self benefits. He has great intuition and is very cordial with others. He confronts his enemies at every step but he is capable of escaping any conspiracy. He has soft corner for the downtrodden people and he devotes his time and energy for the uplift of this section of the society.

3. Education, sources of earnings / profession: He overtakes all hurdles and courage with hard work. He will not be leading a comfortable life up to the age of 32 years. 33 years to 54 years of age will be his golden period. He gets help and reward from unexpected quarters without putting many efforts. The age of 22, 27, 30, 36, 39, 43 and 48 years will be very bad in all respect for him. He may earn as a sculptor or mechanic or as a factory employee or a textile technologist.

4. Family life: He sincerely loves his co-born, and parents. His father leads a separated life. He has a life away from his father. His father has distinct identity of his own and is well known. He cannot stay in the house where he is born. Either he will leave the house or the house where he is born will be sold or destroyed. In other words, he will be settling down at a distant place away from his native house. He cannot enjoy a happy married life. He has to shoulder lot of responsibilities and face lot of criticisms throughout his life.

5. Health: He may suffer from inflammation of kidney, bladder, brain fever and diseases connected with worms. Abdominal tumours or appendicitis also have been noticed.

Female Natives

1. Physical features: She has a beautiful tall body. She has natural long hair.

2. Character: She is proud, veracious, sinful and lazy. She commits sinful deeds.

3. Education, sources of earnings / profession: She will have her education in science subjects. She may be a nurse or a model or film extras. If planets are not well or moderately placed, she may be employed in agricultural field.

4. Family life: There is likelihood of death of the partner, divorce or complete absence of pleasure from her husband. In some cases, childlessness is also noticed.

Positive traits: He has an ability to create, build and appreciation of intrinsic beauty with an aura of polished glamour, refined and attractive mannerisms, power of intuition and ability to inhabit a mystical realm count amongst his positive traits.

Negative traits: He has intrinsic tendency to hide behind a veiling facade, selfishness, indulgence, dubiousness in hisnature and does self service rather than generosity.

Career: He has career related to engineering, gardening, and horticulture, interior decor, glamour and hospitality industry.

15. Swati

General: He/She is endowed with tremendous beauty, jolly, long lived, kind, famous, good-natured, rich, merciful, religious, and soft spoken and receives wealth from the king. He/She cannot be corrupted or manipulated and achieves goals. He/She is broad-minded and may incur a lot of debt. He/She will do well in professions related to medicine and drugs, chemicals and travel industry. He/She tends to focus too much on social work and this may create friction within the family. He/She is rich, renowned and has prosperity, success. His childhood days will be full of problems.

Male Natives

1. Physical features: His under part of the feet is curved and the ankle risen. His feature is very attractive to the women folk. His body will be fleshy type.

2. Character: He is a peace loving but adamant. Once he loses temper, it will be very difficult to calm down. He is the best friend in need and worst enemy of the hated. He does not hesitate to take revenge on the persons who is against him.

3. Education, sources of earning/profession: He suffers, financially and mentally, even if born in a wealthy family, till his 25th years of age. He cannot progress in profession or business up to the age of 30 years. Thereafter he has a golden period up to his 60 years of age. He will earn through the profession as Gold Smith, travellers or drug seller, an actor or dramatist or may join defence (navy)and as astrologer or a mechanical engineer.

4. Family life: His married life is not much congenial and it is not adjustable couple.

5. Health: He has a very good health except piles and arthritis.

Female Natives

1. Physical features: She is very slow in walking.

2. Character: She is sympathetic and loving, religious, truthful, virtuous and enjoys a very high social position and has many friends. She will win over enemies.

3. Education, sources of earning/profession: She gets employment with much fame but does not like to travel. She is forced to accept more travelling due to job.

4. Family life: She will enjoy complete satisfaction from family and her children.

5. Health: Her internal constitution is weak. She may suffer from bronchitis, asthma, breast pain, broken feet ankles and uterus trouble.

Positive Traits: He is very knowledgeable, witty, very much independent, God fearing and religious minded, helpful, truthful in nature and competent; and has a high self respect and posses a strong liking to work in freedom.

Negative Traits: He is stubborn, adamant, and not ready to listen on work with some people. He has restlessness and uncertainty in decision making and is not very popular in the society.

Career: He is excellent and gets success in the later part of his life and not in the earlier stages. He has career interest in the financial and legal domains, gold business, acting, textile, travelling, mechanical engineering and astrology as well.

16. Visakha

General: He has power, position, authority, attractive face, good looking personality, very highly energetic and full of vigour and vitality, intelligence and truthfulness and leads life with his own principles and possesses his belief in Gandhian thoughts of non violence and austerity. He/She is dark in complexion, cheerful, religious, communicator, writer, speaker, unfriendly and will perform rites and rituals. He/She has an attractive personality and very popular, especially with members of the opposite sex. He/She has a strong sense of justice and guards people against sarcasm. His/Her careers will be in journalism, sales and marketing, writing. He/She has a very happy marital life. He/She is very goal oriented, and doesn't give up until achieve success. He/She is patient, persistent and determined. He/She experiences success in the second half of life.

Male Natives:

1. Physical Features: He may be fatty and long structure or very lean and short structure.

2. Character: He does not believe in the orthodox principles or the age-old tradition and is fond of adopting modern ideas. Mostly he lives away from his family. Slavery is suicidal for him. He treats all religions, castes and creed as one and follows of Gandhian philosophy of 'Ahimsa. He accepts Sanyas (saints) when he touches 35 years of age and simultaneously looks after the family and follows Sanyas.

3. Education, sources of earning/profession: He is a very good orator and has the capacity to attract crowd. Hence he is the fittest person to be in the political circle. He does an independent business or job involving high responsibility, banking and religious professions or mathematician or a teacher.

4. Family life: He cannot enjoy love and affection of mother, may be either due to mother's death or unavoidable circumstances. He is always proud of his father. He leads more or less a life of an orphan. There is a lot of difference of opinion between his father and him and due to these reasons he is, right from the childhood, a hard working and self made person. He loves his wife and children very much. He is addicted to alcohol and indulges in too much sex with other ladies.

5. Health: His health is very good, but prone to paralytic attack after the age of 55 years.

Female Natives:

1. Physical features: She is extremely beautiful and faces lot of problems due to this.

2. Character: She has a very sweet tongue and expert in the household activities and in the official activities. She does not believe in pomp and show and is very simple.

3. Education, sources of earning/profession: When Moon and Venus are together, she may become a famous writer. She will have academic excellence in arts or literature.

4. Family life: She treats her husband as her god. Her religious attitude will confer love and affection from her in-laws. She looks after the welfare of all family members and relatives. She may frequently be visiting sacred places.

5. Health: She has good health, but prone to kidney trouble, goitre and weakness.

Positive trait: He is very sharp and keen to learn new things, and do not like orthodox procedures and beliefs. He is full of energy and enthusiasm and quite single minded.

Negative traits: He gets involved with unsocial elements and practice like drug abuse, sex and alcohol habit. He is quite restless in nature and never feels contented with his achievements. This leads him to frustration and continuous worry.

Career: He is natural good orator and has a tendency to enter the political arena. Other careers are independent business, high responsibility jobs like those of administrative arena and teaching profession or mathematician.

17. Anuradha

General: He/She dictates and has energy, propensity in communications, ability to gauge a situation. He/She has commendable complexion and is destroyer of his/her enemies, sensualist, famous, experts in arts, server of king, stationed in countries other than his/her own, truthful and respects his/her mother. He/She is fit for friendship, kind hearted, helping nature, enjoyments, traveler, fun loving, determined, very ambitious, spiritually inclined, dark in complexion, glowing personality and enemy killer. He/She makes a lot of friends. He/She may go far away from place of birth because of profession. He/She may be successful in business. Careers that will suit are sales and marketing, astrology, and counselling. He/She will have a satisfactory marriage. He/She has superb leadership and organizational skills. He/She is sensual and loving. He/She knows how to share and accommodate others. He/She has been known to live far from place of birth. There are many opportunities for travel. He/She has a peculiar frustrated face and a gloomy appearance because he has to confront on several occasions. There will not be peace of mind in his/her life. Even a smallest problem will start pinching his/her mind repeatedly. He/She always thinks of taking revenge whenever opportunity comes. In spite of these drawbacks he/she is the most hard working and ever ready to complete the task. After several reversals he/she ultimately achieves the desired result. He/She is a firm believer of god. His/Her life is full of helplessness but independent.

Male Natives

1. Physical features: He will have a beautiful face with bright eyes. In some cases where combination of planets is not good, he has cruel looks.

2. Character and general events: He is liable to face several obstacles in his life. Even then he has special aptitude to handle the most difficult situation in a systematic way.

3. Education, sources of earning/profession: He has a special calibre of how to pocket his superiors. He starts earning his bread at quite young age say on or about 17 or 18 years of age. He will have good life between the periods of 17 years to 48 years of age. His life after the age of 48 years will be extremely good. It is in this period he can settle down in his life to the desired way and become free from most of the miseries of the life. If Moon is in the company of Mars, he may be a person dealing with drugs and chemicals or a doctor.

4. Family life: No benefit will be derived from the co-born under any circumstances. His spouse will have all the qualities of a good housewife. He likes to provide all necessities of life as far as possible and love and affection to his children. Hence his children reach a high position, much more than that of him.

5. Health: His health will generally be good, but prone to asthmatic attack, dental problems, cough and cold, constipation and sore throat.

Female Natives

1. Physical features: She has attractive face, beautiful body and beautiful waist which attract a man.

2. Character: She is pure hearted. She likes to lead a simple life, a selfless, agreeable and attractive disposition. She can shine well in social and political field. She will respect her elders. Her friends will cordon her as if she is the head of friend's circle.

3. Education, sources of earning/profession: She is interested in music and fine arts. She may obtain an academic degree or high degree in music, dance or other professions.

4. Family life: She is very much devoted to her husband like Radha and observes religious norms. She can be called as a model mother as far as upbringings of her own children are concerned. Her devotion to her in-laws pours further glory in her personal life.

5. Health: She may suffer from irregular menses. Severe pain will be felt at the time of bleeding, as the flow of spoiled blood will be quite intermittent.

Positive traits: He/She has amiability, balance, good conduct, leadership ability, sensitivity to people and situation, and outstanding ability to forge rapport and harmony and is spontaneously efficient in bridging differences and animosity.

Negative traits: He/She has over emphasis on secrecy, frivolousness and unexpected mood swings, idleness, weakness and absence of purpose.

Career: His career opportunities are revolving around mathematics, science and engineering, tourism and travel and administrative job with an insistence on organizational skill.

18. Jyeshtha

General: He/She has material richness, achievement, kindness and hot tempered nature. He/She is attractive, tall and endowed with few children, full of vitality, fond of physical exercises, warrior and dark complexioned. He/She achieves success early in life at the age of twenty-one. Careers that will suit are engineering, Architecture and construction. Married life is happy. He/She will have very supportive spouses. He/She is wise and the patriarch or matriarch of the family and knows how to deal with wealth and power. i) If born in the first pad, he/she is full of lustre and splendour, achieves fame and greatness, and is rich, brave, a hero and an excellent conversationalist; (ii) If born in second pad, he/she is of very cruel nature and quarrelling nature; (iii) If born in third pad, he/she follows the evil path and (v) If born in forth pad, he/she, although, has many sons but because of the influence of this malefic Nakshatra, it causes anguish and pain.

Male Natives

1. Physical features: He has a very good physical stamina, defective teeth bulging out.

2. Character: He is very clean and very sober. He can not maintain secrecy or anything hidden in his mind even if it pertains to his own life. He is hot tempered and obstinate and forms a wall to his progress in life. He takes spot decisions without seeing the opportunity and circumstances, which ultimately leads him to a precarious state. He will not hesitate to cause problems and troubles to those who rendered him all possible help when required. He is prone to drugs and alcohol and goes out of control quickly damaging family life.

3. Education, sources of earning/profession: He will leave his home at a very young age and seek refuge in a distant place. He earns his bread with his own effort. Constant change of jobs or professions is noticed. Life will be full of trial up to 50 years of age and stability starts only after that. There will be maximum trouble from 18 years of age to 26 years of age and he will have to undergo financial problems, mental agony or mental disarrangement. There will be a beginning of progress towards stability from 27th years of age at very slow pace up to his 50 years of age.

4. Family life: He cannot expect any benefit from his mother and co-born because they become his enemies. His near and dear ones generally dislike him. He likes to keep a separate identity and existence. His spouse will always have an upper hand. There are health problems of his wife or separation due to some unavoidable circumstances.

5. Health: There is frequent temperature, dysentery, cough, cold, asthmatic attach and stomach problems. He may have severe pain in arms and shoulders.

Female Natives

1. Physical appearance: She has long arms, height above average, broad face, short and curly hair.
2. Character: She has strong emotions, passionate jealousies and deeper loves. She is intelligent, thoughtful, perceptive and good organiser.
3. Education, sources of income/profession: She is active in sports. She will get medium academic education. She is often contended herself in the home, seeking to enrich her own life through her husband's success. She not involved in any earning job.
4. Family life: She lacks marital harmony and notices loss of children. She is subject to harassment by in-laws. Her neighbours and relatives put poison in her life. She is always worried and her children will neglect her.
5. Health: Her health will not be so good. She suffers from disorder of the uterus. She is also prone to prostrate gland enlargement or pain in arms and shoulders.
Positive traits: He/She is hard working, result oriented, willing to shoulder responsibility and position, and provides supportive protection to the weak and unsheltered persons.
Negative traits: He/She has temper and ego, hot headedness, obstinacy, and unwillingness to compromise.
Career: Career opportunities are military at managerial post, investigation, protection of law and order, and entrepreneurship.

19. Moola

General: He/She has ooze charm and is attractive, fast walkers, endowed with good qualities, unorthodox, strong, determined, fair in complexion, close to politics & power, wealthy, famous, kind hearted and having long life and also very ambitious. He/She is financially successful and leads a materially comfortable life. The work area is gardening, medicine, travel, tourist or agricultural. He/She brings people before the public and gives good judgement. He/She enhances the prosperity and comforts his/her mothers. He/She is sage-like, wealthy and has vehicles, does permanent jobs and defeats his/her enemies. But, if born in the initial three parts of this Nakshatra, he/she causes destruction to his family. (i) If born in the first pad, he/she causes loss to his father, (ii) If born in the second pad, he/she causes loss to his mother and (iii) If born in the third pad, he/she causes loss to the family wealth.
Male Natives

1. Physical features: He has good physical appearance. He will have beautiful limbs and bright eyes. He will be the most attractive person in his family.

2. Character: He has a very sweet nature and is a peace loving person and has a set principle. He can stand and penetrate against any adverse tidal wave and reach the destination. He is not bothered about tomorrow. He keeps all the happenings in the hands of god. There will be frequent changes of profession or trade with no stability. He is always in need of money. He does not earn anything by illegal mode. He believes that all that is taking place on the earth is due to the blessings of God.

3. Education, sources of earning/profession: He has much better success in a foreign land and earns his livelihood in a foreign place in the field of fine arts or a writer.

4. Family life: He cannot have any benefit from his parents whereas he is all self-made. His married life will be satisfactory. His spouse has all the requisite qualities of good wife.

5. Health: He will be affected with is tuberculosis, esonophilia, and paralytic attack. There will be some severe health problems in his 27th, 31st, 44th, 48th, 56th and 60th years of age. He may be addicted to drug and intoxication.

Female Native:

1. Physical features: She will have reddish colour and her principal teeth will not be close, but at great distance, which is wealthy sign.

2. Character and general events: Mostly he is pure hearted, but very much adamant. Since she lacks knack of dealing, she quite often lands into problems.

3. Education, sources of earning/profession: She does not acquire much education. She doesn't show any interest in studies. She spends more than two terms in the same standard or class. Ultimately she leaves further education. The only exception is that if Jupiter is placed in opposition i.e. aspects or placed in Magha Nakshatra, she may be a doctor or employed in an envious position i.e. she will have excellent academic record and reach to the top.

4. Family life: She cannot enjoy full conjugal bliss, but a separated life, mainly due to the death of her husband or divorce. This result cannot be blindly applied as other planetary positions of favourable nature will nullify the bad effects. There may be delay in the marriage and also some hurdles. If the position of Mars is unfavourable she will have to face a lot of problems either from her husband or from children.

5. Health: She is prone to rheumatism, backache or pain in arms and shoulders.

Positive traits: He/She is hard working, committed, and intelligent to innovate ideas to make the road to success, highly optimistic and can come out of the toughest of situations. **Negative trait:** He/She is provoketive, stubborn and adamant. He/She has lack of knowledge to deal with things which often lands him into various problems.

Career: He/She does well in providing religious and financial advice, self business or self employment, in the foreign soil where he has a high possibility of success as compared to the native land. He/She switches his jobs frequently and possesses varied career interests.

20. Purva Ashada

General: He/She is kind hearted, hugely popular in the society for his helping nature, cordial behaviour, tall, lean, attractive physical appearance, dominating personality and possesses extraordinary argument capabilities. He/She has leadership abilities, strong sense of justice, politeness, steadiness and a beautiful wife. He/She is caring, gentle, a hard worker, supportive, generous and has many friends and is an excellent manager. He/She has a forgiving nature and does not hold grudges against others. He/She prefers a simple life above fame and fortune. Careers that will suit are financial planning, social work, NGO's volunteer work. His/Her married life is satisfactory. He/She could suffer from ailments like Tuberculosis and uterine problems. He/She is red in complexion. He/She will travel overseas. He/She is wealthy, tall, good speakers, endowed with attractive face, obedient and lives with respect. He/She is loved by people, and experts in almost all fields. He/She is always successful and success comes at an early age. Great oratory abilities make him/her successful at debates.

Male Natives

1. Physical features: He is lean and tall. His teeth will be very beautiful, eyes bright, waist narrow and arms long. In other words, he has good attractive physical features.

2. Character: Nobody can defeat him in argument. He has extraordinary convincing power. He will not under any circumstance subdue to others, whether he is right or wrong. He can give a lot of advices to others. In decision-making, he is very poor, but too much dominant. He hates external show. He is God fearing, honest, humble and far from hypocrisy. He will be highly religious and devotes much of his time in Puja or others religious act. He is good collector of antiques. He may also take interest in writing poems.

3. Education, sources of earning / profession: Even though he can shine in almost all the fields, he is particularly fit for doctor's profession or fine arts. He is a not fit for any business. It will be period of trial and error up to 32 years of age. Thereafter he slowly starts climbing ladder of success. It is very good Period between 32 years to 50 years.

4. Family life: He cannot enjoy any benefits from his parents. However, he will be lucky to have benefits from his co-born, particularly from his brothers. He will be spending most of his life in foreign land. His married life is happy. His marriage may be delayed. He is more inclined towards his wife and in-laws. He has the most talented and respectful children, who will bring name and fame to his family. There may be at the most two children.

5. Health: His health is not good. He is prone to severe whooping cough, breathing trouble, bronchitis, tuberculosis, heart attack and malaria or esonophilia.

Female Natives

1. Physical features: She is extremely beautiful and magnetic. She will have long nose and graceful look, fair complexion, brown coloured hair.

2. Character and general events: She has energy, enthusiasm, vigour and vitality. She is greedy and is not obstinate. She will come to a final decision after deep consideration. She is straight forward. She will be fond of dogs and other pet animals. She makes promises but will not be fulfilled. She hates to her parents and brothers. She will be leader among her relatives. She is a determined, truthful character.

3. Education, sources of earnings / profession: She is educated. She may be a teacher, bank employee or attached to religious institutions. If mercury is placed with Moon, she may earn as a publisher or writer.

4. Family life: She is very good to her family and more attached to her husband. Benefit from the children will be to a limited extent only.

5. Health: Her health will be good. She will have acute disorder of the womb and uterus.

Positive trait: He/She possesses helpful and empathetic behaviour and is confident, self made and true to himself and the world. He/She provides with a lot of advice and ideas, which are helpful to people many a times.

Negative traits: He/She is arrogant, poor decision maker and cannot be deterred from doing what he has decided to do once.

Career: He/She has a variety of career interests, like medicine and fine arts. He/She is not recommended for ventures where important decision making is needed. Fields of science and philosophy are good career interests from the success point of view.

21. Uttara Ashada

General: He/She possesses good leadership skills and goes up to a respectable position in administrative services and has a beautiful and supportive life partner. He/She is good looking and attractive with fair complexion charming and gracious personality. He/She is very honest, plain hearted, very simple, straight forward and do not like hypocrisy and show off. He/She has political leadership ability. Careers that will suit are research, teaching, counselling, banking, notoriety and publishing. He/She will have a happy marriage and prove to be a devoted spouse. He/She may have more than one marriage. He/She is happy with his sons. He/She cannot lie is successful, and victorious. He/She does physical exercises, loved by people, and travels a lot. He/She is giver, gets enjoyment from wife and endowed with many children.

Male Natives

1. Physical features: He is a well proportioned body, broad head, tall figure, long nose, bright eyes, charming and graceful appearance with fair complexion.

2. Character: He gives respect to all and does god fear. There will be black mole around his waist or on the face. He will not deceive or cause any trouble to others. He will not take any hasty decision. Even in the state of conflict he cannot utter harsh words directly to any person. He i shoulders many responsibilities at a young age. He is subjected to maximum happiness at one stage and the maximum unhappiness at the next.

3. Education, sources of earning/profession: He has to be very careful in making any collaboration, otherwise, failure is certain. There will mark all round success and prosperity after the age of 38 years.

4. Family life: His childhood is better. His married is very good between the age of 28 and 31 years with a responsible and loving wife. The health of the spouse will be a cause of concern for him. His wife will be having problems of acidity or uterus disorder. He lacks happiness from the children, who will be main cause of concern.

5. Health: He is prone to stomach problems, paralysis of limbs, pulmonary diseases.

Female Natives

1. Physical features: Her forehead will be much wider, nose larger, eyes attractive, teeth beautiful, body stout but not so beautiful hair.

2. Character: She is the "obstinate daredevil". Her utterances are not good and jumps into conflict with others, but leads a very simple life. She is very well fit to the proverb "There are two horns for the rabbits I have got".

3. Education, sources of earning/profession: Generally, she is educated. She may be a teacher, bank employee or attached to religious institution. If Mercury is placed with Moon, she may earn as a publisher or a writer.

4. Family life: She does not enjoy married life due to either separation or some other problems connected with husband. She observes all the religious formalities. She derives complete happiness and contentment by marrying a Revati or Uttara-Bhadra Pad boy.

5. Health: She will have wind problems, hernia or uterus problems etc.

Positive traits: He/She is very much humble and respects others regardless of his social or financial positions. He gives utmost respect to women. He is very modest and doesn't utter anything bitter against anyone in spite of having strong opinion difference.

Negative traits: He/She lacks motivation to do things in life and expect appreciation of his work, failing which he feels unhappy and depressed.

Career: He/She achieves success in career after 38 years of age. He may have to face initially some struggle. The fruitful career interest includes engineer, architects, mechanical engineering jobs and working with maps and planning.

22. Shravana

General: He/She is truthful, stable minded, endowed with many children and friends, winning over his enemies, large-hearted, knowledgeable, brave, learned, wealthy, religious, and famous, and pure in heart. He/She is loved by all. Shravana bestows with immense wealth of knowledge, a lot of respect to their parents and caring of them with the utmost sincerity to duties and responsibilities, educational proficiency and intelligence. He/She is experts in various art forms like music, dance, and drama, and acting. They do not ever hurt or pose problem to anyone wilfully. They are extremely religious, pious and honest which is very much revered by the people around them. They always maintain peace with their surrounding situations and people. He/She is good speaker, enjoys life and has many sons, organizational leadership skills. He/She is financially successful, and leads a materially comfortable life. He/She will be a good manager or project leader and will do well in business. He/She works for charitable organizations. Careers that will suit are engineering, management related jobs, charitable and social institutions. Marital life will be extremely happy and blissful. He/She is egoistic by nature. He/She can get service in education field and may be great teacher, or perpetual students. Counselling is a gift and he/she has the ability to truly listen.

Male Natives

1. Physical features: He has a very good attractive physical feature. His height is small. His face may have a mole or some other marks appearing to be a kind of disfigurement. He has black mole beneath his shoulder.

2. Character: He is very sweet in the speech and principled and expects his surroundings to be very clean and neat. He takes pity on the condition of others and tries to help them as far as possible. He is a very good host. He is God fearing and has full Guru Bhakti. He is a believer of 'Satyameva Jayate' (Truth only will win). Neither he will reach to the top nor will he be at the bottom. He will shine well. Since he has to shoulder many responsibilities and spend for fulfilling his responsibilities he will always be in need of money.

3. Education, sources of earnings / profession: He will undergo several changes up to 30 years of age. He will mark stability in all walks of life between 30 years to 45 years of age. He can expect remarkable progress both economical and social beyond 65 years. He will take up mechanical or technical work or engineering connected with petroleum.

4. Family life: His married life will be filled with extra ordinary happiness. He will have most obedient wife with all good qualities. He has sex relationship with other ladies.

5. Health: He suffers from ear problem, skin disease, eczema, rheumatism & tuberculosis.

Female Natives

1. Physical features: She is tall and lean. Her head is comparatively big with broad face. She has large teeth. There will be distance between the front teeth with prominent nose.

2. Character and general events: She believes in charity. She is highly religious and visits several holy places. She is internally very cunning. However, sympathy towards the weak and generosity towards the needy are the notable features. She is a 'chatter-box' without having any control over her tongue.

3. Education, sources of earning / profession: She is Illiterate or medium educated and will be engaged in agricultural field, or employed as typists, clerks or receptionist etc.

4. Family life: She is a pride in family but has frequent friction with her husband.

5. Health: She has skin disease and prone to eczema, filarial, pus formation, and tuberculosis, leprosy of low intensity and ear problem.

Positive Traits: He/She is large hearted and compassionate, and helps people with his knowledge, which makes him lovable among people. He/She has courage, forgiving gratitude and a comprehensive personality in the society and respected for that.

Negative Traits: He/She is shrewd, selfish and adamant and has wrong steps or unorganized efforts and harms others for the sake of accomplishing their target.

Career: His career spans through services to having his own business and is successful in career. He/She is interested in engineering, medicine, education, science and many other forms of arts. He/She is financially sound unless he goes to the wrong path.

23. Dhanishta

General: He/She has symphony, prosperity and adaptability, ability in music dance, confidence, stability, dependability, hard work, energy, exceptional sharpness, commercial skills and benevolence, fame, success and prosperity in abundance, striking a comfort zone with the given surrounding, luxury and good life and group centric activities like an artist. He/She is difficult to win, very famous, serves the elders and always protect others. He/She is religious, endowed with many good qualities, wealthy, kind hearted, sad and suffers from diseases like tuberculosis. He/She is after other sex as is loved by them. He/She is a glib talker, skilled in business and astrology. Marital life will be happy and satisfactory. He/She is dark in complexion, cheerful, have all materialistic pleasures, soft-spoken and well mannered. The work area is Ayurveda, mining and engineering in underground works, land, and business of any product, commission agent, metal related work and machinery. He/She is visionaries and good for public relations. Marriage may be delayed or denied. He/She gains fame, recognition and loves travel.

Male Native

1. Physical features: He has a lean body with lengthy figure and stout figure.

2. Character: He has extremely intelligent mind and all-round knowledge. He does not cause any trouble to others and has religious spirit. He is very revenge taking and, if any person causes trouble to him, he waits for an opportunity to bounce upon him.

3. Education, sources of earning/profession: He is the born scientists and historians. He has an inherent talent of keeping secret; he is quite suitable for secret service, private secretaries to the senior executives. His intelligence is beyond questionable. Hence lawyer's profession is the best for him. It will show progress in the earning field from 24th years of age onwards. He should be very careful before putting trust on others.

4. Family life: He will be the uppermost administrator. His relatives will cause a lot of embarrassment and problems. He is more inclined to his brothers and sisters. He will have ample inherited property subject to the placement of planets in beneficial position. He cannot have much benefit from his in-laws. His wife will be an incarnation of 'Lakshmi' (goddess of wealth). There will be improvement in his finance only after the marriage.

5. Health: His health is not so good. He is prone to whooping cough, anaemia.

Female Natives

1. Physical features: She is beautiful and ever sweet seventeen while she crosses her forty or may have ugly appearance due to teeth protruding outside the lips.

2. Character: She has sympathy towards the weak. She is an enforcement master like Shravana. She has congenial atmosphere in the home front.

3. Education, sources of earning/profession: She has mixed talents and education. Hence usually, she will be teachers or lecturers or research fellows.

4. Family life: She will be an expert in the household administration.

5. Health: Her health is not good. She is prone to anaemia, uterus disorders of acute intensity and spoiled blood

Positive traits: He/She has vibrant mannerisms, frankness and easy adaptability, and striking a bond of geniality and harmony with the immediate surroundings, hopefulness, joy and sympathy.

Negative traits: He/She is susceptible to the society and may manifest subsequent negativity in his behavioural traits and has aggression, talkativeness, materialistic ways, covetousness, lust for success and susceptibility to select incompatible life partners.

Career: Careers are related to performing arts, exceptional niche in managerial positions and catering to group activities. Thus career opportunities based on management and entrepreneurship are also suitable for them. They are found to be equally prosperous in medical profession; particularly in specialized branch of surgery, military bands, real estate and scientific research.

24. Shathabisha

General: He/She has care, remedy and healing, intellect and power of intuition, reclusive loneliness, scientific attitude, meditation and reflection, depression and mood swings, strict adherence to discipline and rigid compliance with norms and seldom going beyond the norms. He/She lives in foreign lands and is most sexy. He/She is respected by the world, long lived, well in transactions and suffers from various diseases. He/She is endowed with many children, most selfish, gambler, questionable character, experts in magic, stubborn, especially knowledgeable, and cruel, eat little, wealthy, server, and destroys his enemies. He/She has a noble and an aristocratic demeanour, and can become dangerous when provoked. Careers that will suit are Engineering, Accounts, and medicine and surgery. Marital life is not without friction. He/She is dark in complexion, weak body and has saving habits. He/She may be healers or doctors. He/She can be astronomers as well as astrologers. He/She may be alcoholic and may suffer from diseases difficult to heal or hard to cure

Male Natives

1. Physical features: He will have excellent memory power, wide forehead, attractive eyes, bright countenance, prominent nose and bulged abdomen. He appears to belong to an aristocratic family at the first sight itself.

2. Character and general events: He is of the type 'Satyameva Jayate'. He will not hesitate to sacrifice his own life for upholding the truth. He is born with certain principles; he has to quite often confront with others as he cannot deviate from his principles of life. Selfless service is his motto. He does not believe in pomp and show and has interesting and attractive conversation, which will be highly instructive and educative.

3. Education, sources of income/profession: It will be a trial period up to 34 years of age and after 34 years, it will be the period of constant progress. He practices astrology, psychology and healing arts. His literary capacity and greatness will come to limelight even when he is very young. He is capable of acquiring very fine and high education. He will be eminent doctors and research fellows in medicines.

4. Family life: He faces a lot of problems from his dear and near ones. He always helps to his near and dears. He undergoes to maximum mental agony due to his brothers. He cannot also enjoy much benefit from his father, whereas full love and affection is derived from his mother. He does not have a happy married life. In some cases, when there is severe affliction of Saturn and Jupiter, it has been found that he remains a chronic bachelor throughout his life, but his wife will have all the good qualities of a companion.

5. Health: He is prone to urinary diseases, breathing trouble and diabetics. He is too much inclined in the sexual pleasures and may have sexually transmitted diseases. He may keep illicit relationship with other females. He may have problems with his jaws. He is also liable to suffer from colic troubles.

Female Natives

1. Physical features: She is tall, lean, fairly beautiful, elegant disposition, fleshy lips and broad cheeks with prominent temples and buttocks.

2. Character: She is religious and god fearing. She has hot-temperament and mostly confronts with family quarrel and lacks mental peace. She has very good memory. She is highly sympathetic and generous.

3. Education, sources of income/profession: She has scientific study as a doctor.

4. Family life: She loves her husband, but life is full of problems due to long separation from her husband or widowhood.

5. Health: Her health is not good. She suffers from colic, chest pain, Urinary and uterus.

Positive Traits: He/She has hard work, determination, discipline, ability to cure and heal; powers of intuition and meditation, educational excellence, methodical approaches together with pleasing skills of presentation and powerful memory.

Negative Traits: He/She is susceptible to obstinacy, adamant and head strong, and have farfetched seriousness and flair for loneliness, bouts of angry outbursts and depression, unfriendly mannerisms, inadaptability, long drawn insistence on tradition and inability to innovate and has laziness, rudeness and absence of social skills.

Career: He/She shines exceptionally in careers, on account of his academically oriented nature, memory and skilled faculty of intuition, revolving around medicine, psychology, astrology and astronomy and due to owing to the predominating influence of Rahu, natives are prone to making careers out of politics, business and in those areas where leadership skills are required.

25. Purva Bhadrapada

General: He/She has mystery, supernaturalism and occult phenomena, honesty, principle and benevolence, sincerity, determination, discipline and ability to confront all kinds of physical and mental hardships. He/She is worshiper, teachers, broad minded, respectable, above controversies, soft spoken, and loving his/her relatives. He/She is endowed with children, sleeps excess, helping nature, and has the knack of trade. He/She, most likely, will be employed in a government organization and will gain good promotions. He/She has quite independent life both socially and financially. It will mark remarkable all-round progress in his life between 24 and 33 years of age. It will be his/her golden period between 40 years and 54 years of age, when he/she can establish fully. He/She keeps much restriction on the spending activities. He/She can shine in the field of business, banking, government job, or as a teacher, actor or writer, research worker and astrologer or astronomer.

Male Natives

1. Physical features: He has a lifted ankle of the foot, medium size with fleshy lips.

2. Character and general events: He is peaceful, loving, very simpleton type, very much principled, and likes good food and is a voracious eater. He normally expresses impartial opinion. He does not believe in the blind principles of religion. He is ever ready to lend a helping hand to the needy. Even so, hatred and resistance will be his reward in return. He is financially weak. He is God fearing and performs religious rites. He is moderately rich, but likes to have respect and honour from the public rather than accumulating money.

3. Education, sources of earnings / profession: He has the knack of trade and shines in any type of job he undertakes. He is employed in a government organization with unexpected gains or promotions in the revenue collection department or where cash transactions take place.

4. Family life: He cannot enjoy full love and affection from mother. Mother is away most of the time and he is automatically separated. He can be proud of his father. His father has fame in the field of fine arts, oration or in the writing field and possess a very good moral character. In spite of these good qualities of his father, he quite often confronts with him.

5. Health: He is prone to paralytic attack, acidity and diabetes. He may have problems with the ribs, flanks and soles of the feet.

Female Natives

1. Physical features: She is very beautiful in appearance and has slim figure.

2. Character: Honesty and sincerity are her main characteristics. She believes that let her head is cut down, but she will not deviate from the right path and principles. She is a born leader. She is capable of extracting work from others. She will be successful when she gets power and authority. She has the humanitarian doctrine, and politeness.

3. Education, sources of earnings / profession: Her education will be in the scientific or technical field and becomes teacher, statistician, and astrologer or research worker.

4. Family life: She is more attachment towards her husband and will be blessed with children. She has good ability in the house hold administration. She will enjoy lots of benefits from her children. She will have many children if she marries a Rohini boy.

5. Health: She is prone to low blood pressure, dropsy or swollen ankles, apoplexy and palpitation, perspiring feet and enlarged liver.

Positive traits: He/She has sincerity, dutifulness, endurance and strength, helpfulness, protectiveness and concern for acquaintances and family.

Negative traits: He/She has innate tendencies to cause harm and malice. He/She is prone to socially destructive ways and terrorism. Being misled into dishonesty, negativity and immorality also count amongst his negative traits.

Career: His careers relate to administrative service, business, and governmental responsibilities and teaching profession; in scientific research, writing and acting. Lines related to astrology, astronomy and analysis of scriptures are also beneficial for him.

26. Uttara Bhadrapada

General: He/She is wise, prominent amongst his/her clan, wears jewels, a lawyer or arbitrator, always does constructive jobs, earns wealth, giver, endowed with children, religious, winners of his/her enemies, happy, determined, sexy, learned, great sacrificed, loved & respected in all circles, fickle, and suffers from fluctuation of funds. He/She is rich, famous, attractive, charming, loyal to friends, follows the virtuous path, always willing to come to aid and will always stand out in a crowd. He/She learns to face the world with confidence. He/She has the 'possession of lucky feet', and represents prosperity, strength and martial qualities, material bondage and rainfall. Malefic or the benefic aspects of ruling Saturn plays an important role in determining his personality. He/She has determination, wisdom and experience, changeability, attracting virtue and mannerisms and highly ethical standard. He/She is intelligent financially successful and self-sufficient. He/She is a loving and merciful person and always willing to reach out to others and spend a lot of time and money on charity. He/She is good at solving problems. Marital life will be happy, harmonious and satisfactory and children will be a source of joy and happiness. There is wealth usually as a gift or inheritance, usually latter in life. He/She is extremely protective of his/her loved ones. His/Her wisdom seems to have magical powers. He/She has happy home life and is blessed with good children.

Male Natives

1. Physical features: He is most attractive and innocent looking person. There is an inherent magnetically force in his look. If he looks at a person with a mild smile, rest assure, that person will be his slave.

2. Character: He keeps equal relationship with high and low people status of the person. He has a spot-less heart. He does not like to give troubles to others. He has temper always on the tip of his nose. However, such short-temper is not of a permanent nature. He will not hesitate to sacrifice even his life to those who love him. At the same time once he is hurt he will become a lion. He has wisdom, knowledge, and personality. He is expert in delivering attractive speeches. He is capable of vanquishing his enemies and attains fairly high position. He is sexually inclined always and desirous of being in the company of other sex.

3. Education, sources of earning/profession: He can attain mastery over several subjects at the same time. He is academically much educated and his expression and knowledge put forward to the world will equal to that of highly educated and learned persons. He is much interested in fine arts and has ability to write prolonged articles or books. Laziness is a remote question for him. Once he opts to undertake a job he cannot turn back till that job is completed. He is employed and reaches to the top. Even born in poor family, he is employed and reaches to a good position and he always receive reward and praise from others. His stability in life or the slightest upward movement begins after his marriage. He starts his livelihood at a very young age say 18 or 19 years of age. He will have important changes in the professional field at his 19th, 21st, 28th, 30th, 35th and 42nd years.

4. Family life: He cannot virtually derive any benefit from his father. He leads a neglected childhood. He is normally subjected to a life away from his home town. His married life will be full of happiness. He will be blessed to have a most suitable wife. His children also will be an asset, most obedient, understanding and respecting. He will be blessed with grandchildren also. He is an ornament in his family.

5. Health: His health will be very good. He is non-care about his own health. Hence he will search for a doctor only when he is seriously ill. He is prone to paralytic attack, stomach problems, piles, and hernia.

Female Natives
1. Physical features: She is of medium height with stout body. She will have large and protruding eyes.
2. Character: She is a real 'Lakshmi' (goddess of wealth) in the family. Her behaviour is extremely cordial, respectful and praise worthy and has adaptability.
3. Education, sources of earning/profession: Employed females can attain good positions due to her own effort. She is best suited to the profession of a lawyer or arbitrator. She is also a good nurse or a doctor.
4. Family life: She will be a gem in any family, where she is born or married. In other words, their footsteps are sufficient to bring in Lakshmi (goddess of wealth).
5. Health: She is prone to rheumatic pains, acute indigestion, constipation, hernia and in some cases tuberculosis of low intensity.
Positive traits: He/She is kind, logical, and deft at skills of calculations, and has serenity and sense of equality and justice. He/She has intrinsic innocence along with an innate capacity for mesmerizing people and a unique potential for gaining with experience.
Negative traits: He/She has susceptibility to fits of anger and is unnecessarily drawn into controversy and disputes and has temperamental, laziness, inertia and loss of control are their other negative traits.
Career: With wealth of knowledge gained by experience, he makes favourable impression on business, entrepreneurship and managerial positions. Career opportunities centring on consultancy, justice and counselling are also meant for him.

27. Revati

General: He/She as power to transcend, idealism, benevolence and sustenance, helpfulness, elements of graceful sociability; streaks of knowledge and educational excellence. He/She reaches to people and has sympathetic approach to people and situations which gives commanding awe, reverence and love. He/She is endowed with good qualities, always thinking about home, wise, wealthy, enjoys life, learned, fickle, earn good money, sexy, givers, pure minded, capable, bright, helps others and moves in foreign lands. He/She is extremely likeable and easy to get along with. He/She will enjoy great prosperity.

He/She has amicable nature, controls his senses, acquires wealth and possesses sharp intelligence. He/She enjoys being a robust physique and good health. He/She should be careful with finances or could incur a lot of debt. His/Her careers will suit in Counselling, Psychology, Psychiatry, Social work and volunteer work in spare time. Marital life will be very harmonious and spouse is very compatible. He/She is fair in complexion, peaceful, Pandit, good working, soft spoken, famous in family, wealthy, and enjoys family values. He/She gets very angry. He/She has an affinity or love towards small animals. He/She may have disappointment early in life. He/She is a bit weak and may be prone to childhood illnesses. He/She loves water, and usually benefits from living by the water. There is a deep devotion and faith to God.

Male Features

1. Physical features: He will have very good physique, moderately tall and symmetrical bodies with fair complexion.

2. Character: He is very clean in his heart, sincere in his dealings and soft-spoken. He cannot keep anything secret for too long time. He will not blindly believe even his loved ones. But once he takes somebody into confidence, it is not easy to keep away them out of his attachment. He is very hot tempered. He tries to observe the principle, which he feels, is correct and resist tooth and nail till the end. He is God fearing, superstitious, religious and rigid in the observance of orthodox culture and principles. Hence he also enjoys the maximum blessings of the Almighty.

3. Education, sources of earnings / profession: He studies historical research and ancient cultures like astrology and astronomy. He is a good physician or poet. He is employed in a government organization with the most success. He is mostly settled in foreign countries at quite a reasonable distance from one's own birthplace. He will come up in his life with his own efforts. His intelligence and abilities are the inherent qualities at birth. He cannot stick to any particular field of job for long time. He cannot gain much till his 50th years of age. It will be a good period between 23 years and 26 years whereas period between 26 to 42 years of age will mark a lot of problems financially, mentally and socially. It is only after his 50th year he can think of worry less and stable life.

4. Family life: He cannot get help from his relatives or father. He is unlucky to enjoy the benefits from his near and dear ones. However, his married life will be moderately good. His spouse will be quite an adjusting type.

5. Health: He is prone to fever, dysentery or dental diseases, and intestinal ulcers.

Female Natives

1. Physical features: She will be extremely beautiful. She can be recognized easily even out of thousand ladies due to her magnificent attractive personality.

2. Character and general events: She is somewhat stubborn. She likes to exercise authority over others. She is God fearing, religious and rigid in the observance of orthodox culture and principles and highly superstitious.

3. Education, sources of earnings / profession: She will have her education in the field of arts, literature or mathematics. She will be in the general line i.e., she may be a telephone operator, typist, teacher or a representative of companies. When good aspect of benefice planet is received, she may be an ambassador or a person representing her country for cultural or political matters.

4. Family life: She will enjoy a most harmonious married life.

5. Health: She may have some deformities of the feet, intestinal ulcers or abdominal disorders. In some cases, deafness has also been noticed.

Positive traits: He is pillar of strength and support to people and society at large, oriented towards religion, intrinsically constructive by nature and has ethics and principles, appreciable sense of decorum, sophistication and culture, and optimism. Ability to reach out and serve others being the major fulcrum of his behavioural characteristics, he is often seen holding magical sway over others.

Negative traits: He is susceptible to idleness and depression and has strict adherence to principle and unexpected norms of behaviour obstinacy, temper, head strong ways, orthodoxy and superstition.

Career: Career involves scientific research, archaeological survey; poetry, literature in ascendance. He can also be successful in careers related to administration, astrology and astronomy due to his wisdom and social acumen.

5

House (Bhava)

The Sanskrit name of the House is "Bhava" or "Sthana". The Houses govern all the "events" of our lives. If we look directly east and point our hand at the Eastern horizon where the Sun rises, we are pointing to the first House, which is called Lagna or Ascendant. Each House/Ascendant/Lagna occupies the 30-degree span of space below the Eastern horizon. Directly across from it or 30-degree span of space above the Western horizon is the 7th house of the Natal Chart. From the first to the 7th are the 2nd through 6th houses. Directly over our head is the 10th house, which is called MC. That is the straight up into space where the high noon Sun beams down on us. Straight below our feet is the 4th house. Suppose 9 O' clock position of the clock represent the Eastern horizon. So, the first house governs at 9.00 to 8.00 O' clock position of the clock. The 2nd house is just below it, i.e. at 8.00 to 7.00 O' clock position of the clock and so on. Thus, the 12th house governs at 10.00 to 9.00 O' clock position of the clock, and this brings us back around to the first house of the Natal Chart. Thus, the entire 360 degree circular span of space surrounding us are divided into 12 equal part of 30 degree sections

5.1 Concept of House & Lord of House working

In Vedic astrology, one sign always occupy one house. Whatever sign is raising the entire portion of sign is considered in the first house. In this way there is an exact correlation between Sign, Sign Lord and House. The Lord of the sign is the Lord of that House. The lord of the first House and its position in the other house will together determine the body. The body gets shape of the house in which the First House ruler has come to stay in and that house greatly affects the body. The matter related to the body is affected during periods ruled by the lord of 1st House in the Vimshotari Dasa. Secondly, the planets associated with the 1st lord greatly affect the affairs or events of body of the native. **Example:** If the lord of the first House is placed in the first house; the body becomes the focus of the life. If the lord of the first house is placed in the second house then the body will be greatly influenced by the second house, i.e. possession and wealth. The native is busy with accumulating wealth whether he is coming from a strong family background or otherwise. If the lord of the first House is placed in the third house, then the body is found to be connected with younger brothers, sisters, and the intelligence, good education, college degrees and other things that the third house rules. If the lord of the first house is placed in fourth house then the person will be greatly concerned with family affairs, home, parents and house & land properties in his life. The native will spend a lot of time at home, rather than on the road or at work. Since the fourth house is tenth from the seventh house, it also means that the native will have some connection with the career of the spouse. This means that the person might work for spouse or work with her spouse or perhaps a home-based business. If the lord of the first house is placed in the fifth house then the children, creativity, romance and helping nature becomes the focus of his bodily activities. If the lord of the first house is placed in the sixth house, then the deaths, diseases and enemies overwhelm the person. If the lord of the first house is placed in the seventh house, then the person is heavily focused on his partner, his spouse during his life. If the lord of the first house goes to the eighth house in a horoscope, then serious physical harm; unforeseen difficulties and serious problems comes to the native. If the lord of the first is placed in the ninth house then there is an overall fortunate protective cover on the native throughout his lifetime. If the lord of the first is placed in the tenth house, then the native is heavily focused upon the attainment of career, status position and success. If the lord of the first is placed in the eleventh house, then the person is heavily focused upon the achievement of desires and certain gainful things in life. If the lord of the first is placed in the twelfth house, since this house rules charity, donations, and losses, the person will be busy in these ways throughout his life. In other words, he may donate himself to these causes. Or, there may be a lot of loss in his life.

Thus, the placements of the lords have their first level meanings. Then the second, third, fourth and deeper and deeper meanings depend on the astrologer's ability to read the complexity of the house relationships with the planets and signs.

5.2 House-to-House Relationship

Houses are related to each others. The second house to any of the other house will indicate the wealth or money acquired from that house. The eighth house is second to the seventh house and will indicate the wealth and money of the spouse. The sixth house is the 2nd to the fifth house and will indicate the financial prosperity of the children. The eleventh house is the second house to the tenth house and will indicate financial gain through the career. Similarly, the twelfth house to any house is the end or loss to the house in question.

Example: The sixth house is the twelfth to the seventh (marriage), so it represents the end of the marriage or wife. The third house is the end of the mother because it is the twelfth to the fourth (mother), so it represents the end of the mother. The third house is our energy, will, and life force and the second house is the twelfth house from our life force and so is the loss of our life force. Thus, Maraka house (2nd) derives its meaning from this principle. It is the twelfth from the third. Similarly the eighth house is our length of life so the twelfth from the eighth, i.e. the seventh house would be the loss of life, i.e. the death house. So, 2nd and 7th houses are called Maraka houses.

The 4th house indicates about "mother". 5th house will indicate the mother's money as it is the 2nd house from 4th house. 6th house will indicate the mother's younger siblings as is the 3rd house from 4th and so on. 1st house will indicate the mother's Career, because this is the 10th house from mother's house. These are called the "compound" or "secondary" houses. The 12th house rules loss, and therefore, the 12th house from any house is the loss of that house. Take the 9th house which rules fortune, religion or Dharma, the father, the spiritual master and guru, and God's grace. The 9th house is 12th from the 10th house. This means that the house of Dharma or religion is the house of loss to the house of career, profession and position. The 10th house rules not only career, but mainly it rules rise and status in material life. It rules standing up tall and straight and getting some position, some fame, and some power in this material life.

Similarly, the 5th house is 12th to the 6th house. Amongst other things, the 5th house rules winning at the lottery and the 6th house rules debts, therefore, it is easy to understand that if we win at the lottery, we can cure all our debt problems. So, the 5th house, which is a money house, puts an end to the 6th house, the house of debts. The 6th house is 12th to the 7th house. The 7th house rules marriage, the spouse and partners in our life. Therefore the 6th house rules the loss of the partner or loss of the spouse.

The 7th house is 12th to the 8th house. The 8th house rules the vital source of energy, which is longevity and so, the 7th house rules the end of vital energy. The 8th house is 12th to the 9th house, which rules father. The 8th house, therefore, represents loss of father.

Therefore, if we find the 9th lord in the 8th house in a chart, it is often found that the person lost his father.

5.3 Classification of House/Bhava:

Houses Categories	Houses	Effects of houses
Trikona (Trine) or Kona or Auspicious houses	1, 5, 9	They are the most auspicious or Benefic houses. They give fortune, luck, bring spirituality and well being if, unaffiliated.
Upachaya (Pratipas/Trika)	3, 6, 11	Upachaya means "improvement" and are considered auspicious. Life improves and gets better over time with these houses if, unaffiliated.
Dustasthana (Trikas) or Malefic houses	6, 8, 12	Trikas are the most in-auspicious or malefic and deal with suffering, ill health, disease, death, loss and sorrow. The rulers of these houses will inflict this type of suffering.
Bhoga	2, 4, 10	Bhoga are considered auspicious and deal with the pleasure and luxury in all respect such as Cars, furnished sweet home and other part of life.
Kendra	1, 4, 7, 10	Kendra is considered auspicious. The planets in Angles give effects in one's boyhood.
Panapharas	2, 5, 8, 11	The effects of planets in Panapharas are felt in the middle age.
Apoklima (Cadent)	3, 6, 9, 12	The planets in Apoklima give result at the conclusion of the life, i.e. old age.
Maraka	2, 7	Maraka means "killer". The rulers are considered the killer sometimes. They are prominent when death or injury occurs.
Lakshmi	4, 10	The planets in Lakshmi gives home, happy life, conveyances, happiness, treasure, lands and buildings, heritage, real estate, good profession or livelihood or honour, living in foreign lands, reputation, business and social activities.

| Vishnu | 5, 9 | The planets in Vishnu provide pleasures through children, love affairs and romance, knowledge, royalty or authority, and fun, long distance travel and make fortunate, and gain of spiritual knowledge, |

Aims of Life	Houses Number	Effects of houses
Dharma	1,5,9	They relate to our sense of purpose and the spirit that moves us such as self through 1st; creative expression through 5th; and our spiritual beliefs and truths through 9th house.
Artha	2,6,10	They relate to our achievements, such as money and material through 2nd; hard work through 6th; and the public recognition and career through 10th house respectively.
Kama	3,7,11	They relate to our sense of conveying our ideas, needs, and desires, such as, the need of a life through 3rd; partnership through 7th; and feel connected to every one through 11th.
Moksha	4,8,12	They relate to our liberation or freedom of soul through 4th; past essence of the soul through 8th; and about releasing all attachments to the world through 12th.

5.4 House Signification (Karaktva):

The Karaka of the twelve Houses and their Signification are as given below. If the Karaka planet is in the house by position or aspect that house, it gives very good effect to the native.

1st House (Thanu/Tanu Bhava): The First house or Lagna called Tanu Bhava represents body looks and soul, head/Brain, personality traits, longevity, health, character and nature, life style, complexion, inherent disposition, vitality, temperament, ego, paternal grandmother's wealth, maternal grandfathers' wealth, residence abroad, livelihood and pride. Benefic in the Ascendant will endow with servants, happiness and robes, while malefic give adverse effects in regard to these. It is one of the most important and auspicious house in the horoscope.

2nd House (Dhan Bhava): The Second house called Dhana Bhava represents wealth, money, quality of speech, family, face, right eye, mouth, food, charity, death, primery education, one's deposits, income, friends, sanyas, and security. It predicts material and financial resources and the ability to earn money, income, and inflow of finance, wealth possessions, precious stones possessions, domestic life, bank position, and understanding with family members, law suits and domestic comforts in general. The second house also projects any accident, nature of accident and how death will come. A benefic in the 2nd will, however, cause his/her death in during its mahadasha.

3rd House (Sahaja / Bhatru / Vikrama / Parakrama / Sahottha / Yodha Bhava): The Third house called Bhatru Bhava represents younger brothers and sisters, sibling, co-born, Business, sports, throat and singing, voice, music, Servants and subordinates, communications, higher secondary level education, talents and skills, short distance travels, neighbour, surroundings, relatives and relations with them; boldness, parent's death, arts such as theatre or filmy arts, filmy direction, painting, drawing, and success through own efforts, competition, hearing and father in law. The third house rules menial work, cough, respiratory system and partition of property.

4th House (Bandhu Bhava): The Fourth house called Matru Bhava represents mother, home, relatives, office or factory, emotions, domestic and house related happiness and luxury, landed and house property, mental peace, chest and lungs, higher education such as master degrees, or doctorates, home affairs & home pleasure, possession of vehicles/conveyances, treasures, conditions at the old age, matters of the heart, inheritance and false allegations, pleasure trips, savings, cattle and pets.

5th House (Santana/Putra Bhava): The Fifth house called Putra Bhava represents children, intelligence and knowledge and intellect, creativity, mantra, tantra and pooja, speculation, love affairs, recreation, romance, creativity and stomach. 5th house represents teaching, principal and gynecologists, sports, relationships, luck, legal or illegal amusements, authority, and pregnancy. It predicts abortions, politics, good karma, destiny, lotteries, and self-projection in order to please prominent people or the public or a boss at a job. The fifth house indicates wanderings, arbitration, and higher education. It covers number of children, their longevity and their character, status, sex of children and education of children. It covers intuition, previous karmas and father's side of influence.

6th House (Ari Bhava): The Sixth house called Shatru Bhava represents health, illness, injuries, loans, sports, enemies and opposition, digestive system, quarrels, court cases, litigation, maternal uncle and aunt, servants, work environment, jobs & service, step-mother, imprisonment, medical profession, food, restaurants, subordinates, obstacles in life, mental worries, calamity, employee and hard work. The sixth house denotes fear from thieves or enemies, fighting, misery and success over enemies, loss of moneys, cheating, danger and calamities (troubles) through women.

7th House (Yuvati/Kalatra/ Jaya Bhava): The Seventh house called Kalatra or Juvati Bhava represents spouse, partners, sex, marriage, business, trade, employment in a private firm, reproduction and genital organs, sexual enjoyments, Kundalini Shakti, sexual organs and diseases thereof, death, relationships and signifies kidneys. 7th house represents married life, travel, conjugal happiness, loved one, divorce, honour, residence and reputation in foreign country, interaction with others, attitudes, sexual passions, open enemies, impotency, desire, disputes, relationships with wife, family life, age of getting married, wife age, health and her nature, journeys to distant places, loss through females, relationship and freedom.

8th House (Randhra/Ayusthan Bhava): The Eight house called Ayusthan Bhava represents longevity, destruction, accidents, physical pains, inheritance, legacies, death and reasons of death, underground wealth, historical things, monuments, parental property, longevity, failure, family of spouse, needs, life's secrets, joint resources, anus and sex power. 8th house represents financial windfalls, lottery winnings and physical pains. It represents parental property, worries, finances through unfair means, obstacles, gain from in-law, mode of death, imprisonment, enemies, support from others, the intimacy with other; struggles, Mafia & underworld, bankruptcy, obstacles, surgery, research, intuition, and long-term sickness, monetary gains from partner, misfortunes, trouble from enemies and loss of property. Malefic in the 8th will cause loss of spouse in the Dasa periods of the Lord of the Navamsa occupied by the 8th Lord.

9th House (Dharma Bhava): The Ninth house called Bhagya Bhava represents luck, prosperity, guru, father, religious and spiritual progress and knowledge of the scriptures, sadhana, pilgrimages, long journeys like foreign travel and foreign trade and dealing with foreigners, grandchildren, higher studies like doctorate, knowledge of foreign languages, grandparents, and signifies hips. 9th house represents fortunes, prosperity, writing books, powers of foresight, religious institutions, teaching, lawyers, fame, happiness, and unexpected gain of wealth, gain from lottery and affluence, association with good people, inclination toward God, religious and social work and fame in it, and satisfaction and fulfilment of desires.

10th House (Karma Bhava): The Tenth house called 'Karma' represents profession, business, authority, power, honours and achievements, acts (karma) one does, father, government service, politics, management, career, status, mother-in-law, prestige, reputation and signifies knees. It increases public image and makes the native's parent of greater influence, and provides power or authority like bosses, judges, and big stars. 10th house signifies popularity, status, activities outside house, pleasures, government favour, command, adopted son, worldly activities and moral responsibilities, livelihood, living in foreign lands, debts, reputation, social activities, social position, fame, wealth of the father, earned money, meritorious deeds, hardship in work, service, and foreign place of settled life, promotion and number of promotion.

11th House (Labh Bhava): The Eleventh house called Labha Bhava represents gains, sources of income, money gain, elder brothers and sisters, friends, long distance travels, air plane travel, entertainment, friend's circle, daughter/son-in-law, associations, accumulated wealth, ambition, social life, association and club, emotional attachments, son's wife, quadrupeds and attitude towards them, desire in life and colleagues or co-workers, love affairs and girl friends, honour and social success, clothes, the staple food, gold, and gain of wealth, pleasure, followers, dependent, insight, power of overcoming obstacles, redemption, worth of garments and signifies legs.

12th House (Vyaya Bhava): The Twelfth house called Vyaya Bhava represents losses, waste, expenses, long journeys like foreign countries and residence in foreign countries, imprisonment, death, sadhana and Moksha or final liberation, bad habits, hospital, quests, export and import, feet, sleep, donation, foreign stay, subconscious, psychological issues, secrets disputes, and signifies mental agony, bodily injury. 12th house covers one's own death, good food & comforts, bed and couch pleasure, donations, miseries, sufferings, troubles, betrayals, law suits, imprisonments, hospitalisation, conjugal relations with opposite sex other than wife, contacts, misfortunes, secret enemies, spiritual liberation, sea or ocean travel, interest in arts & films, renunciation and enjoyment. The twelfth house also indicates the foreign trips, number of the foreign trips, benefits or loss due to the foreign trips and settled in the foreign lands, divine favour and travels.

6

Zodiac Sign

The Zodiac is a band of group of stars or the positions of celestial bodies. The Zodiac is divided into twelve divisions of 30 degrees each called "Sign". Each segment is called a Sign (Rashi).

6.1 Sign Characteristics

Zodiac has twelve signs having different characteristics. They act in different ways and are also known for its different nature. These are given in the following tables:

Table 1: Sign Characteristics

Ascendant	Benefic Planets	Malefic Planets	Most Malefic	Neutral Planets
Aries	Mars, Sun, Jupiter,	Venus, Saturn	Mercury	Moon
Taurus	Venus, Sun, Mars, Mercury, Saturn	Moon	Jupiter	--
Gemini	Mercury, Venus, Saturn	Sun, Jupiter	Mars	Moon
Cancer	Moon, Mars, Jupiter	Mercury, Venus	Saturn	Sun
Leo	Sun, Mars, Jupiter	Moon, Mercury, Venus	Saturn	--
Virgo	Mercury, Venus, Saturn	Sun, Moon, Jupiter	Mars	--
Libra	Venus, Mercury, Saturn	Sun, Moon	Jupiter	Mars
Scorpio	Mars, Sun,	Venus	Mercury	Saturn

		Moon, Jupiter		
Sagittarius	Jupiter, Sun, Mars	Moon, Mercury, Saturn	Venus,	--
Capricorn	Saturn, Mercury, Venus	Moon, Mars, Jupiter	Sun	--
Aquarius	Saturn, Sun, Mars, Venus	Moon, Mercury	Jupiter	--
Pisces	Jupiter, Moon, Mars	Sun, Mercury, Saturn	Venus	--

Table 2: Sign Characteristics

Lagna (Ascendant) Sign	Death inflictor (Maraka Graha)	Raja Yoga Karaka Graha	Neutral Planets	Badhaka Rashi (Signs)
Aries (Mesha)	Mercury, Venus	Jupiter	Venus	Aquarius
Taurus (Vrishabha)	Jupiter, Venus, Moon	Saturn	Sun	Scorpio
Gemini (Mithuna)	Mar, Sun, Moon	Saturn	Moon	Leo
Cancer (Karka)	Venus, Saturn, Mercury	Mars	Venus	Taurus
Leo (Simha)	Saturn, Venus, Moon	Mars	Mercury	Aquarius
Virgo (Kanya)	Jupiter, Moon, Sun	Venus	Moon	Scorpio
Libra (Tula)	Jupiter, Sun	Mercury, Rahu	Mars	Leo
Scorpio (Vrischika)	Mercury, Venus, Saturn	Moon	Saturn,	Taurus
Sagittarius (Dhanu)	Venus, Moon, Mercury	Sun, Ketu	Mercury	Aquarius
Capricorn	Mars,	Mercury	Mars	Scorpio

(Makara)	Jupiter				
Aquarius (Kumbha)	Sun, Jupiter, Moon	Venus	Sun		Leo
Pisces (Meena)	Saturn, Venus, Sun, Mercury	Mars	Mercury		Taurus

Table 3: Sign Characteristics

Lagna Sign	Sign Lord	Sign Gender	Nature/ Mode	Sign Element (Tatva)	Affected Part of Body
Aries (Mesha)	Mars	Male	Odd	Fire	Head, Face, Brain
Taurus (Vrishabh)	Venus	Female	Even	Earth	Throat, Gland Right Eye
Gemini (Mithuna)	Mercury Rahu	Male	Odd	Air	Neck, Nose, Lungs, Blood, Hand's Bone Ear,
Cancer (Karka)	Moon	Female	Even	Water	Chest, Breast, Stomach, Shoulder's Bones,
Leo (Simha)	Sun	Male	Odd	Fire	Upper Stomach, Back, Waist, Heart, Spinal Bones
Virgo (Kanya)	Mercury	Female	Even	Earth	Digestive Organs, Kidney, Lever, Back Bones, Tili, Gurda
Libra (Tula)	Venus	Male	Odd	Air	Skins
Scorpio (Vrischika)	Mars Pluto	Female	Even	Water	Uterus, Ovary, Penis, Nectar
Sagittarius (Dhanu)	Jupiter Ketu	Male	Odd	Fire	Thighs, Waist, Veins

Capricorn (Makara)	Saturn	Female	Even	Earth	Knees, Feet, Digestive Organs
Aquarius (Kumbha)	Saturn Uranus	Male	Odd	Air	Calf Mussels, Feet, Digestive Organs
Pisces (Meena)	Jupiter Neptune	Female	Even	Water	Left Eye, Ankles, Palms

Table 4: Sign Characteristics

Planet	Exaltation Sign	Debilitation Sign	Max. Exaltn. Debilitn.^{0}	Moola-Trikona Sign	Moola-Trikona.^{0}
Sun	Aries	Libra	10	Leo	0 – 20
Moon	Taurus	Scorpio	3	Taurus	4 – 30
Mars	Capricorn	Cancer	28	Aries	0 – 12
Mercury	Virgo	Pisces	15	Virgo	16 – 20
Jupiter	Cancer	Capricorn	5	Sagittarius	0 - 10
Venus	Pisces	Virgo	27	Libra	0 – 15
Saturn	Libra	Aries	20	Aquarius	0 - 20
Rahu	Taurus Gemini	Scorpio Sagittarius	--	Virgo	--
Ketu	Scorpio Sagittarius	Taurus Gemini	--	Pisces	--

Table 5: Sign Characteristics

Lagna Sign	Person's Characteristics	Person's Profession
Aries (Mesha)	Hasty, impulsive, restless, short-tempered	Govt. job, surgeon, mechanics, industrialists, athletes, Police, Military Service, Fire Service, Sports, Engineering, arm manufacturing, trade union leader
Taurus (Vrishabh)	Slow in movement, inclined to ease	Musician, singer, actors, banking, tailors, property dealing, Jewellery business,

	and luxury, faithful & obedient	money lending, commission agent, financial institutions, handicrafts, fancy articles, scented materials, five star hotels, drama, cinema, music, poet, story writer.
Gemini (Mithuna)	Good speakers, witty and humorous, inquiring and curious, fond of knowledge, fun seeking,	media and journalism, accountants, translators, writers, Information and broad casting, space department, education department, book publishing, mathematics department, auditors, law and order councillor, ambassador.
Cancer (Karka)	Emotional, forgiving, sensitive	Export and Import, naval and marine, fishing, nursing, interior design, food, petroleum, historians, shipping, transport department, agriculture, hotel business.
Leo (Simha)	Dominative, behaves like ruler	Govt. Job, Politics, Administrator, Social Services, Charitable institutions, Engineering, Industry, religion, investing, diplomacy.
Virgo (Kanya)	Intelligent, good speaker, tactful	Auditing, Accounting, Business, Teacher, writer, retail shops, computing, astrology, media, doctors, healing.
Libra (Tula)	Good talker, judicious in dealings	Shop, commission agents, bank, Life insurance, law department, hotel business, bar and Restaurant, Dancing Hall, Beauty parlour, Music, Dance, Cinema, judges, artists, cosmetics, fashion, receptionists, advertising, interior decorating, prostitutes.
Scorpio (Vrischika)	Peevish, straight forward, likes to	Iron Industries, Engineering, and Instrument Manufacturing,

		hide or run away from people and crowds	raw materials, priest, astrology, mantra and tantra, occult practices, chemicals, drugs, liquids, insurance, doctors, nurses, police, occult.
Sagittarius (Dhanu)		Honest, easy going, even-tempered, kind hearted, gambles	Forest department, law, religion, banking and finance, entrepreneurs, athletes, law department, temple, financial institutions, education department, ordnance depot, military training, social service, charitable institutions.
Capricorn (Makara)		Witty, and changeable, good organizer, cautious, secretive, ambitious, preserving, pragmatic	Hotels, food products, engineer, doctor, business, building work, Granite stone and sand business, Labourer like porters, coolies, drivers, shoe polishing, shoe makers, plumber and mining.
Aquarius (Kumbh)		Studious, philosopher, honest, benevolent	advisors, consultants, philosophers, astrologers, engineers, computer, Psychology, Religion, Teaching, Research and Development, Administration, Service in Space Dept., Defence, Fire, Jail, Bomb manufacturing, tourist guide, central excise CBI Dept.
Pisces (Meena)		Lazy, emotional, timid, honest, talkative, intuitive, psychic, fond of good food and company	doctors, captain, hospital, prisons Education Department, Religious Institutions, Medicine, Financial, Law Department, External Affairs, Bank, Navy, shipping, temple worker, priest

6.2 Predictions by Sign Element

Every Zodiac Sign falls into one of four elements. There are four Elements, such as earth represents common sense; fire represents action, air represents thinking and communication skills, and water represents the ability to feel and intuitively know. Each Element is assigned to each sign depending to their orientation in the zodiac. Many astrologers consider the element of each of the planets when determining which of the elements may be more significant in a horoscope.

Fire (Aries, Leo, and Sagittarius): Fire is active and masculine. People of the Fire element are outgoing, quite moral, very creative, courageous, passionate, impulsive, hot, dynamic, progressive, action oriented, and direct. Their essence is spirit. They are enthusiastic, optimistic, confident, naive, self-centred, open, confronting, loyal, tactless, impatient, honest, trusting, and independent and feel free.

Earth (Taurus, Virgo, and Capricorn): Earth is a receptive, feminine sign. People are practical, cautious, and pragmatic approach to life and build solid, 'real' material success, i. e. car, home, career success and have long range planning and strong determination to succeed. They are safe/secure, suspicious, sensual, organised, dependable, introvert, and efficient and strong survival instinct.

Air (Gemini, Libra, and Aquarius): They are active, curious, idealistic, unemotional, conceptual, devoid of feeling, good to communicate, social, objective, impersonal, distant, masculine, intellectual, changeable, and impractical, good speech, and natural communicators, extroverted, social, charming, and logical and air has least obvious bad qualities, theoretical, abstract, needs to socialise, needs to share ideas.. The lack of Air Element in a native birth chart indicates difficulty in the expression of that person. Communication of ideas and the ability to conceptualise may prove difficult.

Water (Cancer, Scorpio, and Pisces): People of Water element dissolve everything in them coolly. They take the shape of who they are with, and are quite emotional, sustaining, emotional, sensitive, imaginative, protecting, compassionate, caring, artistic, moody, soulful, subconscious, irrational, introverted, but strong/powerful, vulnerable to hurt, intimate, defensive, psychic, past, suffering, suspicious initially, self-contained, picks up impressions and associated with healing.

Calculation of Element Strength: My method for evaluating the strength of an element in a birth chart is to assign a value of 4 to the element associated with the Sun; the Moon element is assigned a value of 3; Mercury, Venus, and Mars sign elements are assigned a value of 2 each, and Jupiter and Saturn each have a value of 1. Uranus, Neptune and Pluto are disregarded because their element is more societal affecting large groups of individuals born during a period. Using this approach, if as many as 8 points are concentrated in one element, it is considered "Preponderance" in that element. If we get less than 8 points with this approach in one element, it is considered "Absence" in that element.

Example: In one chart, the planets are placed as is given below: For Sun in Aquarius, an Air sign, we assign = 4 Points. For Moon in Libra, an Air sign, we assign = 3 Points. For Mercury in Aquarius, an Air sign, we assign = 2 Points. For Venus in Capricorn, an Earth sign, we assign = 2 Points. For Mars in Sagittarius, a Fire sign, we assign = 2 Points. For Jupiter in Pisces, a Water sign, we assign = 1 Points. For Saturn in Fire, a cardinal sign, we assign = 1 Points.

Thus, we have Element strength, such as, 9 in the Air Element, 2 in the Earth Element s, and 3 in the Fire Element and 1 in Water Element in this chart. This shows a preponderance of Air element in this chart. The preponderance reading of Air element would be appropriate in the above chart.

The preponderance readings of all elements are given below:

A Preponderance of the Fire element: A preponderance of Fire Element indicates high spirits, great faith in self, enthusiasm, direct, honesty, intensely assertive, most daring, individualistic, active and self-expressive, good natured, fun loving, natural leader, having a good time than on material possessions, big egos. He believes so strongly in his own powers and abilities that he overlooks and frequently fails to take advantage of the talents and abilities of others. He tries to do it all himself and don't delegate well. He is constantly "out front" or "on stage", such as an <u>Artist</u> and they need to be recognized and admired for his attainment and accomplishments. Appreciation is more important than money in his estimation. Nothing hurts him more than being ignored. The fire sign sense of honesty is straightforward and often child-like. Thus, he believes everyone is, like himself, an open book.

A Preponderance of the Earth Element: A preponderance of Earth Element indicates cautious, conventional, dependable but quite responsible, methodical, organizer, a builder, and a hard-worker. It provides the skills and attitude necessary to succeed readily in the world of business and never gamble or take unnecessary chances. They understand the reality of a situation and value, reliable and steadfast. They are dependent, diligence and a pragmatic, no-

nonsense approach to life. Lack of ideas or imagination, dullness, rigid, conservatism, extreme materialism, and blind adherence to rules and regulations are their potential faults.

A Preponderance of the Air element: The preponderance of Air Element suggests a strong emphasis on thought, ideas and intellectual and they communicate and express ideas with mental agility and become the impractical dreamers, constantly thinking, people-oriented, but more inclined toward the group than the individual. Your interests are varied, and you're apt to be a life-long student.

A Preponderance of the Water element: The preponderance of Water Element indicates close emotional relationships, romantic, sentimental, affectionate, secure bond with partner, communicate best in non-verbal ways; emotionally, psychically, or through forms as art, dance music, poetry and photography. They have a natural feel and sense for the arts and are apt to let the heart rule the head, highly impractical and impressionable.

6.3 Predictions by Sign in House
Prediction by Sign in 1st House (Thanu/ Body):

Aries (Mesha): He/She is proud, wealthy, having excessive anger, dependent on others. He/She will be medium height, white complexion, smiling face, clever, and lean. He/She suffers from abdominal problems. He/She relishes helping the poor and has faith in God. He/She thinks very high but implements little. He/She inherits huge paternal property. He/She is prone to drowning and accidents. He/She is selfish and forgets a person who is no longer of use. He/She is prone to be cheated by friends and partners. He/She will improve in life and will earn money and believes in donation. Even born in medium income group, he/she earns good money of his/her own. He/She has a personality that is positive, aggressive, and competitive. He/She has leadership qualities, is bold, and empowered with more physical strength. He/She is best in sports, games, trekking, summer camp, and any other outdoor activities.

Taurus (Vrishabh): He/She is a pleasant talker, scholar and loves all. He/She is tall, luxurious, clean-hearted, strong built, good personality, whitish complexion, smiling face, clever and attractive personality. He/She improves in life and earns money and believes in donation. He/She is prone to Litigation or imprisonment because of personal or property disputes. He/She is financially well off, earning much more.

He/She is educated and enjoys a happy married life with educated and glorious progeny. He/She has differences with near relatives and is thus socially unpopular. He/She is prone to accidents. He/She has a tendency toward being heavy by both bone structure and the self-indulgence. Lord Krishna, Mata Amritanand Mayi and also Shri Basaveshwara were born in Taurus, but in Rohini Nakshatra. The second Drekana of Taurus gives the skill of fine arts, music and dance to the native.

Gemini (Mithuna): He/She is proud, loves his friends, charitable, and wealthy. He is annihilator of his enemies and progresses slowly in life. He/She will be medium height, whitish complexion and a faithful friend. He/She is sweet-voiced, jolly and humorous. He/She is well-wisher for everyone and gets cooperation from parents. He/She seldom seeks help and doesn't work under any one. He/She has his own successful business set-up - big or small. He/She is considerate to subordinate and weaker people. He/She has long life, lean body. He/She improves in life; earns money and believes in hard working. He/She, even, born in medium income group, earns money of his/he own. He/She does not get help from his/her spouse. 3rd Drekana of Gemini blesses him/her with fine art, music and dance.

Cancer (Karaka): He/She is religious, handsome, long, has good personality, whitish complexion, and donating. He/She is good-looking, rich and famous. He/She respects his elders and teachers. He/She is prone to head injury during childhood. He/She excels his business away from birth place. He/She is prone to be cheated by partners and close relatives. He/She gains from business abroad in white coloured items. He/She is intelligent and heads an organisation or society. People flock to him/her for advice. He/She may undergo political imprisonment. He/She, even, born in medium income group, earns money of his own. He/She does not enjoy his life due to hardship. His family life is not happy and always difference of opinions between husband and wife. He/She is very protective of those who are close to him/her. He/She is affectionate, emotional, home loving and lovely in their approach.

Leo (Simha): He/She is annihilator of his enemies, has few children, is tall, strong built, good personality, whitish complexion, and smiling face, clever and has attractive personality. He/She is efficient, undertakes tough tasks, and is hard-hearted and always successful. He/She is self-dependent and doesn't trust others. He/She spends as quickly as he earns. He/She crushes his/her enemy, is religious and donating. He/She doesn't forget or forgive his enemy and takes revenge. He/She has differences with father. His wife is long-lived but he/she keeps quarrelling with his/her. He/She is a devoted friend who will remember and repay a kindness. He/She has royal tastes and a sense of luxury. He/She is angry but vents the anger quickly. He/She goes to any authority to prove that he/she is right.

Virgo (Kanya): He/She is endowed with beauty and has a good fortune, medium height, broad chest, whitish complexion, smiling face, clever, very fast in doing the job and is selfish and harm too much out of his selfishness. His/Her young age is very happy. He/She is a successful in politics because he/she has something inside and speaks something else in public. Nobody can measure his political capability. He/She is never crude or coarse. He/She prefers the role of researcher, observer, critic, or teacher. He/She is fault finding type and hypercritical. He/She feels proud in finding the fault and drawback in others.

Libra (Tula): He/She is a scholar, earns his livelihood by virtuous means, wealthy and is respected by everybody. He/She is tall, fair complexion and healthy. He/She is sweet-voiced and benevolent. He/She doesn't stick to one profession and keeps spending too much on research. He/She enjoys little reputation at home but is reputed outside. He/She excels in occupations related to iron. He/She has problem with brothers/sisters. He/She is talkative, is not hard-working and depends on fate, and thus leads insecure life. He/She is unlucky for father's business and gives setback at the age of 12. Initially he/she begins with service but later settles down in own business. He/She leads happy married life with kids.

Scorpio (Vrischika): He/She is wealthy and a scholar. He/She is tall, lean, either very rich or very poor. He/She lives away from home since very early in life and dominates over his family members as well as outsiders. He/She doesn't forget or forgive his enemy. He/She has his own business. He/She has financial problems up to 30 years but later he/she earns money and supports others. He/She cannot sit idle and is world-famous and heads an organisation or society. He/She is prone to injury during fight or accident. He/She gets married more than once. He/She will be rich, famous, popular and smart in love matter. He/She is successful in politics. He/She earns sufficient money in the life but not much savings. This Sign is not good for domestic happiness.

Sagittarius (Dhanu): He/She is an expert in policy matters, religious, important person in his family, medium height, whitish complexion, smiling face, and attractive personality. He/She has strong faith in God, is vegetarian, simple living, believes the people very easily, and is a businessman. He/She does his work with well planning. He/She likes discipline, truth, justice, kindness and independence in his/her life. He/She takes everything and everyone for granted. His/Her reasoning powers are superb. He/She enjoys the good fortune of having thought patterns that remain young and fresh throughout life. He/She makes lots of promises but fail to maintain them. He/She has a great sense of fairness and adopts only fair means to handle the job.

Capricorn (Makara): He/She is inclined towards evil deeds, is greedy and has many children, but hard working. He/She is tall, strong built, good personality, whitish complexion, clever and but selfish, changing his faces frequently as per situations, and very talkative. He/She works under someone and subjects to heavy ups & downs. He/She spends immediately what he/she earns in bad deeds. He/She dominates spouse and quarrels with other family members on that account. He/She stays away without information. He/She is abusive and short-tempered. He/She fails in business and has to go in for service. He/She is attached to mother and has differences with father. He/She is helpful to brothers and sisters and has more daughters. His/Her expense is more than earning and hence always faces shortage of money. The conjugal life is not happy and there is always difference of opinions between them.

Aquarius (Kumbha): He/She leads a happy and contented life. He/She will be well educated, gentle, peaceful, always ready to help others, having good thinking; tall, whitish complexion and attractive personality and straight forward in nature. He/She successfully tackles early age problems and heads for good time later. He/She would spend any amount of time and money to crush his/her enemy or achieve his/her aim. He/She likes to gossip and interact with women and has interest in astrology. He/She is financially well off and is prone to chest problems. He/She will be very hard working and faces difficulties in life. He/She may suffer with the stomach and heart diseases in old age. He/She is strong willed, detached and unyielding in nature. The child of this Sign is unpredictable regarding his/her behaviour and can change frequently himself/herself to any extent during his childhood.

Pisces (Meena): He/She is wealthy, educated, gentle, peaceful, religious, and always ready to help relatives, medium height, and beautiful curly hair and have self confidence. He/She is famous and heads an organisation or society. He/She studies very hard, but he/she is not a high scorer. He/She helps friends and serves society physically but without spending money. He/She works overtime to finish the work same day. He/She becomes favourable of family members and outsiders. He/She just cannot work under any one and leave his/her service very soon for own business. He/She loses temper beyond control but calms down very quickly. He/She is prone to cheating by partners and should better work alone. He/She is likely to break first marital relation (matured) or otherwise, be unhappy with spouse but happy with progeny. He/She may have great interest in writing, music and acting. He/She is likely to set high goals. Drug, alcohol and false promises attract him/her easily.

Prediction by Sign in 2nd House (Dhan/Wealth)

Aries in 2nd Bhava: The Aries in the 2nd Bhava indicates the uncertainty about his/her earning. Sometimes he/she earns more money and sometimes very less money. He/She spends more than the earning on his/her show business and is sometimes harmed by his/her enemies. His/Her good luck starts after marriage.

Taurus in 2nd Bhava: The Taurus, in the 2nd Bhava, indicates his/her good earning. He faces ups and down in life and is harmed by his/her spouse or partners in the business. His/Her good luck starts after 18, 22, 24, 33 and 35 years of age. He/She has a strong drive to earn money; to build and hold financial worth and material possessions. Venus, the planet of love ruling the second house suggests a real love of money. He/She is good at business affairs. He/She is a natural for making and accumulating money.

Gemini in 2nd Bhava: The Gemini, in the 2nd Bhava, indicates his/her weakness in the earning. He/She wastes money in bad relation with the other woman/man. His/Her enemies sometimes harm him/her financially. His/Her good luck starts in business; the private services or in life insurance, small industries, electrical parts industry etc. The Gemini influence in the second house focuses the mind on material matters and on making money. He/She has an active interest in financial affairs. His/Her resourcefulness in accumulating money may result in holding more than one job simultaneously.

Cancer in 2nd Bhava: The Cancer, in the 2nd Bhava, indicates the less earning as compared to his/her hard work. He/She is miser in nature and spends less than the earning and saves more. He/She faces ups and down in the business and is harmed by his/her partners in the business. His/Her good luck starts after 20, 26, 33, 44 and 54 years of age. The influence of Cancer in the second house shows that he/she is protective of his/her financial assets and possessions.

Leo in 2nd Bhava: The Leo, in the 2nd Bhava, indicates the uncertainty about his/her earning in his middle age. He/She spends very happy childhood and does not face any financial crisis in childhood. He spends more than the earning on his/her show business or in uncertain business and is sometimes harmed by his/her enemies. He/She has good luck always and finishes all his/her work successfully. He/She earns money through political works or government job. The Sun denotes his/her self-esteem, so his/her earning capacity may have a lot to do with his/her sense of self-worth.

Virgo in 2nd Bhava: The Virgo, in the 2nd Bhava, indicates he is financially strong. He/She earns money by hard work in business, especially ready-made shop or fancy store shop. Initially he/she faces financial difficulty due to not taking right time decision or due to hot temper. He/She appears very careful with financial affairs, but often he/she can become somewhat penny-wise and pound-foolish.

Libra in 2nd Bhava: The Libra, in the 2nd Bhava, indicates the uncertainty about his/her earning. Sometimes he/she earns more money and sometimes very less money by business. He/She spends more than the earning on his show business or luxuries life and is sometimes harmed by his/her bad habits. He/She has good luck in business like hotel or restaurants. There is balance and harmony in material affairs. This sign suggests the accumulation of possessions is highly dependent on ventures with a partner, normally the marriage partner.

Scorpio in 2nd Bhava: The Scorpio, in the 2nd Bhava, indicates the uncertainty about his/her earning. He/She spends more than the earning on his/her show business or on his/her big planning and is sometimes harmed by his/her friends or relatives. He/She has good luck in business like small industries rather than service.

Sagittarius in 2nd Bhava: The Sagittarius, in the 2nd Bhava, indicates very lucky and good earning. His/Her good luck starts after 24, 28, 33, 37, 48 and 55 years of age. He/She should be very careful while signing any paper. He/She is lucky in this regard. His/Her personality attracts financial success naturally. He/She is a risk taker who may get burned from time to time.

Capricorn in 2nd Bhava: The Capricorn, in the 2nd Bhava, indicates very lucky and god earning and is financially strong. He/She earns good money by business in mines or stone query as compared to service. He/She may serve in planning commission or plans making organization because he/she is very good in big planning. He/She has prudence and is practical in the handling of money. He/She has a 'poor' complex and does not know his/her earning potential and so he/she continues to live as though he/she had very little. When buying investments, he/she is inclined toward the blue chips and sure bets.

He/She is very cautious and practical. He/She may reject luxury or expensive living.

Aquarius in 2nd Bhava: The Aquarius, in the 2nd Bhava, indicates very lucky and good earning and is financially strong. He/She earns good money by many source of income particularly by reporter, writing, publications or businesses as compared to service. He/She does not get benefits from brothers or relatives and gets harmed on good faith by the people. He/She is benefited in partnership and latter life is better than the middle age. He/She is not afraid to take chances financially, and he/she looks for unusual and inventive ways to invest.

Pisces in 2nd Bhava: The Pisces, in the 2nd Bhava, indicates very lucky and good earning and savings and is financially strong. He/She earns good money by many source of income particularly by doctor professions, by sale of medicines, by share business or small industries. He/She controls his expenditures and earns money like anything and does not have peace of mind in life but always busy in earning money. His/Her good luck starts after 24, 28, 33, 37, 48, 55 and 60 years of age. He/She has a sense of timing and intuition that may be an asset financially.

Prediction by Sign in 3rd house (Relations/Sibling):

Aries in 3rd Bhava: The Aries, in the 3rd Bhava, indicates that he/she will be strong mussel and good built, strong built wider shoulders, good personality, very courageous. He/She earns money and believes in making the situations in his/her favours and obedience to his/her seniors. He/She will be talkative and artist and will save money in life. He/She has a strong need to communicate with others, and to communicate forcefully. His mind is active, alert and capable of making quick decisions. He/She is capable in expressing himself/herself. He/She is mentally competitive. He/She is an aggressive learner, always seeking new ideas and new knowledge.

Taurus in 3rd Bhava: The Taurus, in the 3rd Bhava, indicates that he/she will be strong built, good personality, very courageous. He/She earns money by writing, poetry, portrait making and other arts and believes in making the situations in his/her favours and admired by the family members. He/She will be artist. He/She will be able to save money in life. He/She is a slow learner, but once an idea is lodged in his/her brain, it's there to stay. In early education he/she may have been an indifferent student, but as he/she matures he/she accumulates a wealth of knowledge and understanding. He/She has a strong interest in the arts, especially music.

Gemini in 3rd Bhava: The Gemini, in the 3rd Bhava, indicates that he/she will be strong built, good personality, very lucky. He/She earns money and believes in making the situations in his/her favours and honoured by the government. He/She will enjoy the happiness of luxurious vehicles. He/She will be able to save money in life and has very happy family life and good respect and cooperation from spouse. He/She has quickness both in thought and speech. Mercury, ruling Gemini, is the planet of intellect, thought, speech, and wit. He/She is a good conversationalist, a fact collector, and a mentally stimulating person. He/She can be a good diplomat because he/she can agree with divergent views and present rational alternatives.

Cancer in 3rd Bhava: The cancer, in the 3rd Bhava, indicates that he/she will be strong built, good personality, very lucky. He/She earns sufficient money by good business and real estate or construction work and believes in making the situations in his/her favours by his/her noble nature and admired by the society. He/She will enjoy the happiness of friends. He will be able to save money in life and has very happy family life and is religious minded. He/She is very protective of brothers and sisters. He/She has close and emotional ties to the immediate family.

Leo in 3rd Bhava: The Leo, in the 3rd Bhava, indicates that he/she will be strong built, good personality and very courageous and has excellent imaginative power. He/She earns money by writings, poetry, and publication and believes in making the situations in his/her favours and has interest in the great music. He/She will be artist. He/She will face difficulties in education in childhood but finally gets good educations in life and be able to save money in life. His/Her powers of self-expression are outstanding.

Virgo in 3rd Bhava: The Virgo, in the 3rd Bhava, indicates that he/she will be good personality, has very good knowledge of Vedas. His/Her friends are helpful but not the family members. He/She will be short tempered. He/She will be suffering from inferiority complex on communication skills, especially in the early years. He/She can be overly critical of brothers, sisters or neighbours.

Libra in 3rd Bhava: The Libra, in the 3rd Bhava, indicates that he/she will be flicker minded and has relations with low status people. He/She will not be able to take fast decision but too much talkative and does mistake while talking for which he/she repents later on. He/She has differences in family life always. He/She gets along easily with family members because he/she dislikes conflict and argument.

Scorpio in 3rd Bhava: The Scorpio, in the 3rd Bhava, indicates that he/she will be flicker minded and has relations with low status people and is addict of bad habits. He/She will not be able to take fast decision. He/She has differences in family life always and live medium life. He/She is very angry man and does mistake in his/her angriness. He/She does not get help from brothers. There can be friction between him/her and family members, as well as friends and acquaintances.

Sagittarius in 3rd Bhava: The Sagittarius, in the 3rd Bhava, indicates that he/she will lose money in business and will not get success in business. He/She will be able to save money in life in military, police or other government service. He/She is very talkative and has a very cheerful outlook on life, natural exuberance and optimism. He/She has a natural ability to communicate, especially important issues such as politics, education, and religion. He/She has an executive type of mind.

Capricorn in 3rd Bhava: The Capricorn, in the 3rd Bhava, indicates that he/she will be strong mussel and good and strong built handsome body & good attractive face and personality. He/She is very lucky in respect of children and famous among friends and is religious minded. He/She has careful expression of thoughts and ideas. He/She refrains from writing or saying anything unless there is a reason for doing so. He/She has little capacity for small talk, and many people may find it difficult to communicate with him/her. In his early years, he/she may not have good education, but later he/she changes, and he/she is apt to turn to studies to attain his/her ambitions.

Aquarius in 3rd Bhava: The Aquarius, in the 3rd Bhava, indicates that he/she will be peaceful nature and has good relations with brothers but does not get help from them. He/She will get respect in the society and has interest in music. He/She has a sparkling intellect and an inventive mind. He/She is well ahead of time, and sometimes radical. He/She seeks education simply for the sake of learning instead of just for preparing to earn a living.

Pisces in 3rd Bhava: The Pisces, in the 3rd Bhava, indicates that he/she will be good built, handsome body & good attractive face and personality. He/She will be a wealthy man and lucky in respect of children and get full help from children in old age. He/She is famous in society, keeps everybody happy in the society and is religious minded and interested in religious work. In communicating, he/she can become over emotional and he/she may experience problems articulating, especially in his/her youth. He likes to be alone when he/she is performing any type of mental work.

Prediction by Sign in 4th House (Bandhu/Pleasure)

Aries in 4th Bhava: The Aries, in the 4th Bhava, indicates that he/she will be having many cattle. He/She will have relations with many women simultaneously but still he/she enjoys the life happily in different ways and peacefully. He/She is more benefited by agriculture and business. He/She has an aggressive attitude toward the home and family. He/She always takes an active interest in the affairs of the home. He/She has a tendency to force issues and demand too much. The fourth house denotes activities in the latter part of life, the later years will be very active and more daring and outgoing, even youthful personality emerges, as he/she grows older. He/She is more assertive and physically active in the latter part of life.

Taurus in 4th Bhava: He/She is very lucky in respect of children who help him/her in old age and is religious minded. He/She does the social work in big ways. He/She is famous in society, has patient, peaceful, keeps everybody happy in the society and is religious minded and interested in religious work, celebrates worships in big ways. He/She is more benefited by worshipping the lord Shiva. He/She enjoys the happy family life. He/She has strong instincts to provide materially for his/her family. He/She has a very pleasant, easygoing home environment, and harmony, serenity and graciousness in his/her latter years.

Gemini in 4th Bhava: The Gemini, in the 4th Bhava, indicates that he/she will be handsome body & good attractive face and personality. He/She has medium luck in life. He/She is very sexy and happiest person and enjoys the sex in relation with the most beautiful women and spends lots of money on them and fashion and cosmetics. He/She has to work hard to earn money. He/She suffers a lot in his/her old age due to loss of heavy money. His/Her ties to home are not particularly strong.

Cancer in 4th Bhava: He/She will be the luckiest person and has many friends. He/She is very happiest person and enjoys the married life with the most beautiful good nature and fortunate spouse and spends lots of money on fashion and cosmetics. There is a deep attachment to family traditions and home relationships. He/She may be especially protective of his/her parents and assume a role of responsibility for them. He/She may depart his/her early shelter at a young age.

Leo in 4th Bhava: He/She will be the luckiest and very happiest person and enjoys the married life with the most beautiful good nature and fortunate wife and spends lots of money on fashion and cosmetics. He/She is the most angry and irritated person. His/Her relation with the brothers and sisters are not good and does not get help from them. Children are not so happy with him/her. He/She has more female child than male child. He/She has properties and owns a piece of land and lives in a reflection of his/her ego. He/She is apt to be the one that 'rules the roost.' In either sex, this placement often shows the one who is the boss and makes the decisions in the family.

Virgo in 4th Bhava: He/She will be the luckiest person and most fortunate. He/She is very happiest person and enjoys the full and happiest married life with the most beautiful good nature and fortunate spouse and spends lots of money on fashion and cosmetics. He/She gets married in young age and has more male child than female child. He/She has sufficient money to enjoy the life. The childhood is difficult financially but becomes wealthy man after the age of 28 years and is the richest man at 36 years. He/She is learned, educated good nature person.

Libra in 4th Bhava: The Libra, in the 4th Bhava, indicates that he/she will be handsome body & good personality. He/She will be the successful businessman and will spread his business all-around and is most fortunate even though he/she will be poor in childhood. He/She is very happiest person and enjoys the full and happiest married life till old age. The childhood is difficult financially but becomes wealthy man after he/she starts working after marriage. He/She is educated, kind, peace loving and good nature person who helps others. He/She spends some of his/her wealth on religious matters and works and keeps at the distance from the bad people. He/She wants to own his/her home and possessions, free and clear.

Scorpio in 4th Bhava: The Scorpio, in the 4th Bhava, indicates that he/she will be very anxious and worried person; his/her mind will not be peaceful and will face many difficulties in life. He/She is afraid of enemy as they harm him/her frequently. Every time, he/she faces difficulties while starting any job but gets success at the end. He/She starts life from poor childhood but earns money as he/she grows older and saves money to enjoy the family life till old age. He/She has extremely strong feelings about the home and family life. He/She has a strong loyalty and protective tendency shown toward family members. He/She has a sense of royalty, splendour, and space in the home environment.

Sagittarius in 4th Bhava: The Scorpio, in the 4th Bhava, indicates that he/she will be fighting nature and always fighting with people and will be successful in war. He/She will earn money and gets success in business of lending money & interest as compared to service. He/She is involved in litigation and court cases. He/She gets success in military or police job and is his/her own fortune maker and enjoys a medium happy family life. He/She has a strong urge to control and direct family matters. His/Her latter portion of life will be very beneficial. He/She will get wealthier as grow older, both in material ways and in spiritual ways.

Capricorn in 4th Bhava: The Capricorn, in the 4th Bhava, indicates that he/she will be the owner of gardens and vegetations and serve in the same field. He/She has many good and helping friends and is fortunate in respect of friends. He/She has differences with wife but the children are helpful to him/her and beneficial to him. He/She has much responsibility in house of home and family. With the planet Saturn ruling the home, issues of security can be of paramount importance. There is also ambition linked with this placement. He/She dominates his/her home with a practical and no nonsense outlook. He/She is a strong disciplined man, with strict, old-fashioned principles and a lofty code of ethics.

Aquarius in 4th Bhava: The Aquarius, in the 4th Bhava, indicates that he/she will be happiest person and enjoys the married life with the most beautiful, good nature, educated and fortunate spouse who is helpful in making the fortune in the life and spends lots of money on fashion and cosmetics. He/She gets wealth from father in-law. His/Her fortune starts after marriage. He/She has a strong demand for freedom in the affairs of the home. He/She is likely to find it necessary periodically to change his/her address, and he never likes the idea of moving. His/Her home environment is distinctive, perhaps even a bit unusual. He/She has many original ideas and wants the latest innovations as a part of his life style.

Pisces in 4th Bhava: The Pisces, in the 4th Bhava, indicates that he/she will be a captain on the ship or will be in service connected to water transport. He/She will be educated and peaceful person and will get respect in the society due to his/her good nature. He is educated, kind, peace loving and good nature person who helps others. He/She spends some of his/her wealth on religious matters and works and keeps at the distance from the bad people. He/She has sufficient money to enjoy the life. The childhood is difficult financially but becomes wealthy and enjoys the happiest family life. He/She has an emotional tie to the home. He/She is sentimental about his/her family and willing to make sacrifices for his/her loved ones. He/She has a strong need for domestic peace and seclusion.

Prediction by Sign in 5th House (Santana/Children)

Aries in 5th Bhava: The Aries, in the 5th Bhava, indicates that he/she will be very angry and foolish man and does mistake in haste for which he/she repent later. There is always difference of opinions among the husband, wife and children. He/she is always restless and can't take a decision on any matter immediately. He/she never minds spending his/her money for leisure time activities, whatever he/she may be. He/she loves the outdoors and a need to stay on the move. He/she has a sporting attitude that makes him/her fun to be around. Physical activity is necessary for his/her well being and happiness.

Taurus in 5th Bhava: The Taurus, in the 5th Bhava, indicates that he/she will have no issue or will have issue after late or the issue will be weak, unhealthy or mentally retarded. He/she will have better understanding, peaceful, kind and patient. He/she will earn money easily and sometimes from lottery or gambling. He/she will be marrying to a beautiful and fortunate spouse who will be taking all the care of the family. He/she has a very loving nature toward his/her offspring, and has very strong and fixed views regarding their behaviour and the proper upbringing. He/she has personal artistic talents, and he/she may have a natural artistic ability leading to self-expression in some form of the arts.

Gemini in 5th Bhava: The Gemini, in the 5th Bhava, indicates that he/she will have issue and will have all happiness from the issue who will be healthy and educated. He/she will have better courageous, rigid, strong will. He/she will be well educated even though faces difficulties during education. He/she gets angry soon and cooled down soon too. He/she has a cool and intellectual approach to romance. Gemini is the sign of mental energy and the fifth house denotes creativity. He/she is writers and otherwise talented people.

Cancer in 5th Bhava: The Cancer, in the 5th Bhava, indicates that he/she will have more daughters than son in number or the son may take birth at latter age or very late. However, he does not get happiness from the children but pain. He/she will be lazy and has faith on others easily. He/she is very protective of his/her family, and is very maternal toward offspring. His creative work may tend toward the artistic, especially theatre. He/She has good writing abilities and he/she can communicate more easily in writing than in speech.

Leo in 5th Bhava: He/she will have more daughters than son in number or the son may take birth at latter age or very late. He/she will be working away from the native place to earn money for the family. He/she will be courageous, rigid, strong will and hard working. He/She devotes fully to whatever creative activity has his/her interest at the moment. He/She identifies strongly with his/her children and he/she is very proud of his/her accomplishments. He/she is a born gambler and speculator, with more than his/her fair share of luck.

Virgo in 5th Bhava: The Virgo, in the 5th Bhava, indicates that he/she believes in unnecessary show business other than the reality, and does not have issue or does not get help or happiness from the issue. He/she take fast decision and misuse the power immediately after getting it. Relationships with offspring can be strained as he/she lacks the patience for properly disciplining them and understanding their needs.

Libra in 5th Bhava: The Libra, in the 5th Bhava, indicates that he/she will be a gentle man, peace loving, and effective personality, very handsome and educated. Romantically he/she is charming, but very inconsistent and fickle. He/She is lucky at love. Libra is naturally suited to the raising of children, although he/she is somewhat prone to spoiling them. He/She has many child-like qualities in his/her make-up.

Scorpio in 5th Bhava: The Scorpio, in the 5th Bhava, indicates that he/she will be a gentle man, peace loving, effective personality, very handsome, religious minded and educated. He/she will suffer from venereal disease. He/She will get happiness from the children. He/She has a good planning and completes the job taken in hand successfully. He/She has attachments and romantic encounter. He/She is concerned for his/her children, almost to an extreme degree.

Sagittarius in 5th Bhava: The Sagittarius, in the 5th Bhava, indicates that he/she will be interested in horse riding or horse driving like horse race man, or horse cart man. He/She respects the people and enjoys the full happiness from his/her children because all are obedient. He/She has a constant need to show off, and he/she enjoys the thrill of any adventure. He/She is a natural gambler who will speculate on just about anything. He/She has a good understanding of young people, and he/she can get along with them so well because he/she treats them with true respect as individuals.

Capricorn in 5th Bhava: The Capricorn, in the 5th Bhava, indicates that he/she will a bad man, which enjoys the bad work or bad politics, very clever to get job done by the enemies too, principle less. But, if she is a female, she will be a big Guru Matta and worship able by the people, knowledgeable of all the Vedas and religious books. He/She is a hard worker, good with details, and can make a scrupulous teacher and disciplinarian of young people. In romantic affairs, he/she may be prudish, and may even appear cold.

Aquarius in 5th Bhava: The Aquarius, in the 5th Bhava, indicates that he/she will a fixed mind, good in education and get all happiness from the children. He/She talks truth and help and gets respect from others. He/She is courageous and never gets afraid of problems in life and face with courage. He is cool and detached concerning romance. His/Her children are apt to display a rebellious nature, and they need to be taught discipline at an early age.

Pisces in 5th Bhava: The Pisces, in the 5th Bhava, indicates that he/she will be very sexy, popular in a female society and spends a lot for them, always smiling face, sentimental. He/She has differences with spouse and children and hence has medium level happiness from children. He/She has frustration with his/her offspring. He/She may sometimes be disappointed and disillusioned. He/She may dream about creating something so significant that he/she will be remembered for his/her efforts long after he/she is gone. Raising children can be difficult and confusing.

Prediction by Sign in 6th House (Dukha/Sadness)

Aries in 6th Bhava: The Aries, in the 6th Bhava, indicates that he/she will be bad and corrupt man, hard working but always complaining to someone and telling badly about others. He/She is a very hard working member of society. He/She may run into difficulties with co-workers and employees because of his/her critical nature.

Taurus in 6th Bhava: The Taurus, in the 6th Bhava, indicates that he/she will be a frustrated man; hard working but always having differences with family members and his/her children are aggressive type. Teaching is very attractive to him/her because he/she gets the time for his/her varied extracurricular activities.

Gemini in 6th Bhava: The Gemini, in the 6th Bhava, indicates that he/she will be a frustrated man; hard working but always having differences with family members and his/her children are aggressive and fighting nature. He/She does business rather than service but always loses money in life. He/She dislikes repetitious tasks. He/She is likely to succeed in fields related to scientific research, business or finance.

Cancer in 6th Bhava: The Cancer, in the 6th Bhava, indicates that he/she will not be taking rest in life and will not allow his/her relatives or subordinates to take rest. He/She will always be fighting with others and in very much love and affection with his/her children. As a supervisor, he/she is very understanding and concerned about the welfare of his/her employees.

Leo in 6th Bhava: The Leo, in the 6th Bhava, indicates that he/she will be angry-man, infighting and very sexy and will suffer financially due to entangle with other man/women. He/She will suffer and loose relation with spouse and children due to other man's/women's relation. He/She can lose himself/herself completely in work and service dominates over co-workers and subordinates. He/She has a feeling of authority where work and services are concerned.

Virgo in 6th Bhava: He/She will lose huge money due to contact and connection with low status/ level man/women. He/She seldom eats or drinks too excess and he/she is health conscious. History is especially interesting to him/her because of his/her ability to accumulate and relate details.

Libra in 6th Bhava: The Libra, in the 6th Bhava, indicates that he/she will be very rich man but will be selfish and will not believe in god. He/She is cooperative, tactful, and diplomatic in the work place.

Scorpio in 6th Bhava: The Scorpio, in the 6th Bhava, indicates that he/she will be sexy and luxurious. He/She will be hard working and will make money with hard work. He/She will not be so fortunate. He/She has commitment and seriousness about work and gets intensely involved in his/her job. His/Her work often involves matters of investigation such as journalism, research, laboratory science or psychology.

Sagittarius in 6th Bhava: The Sagittarius, in the 6th Bhava, indicates that he/she will be poor and has to work hard and struggle for bread and food whole life. He/She has a tendency to overwork and to over-extend himself/herself in his work. He/She eats or drink too much, or to have an extravagant taste in food and drink.

Capricorn in 6th Bhava: The Capricorn, in the 6th Bhava, indicates that he/she will be poor and has to work hard and struggle for bread and food whole life. His/Her expenditure is more than earning because of children is always sick. As a supervisor, he/she is a disciplinarian. Some Capricornia avoid work altogether because of disinterest in his/her job.

Aquarius in 6th Bhava: The Aquarius, in the 6th Bhava, indicates that he/she will work on ship or in navy. He/She is quarrelling with his/her boss and colleague. He/She has danger of drowning in water in childhood and also at the age of 41 years. He/She can function especially well within group situations.

Pisces in 6th Bhava: The Pisces, in the 6th Bhava, indicates that he/she will be infighting and so does not has good relations with children, spouse and other family members and so feel loneliness in whole life. He/She may be spending in litigation. He/She has excessive worry about the job.

Prediction by Sign in 7th house (Jaya/Wife)

Aries in 7th Bhava: The Aries, in the 7th Bhava, indicates that his//her spouse will be cruel, angry-nature, rigid, very frequently developing bad relation with husband and habituated of making disturbed family life, even though she will be educated. His/Her partner is energetic, aggressive, sexy, innovative, and pioneering.

Taurus in 7th Bhava: The Taurus, in the 7th Bhava, indicates that his//her spouse will be very beautiful, intelligent, sweet-talking, good mannered, better-understanding and talkative but proudly of her beauty. He/She will be good in conjugal life. He/She will be a loyal marital partner. There is much stubbornness in both him/her and his/her partner.

Gemini in 7th Bhava: The Gemini, in the 7th Bhava, indicates that his//her spouse will be beautiful, intelligent, sweet-talking, good mannered, better-understanding and talkative. He/She will be knowledgeable of stitching, dance, music, good cooking. He/She will be fortunate but always unsatisfied in conjugal life.

Cancer in 7th Bhava: The Cancer, in the 7th Bhava, indicates that his//her spouse will be very beautiful, intelligent, attract others due to her good manner, imaginative, thinker, better understanding and excellent in beauty. He/She will be knowledgeable of stitching, dance, music, good cooking and will be fortunate. He/She will be sentimental and good in conjugal life.

Leo in 7th Bhava: The Leo, in the 7th Bhava, indicates that his//her spouse will be cruel, angry-nature, rigid, very frequently developing bad relation with others and stay separately even in the friend's party. He/She will be too selfish and will be attached too much to ornaments and money, even though he/she will be educated, intelligent and good thinker. He/She will have happy married life. His/Her partner is dynamic, dramatic, strong, and vital.

Virgo in 7th Bhava: The Virgo, in the 7th Bhava, indicates that his//her spouse will be very beautiful, intelligent, always helpful to his/her spouse in difficulty and painful time, good mannered, better understanding and excellent in beauty. He/She will have good conjugal life. Marital partner is hard working and effectual, and assumes most of the responsibility of helping him/her and takes care of his/her practical affairs.

Libra in 7th Bhava: The Libra, in the 7th Bhava, indicates that his//her spouse will be very beautiful, intelligent, attract others due to her excellent beauty, helping attitude, interested in religious work, donor and good mannered, imaginative, thinker, better understanding and excellent in beauty. He/She will be helpful to his/her spouse and will be fortunate. He/She will have good conjugal life.

Scorpio in 7th Bhava: The Libra, in the 7th Bhava, indicates that his//her spouse will be less educated or uneducated, does not know stitching or any ladies' art and not even interested in the same, very unfortunate in life since childhood, always facing difficulties in whatsoever work he/she undertake. He/She will be suffering from diseases like headache and stomach. His/Her mate may be inwardly powerful and dynamic.

Sagittarius in 7th Bhava: The Sagittarius, in the 7th Bhava, indicates that his//her spouse will be cruel, angry-nature, rigid, very frequently developing bad relation with others. She/He will be less educated and will have no interest in stitching or any other ladies' arts. He/She will have happy married life, fortunate spouse, who will bring monetary help from his/her parent and will give good children, but reluctance to be hemmed in formal partnership entrapments.

Capricorn in 7th Bhava: The Capricorn, in the 7th Bhava, indicates that his//her spouse will be cruel, angry-nature, rigid, very frequently developing bad relation with others and educated and will have no interest in stitching or any other ladies' arts. He/She will have happy married life. He/She is restricted by the duties of wedlock or of any partnership. He/She needs partners to be his/her nurturing support and a mate as the parent figure to give a solid base, be supportive and responsible.

Aquarius in 7th Bhava: The Aquarius, in the 7th Bhava, indicates that his//her spouse will be very beautiful, intelligent, always helpful to her husband in difficulties, hard working, ready to face strongly in difficulties, fearful of god, respecting elderly people, interested in sacred work. He/She will have good conjugal life and will give good children.

Pisces in 7th Bhava: The Pisces, in the 7th Bhava, indicates that his//her spouse will be beautiful, intelligent, helping attitude, interested in religious work, donor, good mannered, better understanding and very active in social work. He/She will be helpful to his/her spouse and will be fortunate. He/She will be good swimmer and have good conjugal life and deliver intelligent and good children. He/She is willing to make sacrifices to benefit the union. His/Her partner helps him/her expand his/her horizons to new fields.

Prediction by Sign in 8th house (Mrityu/Death)

Aries in 8th Bhava: The Aries, in the 8th Bhava, indicates that he/she will be mostly settled in foreign country. He/She will be suffering from the disease of talking or walking in sleep. He/She will be unhappy by remembering his/her past incidents of life. He/She will be rich and die in foreign country. In joint financial affairs, there can be conflict and disagreement about money. He/She is decisive in this regard. Sometimes such conflicts may result in litigation.

Taurus in 8th Bhava: The Taurus, in the 8th Bhava, indicates that he/she will be seriously suffering from the disease of cough and die due to serious cough congestion of respiratory system. He/She has a good head for business and a sense for sound investments. There is a tendency to marry for money or security, as well as for love. There may be ups and downs in his/her financial affairs, but eventually matters turn out profitably.

Gemini in 8th Bhava: The Gemini, in the 8th Bhava, indicates that he/she will be seriously sick of Prameha and Gurda Roga in old age. He/She will die due to fighting with enemies. He/She is full of ideas about how properly to handle joint finances.

Cancer in 8th Bhava: The Cancer, in the 8th Bhava, indicates that he/she will die due to drowning in water or in the house peacefully at old age. He/She rarely exposes emotional needs and ceases to be the go-getter that usually marks his/her style.

Leo in 8th Bhava: The Leo, in the 8th Bhava, indicates that he/she will die due to snake biting or due to fighting with thief He/She has a large capacity for romance. There is also a much creative energy directed toward business and large-scale investments. Moneymaking becomes a game for him/her to be enjoyed. This position usually promises living to a ripe old age.

Virgo in 8th Bhava: The Virgo, in the 8th Bhava, indicates that he/she will die due venereal diseases or due to consumption of poison. This sign suggests a restrained or constrained sex life.

Libra in 8th Bhava: The Libra, in the 8th Bhava, indicates that he/she will die due to hunger or due to fighting with shoulder or due to anger and blood pressure and has financial gain through marriage and partnerships. He/She gets just everything he/she really wants with a diplomatic approach.

Scorpio in 8th Bhava: The Scorpio, in the 8th Bhava, indicates that he/she will die due to skin diseases or due to consumption of poison.

Sagittarius in 8th Bhava: The Sagittarius, in the 8th Bhava, indicates that he/she will die due to drowning in water or in the house peacefully at old age among the family members. He/She has a natural flare for business and good fortune when it comes to money. He/She has good fortune to benefit from large-scale enterprises that grow and prosper. Jupiter, the ruling planet of Sagittarius, provides good fortune and abundance in the part of the chart it controls. He/She gets financial help throughout his/her life, due to Jupiter influence. He/She can usually get what he/she wants out of life and is good-humoured nature and generous, outgoing demeanour.

Capricorn in 8th Bhava: The Capricorn, in the 8th Bhava, indicates that he/she will die due to old age and worshipping god in the house peacefully among the family members. He/She does not like to borrow and being in debt. He/She would never stoop to cheating or deceptions in business dealing. He/She handles other people's money and does so with serious concern.

Aquarius in 8th Bhava: The Aquarius, in the 8th Bhava, indicates that he/she will die due to burning or due to respiratory problem or due to saviour wound problem. He/She has a strong interest in the spiritual side of life. He/She is very unorthodox on sex, birth control and abortion.

Pisces in 8th Bhava: The Pisces, in the 8th Bhava, indicates that he/she will die due to lever problem or due to blood diseases in old age among the family members.

Prediction by Sign in 9th house (Bhagya/Fortune)

Aries in 9th Bhava: The Aries, in the 9th Bhava, indicates that he/she will be fortunate. He/She is always worried for money. His/Her expenses are more than the earning. He/She makes profit by purchase and sale of animals, agricultural lands and house properties, & product of religious work. He/She loves to travel.

Taurus in 9th Bhava: The Taurus, in the 9th Bhava, indicates that he/she will be fortunate at the age of 28 and 36 years. His/Her child hood is difficult and face problem in education but is successful in completing his/her education. He/She makes profit in business of fancy store but progress slowly in service. He/She earns wealth in second part of life and earns good name & fame and enjoys the life luxuriously. Venus makes travel enjoyable, and lives far from his/her native home.

Gemini in 9th Bhava: The Gemini, in the 9th Bhava, indicates that he/she will be fortunate at the age of 27 years, is simple, vegetarian, and gentleman and talks sensibly. He/She is successful in completing his education and helps poor people. He/She makes profit in business and thinks of always about business. He/She earns wealth in middle part of life and earns good name & fame and enjoys the life as a respected person. He/She especially enjoys travel mostly for observing different people and places.

Cancer in 9th Bhava: The Cancer, in the 9th Bhava, indicates that he/she will be simple, vegetarian, and gentleman and talks sensibly. He/She is successful in teacher, reporter, publisher and writer and helps poor people. He/She suffers from stomach problems and gastric. He/She takes too much time in taking the decisions and due to this, he/she suffers. He/She earns wealth but his/her middle part of life is full of struggles and loss of wealth.

Leo in 9th Bhava: The Leo, in the 9th Bhava, indicates that he/she will be totally against the religions and will be arguing against the religions. His child hood is difficult and face problem in education but is successful in completing his/her education. Whole life he/she will be struggling and will do hard work for survival. He/She earns wealth in second part of life and earns good name & fame and enjoys the life luxuriously. His/Her nature of job will be touring type.

Virgo in 9th Bhava: The Virgo, in the 9th Bhava, indicates that he/she will be luxurious at the young age and gentleman and talks sensibly. He/She earns wealth in middle part of life but fortune never favours him/her. He/She will be fortunate at the age of 23 years and enjoys his/her life at old age with grand children.

Libra in 9th Bhava: The Libra, in the 9th Bhava, indicates that he/she will be totally religious minded and will be arguing always in favour of the religions. His/Her child hood is difficult and face problem in education but is successful in completing his/her education. Whole life he/she will be struggling and will do hard work for survival. He/She will be more successful in business than service. He/She earns wealth from business and earns good name & fame in the society as a leader and enjoys the life luxuriously in leadership. He/She gets wealth from his/her in-law. He/She will be fortunate after the age of 24 years.

Scorpio in 9th Bhava: The Scorpio, in the 9th Bhava, indicates that he/she will be mean mind. His/Her child hood is difficult and face problem in education, will be struggling and will do hard work for survival. He/She will be more successful after the age of 28 years.

He/She earns wealth after 28 year of age and earns good name & fame in the society and enjoys the life luxuriously. His/Her nature of job will be touring type.

Sagittarius in 9th Bhava: The Sagittarius, in the 9th Bhava, indicates that he/she will be clever, peaceful and gentle man. His/Her child hood is good and does not face problem in living. He/She will be struggling and will do hard work for survival at the middle of age. He/She will be more fortunate and successful after the age of 45 years and achieve very high position and wealth. He/She earns wealth after 45 year of age and earns good name & fame in the society and enjoys the life luxuriously. His/Her nature of job will be related to water like captain, sailor or any other service with irrigation department where he/she gets lot of success. He/She will have much more journey in life.

Capricorn in 9th Bhava: The Capricorn, in the 9th Bhava, indicates that he/she will be clever, peaceful and gentle man. His child hood is not good and faces problem in living. He/She will be struggling and will do hard work for his survival. He/She will be more fortunate and successful after the age of 45 years. He/She earns wealth after 45 year of age and achieves very high position and wealth and earns good name & fame in the society and enjoys the life luxuriously.

Aquarius in 9th Bhava: The Aquarius, in the 9th Bhava, indicates that he/she will be clever in politics, a leader. His/Her child hood is not good and faces many problems in living. He/She will be struggling in education and will do hard work for his/her survival. He/She will be more fortunate and successful after the age of 45 years and 36 years and achieve very high position and wealth and earns good name & fame in the society and enjoys the life luxuriously. He/She starts earning wealth after 28 year of age.

Pisces in 9th Bhava: The Pisces, in the 9th Bhava, indicates that he/she will be totally religious minded and child hood will be good but is successful in completing his/her education. He/She will be more successful in business than service, particularly in yellow metals business. He/She will be fortunate after the age of 27 years and earns wealth from business and earns good name & fame in the society as a leader and enjoys the life luxuriously in leadership.

Prediction by Sign in 10th house (Karma/Profession)

Aries in 10th Bhava: The Aries, in the 10th Bhava, indicates that child hood is difficult and face problem in education but is successful in completing his/her education. Whole life he will be struggling and will be busy in more than one work and will do hard work for survival. He/She earns wealth in second part of life but does not earn good name & fame and does not enjoy the life luxuriously. He/She has tendency toward shyness and to disappear on the public stage, or even in public view.

Taurus in 10th Bhava: The Taurus, in the 10th Bhava, indicates that child hood is good but will spend more than his/her earning in his life. Whole lives he/she will be struggling and will be busy in earning and will not be able to save much. He/She will be patriotic to his/her father. He/She chooses career, such as art, theatre, and music.

Gemini in 10th Bhava: The Gemini, in the 10th Bhava, indicates that he/she will be patriotic to his/her father and respect and obey the elders and will be agriculturist and earn money from agriculture or real estate. He/She may earn more money from business rather than service.

Cancer in 10th Bhava: The Cancer, in the 10th Bhava, indicates that he/she will be child hood is good but will spend more than his/her earning in his/her life. Whole lives he/she will be busy in helping poor people and doing social work such as making temples and drinking water facilities for the people and will be busy in earning.

Leo in 10th Bhava: The Leo, in the 10th Bhava, indicates that child hood is difficult and face problem but is successful in completing his/her education. Whole life he/she will be struggling and will do hard work for survival. He/She earns wealth in second part of life but does not earn good name & fame and does not enjoy the life luxuriously. He/She does not get help from the family. He/She is likely to get a position of leadership and authority, and be admired and is happiest running his own business.

Virgo in 10th Bhava: The Virgo, in the 10th Bhava, indicates that he/she will be patriotic to his/her father but does not enjoy the life luxuriously. He/She will be self respected person and works to maintain his/her respect at any cost. He/She will not do buttering of any other person in life, even though he may lose something. He/She will earn good money from business rather than service. He/She finds employment in a large, well-established organization such as civil service, a church, or an educational institution.

Libra in 10th Bhava: The Libra, in the 10th Bhava, indicates that he/she will be patriotic to his/her father and respect and obey the elders and enjoy the life luxuriously. He/She will earn more money from business rather than service. Business will be lucky for him/her and will progress in life by business. He/She will be rich of his/her speech and fulfil whatever he/she will promise. He/She will reach to the highest position in the middle age. He/She will enter a business with a partner, or in cooperation with others. He/She will be an excellent administrator. His/Her public standing may rise well above that of his parent.

Scorpio in 10th Bhava: The Scorpio, in the 10th Bhava, indicates that he/she will be totally religious minded and get popularity and respect in the society and will be clean and honest. His/Her child hood is good but will spend more than his/her earning in his/her life in religious work. Whole lives he/she will be busy in helping poor people and doing social work and will be busy in earning and will not be able to save much. He/She will reach to the highest position in the service. He/She make his/her mark in the world. There has drive to succeed and is naturally able to impress others with the forcefulness of his/her mind.

Sagittarius in 10th Bhava: The Sagittarius, in the 10th Bhava, indicates that he/she will be patriotic to his father and respect and obey the elders and enjoy the life luxuriously. He/She will earn more money from business rather than service. Business will be lucky for him/her and will progress in life by business. He/She may face a lot of problem in service but will be hard working and makes everything favourable in life. He/She will get popularity and respect in the society. He/She has the abilities to be a leader and never hesitates using his/her influence to attain goals. A good deal of travel is likely to be associated with his/her career. He/She may have more than one career in his/her life.

Capricorn in 10th Bhava: The Capricorn, in the 10th Bhava, indicates that he/she will be totally religious minded and get popularity and respect in the society and will be clean and honest. His/Her childhood is good but will spend more than his/her earning in religious work. Whole lives he/she will be busy in helping poor people and doing social work. He/She will be able to make favourable conditions and will reach to the highest position in the middle age and people will be astonished with his/her rise. He/She has a very strong sense of duty, an attitude of dedication. The progress in the career may be slow, but it is consistent. He/She is capable of climbing to the top. He/She can build a solid public image. This is a very achievement oriented placement.

Aquarius in 10th Bhava: The Aquarius, in the 10th Bhava, indicates that he/she will be political minded and get popularity and respect in the society and will be talkative. His/Her child hood is good but will be busy in helping people. He/She has followers in politics. He/She makes favourable conditions and will reach to the highest position.

He/She is a team player and function well as a part of the team. He/She may be involved in many humanitarian ventures.

Pisces in 10th Bhava: The Pisces, in the 10th Bhava, indicates that he/she will be patriotic to his father and respect and obey the elders and enjoy the life luxuriously. He/She will earn more money from service related to water. He/She may face a lot of problem in life but will be hard working and makes everything favourable in life. He/She will get popularity and respect in the society. He/She can make money by business too.

Prediction by Sign in 11th house (Gain/Income)

Aries in 11th Bhava: The Aries, in the 11th Bhava, indicates that he/she will be patriotic to his/her father and respects and obeys the elders and does not enjoy the life luxuriously. He/She will be self respected person and will not do buttering of any other person in life, even though he/she may loose something. He/She will be hard working person and does not get help from the father and family. Whole life he/she will be struggling and will be busy in more than one work and will do hard work for survival. He/She is fortunate after 28 years of age. He/She is active in associations, club or other such organizations. He/She attracts a wide and varied circle of friends. He/She makes the most of his contacts. Often, friends are the key to helping him/her attain his goals.

Taurus in 11th Bhava: The Taurus, in the 11th Bhava, indicates that he/she will belong to the medium family and does not enjoy the life luxuriously. He/She will be self respected person and work to maintain his/her respect at any cost. He/She will be hard working person and does not get help from the father and family. Whole life he/she will be struggling and will be busy in more than one work and will do hard work for survival. He/She will be known to the great personalities and will live in his/her surroundings. He/She gets benefited with the opposite sex. His luck starts from his/her middle age and reach to the highest post. He/She will be making money, accumulating a comfortable standard of living, and establishing a secure situation. His/Her interest is not just money and possessions for the sake of wealth. Instead he/she wants to gain a sense of security, and status to overcome a basic insecurity. He/She has well-to-do friends, who may be called on to help him/her attain his/her goals.

Gemini in 11th Bhava: The Gemini, in the 11th Bhava, indicates that he/she will be hand to mouth since child hood. He/She will be self respected person and work to maintain his/her respect at any cost.

He/She will be hard working person and does not get help from the family. Whole life he/she will be struggling and will be busy in more than one work and will do hard work for survival. He/She is fortunate in business as compared to service. Up to 42 years, his financial conditions are miserable. He/She will start earning from many sources after 42 years of age. He/She has many personal connections. He/She is naturally attracted to people who are witty, intelligent, and verbal. He/She has a good sense of humour and he can laugh, even at himself/herself. He/She is not a loner, and he/she needs constant, mentally compatible companionship to be happy and fulfilled.

Cancer in 11th Bhava: The cancer, in the 11th Bhava, indicates that he/she will be patriotic to his/her father and respects and obeys the elders and enjoys the life luxuriously. He/She will be self respected person and work to maintain his/her respect at any cost. He/She will be hard working person and does not get help from the father and family. Whole life he/she will be struggling and will be busy in more than one work. He/She will progress in the service as compared to business. He/She will be more attached to his/her family members due to his/her sentiment and he/she may sacrifice for his/her family or friends. He/She can earn money from lottery. He/She will have a successful life.

Leo in 11th Bhava: The Leo, in the 11th Bhava, indicates that he/she will be business minded and calculate everything in term of profit and loss before doing the work. He/She will be self respected person and work to maintain his/her respect at any cost. He/She will be hard working person and does not get help from the family. Whole life he/she will be calculating and will be busy in more than one work and will does hard work for survival. He/She is fortunate in business as compared to service. He/She will start earning wealth in business from many sources and makes lot of money. He/She is kind and respects everybody. He/She will never spend money on luxury and live a simple life. He/She has leadership qualities displayed within groups and organizations. He/She has a very deep need for friends and associations. He/She takes pride in his/her friends and associates; some may be rich and famous. He/She may tend to draw strength from his/her friends as if they fulfilled a special need in his/her life. He/She is always careful to dress in good taste, and notices what others are wearing as well. He/She has a very wide circle of friends, most of who have much respect for him/her.

Virgo in 11th Bhava: The Virgo, in the 11th Bhava, indicates that he/she will be business minded and calculate everything in term of profit and loss before doing the work. He/She will be far sighted and self respected person and work to maintain his/her respect at any cost. He/She will be hard working person and does not get help from the family rather he/she will be helping his brothers and other family members. Whole life he/she will be calculating and will be busy in more than one work and will do hard work for survival. He/She is fortunate in politics as compared to business and service. He/She will start earning from many sources and makes lot of money from politics. He/She is kind and respects everybody. He/She will never spend money on luxury and live a simple life. He/She has a readiness to serve close friends.

Libra in 11th Bhava: The Libra, in the 11th Bhava, indicates that he/she will be calm, quiet, intelligent and gentle person. He/She will be able to identify the right or wrong timing and to do the miracle even in the miserable conditions. Family and friends will be helpful to him/her. He/She has an amiable demeanour with friends and working in group situations. In this context, he is diplomatic and even handed. He/She is likely to be selected to head a group just because of he/she is acceptable to those with divergent interests. He/She has a highly social attitude. He/She really loves to be with friends. Often this sign denotes marriage to a friend of long standing.

Scorpio in 11th Bhava: The Scorpio, in the 11th Bhava, indicates that he/she will be so intelligent that he/she will speak anything suitable to the time, person and place. He/She will be able to do many things at a time and will be successful in all the works undertaken. He/She will be doing the business related to the land and spends a lot for this. He/She will be benefited by agriculture. He/She develops close relationships with intense and aggressive friends who are influential and powerful. He/She never accepts weak individuals as friends.

Sagittarius in 11th Bhava: The Sagittarius, in the 11th Bhava, indicates that he/she will be business minded and calculate everything in term of profit and loss before doing the work. He/She will be self respected person and work to maintain his/her respect at any cost. He/She will be hard working person. Whole life he/she will be calculating and will be busy in more than one work and will do hard work for making lot of money. He/She will have popularity among service class officers and managers and will make use of them to earn more and more money. He/She is fortunate in business as compared to service. He/She will start earning wealth in business from many sources and makes lot of money. He/She is kind and respects everybody. He/She will never spend money on luxury and live a simple life. He/She has a wide circle of friends and associates and enjoys such casual relationships.

Capricorn in 11th Bhava: The Cancer, in the 11th Bhava, indicates that he/she will be a self made person, active and intelligent to take right decision at the right time, right place. He/She will be self respected person and work to maintain his respect at any cost. He/She will be hard working person and get help from the family. He/She is fortunate in service as compared to business. He/She will start earning wealth in service and from many other sources and makes lot of money. He/She is kind and respects everybody. He/She will never talk lye and fulfil his/her promises at any cost.

Aquarius in 11th Bhava: The Aquarius, in the 11th Bhava, indicates that he/she will be a self-made person, active and intelligent to take right decision at the right time, right place. He/She will be self respected person and work to maintain his/her respect at any cost. He/She will be hard working person and does not get help from the family. His/Her childhood is very troublesome but as he/she grows up, he/she starts earning and his/her life is comfortable. He/She is fortunate in service as compared to business. He/She will start earning wealth in service and from many other sources and makes lot of money. He/She is kind and respects everybody. He/She will never talk lye and fulfil his/her promises at any cost. He/She spends money on luxury and lives a simple life. He/She is successful in politics too. His/Her friendships are intellectually motivated rather than sentimentally stimulated.

Pisces in 11th Bhava: The Pisces, in the 11th Bhava, indicates that he/she will be a rich man, self made person, active and intelligent to take right decision at the right time, right place. He/She will be self respected person and work to maintain his respect. He/She will be hard working person and get help from the family. He/She is fortunate and shins in a particular line such as singer or scientist or in arts and in service as compared to business. He/She will start earning wealth in service and from many other sources and makes lot of money. He/She is kind and respects everybody. He/She will never talk lye and fulfil his promises at any cost. He/She spends money on luxury and lives a simple life.

Prediction by Sign in 12th house (Loss/Expenditure)

Aries in 12th Bhava: The Aries, in the 12th Bhava, indicates that he/she will enjoy the life luxuriously since childhood. He/She will earn more money from business rather than service in beginning but will become bankrupt in the middle age and there will be financial crisis. Business will be lucky for him/her in beginning or early age and will progress in life by business up to middle age. He/She may face a lot of problem in business at middle age but will be hard working and makes everything favourable in life. He/She will get popularity and respect in the society. His/Her eyesight will be weak and will get operated in the old age. Though he/she is very slow to anger, when he/she does cross the line, his/her temper can be irrational.

Taurus in 12th Bhava: The Taurus, in the 12th Bhava, indicates that he/she will be a rich man, self made person, active and intelligent to take right decision at the right time, right place. He/She will be self respected person and work to maintain his/he respect in the society. He/She will be hard working person. He/She is fortunate and shins in a particular line such as administrator or tourist or salesman or writer and in service as compared to business. He/She will start earning wealth at the age of 30 years and from many other sources and makes lot of money. He/She is kind and respects everybody. He/She will never talk lye and fulfil his/her promises at any cost. He/She does not spend money on luxury and lives a simple life. He/She has more worry about financial affairs than his/her happy-go-lucky demeanour would imply.

Gemini in 12th Bhava: The Gemini, in the 12th Bhava, indicates that he/she will be imaginative and sentimental type. He/She believes any person and get hurt and cheated in due course of time by that person, particularly by relative, brothers, sisters and others. He/She will never talk lye and fulfil his/her promises at any cost. He/She spends too much money on luxury and lives a luxurious life. Whole life he/she will be struggling and will be busy in more than one work and will do hard work for survival. He/She is fortunate in business as well as service. Throughout, his financial conditions are miserable. He/She will start earning from many sources and will spend too much money. His/Her eyesight will be weak and will get operated in the old age. He/She will find it very difficult to live with or near his brothers, sisters or other close relatives.

Cancer in 12th Bhava: The Cancer, in the 12th Bhava, indicates that he/she will be rigid and angry promise fulfiller type. He/She makes mistake due to his angriness. He/She will never talk lye and fulfil his/her promises at any cost. He/She spends too much money on luxury and lives a luxurious life. Whole life he/she will be struggling and will be busy in more than one work and will do hard work for survival. He/She is fortunate in business as well as service. Throughout, his/her financial conditions are miserable. He/She will start earning from many sources and will spend too much money. He/She believes any person and gets hurt and cheated in due course of time by that person, particularly by relative. He/She is a generous person. He/She gets hurt when those near and dear are beginning to take him for granted.

Leo in 12th Bhava: The Leo, in the 12th Bhava, indicates that he/she will be imaginative and sentimental calm and quite, kind and sweetly spoken type. He/She believes any person and gets hurt and cheated. He/She will never talk lye and fulfil his/her promises at any cost. He/She will not like to spend too much money on luxury and lives a simple life. Whole life he/she will be struggling and will be busy in more than one work and will do hard work for survival. He/She is fortunate in service rather than business. He/She will start earning from many sources and will not spend much money on decorative items. He/She is a power behind the scenes. He/She plays a back room manoeuvring role in matters. He/She is never one to blow his/her horn.

Virgo in 12th Bhava: The Virgo, in the 12th Bhava, indicates that he/she will be sentimental calm and quiet, and sweetly spoken type and successful businessman. He/She will be able to take decision very fast in any matter. He/She will be in business and find out many source of income in it. He/She will not like to spend too much money on luxury and lives a simple life and teach his/her children the same. Whole life he/she will be struggling and will be busy in more than one work and will do hard work for survival. He/She is fortunate in business rather than service. He/She will start earning from many sources and will not spend much money on luxury items. He/She may have to face litigation and charges and due to that, he may face problem. He/She will be self respected person and work to maintain his/her respect at any cost and will do the religious and social work.

Libra in 12th Bhava: The Libra, in the 12th Bhava, indicates that he/she will be a common person, self-made person, active and intelligent to take right decision at the right time, right place. He/She will be self respected person and work to maintain his/her respect. He/She will be hard working person and does not get help from the parent. He/She will be able to make any person in his/her favours and become popular in the society. He/She will have many friends and known personalities and will get benefited from them. He/She is fortunate and shins in service as compared to business. He/She will start earning in service and from many other sources. He/She is kind and respects everybody. He/She will never talk lye and fulfil his promises at any cost. He/She spends money on luxury and lives a simple life. He/She is a loner with some apprehensions regarding reliance on others. His/Her partnership arrangements, including the marriage, may be fated in some way and acting in collaboration with others, rarely works out well for him/her.

Scorpio in 12th Bhava: The Scorpio, in the 12th Bhava, indicates that he/she will be self made person, active, religious minded person and intelligent to take right decision at the right time, right place. He/She will be self respected person and work to maintain his/her respect in the society and will do the religious work. He/She will be hard working person. He/She is fortunate and shins in a particular line such as writer or publisher and in service as compared to business. He/She will start earning wealth at the age of 16 years and from many other sources. He/She is kind and respects everybody. He/She will never talk lye and fulfil his promises at any cost. He/She does not spend money on luxury and lives a simple life.

Sagittarius in 12th Bhava: The Sagittarius, in the 12th Bhava, indicates that he/she will be intelligent. He will be able to take decision very fast in any matter. He/She will be in business and find out many source of income in it. He/She will not like to spend too much money on luxury and lives a simple life. Whole life he/she will be struggling and will be busy in more than one work and will do hard work for survival. He/She will start earning from many sources and will not spend much money on luxury items. He/She will be self respected person and work to maintain his/her respect at any cost and will do the religious and social work and will be popular in the society. He/She will be able to make any person in his/her favour and become popular in the society. He/She will have many friends and known personalities and will get benefited from them.

Capricorn in 12th Bhava: The Capricorn, in the 12th Bhava, indicates that he/she will be a common person, self-made person, active and intelligent to take right decision at the right time, right place. His/Her childhood will be troublesome. He/She will be self respected person and work to maintain his/her respect. He/She will be hard working person and does not get monetary help from the parent. He/She will be able to make money by his/her own hard work and become popular in the society. He/She will start earning wealth at the age of 30 years and from many other sources. He/She will have many friends and known personalities and will get benefited from them. He/She is fortunate and shins in service as compared to business. He/She will start earning in service. He/She is kind and respects every body. He/She will never talk lye and fulfil his promises at any cost. He/She spends money very cautiously and makes balance between income and expenditure and lives a simple life. He/She has subconscious feelings of limitation and inadequacy deeply ingrained in the psyche.

Aquarius in 12th Bhava: The Aquarius, in the 12th Bhava, indicates that he/she will be intelligent, a common person, self-made person, active, religious minded and intelligent to take right decision at the right time, right place. He/She will be self respected person and work to maintain his/her respect. He/She will be hard working person. He/She will be able to take decision very fast in any matter and will be in service and finds out many source of income in it. He/She will spend too much money and does not have control over it and lives a simple life. Whole life he/she will be struggling and will be busy in more than one work and will do hard work for survival. He/She will start earning from many sources and will not spend much money on luxury items. He/She will be self respected person and work to maintain his/her respect at any cost and will do the religious and social work and will be popular in the society. He/She will be able to make any person in his/her favour and become popular in the society. He/She will have many friends and known personalities and will get benefited from them.

Pisces in 12th Bhava: The Pisces, in the 12th Bhava, indicates that he/she will be a common person, self-made person, active and intelligent to take right decision at the right time, right place. His/Her childhood will be simple. He/She will be self respected person and work to maintain his/her respect. He/She will be hard working person and does not get monetary help from the parent. He/She will be able to make money by his/he own hard work and become popular in the society. He/She will start earning wealth at the age of 32 years and from many other sources. He/She will have many friends and known personalities and will get benefited from them. He/She is fortunate and shins in service as compared to business and reaches to the highest position by his/her hard work. He/She will start earning in service. He/She is kind and respects everybody and will never talk lye and fulfil his promises at any cost. He/She spends money very cautiously and makes bank balance and lives a simple life. He/She will be helpful to others.

7

Moon-Sign

7.1 Predictions by Moon Sign (Rashi):

The waxing (rising) moon is a very auspicious planet, capable of causing Neechabhanga of other planets by his mere aspect. The position of Moon in a Sign is at the time of birth is called the Moon-Sign or Birth Rashi or Rashi. **Example:** If the Moon is in the Sign of Mesh (Aries), the Moon-Sign or the Birth Rashi or Rashi (Janma Rashi) is Mesh. It is also called the Moon-Sign Chart, in which Moon in the first House. Accordingly, the distance of the house of all the planets from the Moon is accessed, which is essential to predict the effects of the Maha Dasa and Antar Dasa of planet.

7.1 Aries Rashi (Moon in Aries)

Name starts with phonetic (Chu, Hey, Cho, La, Li, Lu, Ley, Lo, Ae): The Moon in Aries is not congruent to her nature. He/She is restless, eyes, inflicted with diseases, unfaithful, gives pleasures to his wife, fears of drowning into water, hard working and is full of tranquillity in his old age. He/She is adventurous and is too changeable, moody, whimsical and flirtatious. He/She is likely to meet with all sorts of disappointments and even disillusion.

7.2 Taurus Rashi (Moon in Taurus)

Name starts with phonetic (Ee, U, Aye, Oh, Va, Ve, Vo, Vay, Vo): The Moon in Taurus is a natural domicile. He/She is charitable, pious, virtuous, wealthy, and full of radiance, good health and long lived. He/She has determination, loyal friend and is emotionally very strong and seldom changes his/her mind. He/She attracts opposite sex for strong romance. He/She takes up occupations of real estate, property, art, design, jewellery and business. He/She is an ambitious, selfish, has a goal for a luxurious home, plenty of money and intellectual but has a very practical, astute, shrewd ability to judge the average conditions in life.

7.3   Gemini Rashi (Moon in Gemini)

Name starts with phonetic (Ka, Ke, Koo, Gha, Jna, Cha, Kay, Ko, Haa): The moon in Gemini is considered weak. He/She has a melodious voice, talks sweetly, is kind hearted, very lusty and is prone to throat diseases, famous, wealthy, fair complexioned, tall, clever, genius, of firm resolution, efficient in work, and remains judicious in every situation. He/She thrives on communication. He/She prefers job in media, travel as well as sales. He/She chooses his/her partners. He/She is highly imaginative, educated, and is both, a good teacher and a sharp student. He/She cannot limit to one activity and business at a time and would do well with a strong & practical partner. He/She will be entrusting the partner with most of the decision-making in business. He/She is very successful as news person, advertising agency, writers, authors or any other creative field. He/She is likely to make plenty of money, and enjoy great popularity. He/She is good in the business world. He/She will retain a youthful look and behaviour and succeeds at "staying young forever."

7.4 Cancer Rashi (Moon in Cancer)

Name starts with phonetic (He, Hoo, Hey, Ho, Daa, Dee, Doo, Dey, Do): The Moon is considered royal in her own Cancer. He/She is wealthy, has patience, serves his teacher, very clever, lives in a foreign land, keeps good company, and has a high degree of intelligence. He/She is sympathetic, kind, compassionate and sensitive to others' feelings. He/She is with excellent memory, fond of home and parents, peaceful, gentle, affectionate, and romantic. He/She may get delayed marriage. He/She may be excellent artists, musicians, and poets and home life is very important to him/her. As a parent, he/she is quite nurturing and lavishes loved ones. He/She does not hesitate to use tears or a self-sacrificial attitude to get his/her point across. Real estate is an especially good area to invest in for him/her. He/She would also do well in a business run from home and has natural tendency to put on weight.

7.5 Leo Rashi (Moon in Leo)

Name starts with phonetic (Ma, Me, Moo, May, Moo, Ta, Tee, Too, Tay): The Moon in Leo is the natural domicile of the Sun. He/She forgives easily, loves to travel, likes to eat non-vegetarian food, is full of fear, keeps good company, is humble, has excessive anger, devoted to his parents and achieves fame. He/She is self-sacrificing, generous, conservative, discriminating, encouraging, romantic, optimistic, brilliant robust, strong, and decisive and a natural leader and exudes energy and drive. He/She loves excitement and action.

He/She may sacrifice everything in the cause of righteousness and justice. He/She makes others dependent on him. He/She cannot be convinced against his/her will, nor be swayed against emotions. He/She is strong-minded and determined. He/She is warm, loving and outgoing. He/She doesn't settle for less than what he/she wants and is too much commanding. He/She will be smothered with love and care, and may be henpeck.

7.6 Virgo Rashi (Moon in Virgo)

Name starts with phonetic (Too, Pa, Pee, Pu, Sha, Na, Tha, Pay, Poe): The Moon in Virgo is unimpeded mind, passion and flesh. The keyword is "criticism". He/She is a sensualist, respects the virtuous people, religious, clever, charitable, poet, follower of the Vedas, lover of humanity, interested in dance and music, likes to travel, and is troubled by his/her Spouse. He/She has no confusion in the mind, strong social conscience, good communication and is sociable, logical, back-seat drivers and clear-headed. He/She has a strong personal code of conduct and set high standards. He/She frequently takes nursing, dietetics, teaching, and secretarial work as careers. He/She is considered to be cold-blooded and overly ambitious. He/She is very conscious of keeping fit, both physically and mentally. He/She also enjoys taking part in national politics or social issues. He/She makes well, stable business partners and a successful professional. He/She is grave, sexless people with no sexual curiosity, and doesn't understand the meaning of sex. He/She is exceedingly active, and will put more energy into house cleaning, attention to business, and personal doctoring than any other type of person.

7.7 Libra Rashi (Moon in Libra)

Name starts with phonetic (Ra, Ree, Ru, Ray, Tha, Thee, Thoo, They): The Moon in Libra is the best positions. The keyword is "decision". The moon in this position gives artistic temperament, creative ability, good mental understanding, but no executive ability. He/She gets angry unnecessarily, talks sweetly, has restless eyes, mixed fortunes, authority inside the house but powerless outside, a devotee and likes to travel. He/She is charming, creative, and diplomatic. He/She may be romantically amorous, notoriously fickle and wavering in romance. He/She is financially motivated, can be reckless, careless, and/or squandering. He/She is known for charm and social grace, and presents an image of total balance and harmony, and has problem of the kidneys and allergies. He/She takes professions as law, architecture, politics, the arts, and even homemaking. He/She is attracted to very gracious partner and finds a good one in life, although it may not be the first one. He/She is voluptuous, deceptive in habits, with a voracious physical appetite. Usually he/she is so pretty and has so much charm that marriage is a foregone conclusion.

7.8 Scorpio Rashi (Moon in Scorpio)

Name starts with phonetic (Tho, Na, Nee, Noo, Ney, No, Yaa, Yee, Yoo): The Moon in Scorpio is favourable position. The keyword is "Ulterior Motivation". He/She is a traveller from his childhood, has yellow eyes, lusty, proud, behaves roughly with his relatives, acquires wealth through hard work, and is wicked towards his mother. He/She is intelligent, cold, sensual, emotional, materialistic and secretive. He/She can become superb occultists and astrologers. The position is favourable for jobs in medicine, surgery, chemistry and investigative work. He/She can be very possessive, jealous, very cruel and vindictive. He/She never forgets a wrong that someone has done, and will plot and plan for years or decades, if necessary, to seek revenge. He/She is intensely emotional but projects a perfectly cool exterior at all times. His/Her frenzied desires are never satisfied within home. He/She is constantly seeking outside satisfaction. There are few women who have a good deal of scandal running through the life.

7.9 Sagittarius Rashi (Moon in Sagittarius)

Name starts with phonetic (Yey, Yo, Ba, Bee, Bu, Dha, Pha, Dha, Bay): The Moon position in Sagittarius is risk-taking. The keyword is "enthusiasm". He/She is pious, wealthy and virtuous, has a loving nature, knowledge of fine arts, likes drawing and painting, has a wife full of good qualities, sweet-talker, has a heavy physique and in some rare cases, a destroyer of his family. He/She is eternal students with an urge to higher education, impulsive, blunt and outspoken, magnetic and forceful, actively philosophical and believes in justice and fair play and helps out anyone who is in need. He/She likes astrology and prophecy. He/She hates anything hidden or secret. He/She has a strong sense in gambling and believes that he/she simply cannot lose and therefore take unconsidered risks. If it is necessary to terminate a relationship, he/she does so quickly and cleanly without looking back. He/She knows something better is waiting just around the next corner. He/She is great spenders and enjoys a jolly, pagan sort of social life, uninhabited and full of romantic interest. He/She is incurable opposite sex partners' chasers and has put a great deal of enthusiasm into his/her dashing love affairs.

7.10 Capricorn Rashi (Moon in Capricorn)

Name starts with phonetic (Bo, Ja, Je, Ju, Jay, Jo, gha, Ga, Gee): Moon in Capricorn is one of the least desirable positions. The key word is "Management". He/She has values in his family, is under the influence of his wife, scholar, undertakes charitable work, respects his mother, is wealthy, has obedient servants, is kind hearted, has a large family, and lot of worries also. He/She lacks sympathy, and has innate selfishness, self-preoccupation, dignity, tenacity and a realistic vision of the world. He/She has defeated ambitions and dreams, misfortunes, occupational and financial troubles, credit difficulties and all sorts of other misfortunes. He/She will have positions of executive, administrative, public and organizational positions and commercial pursuits. He/She has a natural desire to rise to a position of power and fame and is willing to work hard for accomplishments. He/She is not fortunate enough to achieve fame and fortune and becomes terribly frustrated and may even develop ill health as a result. He/She is very selfish, cautious, and thrifty and lays own aims and ambitions.

7.11 Aquarius Rashi (Moon in Aquarius)

Name starts with phonetic (Goo, Gay, Sa, See, So, Say, Da): The Moon in Aquarius is fixed. The keyword is "disinterestedness". He/She is lazy, owns the most expensive vehicles, wealthy, is blessed with beautiful eyes, and has a simple nature. He acquires wealth and knowledge, achieves fame on account of his virtuosity and kindness, is fearless and enjoys his wealth. He/She is idealistic, caring for the global village, a little detached to home and paradoxically. He/She possesses integrity and honesty, and is not likely to ask for help when in trouble, but ready to help others. He/She is well liked and have strong religious and philosophical instincts coupled to a humanitarian urge. He/She possesses absence of jealousy and possessiveness, and favours all forms of humanitarian, political, and educational pursuits, exploration in all fields, authorship, and astrology too. He/She has a tendency to gossip and spread rumours. He/She loves the business and professional world, and has plenty of patience, kindness, intelligence and understanding.

7.12 Pisces Rashi (Moon in Pisces)

Name starts with phonetic (Dee, Du, Tam, De, Do, Cha, Che): The Moon in Pisces is not favourable position. The key word is "anxiety". He/She is brave, talks cleverly, but often has excessive anger, loved by his family, a devotee, a very fast walker, efficient in charity and knowledge, virtuous, sacrificing and receives affection and love from his friends and family members. He/She has strong creativity, powerful imagination and impressionability. He/She is natural worshippers of beauty, very loyal to friends and is more inclined to romantic attitudes. He/She succeeds best in intuitive judgement, discretion, assiduity, and detailed work. He/She does well as entertainers, dealers with liquids of all sorts, promoters, seafarers, and detectives. He/She has strong intuitive and psychic qualities. He/She should generally follow his intuition, and cares a great deal about others and seek to serve society as a whole in own way. He/She gives love freely to others, and may get deceived by others. He/She may feel like rejecting the world altogether. His/Her life probably will not be one of the Cinderella or Prince Charming of his/her dreams. He/She is not afraid to work hard and is subjected to wild swings in mood. He/She can give pure, unselfish, transforming love and compassion. He/She can often find satisfaction and relief in religion and art.

8

Sun-Sign Predictions

8.1 Predictions by Sun-Sign

The date of birth determines the Sun-Sign and gives special attributes to the native as mentioned below:

8.1 Aries (March 21 - April 20)

Who works from morning to evening, and never likes to be outdone? Who is outspoken, alert, and ambitious and whose walk is almost like a run?

Personal Quality: The Arian is born leader, adventurous, brave, fearless, highly dominating, and full of energy, aggressive, argumentative, good athletes and soldiers. He/She is outspoken, alert and quick to act and speak. He/She is always willing to help the persons in need. He/She is not a follower. He/She is large hearted and speak straight forward for what he/she feel about the person. He/She is childishly egocentric, extremely demanding and liable to throw tantrums if denied. He/She is quick to anger and known for his impatience, and is prone to be arrogant. Under planetary afflictions he is subject to brain fever, dizziness, nosebleed, neuralgia, inflammation of the cerebral hemispheres, and diseases of face.

Positive Quality: He/She is generous, a lover of justice, and wishes to earn by own efforts and never looks at others wealth. He/She gives time, effort, money and sympathies to others. He/She likes to be challenged and enjoys solving any obstacle. He/She has both moral and physical courage.

Negative Quality: He/She is not very tactful in communicating and will never bend, and is not diplomatic and is sharp tongue and shows anxiety. He/She has a spending nature to maintain the image. He/She gets nervous when things are not moving his/her way. He/She is quick-tempered, violent, impatient, egotistical and intolerant.

Physical Appearance: He/She is angular, slim in early life, although may fill out later. He/She has sharp elbows and knees.

Relationships: He/She is possessive, jealous, faithful and idealistic, passionate and incurably romantic. He/She tends towards joyous sex and close relationships throughout the lives. His/Her life partner will be Leo and Libra. Aquarius and Sagittarius might be very helpful to him/her in business.

Career: His/Her income and social status will rise at the age of 48-52 and will get promotion at the age of 30, 36 and 45. He/She will deal with 'futures' on the money market or hacking through the Amazonian forest. He/She can be Dentist, Director, E M T, Entertainer, Entrepreneur, Landlord, Lawyer, and Make-up artist, Optometry, Producer, Sports person and Stockbroker.

Health: He/She is always in a tearing hurry and often has a fast metabolism, which keeps the weight down. He/She is prone to stress and suffer from tension headaches.

Lucky stone: Coral & Pearl. The glittering Coral will gives all the courage, makes him/her rich and gives a comfortable future. Topaz and Moonstones will be auspicious too.

Lucky Number: The number 1, 9 & 14 can bring luck in life.

Lucky Colour: Revel in the magic of peacock blue or Shades of Red will be lucky.

Lucky Day: Dig gold on Tuesday.

Lucky Flower: Sweet Pea.

8.2 Taurus (Vrishabh) (April 21 - May 21)

Who loves good things and smiles through life except when crossed? Who is stolid, tenacious and determined and thinks he knows the most?

Personal Quality: The Taurus is stubborn, generous, highly reliable, and good in the position of managers and achieves almost everything in life. He/She makes friendships very rarely but, once made, he/she is faithful. He/She is reliable, responsible, affectionate and loyal. He/She is easy to get along with and good team player. He/She can reach to the desired height with hard work, devotion and patience. He/She is practical, reserved and is possessing tremendous willpower and self-discipline. He/She is incredibly and uncompromisingly loyal. His/Her economic position will be good from the age of 35-46 and becomes a rich man in the society. His/Her early part of life is very struggling. His/Her children are generally intelligent and bright and he/she has a pleasant married life.

Positive Quality: He/She is helpful and does a lot of things for family and considers it as a sacred duty. He/She has ability to concentrate and never leaves anything unfinished and does not believe in shortcut of anything. He/She is warm, loving, gentle and charming most of the time. He/She is honest, reliable and loving.

Negative Quality: He/She very seldom changes the mind once made-up and does not bother at all for the result. He/She is expressed in dullness, stubbornness and resistance to change. He/She is very suspicious and is afraid of getting deceived. He/She believes to the person so easily that any body can cheat him/her. He/She cannot forget or forgive people so easily.

Physical Appearance: His/Her stature varies from short to medium to stocky. His/Her eyes are bright and soulful and he/she carries himself gracefully.

Relationships: He/She is intense and passionate. He/She demands perfection from mate and is exacting. This makes him/her ardent and fascinating lovers. He/She makes charming company and is loyal and devoted.

Career: He/She has a wide spread of potential careers, right from banking to the fine arts. He/She will hold on to one job for the rest of his/her working lives. Here are some occupations that a he/she might consider such as Advertising director, Antique dealer, Business person, Cashier, Clothing designer, Financial advisor, Florist, Patron of the arts, Perfumer, Real estate agent, Singer, Venture capitalist and Woodworker.

Health: Traditionally, he/she is endowed with a vigorous constitution and splendid health. If there is a weak point, it is usually his/her throat or neck. He/She is hopelessly addicted to food and alcohol.

Ideal Partner: He/She vibes best with one of Taurus.

Lucky stone: Diamond, Emerald & Blue Sapphire. Wearing Diamond or Emerald or Blue Sapphire can bring wealth and makes a better person and can give him/her strength.

Lucky Number: Number2, 3 & 8 are best for good fortune.

Lucky Colour: Lotus pink or Shades of verdant green will do a world of good.

Lucky Day: Monday is the day of new beginnings

Flower: Daisy

8.3 Gemini (Mithuna) (May 21 - June 20)

Who oscillates, communicates and changes often and who is fond of life, fun and pleasure?

Who has a free soul and loves others attentions and exchanging of an intellectual nature?

Personal Quality: The Gemini is adaptable, dual natured, affectionate, courteous, kind, generous, scientists, and talented, bright, witty, entertaining army personnel, thoughtful towards poor and sufferer, adaptable and adjusting nature. He/She has unpredictable temperament. He/She is very well with Aquarius. He/She has wealth in later half of life. He/She will completely change his/her mind like Chameleon.

Positive Quality: He/She is usually quite, creative and has a strong self-confidence. He/She is often the centre of attraction in the gathering and is versatile, adaptable, and inquisitive and always moves along with the times.

Negative Quality: He/She is sharp-tongued and sometimes boring. He/She cannot concentrate well at one point and hence does not finish the job. He/She becomes cynical, biting, moody and quickly angered. He/She is superficial, restless, nervous, lacks concentration and conniving. He/She does not keep their promises.

Physical Appearance: He/She has small, narrow hands and feet and slim. He/She is generally tall. He/She is highly energetic and exudes oodles of charisma.

Relationships: He/She is emotionally undemonstrative but enjoys being in a lively family, and seeks a partner with strong opinions. He/She is keen to make relationships but tends to be too egocentric.

Career: He/She is particularly suited to media work, in sales pitches and can be writer, Broker, Commentator, Concierge, Correspondent, Debater, Impersonator, Journalist, Librarian, Linguist, News commentator, Novelist, Orator, and Playwright.

Health: He/She has the ill effects of smoking since he/she has delicate lungs.

Ideal Partner: He/She is best with Aquarians. Virgo, Libra and Aquarius people may help him/her.

Lucky Gem: Blue Sapphire, Diamond & Emerald. He/She can count on the Emerald, which will bless him with all the intelligence he needs.

Lucky Number: Number 3, 7 & 9 will bring him good news.

Lucky Colour: Sky Blue; Reach for the skies with sky blue.

Lucky Day: Thursday; all his dreams come true on Thursday.

Lucky Flower: Rose

8.4 Cancer (Karka) (21st June - 20th July)

Who cannot stick to any adhesion and changes like a season?
Who most perplexing character and let's go without any reason?

Personal Quality: The Cancer is protective, jealous, sensitive, full of suspicion, not easy to understand for his/her moods, often fluctuate from sweet to cranky and lacks faith in others. He/She can be untidy, sulky, devious, and inclined to self-pity because of an inferiority complex but always ready to cooperate. He/She is very fond of food and is usually hearty eaters. He/She gets ancestral and sudden properties after the age of 35. He/She is the most family-centred persons and fiercely protective of loved ones. He/She possesses strong paternal and maternal instincts.

Positive Quality: He/She is loving, kind, faithful, loyal, honest, hard working and sensitive and leaves his unforgettable impression on others.

Negative Quality: He/She is sulky, devious, moody, clinging, manipulative, overly emotional and insecure and inclined to self-pity and prone to a sense of personal inferiority and believes his/her views, opinions and behaviour to be impeccable, and beyond question or criticism.

Physical Appearance: He/She is small with round faces and possesses a tendency to gain weight. He/She has abundant shiny hair, expressive eyes and is economical with his/her gestures.

Relationships: He/She treasures emotional bonds and doesn't severe tie easily. He/She clings on to a failing relationship and finds it difficult to let go. He/She wears his/her hearts on his/her sleeve, and is prone to emotional excesses.

Career: He/She is best suited for counselling and charity work, good journalists, writers or politicians, archaeologist, caterer, dairy farmer, deep-sea diver, dietician, hotel worker, manufacturer, merchant small businesses. He/She works well with people and often adopts the role of a mediator.

Health: He/She has a weak digestive system and constantly suffers from heartburn and ulcers. He/She tends to become hypochondriacs.

Ideal Partner: He/She seeks steady, stable and practical partners, and usually bonds best with Capricorn.

Lucky Gem: Yellow sapphire, Pearl & Coral. If he suffers with sleepless nights, a pearl could be the perfect cure. Wearing the pearl ensures peace of mind, and brings all the good luck in the world.

Lucky Number: Number 4 & 6 are his/her pick for good fortune.

Lucky Colour: white, Soak in the elegance of white.

Lucky Day: Wednesday; it is time to look ahead on Wednesday.

Lucy Flower: Larkspur.

8.5 Leo (Simha) (July 21st to August 21st)

Who praises all his kindred and expects others to praise them too?
Who possess grace, dignity and generosity but cannot see their senseless view?

Personal Quality: He/She is creative, strong willed, self-confident, generous, broad-minded and faithful. He/She has a powerful presence of mind and power of success in the conquests. He/She is manager because he dislikes subordination. He/She excellent organizers. He/She has self-confidence, alertness and hard struggling power. He/She never forgets the goal and tries hard to achieve it with patience, wisdom and hard labour. He/She is a dominant, always busy with some planning and work and cannot seat idle even for an hour. He/She is able to attain the top most position. He/She achieves full success after the middle age. He/She prefers position and honour to money. For position, he/she can forget any money. He/She always cares for others and other's interest or benefits except, where it is necessary to take care for him/he. He/She is born to lead. He/She is sometimes great saints. There is a reality in his/her love and he/she can do anything to please that he/she likes most. He/She will keep improving day by day after 30 years of age. He/She is straightforward and uncomplicated individuals. He/She is stubborn, and may suffer from short bouts of depression. He/She walks forward always, head held proudly and face turned towards the sun. He/She accumulates good amount of money, wealth and properties. His/Her fortune is good and he/she will never lack money in life.

Positive Quality: He/She is honest possesses a strong positive nature and doesn't shrink from any adverse circumstances. He/She can never bear dishonesty and injustice in life. He/She is witty and sets examples for others to follow. He/She is direct and to the point.

Negative Quality: He/She sometimes thinks himself/herself to be the only capable person in the total world. He/She believes in commanding only and does not care for others feelings. Some Leos think for money and profit and forget their other duties. It is very difficult to face his/her furies. He/She can be too sensitive to personal criticism, and when his dominance is threatened he can go into a sudden rage. He/She is conceited and arrogant.

Physical Appearance: He/She has a distinguished stature and attracts attention easily. He/She normally possesses healthy skin and a well-sculpted body.

Relationships: He/She makes wonderful social companions and is passionate and faithful lovers, but very sensitive and gets hurt easily.

Health: He/She suffers from back and spinal problems, has tendency to be overweight by lack of exercise. He/She gets easily stressed and can suffer from heart ailments.

Ideal Partner: Hence, he/she gets best with Aquarius, but Taurus, Gemini and, Sagittarius women are ideal partner for him/her.

Lucky Gem: Ruby and Topaz. The ruby is a miracle stone, and wearing it will bring him/her health and good fortune. He/She will also acquire the power to make instant decisions. The Ruby and Topaz can give him/her success and power in life.

Lucky Number: Number 1, 3, 4, 5, 6 & 9 will pull him/her out of trouble.

Lucky Colour: Saffron. Let saffron lift his/he spirits.

Lucky Day: Friday. Preen, for he/she will rake in accolades on Friday.

Lucky stone: Peridot.

Flower: Gladiolus

8.6 Virgo (Kanya) (August 22 to September 22)

Who criticizes all she sees and would even analyse a sneeze?
Who is observant, shrewd but hugs and loves her own disease?

Personal Quality: The Virgo is critical, precise, easygoing, reliable, steady, helpful, intellectual, studious, logical, methodical, communicative, sciences, languages, takes a romance to new heights, good followers and the best employee one can ever have. He/She often dislikes delegating. He/She knows how to please the persons in power and position to get his/her work done easily and so, gets promotion very fast. He/She has a pleasant nature, colourful personality and sharp mind with a great sense of responsibility. He/She is precise, refined, and a lover of cleanliness, hygiene and order. It is not so easy to measure the depth of his nature. He/She knows to mould others in his shape by his clever activity. If some body offends him/her, he/she does not show his real feelings on his/her face but act secretly to take revenge and hit back all his/her might when he/she gets the opportunity. The early part of his/her life is spent in struggles. The luck favours him/her at the age of 24, 36 to 42 years of age. He/She gets properties after the age of 40.

Positive Quality: He/She is good planners, practical and hard working, trustworthy and able to do perfect work. He/She is a hard worker, conscientious and perfectionists. He/She is plain spoken and is able to express well. He/She leads a moderate life and does not like excesses.

Negative Quality: He/She is sometimes very critical and thinks himself/herself all in all. He/She has a penchant for turning molehills into mountains, difficulties into stress and cleanliness into obsessive behaviour and a capacity for endless worry. He/She is miser and selfish and wish others to follow them.

Physical appearance: He/She has good bone structure and is often highly photogenic. He/She is attractive, with beautiful eyes that sparkle with intelligence.

Relationships: He/She is truthful and loyal. He/She is tense in close relationships, which could badly upset his/her sex lives and makes it hard for him/her to become truly intimate with those he/she loves.

Career: He/She is usually happy working in a job that calls for precision, a shrewd mind, and logic. Dogged, analytical and intellectual, he/she is makes skilled and inspired research scientists, analysts or even literary critics. He/She can be at the best as a manager, a secretary, a lawyer and a trader.

Health: He/She is often martyrs to stomachs, and may suffer from Irritable Bowel Syndrome, from food allergies. Virgo rules the abdominal region, intestines, and the lower lobes of the liver, the spleen, the duodenum, and the sympathetic nervous system.

Ideal Partner: He/She gets attracted to opposites and vibe well with dreamy Pisces.

Lucky Gem: Pearl, Topaz and Emerald. He/She should wear Emerald to crack a tough problem, to help him. The stone will bless him/her with all the intelligence he/she needs.

Lucky Number: 2, 3, 5, 6, 7 & 15are for all the good luck.

Lucky Colour: Get close to Nature. Wear earthy browns.

Lucky Day: His/Her quest for perfection pays off on Friday.

Lucky Flower: Lavender

8.7 Libra (September 23 to October 22)

Who is easygoing, sociable and keeps you waiting for half the day? Who puts you off with promise gay and compromises all the way?

Personal Quality: The Libra is diplomat, impartial, sociable, cheerful, charming and sensitive to the needs of others. He/She is affectionate, polite in behaviour, cooperative, peace loving and sacrificing. He/She is natural arbitrators and diplomats, justice, honest and hard worker. He/She can win over his/her staunchest enemies with the help of his/her sweet voice. He/She has an idealistic and generally peace loving nature. He/She is the most civilized of the twelve signs. He/She has financial stability in the life. He/She is sure to have properties of his/her own but keep away from others' properties. He/She is more interested in making friends than enemies, and is willing to go along with others and do whatever it takes to maintain a relationship. He/She is a sensual lover and does not like any interference in the matters of love and marriage. Discord makes him/her totally insecure, and uncomfortable. He/She likes harmony in his/her life, and will do whatever it takes to have it.

Positive Quality: He/She always maintains a sweet relation with others. He/She is usually sympathetic, kind, loving nature and artistic. He/She does not like injustice, quarrels and disagreements. He/She is fair minded and loyal and have reach taste or good sense of humour. He/She does not hurt other person's feelings and likes to help the person in need.

Negative Quality: He/She is insincere and jealous and likes self admirations. He/She does not have argument power well even he is right. He/She tries to keep away from truth and painful experience. He/She can be frivolous, flirty and quite shallow. He/She is fickle minded, dependent, indecisive, sulking, and likes peace at any cost.

Physical Appearance: He/She is smart and attractive. He/She has sweet open faces with laughing eyes, and a devastating smile. He/She tends to be plump rather than angular or skinny.

Relationships: He/She highly understands his/her companions and he meets her with his own innate optimism. His/Her married life is delightful with the Gemini and Cancer girls. He/She gets special co-operation from Gemini, Aquarius, Sagittarius and Leo natives. He/She is not very good at handling relationships.

Career: He/She gets success in life as a businessman, engineer, lawyer, chartered accountant or a doctor. He/She is attracted to careers in the luxury trades, including fashion, beauty and design. He/She also makes good diplomats and counsellors. He/She is successful as writers, composers, fashion designers, interior decorators, critics, administrators, lawyers, and in civil services.

Health: This sign rules the kidneys, the lumbar region of the spine, the skin, the urethras, which are the tiny ducts running between the kidneys and the bladder, and the verso-motor system. He/She tends to suffer from nervous stomachs and ulcers. He/She needs to drink plenty of water in order to flush out the toxins from kidneys.

Ideal Partner: He/She gets along with the best ones and the gentle that captivate attention for life.

Lucky Gem: Diamond and Blue Sapphire. He/She does love the real thing, and wearing a diamond can bring him/her wealth and can make him/her a better person.
Lucky Number: Number 1, 2, 4, 7 & 21 will fill with joy.
Lucky Colour: Royal Blue. He/She is can rule over the world with royal blue.
Lucky Day: Tuesday. Pack your bags for a holiday on Tuesday.
Lucky Flower: Aster.

8.8 Scorpio (Vrischika) (October 23 to November 22)

Who has an intense and powerful nature and keeps ready an arrow in his bow?
Who is self centred, wilful, proud, and detective and if you prod him, lets it go?
Who is a fervent friend, a subtle foe?

Personal Quality: Scorpion is ruthless, mysterious, magnetically, attractive and emotional intimacy and is the most intense and passionate. He/She also has immense degree of willpower and is highly tenacious. He/She is of a secretive, timid, retiring nature, one that does not talk of his affairs. He/She is honest and independent nature person with hard struggles in life. He/She does not believe in accumulating the illegal wealth. He/She has strong will power and a natural quality of leadership. He/She does not believe in empty promises. He/She can be vindictive, dangerous enemies and possesses a strong streak of venom. He/She starts good earning at the age of 24 and after 40 years of age acquire properties. He/She is self contained and self centred and seethes and doesn't give up the enmity. He/She may burst any moment. He/She is too demanding, too unforgiving of faults in others. He/She is very jealous. He/She is the symbol of sex and passionate lovers.

Positive Quality: He/She is brave, courageous, sincere and loving, subtle, imaginative, powerful, generous, loyal, passionate, exciting, and magnetic. He/She is the person with the fixed mind and achieve goal directly by his deed. He/She is not afraid of obstacles because he has a strong will power.

Negative Personality: He/She is proudly, over sensitive and careless. He/She is jealous, resentful, obstinate, compulsive, obsessive, brooding, secretive, revengeful, possessive, and extremist and can appear cold and impassive.

Health Concerns: He/She is prone to ailments of the liver and kidneys, stones and gravel in the bladder or genitals, and other genital ills such as pianism, abscesses, boils, carbuncles, fistulas, piles, ruptures and ulcers.

Physical Appearance: He/She is always striking and has a magnetic face and dress elegantly. There is a strange mysticism and magnetism in his personality, which is enchanting to the beholder.

Relationships: He/She likes to stay away from his/her family and leads an independent life. His/Her marital life with Pisces, Taurus, and Virgo and Cancer girls will be pleasant.

Career: He/She is traditionally associated with jobs such as mining and detective work. He/She makes demanding bosses. He/She might consider job such as Analyst, Criminologist, Detective, Doctor, Enforcer, Hypnotist, Insurance agent, Investigator, Lab technician, Private investigator, Psychiatrist, Psychologist, Researcher or Scientist.

Ideal Partner: Scorpios fits best with Taurus. This sign gives him the material and emotional security he craves.

Lucky Gem: Coral, Opal. He gathers courage from the coral. The stone could also make him/her rich and also gets confident.

Lucky Number: Discover the magical powers of Number 2, 3, 7, 8 & 9.

Lucky Colour: Midnight Blue, Maroon. Kiss those blues away with midnight blue.

Lucky Day: Sunday. There will be sudden windfall on Sunday.

Lucy Flower: Chrysanthemum.

8.9 Sagittarius (Dhanu) (November 23 to December 20)

Who loves the dim, religious light and always keeps a star in sight? Who is versatile enterprising and an optimist, both gay and bright?

Personal Quality: The Sagittarius is moralistic, impulsive, full of versatility and eagerness, and has a positive outlook towards life. He/She enjoys travelling and exploration. He/She is ambitious and optimistic, honourable, honest, trustworthy, truthful, generous and sincere with a passion for justice and truth. He/She has very charming voice and benevolent personality. He/She is 'Yes' man and never says 'No' to anyone. He/She doesn't get demoralized in his failure. He/She does not like to harm any person and does social work too. He/She is noted for longevity, intuitiveness and original thinkers. He/She cannot gain or be successful in Gambling, Races and Stock-exchange businesses. He/She can be successful in business of white and artistic gift items or textiles and metal. He/She earns wealth after 36 year of age and also gets parental property.

Positive Quality: He/She is honest, tolerant, and friendly and trusts and respects people. He/She is kind and forgives the people easily is never proud.

Negative Quality: He/She is indiscipline and never learns even from his/her mistake in the past. He/She likes gambling and loose money. He/She does not keep his promises and does not have foresightedness and hence, he/she is unsuccessful. He/She has a quick temper and a biting tongue. Physical Appearance: He/She has darting and piercing eyes that are always likely to flash with laughter. While not particularly fashion-conscious, but he/she looks trendy.

Relationships: He/She has a happy family life and Gemini or Arian girls will be suitable as a partner and also for success in his life. Leo and Libra can be helpful for him. He/She is tactless and can hurt with his/her brutal remarks.

Career: He/She is successful in social administration, in public relations, as scientists and as musicians inquisitive. He/She works best with a tactful, organized business partner. Here are some occupations that he might consider, such as, Academic, Adventure travel guide, Advisor, Astronaut, Consultant, Entrepreneur, Inventor, Humanitarian work, Market researcher, Senator.

Health: He/She often fails to look before where he leaps, and as a result suffer quite a few bruises, pulled muscles and broken bones.

Ideal Partner: He/She needs to spend his life with organized and tolerant people, so his vibe best with Aquarius or Libra.

Lucky Gem: Topaz. If he/she is feeling invincible, thank the Hessonite for it. This is a stone of power, and the world is for you to conquer.

Lucky Number: He should be sure of success with Number 2, 3, 5, 6 & 8.

Lucky Colour: Red. Wear Reds for warmth and energy.

Lucky Day: Thursday. An old friend will brighten up an otherwise dreary Thursday.

Flower: Holly

8.10 Capricorn (Makara) (December 21 - January 19)

Who takes what's due and climbs and schemes for wealth and place? Who is confident, a resourceful and morns his brothers fall from grace?

Personal Quality: The Capricorn is miserly, very ambitious, self-confident, loyal, snobbish, true workaholic and accepts hard work and reaches to greater heights. He/She makes superb administrators, and often raises to very high positions in his/her careers. He/She wants to get everything with money power. He/She can be temperamental and moody. He/She acquires wealth, dominant position and what else due to his/her smooth talking and hard working. He/She never gives up the thing what he/she had decided to get and takes rest only after completion of the work. He/She knows the value of money only and is always ready to face the consequences to grab the money. He/She saves money and does not spend without need. He/She has very wide circle of friends. He/She is resourceful, practical manager, works well in a disciplined environment. He/She can be frugal, possessing the ability to achieve results with minimum effort and expense. He/She manages several projects simultaneously. His/Her bright carrier begins after the age of 24 and between the ages of 35 to 48 years. He/She is among the wealthy persons. He/She moulds himself/herself as per the situation and hence gets success in life.

Positive Quality: He/She is honest, very practical person and can be relied on. He/She does duty sincerely and finishes the work with great responsibility. He/She is goal oriented. He/She has great control and authority. He/She is loyal to intimates. Never impetuous, he/she considers business and personal relationships carefully before becoming involved. He/She is family person, and family usually comes first, except where business is his/her primary concern.

Negative Quality: He/She is sometimes too bossy and very narrow-minded and thinks highly of him. He/She doesn't function well in subordinate positions. He/She can spread gloom and tension in a minute and is quite capable of depressing everyone else around him. Never really up, but often down, he/she needs a positive environment to enliven his/her spirits. He/She thinks as the wisest man in the world and likes to show the others. Sometimes, he/she is very proudly. He/She is stubborn, overbearing, unforgiving, inhibited, fatalistic, condescending.

Physical Appearance: He/She tends to be tall and have sharp, angular features. He/She generally has serene expressions and an air of tranquillity. He/She takes great care of appearance.

Relationships: He/She can be surprisingly passionate behind closed doors. However, he/she takes a while to warm up and is very cautious about relating to others. He/She is very unhappy with emotional scenes and upheaval, gets hurt easily, but thaws just as quickly if he/she finds that his/her partner is genuinely repentant. Cancer and Libra girls will be ideal for his/her life partnership. Taurus, Virgo and Libra people will be helpful to him.

Career: He/She is usually determined and ambitious, with a strong sense of discipline and a good head for business. He/She sees duty and law enforcement as of paramount importance. Here are some occupations that he/she might consider such as Administrator, C E O, Coach, Commissioner, Economist, Governor, Industrialist, Leader, and Manager, Mountain climber, Office manager, Official, Operations manager, Organizer, President, Professional, Programmer, Proprietor or overbearing.

Health: Traditionally, Capricorn problem areas are their joints, bones and teeth. He/She needs to ingest extra calcium, cod-liver oil and evening primrose oil supplements to help his/her flexibility.

Ideal Partner: Always seeking a loving and strong partner, he/she gets along very well with Taurus. Capricorn needs a strong, loving partner and bond best with Taurus.

Lucky Gem: Blue Sapphire, Garnet and Diamond. This blue sapphire promises him/her all the luck he/she could wish for. He/She could soon be in for a lot of money, so better brush up on his/her financial skills.

Lucky Number: Number 6, 8, 9 & 18 are his/her secret weapon for the success.

Lucky Colour: Peacock Blue. Wear peacock blue for luck.

Lucky Day: Friday. Watch things falling into place perfectly on Friday.

8.11 Aquarius (Kumbha) (January 20 - February 19)

Who gives to all a helping hand but bows his head to no command?
Who are inventor, genius and superman and higher laws doth understand?

Personal Quality: He/She possesses ill health and likes to make the world a better place. He/She has spiritual bent of mind. He/She is friendly, humanitarian and original thinkers, but can be rather eccentric. He/She is far sighted and innovative. He/She will go out of his/her way to help when needed, but never gets involved emotionally. He/She gets respect and the dignity in the society due to his/her kind nature and service to the people. He/She is broad-minded and expects others the same. He/She is known by his own name and is not follower but makes his own path. He/She often adopts a life style that goes against the trends, because the odd and unique fascinate him/her. He/She is an active man who is always busy with some kind of mental and physical work. He/She can achieve masterly in artistry, writing, medical, management, police or intelligence work.

He/She enjoys his/her own company and is recharged by this quiet time.

Positive Quality: He/She is kind hearted, honest, kind, tolerant, cool, clear, logical people. He/She always thinks for welfare of everyone. He/She is a dedicated person and never bears injustice. He/She does not like to hurt other's feelings and are very helpful to the people in need.

Negative Quality: He/She is, sometimes, not efficient planners and the work undertaken is seldom completed. Sometimes, he/she thinks himself very clever and intelligent than others. He/She is an enigma. He/She is quite aloof people and, sometimes, does not accept his/her fault.

Physical Appearance: He/She has clean-cut good looks and a ready smile that shows off his/her excellent teeth to advantage. He/She can eat any amount of junk food and still remain slim. He/She likes to wear unusual outfits.

Relationships: Physical relationship is not so important for him/her but he/she is very emotional lover and has in depth love. Gemini, Libra, Leo and Aries are the persons who will be helpful to him/her.

Career: He/She might excel in technical fields linked with electrical and radio industries. Most are hard working, even driven in his chosen field. Many choose careers like Astrology. Here are some occupations that he/she might consider such as Academic, Adventure travel guide, Advisor, Astronaut, Consultant, Entrepreneur, Inventor, Humanitarian work, Market researcher and Senator.

Health: His/Her lungs are particularly sensitive to cigarette smoke and air pollution.

Ideal Partner: The most suitable match for Aquarians is Capricorns.

Lucky Gem: Amethyst, Blue Sapphire and Hessonite. If he/she is feeling invincible, think of hessonite for him. This is a stone of power, and the world is for him/her to conquer.

Lucky Number: He/She can be sure of success with Number 2, 3, 7 & 9.

Lucky Colour: Red. Wear reds for warmth and energy.

Lucky Day: Thursday. An old friend will brighten up an otherwise dreary Thursday.

Lucky Flower: Violet

8.12 Pisces (Meena) (February 19 to March 20)

Who possesses a gentle, compassionate, sensitive and spiritual nature?

Who are friendly and respond to suffering, which others encounter?

Personal Quality: The Piscean is charity, anxious, self-sacrificing, gentle, patient, malleable nature and has strong intuitive powers. He/She has superb observation, concentration while listening and good grasping power. He/She has instinct for nature, beauty, travelling, luxury and pleasure. He/She is good in subordinate positions and heads of small business. He/She is the kindest and most charitable of all the signs. He/She will make many sacrifices for other people. He/She lives life in lonely. As a lover, he/she is faithful and love to dabble in the art of sexual fantasy. He/She has many generous qualities and is friendly, good-natured, kind and compassionate, sensitive to the feelings of those around them, and respond with the utmost sympathy and tact to any suffering. He/She gets ancestral properties but he/she wishes to make money by his/her own efforts. Horseback riding, dancing, skating, swimming or sailing are favoured activities.

Positive Quality: He/She is versatile, intuitive and has quick understanding. He/She observes and listens well, and are receptive to new ideas and atmospheres. He/She readily adapts to change.

Negative Quality: His/Her dominant keyword is "I believe". His/Her nature tends to be too otherworldly for the practical purposes. He/She also dislikes disciple and confinement. The nine-to-five life is not for him/her.

Relationship: He/She tends to bond romantically well with Aquarians. He/She is never egotistical in personal relationships and gives more. He/She can be loving and affectionate partners for life.

Health Concerns: He/She can be threatened by anaemia, boils, ulcers and other skin diseases, especially inflammation of the eyelids, gout, inflammation, heavy periods and foot disorders and lameness. He/She is prone to all kinds of allergies and crippling headaches.

Partner: Aquarians are the best match for Pisceans and two tend to bond romantically.

Lucky Stone: Yellow Sapphire, Topaz & Coral.
Special Flowers: Water Lily, White Poppy & Jonquil
Special Colours: Pale Green & Turquoise
Lucky Numbers: 1, 2, 3, 4 & 6
Lucky Day: Friday

9

Planet

There are nine Graha, namely, Sun, Moon, Mars, Mercury, Jupiter, Venus, Saturn Rahu and Ketu. But after further studies, the Uranus, Neptune and Pluto have been added to the astrology to make it more fascinating subject. Moon, Mercury, Jupiter and Venus are benefic by nature and others are malefic.

Surya: Surya's eyes are honey-coloured. He has a square body. He is of clean habits, bilious, intelligent and has limited hair (on his head).

Chandra: Chandra is very windy and phlegmatic. She is learned and has a round body. She has auspicious looks and sweet speech, is fickle-minded and very lustful.

Mangal: Mangal has blood-red eyes, is fickle-minded, liberal, and bilious, given to anger and has thin waist and thin physique.

Buddha: Buddha is endowed with an attractive physique and the capacity to use words with many meanings. He is fond of jokes. He has a mix of all the three humours.

Guru: Guru has a big body, tawny hair and tawny eyes, is phlegmatic, intelligent and learned in Shashtra.

Sukra: Sukra is charming, has a splendours physique, is excellent or great in disposition, has charming eyes, is a poet, is phlegmatic and windy and has curly hair.

Sani: Sani has an emaciated and long physique, has tawny eyes, is windy in temperament, has big teeth, is indolent and lame and has coarse hair.

Rahu (The Dragon Head/ North Node of Moon): Rahu is the dead body of the lusty demon, killed by Vishnu. So, Rahu is the head of the dragon and symbolizes the trouble. Rahu has smoky appearance with a blue mix physique. He resides in forests and is horrible. He is windy in temperament and is intelligent.

Ketu (Dragons Tail/South Node of Moon): Ketu is the dead body of the lusty demon, killed by Vishnu. So, Ketu is the tail of the dragon and losses and symbolizes the death along with Saturn. Ketu has smoky appearance with a blue mix physique. He resides in forests and is horrible. He is windy in temperament and is intelligent. Ketu is akin to Rahu.

Note: Each planet has been allotted some points, such as Sun - 48; Moon - 49; Mars - 39; Mercury - 54; Jupiter - 56; Venus - 52; Saturn - 39. Thus it is totalling 337 points all together.

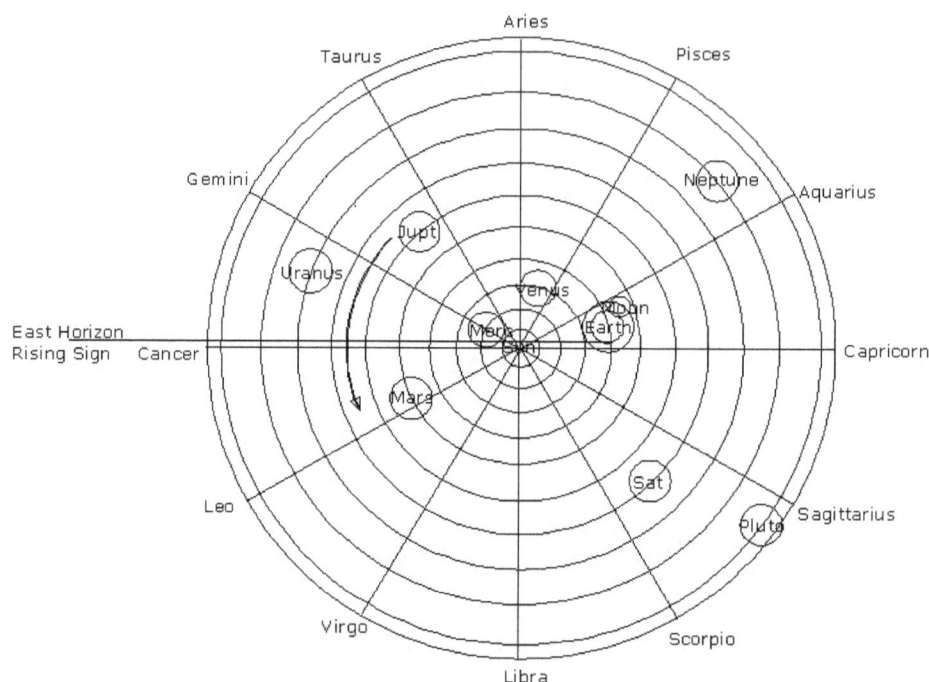

Fig: Planets position with respect to Rising Sign (Lagna) as View from the Earth

9.1 Planets in Transit, Motion, Stationary and Retrograde

Transit refers to the movement or revolution of planets across the Sun. Transit Chart is the position of the Planets at a particular time and place while orbiting the Sun because of their daily motion. The transits Chart show the effect of planets, the ups and the downs brought about by circumstances in our life that we face on daily or monthly basis. The planet's power increases when it is retrograde or stationary. The following table represents the average daily angular motion of the planets and the approximate time they spend in one sign and they need to complete a full circle of the Sun. The symbols in the brackets indicate the direction of their movement.

Table 2: Astronomical Details of Planets

Planet	Average Daily Motion	Time spent in a Sign/House	Cycle Duration to cover 12 Signs
Sun	$0°59'$	30 days (1 month)	1 year
Moon	$13°10'$	2 1/4 days	1 month
Mars	$0°31'$	1 1/2 months	18 months
Mercury	$4°5'$	27 days	1 year
Jupiter	$0°5'$	1 year	12 years
Venus	$1°36'$	28 days	1 year
Saturn	$0°2'$	2 1/2 years	30 years
Rahu (R)	$0°3'$	1 1/2 years	18 years
Ketu (R)	$0°3'$	1 1/2 years	18 years

Saturn Transit (Sadesati): Transit Saturn gives significant results with respect to Moon. When Saturn crosses over Moon, it gives Sade Sati effects. Sade Sati normally gives lots of tension. Similarly Saturn crossing over 8th house from Moon also gives tension and losses. Saturn transit over Jupiter or over 7th house predicts marriage time and Saturn over Rahu gives break in business partnership. During the first Sade Sati of Saturn, it gives a drop in academic results and in 2nd it gives loss in business. When Saturn crosses over 8th house from Moon it gives break in partnership and brings the income to the rock bottom level. It also gives bad health to mother and loss of life of one of the close relatives. Transit of Saturn from Moon is very important.

Jupiter Transit: If transiting Jupiter is conjunct to Venus in the birth chart, it gives enjoy a time of tremendous financial or romantic opportunities. Jupiter transiting over 4th house, over Rahu or over Saturn, it gives finance to buy a new property. In a nut shell, transits seem to be showing their results very effectively; even more than Dasha. Major transits bring major changes in the life of the native. Jupiter and Rahu transit over Lagna as well as Moon give results. If we superimpose the result of Dasha over that of the transit we can predict better.

Motion of the Planets: There are eight kinds of motions to planets, Mangal to Sani. These are Vakra (Retrograde), Anuvakra (entering the previous Rashi in retrograde motion), Vikal (devoid of motion, Stationary, i.e. fixed), Mand (slower motion than usual), Mandatar (slower than the previous motion), Sama (increasing in motion), Char (faster than Sama motion) and Atichar (entering next Rashi in Accelerated motion). The strengths, allotted due to such 8 motions are 60, 30, 15, 30, 15, 7.5, 45 and 30 respectively.

Stationary Motion (Vikal): Some times, planets apparently seem to slow down or stand still or proceed backwards and then again stop, turn around and go forwards again. This is because of the law of retrograde motion. Let us consider the Figures. As Mercury moves along on the other side of the Sun, Mercury comes up alongside the Sun and is soon going to come between the Sun and Earth. There is a point where for a short time, to its cresting on the edge of its orbit in relation to us; it appears to stand still when we judge by looking at the sign behind it, i. e. Zodiac. This is called 'Stationary Motion'.

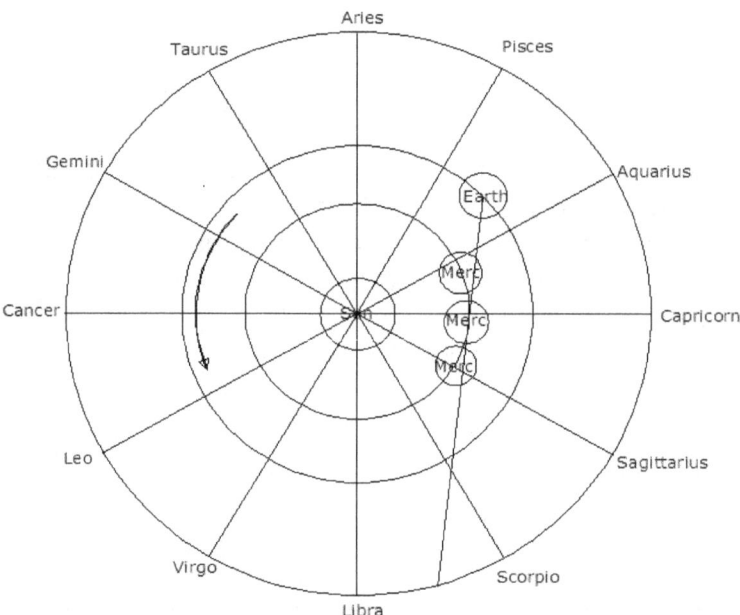

Figure: Planet appears as "stationary" or "standstill" as viewed from Earth

ii) Retrograde Motion (Vakra): The planets move within the belt of the zodiac with a different average of speed. When any planet appears moving apparently in the opposite direction to the Sun, as viewed from the earth, that motion is called Vakra or Saktha Avastha or Retrograde Motion. The two planets, i.e. the Sun and the Moon have steady and direct motion. Except the Sun and the Moon, the other planets, such as, Mars, Mercury, Jupiter, Venus and Saturn change their proper motion through the Zodiac periodically and appear to move backwards for short period and are called retrograde planet and after some time they resume their direct motion. The planet in retrograde is marked in the horoscope with the mark 'R'. A retrograde planet becomes more powerful. It also gives some unusual results and sometimes in the reverse order in the timing of effects etc. But, the lunar nodes, Rahu and Ketu always move in retrograde direction. The rest of the planets, i.e. Mars, Mercury, Jupiter, Venus and Saturn most of the time move in the direct way, but at some times they fall into a retrograde cycle.

The greatest angular distance of Mercury from the Sun is 27^0, while for Venus, it is 47^0. Before retrogression occurs, the planet gets stationary for a certain period of time. The same thing occurs when the planet ends its retrograde cycle. Sun and moon will never retrograde. At this time the planet gives a very strong and steady effect while it is stationary. In the figure shown below, while Mercury moves from position-I to the position-II around the Sun, it cast a shadow of position I and II in opposite direction in the Capricorn Sign, which appears the Mercury moving in Retrograde, i.e. in opposite direction. This is called Retrograde Motion.

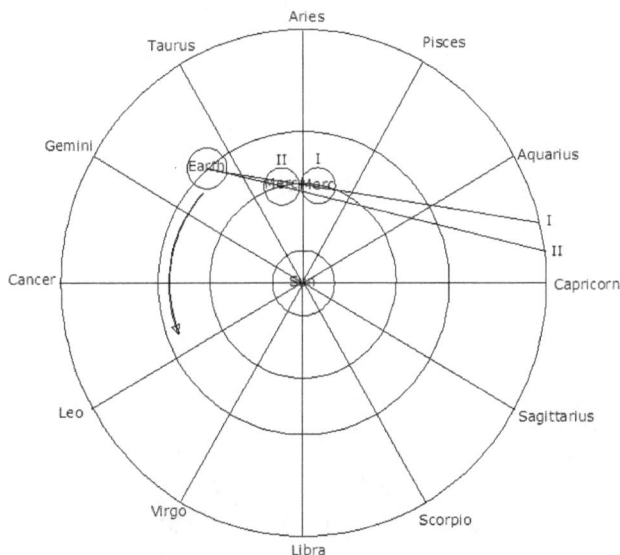

Figure: Planet in "Retrograde Motion"

9.2 Planet's Characteristics

The Planet is the most important character in the horoscope. The planets possess some intrinsic properties when they occupy the sign and house. Planets change their behaviour like Favorable/Benefic or Unfavorable/Malefic according to the position they occupy in Sign and House and lordship.

Once the planet's behavior is ascertained because of its placement, predictions are changed with time, depending upon Dasa operating at that time or by Transit.

Table 1: Planet's Characteristics

Planets	Friends	Neutrals	Enemies
Sun (Surya)	Moon, Mars, Jupiter	Mercury	Saturn, Venus,
Moon (Chandra)	Sun, Mercury	Mars, Jupiter, Venus, Saturn	--
Mars (Mangal)	Sun, Moon, Jupiter	Venus, Saturn,	Mercury
Mercury (Buddha)	Sun, Venus,	Mars, Jupiter, Saturn	Moon
Jupiter (Guru)	Sun, Moon, Mars,	Saturn,	Mercury, Venus
Venus (Sukra)	Mercury; Saturn,	Mars, Jupiter	Sun, Moon
Saturn (Sani)	Mercury, Venus,	Jupiter,	Sun, Moon, Mars

Table 2: Plant's Characteristics

Characteristics	Sun	Moon	Mars	Mercury
Planet's Natural behaviour	Malefic	Benefic	Malefic	Benefic
Karaka	Father, Soul, and 1st house	Mind, Mother and 4th house	Strength, Male's Younger brother, and Female's Husband.	Speech, education, & 3rd house.
Colour	Red-orange	Tawny	Red	Green

Cabinet	King	Queen	General	Prince
Deities	Agni	Varuna	Kartikkeya	Mahavishnu
Sex	Male	Female	Male	Female
Tattwa	Agni (fire)	Jala (water)	Agni (fire)	Bhumi (earth)
Varnas (Deeds)	Kchatriya	Vaishhya (trader)	Kchatriya	Vaishya (trader)
Gunas (Behaviour)	Sattva (goodness)	Sattva (goodness)	Tamas (ignorance)	Rajas (passion)
Abode	Temple	Watery places	Fireplace	Sports ground
Dhatu (Body Part)	Asthi (bones)	Rakta (blood)	Majja (marrow)	Tvak (skin)
Time periods	Ýyana (half year)	Kchana (second)	Vara (day)	Ritu (season)
Taste	Pungent	Saline	Bitter	Mixed
Ritu (Seasons)	--	Varsha (rainy)	Grishma (summer)	Sarad (Winter)
Lord	Lord Siva	Goddess Parvathi	Lord Karthikeya	Maha Vishnu
Dasa Period	6 years	10 years	7 years	17 years

Table 3: Planet's Characteristics

Characteristics	Jupiter	Venus	Saturn
Planet's Nature	Benefic	Benefic	Malefic
Karaka	Knowledge, happiness, Male, 5th house; & 9th house.	Male's wife, Sexual Love, Younger sister, 2nd house, & 7th house.	Misery & Grief, Profession, Elder brother and 10th house
Colour	Tawny	Variegated	Black
Cabinet	Minister	Minister	Servant
Deities	Indra	Sukra	Brahma
Sex	Male	Female	Impotent
Tattwa	Ýknna (ether)	Jala (water)	Vayu (air)
Varnas (Deeds)	Brahmana	Brahmana	Chhudra

		(priest)	(priest)	(worker)
Gunas (Behaviour)		Sattva (goodness)	Rajas (passion)	Tamas (anger)
Abode		Treasure house	Bedroom	Filthy places
Dhatus Body Part		Vasa (fat)	Virya (semen)	Snayu (muscle)
Time periods		Masa (month)	Pakcha (fortnight)	Varsha (year)
Taste		Sweet	Sour	Astringent
Ritu (Seasons)		Hemanta (Dew)	Vasanta (spring)	Sisir (fall)
Lord		Lord Dakshinamurthi	Maha Lakshmi	Lord Yama
Dasa Period		16 years	20 years	19 years

Note: 1) The Dasa period of Rahu and Ketu are 18 years and 7 years respectively.

2) The Lord of Rahu and Ketu are Goddess Durga and Ganesha respectively.

Table 4: Planet's Characteristics

Planet	Ownership	Exaltation	Debilitation	Aspects	Nature	Sex
Sun	Leo	Aries	Libra	7	**Natural Benefic / Malefic Rating**	Male
Mon	Cancer	Taurus	Scorpio	7	Malefic*	Female
Mars	Aries/ Scorpio	Capricorn	Cancer	4,7,10	Benefic*	Male
Mercury	Gemini/ Virgo	Virgo	Pisces	7	Malefic**	Female
Jupiter	Sagittarius/ Pisces	Cancer	Capricorn	5,7,9	Benefic**	Male
Venus	Taurus/	Pisces	Virgo	7	Benefic***	Female

	Libra					
Saturn	Capricorn/ Aquarius	Libra	Aries	3,7,10	Benefic***	Impotent
Rahu	-	Taurus	Scorpio	5th, 7th, 9th	Malefic***	-
Ketu	-	Scorpio	Taurus	5th, 7th, 9th	Malefic***	-

Table 5: Planet's Characteristics

Name of Graha	Deeptaansh	Exaltation Sign	(Debilitation Sign)	Mula Ans
Sun (Surya)	15	Aries	0-10 ^{0}	7
Moon (Chandra)	12	Taurus	0-3 ^{0}	3
Mars (Mangal)	8	Capricorn	0-28 ^{0}	10
Mercury (Buddha)	7	Virgo	0-15 ^{0}	6
Jupiter (Guru)	9	Cancer	0-5 ^{0}	9
Venus (Sukra)	7	Pisces	0-27 ^{0}	5
Saturn (Sani)	9	Libra	0-20 ^{0}	1
Dragons Head (Rahu)	1	Gemini	--	--
Dragons Tail (Ketu)	1	Sagittarius	--	--

Table 6: Planet's Characteristics

Name of Graha	Pindaaya	Dhruva	Karaka of Followings	Affected house	Best karaka

	Dhruva ank	ank	Signs & Bhava	Badly During Transit	Of House
Sun (Surya)	0-10 ansa	19	Aries ; Thanu Bhava	5	1st, 10th, 11th
Moon (Chandra)	0-3 ansa	25	Cancer; Bandhu Bhava	8	1st, 4th, 7th
Mars (Mangal)	9-28 ansa	15	Gemini; Sahaj Bhava & Virgo; Ari Bhava	7	3rd, 6th, 10th
Mercury (Buddha)	5-15 ansa	12	Capricorn; Karma Bhava	2	2nd, 10th, 11th
Jupiter (Guru)	3-5 ansa	15	Thanu Bhava; Dhan Bhava & Leo; Putra Bhava & Sagittarius; Dharma Bhava & Aquarius; Labh Bhava	3	2nd, 5th, 9th and 11th
Venus (Sukra)	11-27 ansa	21	Libra; Yuvati Bhava	6	4th, 7th, 12th
Saturn (Sani)	6-20 ansa	20	Scorpio; Randhra	1	3rd, 6th, 8th, 10th, 12th
Dragons Head (Rahu)	-	-		9	3rd, 6th, 10th
Dragons Tail (Ketu)	-	-		9	2nd, 8th

Table 7: Planets and Education or Professions

Planets	Liberal	Semi-Technical	Technical
Sun	Political Science	Technical,	Physics

		Statistical, Mathematics	
Moon	Music, Fine Arts	Paramedical	Chemistry, Medical
Mars	Logics Related	Mechanic	Engineer
Mercury	Accountancy	Semi-tech Accounts	Higher Accountancy
Jupiter	Vedanta, Classical knowledge, Literature, Preacher	Journalism, Astrology, Philosophy, Psychology	Management
Venus	Music, Dance, Arts, Painting	Hotel Management, Tourism	Computer Animation, Graphics
Saturn	Prachin Vidhya, Labour, Law	Mechanical Work, Geology	Engineering
Rahu	Research, Psycho-analytical work	Pilot, Air-Hostess	Space Engineer
Ketu	Language	Language Specialist	Metrologies, Computer

Table 8: Planet's Characteristics

Graha	Person's Profession
Sun (Surya)	Sun provides gains of education, wealth through authority; Politicians & political leaders like Ministers, President; professionals like Directors, & Physicians & Doctor, Actors, Singers or big Businessman, Government service like I A S officers, Specialist; Social workers, and Philosopher.
Moon (Chandra)	Moon gives gain of finance by profession related to Farmer, nursing, the public health care taker, Advertiser, public relation officer, Journalist, Businessman, marine like sailor, Hawker, restaurants, import/export, Boatman cook, agent, hospitality.
Mars (Mangal)	Mars provides gains of wealth through profession like Engineer, Surgeon, Doctor, Dentist, Police & constable, carpenter, mechanic, blacksmith, hunter, military personnel, butchers, barbers, and related to construction, cook, dealing of land, ,spying and

	Service like, Fire service, Sports, industry, Fire kiln, potter, brick kiln, Instrument manufacturing, stone breaking, granite industries, agriculture and arms industry.
Mercury (Buddha)	Mercury provides gains of wealth through profession like ambassador, lecturer, professor, councillor, teacher, writer, landlord, broker, inspector, publisher, journalist, clerk, audit & accountant, mathematicians, public speaker, imports and exports, Salesman, orator, secretary, traders, poet, scripture, clerk, accountant, Astrologer, Lawyer, editor, mathematician, judge, telephone operator, CBI department officer, legal adviser and businessman.
Jupiter (Guru)	Jupiter gives the status, wealth, gains of finance by profession like Judge, Priest, Lawyer, Manager, Teacher, Chartered Accountant, politics like Ministers, lawyer, bankers, temple workers, Yogasana Teacher, Religious teacher, professor, consultants, spirituality, financial management, mathematics, knowledge of Vedas, philosophy and astrology.
Venus (Sukra)	Venus provides gains of wealth through profession like Artists, treasurer, musician, singer, dancer, dramatist, prostitute, call girl, Minister, Accountant, Auditor, beauty Parlour Saloon, Scented materials business, writer, tourism, sculpture, make-up job, beauty context, money lending, financial organizations, commission agent, cinema theatre & Five-star hotels.
Saturn (Sani)	Saturn provides gains of wealth through Labourer s related to real estate, agriculture, building trades, mining, monk, industry, public dealing, service, industrial workers and in Government Service and service related to Speech, dealing with labour jobs, policeman, renunciation man, cheater, killer, dacoit and thief.
Dragons Head (Rahu)	Rahu gives gains of wealth by profession like Labourers related to Chemical Industry, Electronic Industry, cigarette factory, wine manufacture, bomb manufacture, CBI department, Secret organization (detective), defence department, foreign trader, smuggler, robber, cheater, pick-pocketed, chain snatcher, corruption maker, underground work, researchers, engineers, physicians, medicine/drugs,

	speculators and aviation.
Dragons Tail (Ketu)	Ketu provides wealth through profession like Doctor, priest, astrologers, beggars, saints, Lawyer, Tailor, Choir Manufacture, Cable manufacture, weavers, hand loomer, power loom industries, spinning mills, religious teachers, enlightenment (Guru) and metaphysics.
Uranus	Uranus provides wealth through profession like scientists, inventors, computing, astrologers, lab technicians, electronics.
Neptune	Neptune provides wealth through profession like photographers, movies, marine, oil, pharmaceutical, psychics, and poets.
Pluto	Pluto provides wealth through profession like research, investigators, insurance, death, longevity-related technology, espionage. Mars, Saturn and Rahu are considered the role of technical planets.

Table 9: Planet's Characteristics

Graha	Education Field	Relation	Diseases & Affected Parts of Body
Sun	Political – Science, Social – Science	Father, Son	Heart attack & disease, Bones, Back, Veins, Right Eye disease, loss of appetite, Fever, Indigestion, skin disease, Hysteria.
Moon	Arts, Psychology, Chemistry, Hotel Management	Mother, Mother-in-Law, Maternal Uncle wife	Brain, Lungs, chest, Left-eye, Breast, Venereal disease, Skin disease, Cold cough, Phlegm, smallpox, Mental disorder, stomach ache, Sleeplessness, Indigestion, Ajeerna, Arthritis, Jalaghaat, Animal attack.
Mars	Engineering, Physics	Brother, Husband	Nose, Fore Head, intestine, Bone narrow, Accidents, piles, blood pressure, heart disease, constipation,

			anaemia, Kantha, Rakta-Vikaar, Gathiaa-Bukhaar, Aagajani, poison, blood related, heart attack.
Mercury	Mathematics, Statistics, Drawing	Maternal Uncle, Younger Brother / Sister	Veins, lungs, tongue, arms, Mouth, Hair, Stomach disorder, Leprosy, Childlessness, intestinal complaints, skin diseases, mental disorder, throat, tonsils, leucodrama, Nose, ears, baat, peet, koda, jahar, falling from height, madness, asthma, hakalaanaa & Gungaapan.
Jupiter	Audit and Accounts, Philosophy	Teacher, Self	Lever, digestive organs, Kidney, Hernia, Bronchitis, liver complaints, jaundice Gastric, fever, Plane accident, Liquid from Rt. ears.
Venus (Sukra)	Commerce, Marketing, House-keeping	Wife, Sister, Daughter, Daughter-in-Law	Left ear, glands, sex-organs, sperm, Generative organs, Venereal diseases, uterine disorders, diabetes, diseases of ovaries, Abortion, Koda, vaat, peeta, cough, birya-vikaar, eye, urinal, gudaa, ultras-related, pedu's pain.
Saturn	Mining	Elder Brother	Bone joint, Bones, teeth, Knee, cough, Nerves, gas troubles, rheumatism, arthritis, defective speech, indigestion, ulcer, asthma Stomach, leg, gas related disease; falling from tree, Hit from stones.
Rahu (Dragon Head)	Chemical, Nuclear Physics	Paternal Grand Father	Lungs, Lower part of wind pipe, Cataract, leprosy, suicide, rheumatism, sudden death, murder, accidents,

			sexual perversity, homosexuality, lesbianism, Heart attack, Poison–eating, weakness of heart, Digestive systems.
Ketu (Dragon Tail)	Medical, Theology, Astrology, Law	Maternal Grand Father	Throat, leprosy, suicide, small pox, piles, cancer, sudden death, murder, tumour, Wound, Skin.

9.3 Benefic and Malefic Planet:

There are certain rules in determining a functional Benefic and Malefic Planet for each lagna in the Kundali. Lordship of certain houses in the chart makes the planets deviate from their natural tendencies of doing good or bad. Example: Jupiter, even though it is naturally the best beneficial planet, but it is Malefic Planet for Kanya (Virgo) Lagna as he, being natural benefic, is the lord of 4th and 7th house (Kendradhipatya Dosha) and hence it does not give good results for Virgo Lagna. Similarly, Jupiter is a neutral planet for Simha (Leo) lagna. Dr. Raman also talks about Kendradhipati (Lordship) Dosha, the blemishing acquired by a natural benefic by virtue of lordship of an angle 4th, 7th or 10th house and thus the planets become functional malefic in the Kundaly. The lord of Lagna does not get affected by this dosha and so Lagnesha is always a befic planet in Kundali.

Lord of 1st, 5th, and 9th House: The lords of Trikona are three and are always considered auspicious or benefic and produce good results always. If exalted planets are placed in Kendra to Trikona, it again causes Raja yoga.

Lord of 4th, 7th, and 10th House: The lords of Kona are are considered auspicious or benefic, if they are natural malefic. The lords of Kona are considered inauspicious or malefic, if they are natural benefic planet. Further, lords of 4, 7, 10 houses, if placed in auspicious houses, produce good results. If it is placed in 6th, 8th or 12th houses, it produces bad results. The conjunction of the Kendra lords or the Kona lords or the Kendra to Kona lords is highly auspicious and produces good results. These are called Raja yoga combinations. The Raja yoga combinations give rise to authority, power, position and wealth.

Lord of 3rd, 6th, and 11th House: Sage Parashara states that whosoever planet even including benefic, if owns 3rd, 6th, and 11th are considered inauspicious or malefic, and will produce sinful effects and are trouble maker, such as, (1) the Dasa of 11th lord will be gainful (Laabh) monetary wise, but it will cause obstacles (Vighna) and diseases (Rogi) to the native simultaneously. During the Dasa of 11th lord, one will get evil news. His co-born will undergo troubles. His offspring will fall ill. The native himself/herself be miserable, be cheated and may suffer from ear diseases; (2) The Dasa of the lord of 3rd (Tri), 6th (Shad) and/or 11th (Labha) and the Dasa or Bhukti of planet joining with the lords of 3rd, 6th, and /or 11th will be inauspicious and produce evil effects; however, benefic occupying these "Upachayas" give rise to so called "Vasumati yoga". Some Seers have views that the lord of 11th house or planet posited therein or the planet associated with the lord of 11th house, is very enigmatic and mysterious during its Dasa or Bhukti, and is capable of improving acquisitions but at the same time, during this Dasa or Bhukti, the same planet gives difficulties, litigations and serious diseases to the native too. The reason is that (a) the 11th house is 2nd from the 10th house, the central point for one's gains of money or wealth through the Karmas of native. (b) The 11th house being 6th from the 6th house, an epicentre of one's loss of money or wealth through diseases, litigations, punishments and debts. 11th lord is better placed in the auspicious houses and produce good results, otherwise produce bad results.

Lords of 12th, 2nd and 8th House: The remaining three houses as 2nd, 8th and 12th, according to Parashara, are not as bad as the 3rd, 6th and 11th. The houses 2nd, 8th and 12th tend to modify their attributes according to relationship; the lords of 12th, 2nd and 8th (strictly to the order of these lords) will prove auspicious, if they are associated with the angular or the trinal lords. But they become deadly inauspicious by their association with the lords of 3rd, 6th, and 11th. While grading these two groups the sage states the intensity of malfeasance will increase in the ascending order i. e. 6th lord is more harmful as compared to 3rd lord while 11th lord is most harmful / inauspicious of the three. The relationships with favourable or unfavourable lords to the lords of 12th, 2nd, and 8th houses (in this very order) give good or bad results as detailed here. Sage Parashara gives some relief to the lord of 8th house but with certain conditions, that the said planet should have simultaneous lordship over a trine i.e. Saturn in case of signs Gemini rising in the lagna, Jupiter in case of Leo lagna and Mercury in case of Capricorn lagna. Here the sage is silent about the Mercury's lordship over 8th house and simultaneously its lordship over a trine viz. 5th house in case of Aquarius lagna and the reason for dropping this planet from the above list can be the exaltation as well as Mooltrikona sign of Mercury falling in the 8th house. There is a saying "If a planet is the ruler of the 12th house from his position, he destroys that Bhava of occupancy in the Dasa and Antar Dasa of such planet. Lords of 6, 8, 12 house, if placed in another 6, 8, or 12 houses, produce good and auspicious results, but produce bad results if placed in the good houses. 2nd lord is better placed in the auspicious houses and produce good results, otherwise produce bad results. The conjunction of the Kendra lords or the Kona lords or the Kendra to Kona lords is highly auspicious and produces good results. These are called Raja yoga combinations. The Raja yoga combinations give rise to authority, power, position and wealth. The Lords of 6, 8 and 12 house, if conjunct or joins Kendra lords or the Kona lords; it spoils even the the Raja yoga formed in the Kundali and produce bad results. If two Dustasthana lords, (6th or 8th or 12th), exchange their houses of lordship, then also it is auspicious. However, if a benefic house lord exchanges houses with a malefic house lord, like 9th lord with the 12th lord, then the exchange is inauspicious.

9.4 Shani and Divorce:

Shani occupying house, 1, 5, or 12, gives Shani's negative aspect to the 7th house (marriage house) and makes marriage loveless, unhelpful, or full of grief. But, he does not break up marriages and it also doesn't cause divorce. Shani occupying bhava-7 obtains Dik-bala and hence he gives tremendous endurance in loveless marriages to an unsuitable or oppressive spouse. Shani respects the tradition above all. Unless Shani is L-8 (Mithuna and Karka Lagna), Shani is not a divorce causer. He is just a misery causer.

Shani in Lagna or Shani yuti Chandra are equally difficult for Married life. Both the Shani aspect to yuvati bhava and Shani aspect to 7th from Chandra, it profiles anemotionally cold, unsuitable first spouse. Shani's aspect to the 7th house usually signifies the need for a second marriage, because the first marriage is so unsatisfactory. However Shani does give staying power, and the native who has Shani aspecting either kalatra bhava or 7th from Chandra will generally stay in the difficult marriage for many years in order to exhaust all possibilities for agreement. In the end, this harsh drishti forces marital separation and the native feels cut off from social approval as well.

Shani rising or Sani yuti with Chandra combines heavy labour with emotional stress. It's not an easy life for these native. Nevertheless, they are responsible and no matter how deep their suffering, they will always find the strength to accept responsibility for raising children, caretaking elderly parents, or serving the state. The Shani-afflicted may have lifelong emotional cramping, but they will always be working or ready to work.

Saturn gives the karma of carrying others' burdens, of being the result of others' actions, and of remaining in a state of financial, emotional, physical, or intellectual survival throughout life. Shani-afflicted have a deep and often unshakable sense of resource scarcity; yet, thinking they must always be on the run in a state of bare survival, they can rarely save money or build lasting security.

Shani's suffering is caused by Separation from those people, places, and things to which we are most attached in material life. He gives constant attachment-separation in his life, but never makes the victim. Shani will damage marriage trust through chronic self-doubt projected upon the partner. Shani's drishti upon bhava-10 is the least harmful because bhava-10 does involve the management of scarce resources and distrust of egoism. However the native is perpetually unhappy with his own career because of the self-criticism and negative expectations projected upon the boss. Persistent negativity in career duties, will blockone's superiors from guiding the path for the native to acquire increasing public responsibility.

When Saturn is the lord of 11th for Aries ascendant (a moveable Lagna), it is Malefic or Badhaka for the Native. One of the malefic impacts of Saturn is loss of reputation. In Business losses go on increasing. Governmental enquiries start. The fear of imprisonment and repayment of debt troubles a lot. Life becomes miserable because of loss of health. Problem of sleeplessness and stomach pain starts. The native becomes directionless. The friends turn into enemies. Reliable people around the native turn against him. At home all, including children, wife and parents start preaching but nobody comes forward to share his sorrows. It becomes difficult for one to concentrate in work. The condition of the native becomes deplorable and life appears burdensome.

Good Effect of Saturn for Cetain Lagnas

Libra Lagna: Saturn is a Raja Yoga Karaka planet for Libra ascendant and brings material comforts, peace and prosperity in its major and sub periods.

Taurus Lagna: Saturn is a Raja Yoga Karaka planet for Taurus ascendant also, being lord of 9th for fixed Lagna; it is a little bit Badhaka and does not produce very good results in spite of being the lord of Kendra and Trikona house of horoscope.
Capricorn Lagna: It is auspicious for Capricorn Lagna.
Gemini & Virgo Lagna: It is average for Gemini and Virgo ascendant.
Cancer & Leo Lagna: It is inauspicious for Cancer and Leo Lagna but yields average results for Leo.
Scorpio & Sagittarius Lagna: It produces ordinary results for Scorpio and Sagittarius.
Aquarius Lagna: It is slight inauspicious for Aquarius Lagna.
Pisces Lagna: It is not inauspicious for Pisces Lagna.

Yoga Karaka Effects of Saturn:

If Saturn is yogakaraka for a native auspicious results are experienced during Sadesati. If Yogakaraka Saturn transits through Kendra or Trikona from ascendant or during the period in operation, the results of Sadesati are good and not bad. The second special feature is 'Shani verses Rahu', i.e. Rahu is like Saturn. It is observed that in case of Tula lagna born, for which Saturn becomes Yogakaraka, if its dasa comes very late in the native's life, then Saturn confers its favourable result during the dasa of Rahu. If such an individual has both Rahu and Saturn dasa in his life time, then the results experienced are par excellence. The above features are corroborated by the horoscopes of Emperor Akbar, Adolf Hitler, Mahatma Gandhi, Margaret Thatcher and Atal Behari Vajpayee, Amitab Bachan and others.

Remedies of Saturn:

1. Donate any one of the following things that you can afford on Saturdays. 2. Black cloth, black gram, sesame seeds, mustard oil etc. The ill effects of Saturn are reduced by donating these things on Saturdays. 3. We now know that Saturn is a deep acting planet affecting a human being adversely when is afflicted and bestowing prosperity when is pacified.
Vedic Mantra of Saturn: Chanting the following Mantra by Maharishi VedVyasa pacifies Saturn.
"Visanjana sama busam avputhram yama grajam, chaya marthanda sambhootham tham namami Saneescharam"

The following Mantra is also considered very well for planet Saturn. You may choose to recite one of the Mantra which is easy to remember.

Beeja Mantra: 'Om pram preem praum saha Sanaischaraya namaha'.

Gayathri Mantra: "Om Sanaischaraya vidmahe soorya puthraya dhimahi tanno manda prachodayat".

Recite all these Mantra 108 times on Saturdays. Recite the Mantra 12 times with devotion and dedication if you don't get time.

Rahu and his legendary

Rahu is a legendary master of deception who signifies cheaters, pleasure seekers, and operators in foreign lands, drug and poison dealers, insincere & immoral acts. It is the significator of an irreligious person, an outcast, harsh speech, falsehoods, dirtiness, abdominal ulcers, bones, and transmigration. Rahu is instrumental in strengthening one's power and converting even an enemy into a friend. Rahu in Vedic texts represents the chief, the advisor of the demons, the minister of the demons, ever-angry, the tormentor, bitter enemy of the luminaries, lord of illusions, one who frightens the Sun, the one who makes the Moon lustreless, the peacemaker, the immortal (having drunk the divine nectar), bestowal of prosperity and wealth and ultimate knowledge. Rahu is friends with Ketu, Saturn, Mercury and Venus. It is the enemy of Sun, Moon, Mars, and Jupiter. It is important to note that Rahu is more inimical to the Sun.

When Rahu forms close conjunction/aspect with natal planets, it causes affliction and destroys the significations of these planets. The harm is less when the afflicted planets are strong and more when they are weak. When the planets are weak, badly placed and closely afflicted by Rahu (north lunar node), it is indicative of tragic. Rahu is Dig Bala in 10th House. The Rahu works best in 3rd & 6th and 10th houses and will give success, strength & power and can convert enemies into friends & boost for long life. In other houses it can be a bit too much causing problems. Rahu gets exalted in houses 3 and 6, whereas he gets debilitated in houses 8, 9 and 11. Rahu gives bad results in 1, 5, 7, 8 and 11th houses. 12th house is his 'Pakka Ghar' and he proves highly auspicious in houses 3, 4 and 6. As a node of Moon, Rahu does not provide adverse results so long as 4th house or Moon is not afflicted. He gives good results when Mars occupies houses 3 and 12, or when Sun and Mercury are in house 3, or when he himself is posited in 4th house. Rahu further provides good results if placed together with Mercury or aspect by him. Rahu offers highly beneficial effects if placed in houses earlier than Saturn. But if it is otherwise, Saturn becomes stronger and Rahu acts as his agent. Sun provides very good results when Rahu is aspect by Saturn, but Rahu gives the effects of a debilitated planet when Saturn is aspect by Rahu. If Sun and Venus are placed together in a horoscope, Rahu will generally provide adverse results. Similarly, Rahu will provide bad results if Saturn and Sun are also combined in a horoscope. Mars will also become negative, if Ketu is placed in houses earlier than Rahu. Rahu will provide adverse results, whereas Ketu effect would be zeroed. If there is Rahu in the Fifth (5th) house and has a relation with the sun, such girl does not have Menses.

There is a proverb that, no planet can bless like Rahu and no planet can curse like Ketu. The Dasa of Ketu spoils and pulls him/her down. Some body's father lost his/her fortune in Ketu Dasa period and his/her family was ruined whereas, the Dasa period of Rahu helped the noted actor and chief minister of Tamil Nadu, Sri M.G. Ramachandran. Rahu indicates fame, extremes, foreigners and foreign lands, fulfilment of worldly desires, status, prestige, power, worldly success and outer success with inner turmoil. Rahu becomes strong in a Kendra from Mercury or the ascendant. Our most intense desires are granted under Rahu. It grants us fame and fortune. Rahu is the material world, and gives all the desires. Rahu deals with fear, obsessive and compulsive behaviour. Rahu with Venus or Jupiter can bring wealth. Rahu with Saturn can cause suffering to the house they are in.

Rahu is instrumental in strengthening the power and can convert enemies into friends. It is popularly called the planet of success. Rahu does not affect the materialistic pursuits adversely during the Transit, as he is a very fond of sensual pleasures. So during Rahu Transit over the planets holding the charge of the Bhava/House other than those of the 6th, 8th and 12th is, generally, beneficial. The Rahu energize the planets during his transit over the planets. But while crossing over the lords of the Dustasthana planets, such as 6th, 8th and 12th house, insures to create bad results such as diseases, disgrace, impediment and heavy and wasteful expenditure.

Remedies of Rahu:

Remedies: Feeding the ants is considered one way to propitiate Rahu. In some parts of India feeding ants is considered one of the ways of propitiating Rahu. Rahu Dan (donation like mustard, radishes, blankets, sesame, lead, saffron, satnaja (a mixture of seven grains), and coal on Sunday morning is a remedial measure in astrology.

These articles are to be donated by a person facing the evil effects of Rahu, or if Rahu is not in a good position in one's horoscope.

Worship: There is a dedicated temple to Rahu at Naganatha Temple at Thirunageswaram, Tamil Nadu. There is a milk abhishekam everyday during Rahu Kaalam to appease Rahu. The milk turns light blue when it flows down after touching the statue of Rahu. This practice has been followed for over 1,500 years.

Mantra: Beej Mantra: Chanting of Rahu mantra "Om Bhram Bhreem Bhroum Sah Rahave Namah", 18000 times in 40 days.

Ved Vyaas navagraha Mantra: "Ardha kaayam maha veryam chandraditya vimardshanam, simhika garba sambootham tam rahum pranamamyaham".

Gayathri Mantra: "Om nagadwajaya vidmahe padma hastaya dheemahi thanno rahu prachodayat". Reciting above Mantras with faith, devotion and sincerity will help to ward off the evils of Rahu during the Dasha and Antardasha of Rahu. One can also worship lord Shiva for this. Chanting of Rahu Stotram with Meaning, such as, (1) Rahur dhanavamanthri cha simhika chitha nandana, Ardha kaya, sada krodhi, chandradhithya vimardhana. (Rahu, Minister of Rakshasa, one who makes Simhika happy, Half bodied one, one who is always angry, Tormentor who troubles Sun & Moon). (2) Roudhro rudhra priyo daithya swar bhanur, bhanur bheethidha, Graha raja sudhapayee rakadhithyabhilashtaka. (Angry one, Devotee of Rudhra, Ogre, One who is near the Sun, one who terrifies the sun, King of planets, one who got nectar, One who desires the moon and the sun). (3) Kala drushti kala roopa, Sri Kanta hrudayashraya, Vidhunthudha saimhikeya, ghora roopo, maha bala. (One who has death inflicting sight, one who likes death, One who lives in the heart of Shiva, one who made moon dim, One who is the son of ogress Simhika, One who has terrifying form, One who is very strong). (4) Graha peeda karo damshtri raktha nethro mahodhara, Panchavimsathi namani sthuthwa rahum sada nara, ya paden mahathi peeda thasya nasyathi kevalam. (One who torments planets, one who has big teeth, One who has red eyes and one who has a big paunch, If a man recites these twenty five names and prays to Rahu And as soon as he reads big tormenting troubles vanish immediately). (5) Aarogyam putham athulam sriyam dhanyam pasum sthadha, Dadhathi rahu sthasmai thu ya padeth sthothramuthamam. (Health, incomparable sons, wealth, cereals and animals would be given to him by Rahu, to the one who reads this great prayer). (6) Sathatham padathe yasthu jeeveth, varsha satham nara. (The man who reads this regularly would live for one hundred years)

Gemstones: The gem stone related to Rahu is hessonite and it rules the number 4.

Ketu and his legendary

The Ketu is Dig Bala in 12th House and works best in the 2nd & 8th and 12th houses and will give spiritual influences, spiritually speaking, highly renounced, separations, cuts, loses and deprivations. Ketu gives its exalted effect when in 5th, 9th or 12th house and its debilitated effect in 6th and 8th house. Ketu represents son, grandson, ear and spine. 6th house is considered to be its 'Pucca Ghar.' Dawn is its time and it represents Sunday. Venus and Rahu are its friends, whereas Moon and Mars are its enemies. Forty two years is the age of Ketu. Ketu is also considered to be the bed. So the bed given by in-laws after marriage is considered to be auspicious for the birth of a son and as long as that bed is in the house, the effect of Ketu can never be inauspicious. Ketu has tail but no head and represents broken relationships, cleverness, changing events, accidents. Ketu with the Moon can give psychic abilities. The placement of the Ketu in the 4th house causes mother in debt. He promotes ill will and is the trouble-makers of the zodiac. They attack in a highly clandestine manner. The native pines for happiness and becomes a recluse, always cursing his/her fate. Those ruled by Ketu, are secret adversaries and relates to people in pharmaceuticals industries, doctors and astrologers. Ketu is the most spiritual planet, with Saturn being next in line, and Jupiter takes third place. Sun-Ketu combination may make him/her a spiritual leader, Moon-Ketu may give good ability to separate emotions from logic, and Jupiter-Ketu may give deep intellectual or philosophical abilities. If Ketu is not strong to give spiritual benefits, then he would rather give disappointment in material things, and may even prove harmful, giving injuries, sudden changes. Then with the Sun, he will harm status, with the Moon, the mind and with Jupiter, his/her wisdom or religion, similarly as Rahu does. Ketu usually makes him/her unfortunate, i.e., causes loss or lack of opportunity. When Ketu is in the 10th house and aspect by Saturn, Mars and Rahu form Gemini ruled by Mercury, confer him/her the internationally famous mathematician. The Ketu in 10th house gives him/her a fertile brain, happiness, religious and pilgrimages to sacred rivers. When the Ketu is in 10th and in Gemini, it gives him/her good health, happiness, elevation to responsible and exalted positions, good food and at the old age venereal troubles, death of near relatives and loss of reputation and self respect in its own Dasa. If the Ketu is in 10th house in Sagittarius sign, it makes him/her quite disorganized though he/she is very sharp mentally. He/She will say and do the things in such a way the guru (teacher) will fail to understand him/her. If Ketu is in the 10th house, which is aspect by the Saturn without any beneficial conjunction, he/she will earn the money from government service or municipal office.

Remedies of Ketu:

Remedy: Donate black things like mustard on Sundays.
Worship: Worship lord Ganesha.
Mantras: (1) Beeja Mantra: 'Om shraam shreem shraum saha ketave namaha'. **(2) Ved Vyasa navagraha Mantra:** "palasa pushpa sankaram taraka graha mustakam, raudram, raudrathmakam goram tam ketum pranamamyaham", **(3) Gayathri Mantra:** "Om aswathwajaya vidmahe shoola hastaya dheemahi,thanno ketu prachodayat". Recite the above mantras at least 108 times during the Ketu Dasha or Antardhasa. If time does not permit, chant it 12 times daily with absolute faith and devotion.
Gem: Ketu's gemstone is cat's eye and he rules number 7 in Indian numerology.

9.5 Predictions by Cluster of Planets in Houses

Cluster of Planet & Effects: Further, the planets positions often seem to become clustered in one part of the chart or another. When this occurs, there is a sort of hemisphere preponderance.

Most of the Planets in the Eastern Hemisphere: He/She is likely to be more independent, strong-willed, and more individualistic, leadership ability, a self-starter, self-motivated, often self-employed and a risk-taker. In short, he/she more apt to be the author of his/her own destiny and is likely to have many choices in life.

Most of the Planets in the Western Hemisphere: He/She is highly adaptable to opportunities, passive or subtle in his actions and often permitting others to take the lead. He is also somewhat dependent on others in many ways. He/She works best in a partnership or group employment arrangements, always having others get involved in his/her destiny. He/She has a sense for taking advantage of opportunities that are presented to him/her, and because of this he/she usually makes the most of them.

Most of the Planets in the Southern Hemisphere: He/She is ambitious; career oriented, and generally wants fame and recognition. He/She has an objective view of life that lets him/her enjoy working with others. He/She is very active in the outside world, having broad material values and goals. This more subjective view of matters can produce significant focus and thrust of action. When the planets are in the south; he/she says or does escapes public view.

Most of the Planets in the Northern Hemisphere: He/She has a very subjective view of life with personal privacy noted as important. He/She enjoys working alone because he/she is a little introspective and perhaps even introverted. So his/her accomplishment may often lack recognition. This more subjective view of matters can produce significant focus and thrust of action.

Most of the Planets in the Northeast: He/She places great faith in his/her abilities and he/she is less inclined to trust others with tasks. He/She feels that he/she can do better on his/her own. He/she is naively self-contained and highly subjective in his/her judgement. He/she places a great value on maintaining his/her privacy.

Most of the Planets in the Southeast: He/She is a highly self-contained individual very much in control of his/her destiny, totally willing to consider others in life situations. There is a tendency to seek self-fulfilment through sharing with others and success is measured not so much by his/her standards, but by what opinion others hold of him/her.

Most of the Planets in the Northwest: He/She is an adaptive, reactive and practical individual. He/she will seek fulfilment through the development and expansion of his talents and abilities in an attempt to improve upon some phase of life that he/she considers of particular importance. He/she is apt to be strongly influenced in this regard by others, and he/she does not initiate.

Most of the Planets in the Southwest: His/Her actions in his/her life are open and exposed to others, and his/her activities are very much a matter of public attention and record. The achievement of any degree of self-realization requires that he/she gives of himself. Others will constantly prove to be a drain on his/her energies and resources. Though he/she may find himself/herself in the spotlight, he/she may always seem to lack complete control of matters.

9.6 Predictions by Exalted & Debilitated Planets in 12 Houses
(1) Effects of Exalted & Debilitated Sun in 12 houses:

Sun in 1st Houses:

(i) Exalted Sun in 1st House: A benefic (positive) exalted Sun placed in 1st house generally bless the native with good overall health, good creative skills, intelligence, leadership and concentration power and native can achieve professional success in his life by virtue of his creativity. The native can be successful in the fields like top class government services like IAS, IPS and other such services, medical profession, teaching, coaching and consultation professions and other similar professions. A benefic exalted Sun in 1st house can also bless with healthy and good male children and such male children can provide a good support to the native having such placement of Sun in his horoscope. But, a malefic (negative) exalted Sun placed in 1st house can cause problems in the married life of the native by causing mild to serious arguments between the native and his spouse, can also cause problems with child birth, problems related to stomach disorders due to digestion, and high blood pressure. A malefic exalted Sun in 1st house can also make the native suffer from some skin related problems in some particular cases.

(ii) Debilitated Sun in 1st House: Though a debilitated planet is not likely to give great results due to the weakness of the planet, but it can still give well to very good results. A benefic debilitated Sun placed in 1st house can give good result regarding the financial prosperity; very good results in his marriage and married life and good life partners and likely to lead a happy married life. On the other hand, a malefic debilitated Sun in 1st house of a horoscope can trouble the native with health related issues or permanent diseases; causes problems in married life facing delays, losses and problems in his professional spheres.

Sun in 2nd Houses:

(i) Exalted Sun in 2nd House: A positive exalted Sun in 2nd house can bless the native with very good health throughout his life, can make the native a renowned Surgeons and consultant physicians doctor, a renowned criminal lawyers, a renowned astrologer or consultant and the native can earn good amount of money as well as reputation from his profession. But, a negative exalted Sun in 2nd house can bring serious problems in the married life of the native and it can cause one or more than one divorce and can make him suffer from debts and can face to very big debts throughout his life and waste of money by virtue of the habit of Gambling or wrong decisions, loss of money due to spending it unwisely; ;money wasted due to bad habits of eating and drinking; can also trouble with some diseases; can cause big financial losses as well as health losses and the native under the strong influence of this placement can even die because of a disease.

(ii) Debilitated Sun in 2nd House: A benefic debilitated Sun in 2nd house can bless the native with a post of power and authority as government officers, particularly the officers dealing with external affairs like officers working in embassies of his countries which are situated in foreign countries; gives very good spiritual and religious interests earning very good respect in the society. On the other hand, a malefic debilitated Sun in 2nd house can cause problems in the married life going through a very bad marriage which can end up in divorce in some cases; creates financial problems developing the habit of overspending and resulting in financial crisis many times; taking debts; and afflicting him with some diseases caused due to his bad eating and drinking habits.

Sun in 3rd Houses:

(i) Exalted Sun in 3rd House: A benefic (positive) exalted Sun in third house can bless the native with good reputation, good public image, good relations with brothers and friends; good support from colleagues; happiness in married life; professional success; good health; good communication skills; good leadership qualities to achieve success in life; gain of a position of power in some government organization; very good spiritual interests; and very good power of meditation. A malefic (negative) exalted Sun in third house can make him suffer on account of health issues; problems coming from his bosses, seniors and other persons in authority; resignation from a job or facing a complaint or litigation due to unfavorable attitude of his boss; make many enemies by virtue of his aggressive nature troubling him in his life; and undergoing some injuries in his life.

(ii) Debilitated Sun in 3rd House: A benefic debilitated Sun in 3rd house can make the native successful as a police officer, army officer, and sportsperson, powerful politician with power, fame, and physical activity; renders very good creative abilities, intelligence, intellect and sound judgment. On the other hand, a malefic debilitated Sun in 3rd house can give bad result regarding the married life facing separation or divorce, litigations filed by his wife's side; gives problem in professional life; suffers from permanent disease.

Sun in 4th Houses:

(i) Exalted Sun in 4th House: A positive exalted Sun in the 4th house can bless the native with a long life; good health; good powers of meditation and good spiritual interests; gains from ancestral property or inheritance; and many kinds of luxuries like possession of Vehicles and houses. A negative and exalted Sun in 4th house can cause problems in the married life; serious arguments with his wife leading to a long separation or a divorce; getting cheated by his wife for someone other person; disturbance in comforts and peace of mind for most of his lives; changing of his place of residence many times in his life on account of something unfortunate; and suffering from some diseases and undergoing operations or surgeries many times in his life.

(ii) Debilitated Sun in 4th House: A benefic debilitated Sun placed in 4th house can bless the native with great amount of wealth, property and luxuries and is born in rich and powerful families and enjoys life to its fullest right from his childhood; inherits wealth, property and business from his paternal family and provides good financial security to him under its influence, right from the start; company of very beautiful girlfriends or wife. On the other hand, a malefic debilitated Sun in 4th house can develop extreme materialistic, physical or sexual desires wasting a great part of their time, money and health towards physical and particularly sexual pursuits; causes problems in the married life and getting involved in extramarital affairs spoiling his marriage; makes him very suspicious by nature and such a native can spoil many good relations in his life merely due to false suspicion.

Sun in 5th Houses:

(i) Exalted Sun in the 5th House: A positive exalted Sun in fifth house can bless the native with a successful and famous father; very good higher education; going to foreign country to complete his higher education; professional success; earning very good name and reputation in his professional sphere on account of skills and knowledge possessed by him; very good children and the overall luck; very good carrier after the birth of children; very good spiritual results; very advanced spiritual practice and heading of some spiritual organization. On the other hand, a malefic exalted Sun in 5th house can make the native suffer from many delays in his life; no fast progress in his life; and no success in his life. A malefic placement can make the native suffer from unexpected and sudden problems caused by relatives; problems related to the children and suffering on account of health problems of his children.

(ii) Debilitated Sun in 5th House: A benefic debilitated Sun placed in 5th house can bless the native with great success in his professional spheres like lawyers, judges, doctors, engineers, astrologers, scientists, analysts, researchers, politicians, diplomats; gives very good results regarding education, children, and spiritual progress. On the other hand, a malefic debilitated Sun in 5th house can trouble the native with problems in his professional spheres due to negatively biased attitude of his bosses, seniors and colleagues; gives a doubtful and suspicious nature having an adverse effect on his relations; trouble him with health problems and diseases.

Sun in 6th Houses:

(i) Exalted Sun in Sixth House: An exalted Sun in sixth house is considered bad by many astrologers and it is associated with many kinds of bad results by these astrologers. But in actual practice an exalted and benefic Sun in 6th house can bless the native with very good results. A benefic exalted Sun placed in 6th house of a horoscope can bless the native with very good results in his professional life. The natives can become good doctors, lawyers, bankers, consultants or other people working in similar profession; and earn good money; maintain good reputation and status; and are very skilled at their jobs; and capable of achieving new heights in professional spheres. A benefic exalted Sun in 6th house can also bless with very good health; hale and hearty throughout their lives. On the other hand, a malefic exalted Sun placed in 6th house can make the native suffer from health problems; permanent diseases and treating them for a very long part of their life; defamation in litigation or false accusations levied on the native. This placement can also make the native suffer on account of his children suffering on account of a bad marriage, health problems and court cases; and spending lots of money and time to fix the problems in the lives of his children.

(ii) Debilitated Sun in 6th House: A benefic debilitated Sun in 6th house can make the native benefit from ancestral property, wealth or business obtained from inheritance and inherits great amount of wealth, property or business empires from his ancestors provided this fact is supported by their overall horoscope; blesses him with luxuries and comforts and enjoyment of life to the fullest. On the other hand, a malefic debilitated Sun in 6th house can cause serious problems in the marriage and married life facing divorces; poses serious health related problems getting afflicted with a serious, permanent and fatal disease; litigations and court cases, particularly related to the property matters.

Sun in 7th Houses:

(i) Exalted Sun in 7th House: A positive exalted Sun in seventh house can bless the native with good results from a business in partnership; good reputation without facing any serious blames, allegations or defamation throughout his life; marrying to a rich and good wife; getting good financial support from his wife's family and his wife and her family can play an important role in professional success of the native; going to foreign lands on permanent basis; making profits from business which is settled in foreign lands or which connects to foreign lands. On the other hand, a negative exalted Sun in 7th house can make the native suffer in married life; facing serious issues in his married life; financial losses which can occur due to loss in some partnership business or which can occur due to bad spending habits of the spouse; health problems; bringing bad reputation by virtue of wrong decisions taken; serious financial losses and going to jail for a short duration of time.

(ii) Debilitated Sun in 7th House: A benefic debilitated Sun placed in 7th house can bless the native with a good position of power and status becoming high class government officers in the departments like police force, administrative services, and army or can become doctors, surgeons, engineers, physicians, consultants, advisors; does a love marriage leading a happy married life; name and fame in the society and spiritual progress and travels to foreign countries. On the other hand, a malefic debilitated Sun in 7th house can cause problems with his higher education and some may not be able to complete his education; troubles with problems related to child birth and he may have to undergo delays and medical treatment in order to have his child and he is prone to health problems.

Sun in 8th Houses:

(i) Exalted Sun in 8th House: An exalted Sun placed in the eighth house is considered bad to very bad by many Indian astrologers whereas an exalted benefic Sun present in 8th house can bless the native with a permanent career in some foreign country; good results in spiritual and supernatural fields; and practicing in spiritual fields. On the other hand, a malefic exalted Sun placed in 8th house can pose a direct threat to the life and having a short life span or an accidental death; problems in married life, problems in financial spheres and spending lots of money towards the treatment of diseases, court cases and settlements; losing on account of inheritance or ancestral property through litigation or court case; serious health problems and undergoing surgeries and operations many times in his life and can die in an operation theatre while surgery is being performed on him.

(ii) Debilitated Sun in 8th House: A benefic debilitated Sun in 8th house can bless the native with wealth, prosperity and business by means of inheritance easily or some going through court cases and litigations in order to own such inheritance; long life span, a good spiritual progress, interests and progress towards paranormal or occult and practicing of hidden faiths achieving good results and success in such professions. On the other hand, a malefic debilitated

Sun in 8th house can cause serious problems in the marriage and married life like late or a very late marriage, a disturbed marriage, divorce or a broken marriage; the divorces finalizing after much drama, litigations and court cases and parting with a substantial part of his wealth and assets towards the settlement of such divorce or divorces; permanent disease keeping on bothering him throughout his life and may claim his life in the end.

Sun in 9th Houses:

(i) Exalted Sun in 9th House: A positive exalted Sun in ninth house can bless the native with great amount of wealth, good health and leadership or position of power and prosperous life; chances of settling in a foreign country and making money, name and fame there; wealth or a position of power passed on by his father or grandfather; very good gains from father or grandfather; good spiritual and religious interests and becoming the heads of spiritual or religious organization and holding good posts of power in government or becoming a powerful politicians; a powerful position in government office and being successful in government jobs which hold a position of power and authority. On the other hand, a malefic exalted Sun placed in 9th house can make the native suffer from health problems; suffering from debts of his father and forefathers; having a passive and pessimistic attitude and not try to work hard to achieve his goals and then blaming his failures on destiny.

(ii) Debilitated Sun in 9th House: A debilitated Sun placed in 9th house can bless the native with very good all around professional success, name, fame, religious and spiritual interest; gets a wise and financially sound life partner providing more strength and stability to him; can settle in foreign country permanently after the marriage. On the other hand, a malefic debilitated Sun in 9th house can delay the marriage or waiting very long for his marriage; some may walk out of his marriage as he feels suffocation in his marriage or wants to break free from the bonds of marriage; makes him very argumentative and outspoken causing problems in maintaining any kind of relations whether it is a marriage or other relations; affects the sexual drive and sexual pleasures for him in a negative way and depriving of his fair share of sexual pleasures due to one reason or the other.

Sun in 10th Houses:

(i) Exalted Sun in 10th House: A positive exalted Sun in tenth house can bless the native with very good professional skills and a very good professional career; making great progress in his professional life; quality for setting examples for the others; name and fame by virtue of some new discoveries, inventions or ideas formulated;

making lots of wealth through professions; good married life; achieving professional successes with the help of his family members and his spouse and possessing good communication skills; good reputation in the society; a position of power and authority in some government or non government organizations and becoming the head of a big organization. On the other hand, a malefic exalted Sun in 10th house can make the native suffer from financial losses in his professional sphere; undergoing frequent job changes or profession changes throughout his life; serious concerns related to the married life; separation from his wife for a long period of time; health problems caused due to workaholic nature; and working without taking sufficient rest and relaxation.

(ii) Debilitated Sun in 10th House: A benefic debilitated Sun in 10th house can bless the native with very good professional becoming successful as police officers, army officers and other people working in police or paramilitary forces, lawyers, judges and other people working with some government organizations; gives sudden and sharp financial gains through profession or through some other means or unexpected financial profits many times in his life depending upon the overall tone of their horoscopes. On the other hand, a malefic debilitated Sun in 10th house can cause serious problems and troubles in professional sphere facing much delays or facing big losses, financial crisis and even defame or disgrace; very bad results in married life arising due to lack of mental compatibility between him and his wife and not sparing proper time for his marriage which can ultimately break his marriage.

Sun in 11th Houses:

(i) Exalted Sun in 11th House: A positive exalted Sun in eleventh house can bless the native with very good financial gains from his profession and also other than his profession; getting lots of money on regular basis from his brothers, sisters or friends; becoming very successful in politics; getting good position of power with his own as well as his sincere friends and relatives efforts; good friends; getting success and profits with the help of his friends; very good communications skills and leadership qualities and successful in getting a position of power through politics or through some other means. On the other hand, a malefic exalted Sun in 11th house can trouble the native on account of delays in his professional life; putting long and sustained efforts in order to see significant results in his professional life; financial losses on account of gambling, lottery, stock trading and other such practices and health problems.

(ii) Debilitated Sun in 11th House: A benefic debilitated Sun in 11th house can render great name, fame and power which may come to him as inheritance or with the help of some member of his father's family; profiting from professions in foreign countries or some may travel to foreign countries on frequent basis; not likely to settle in these foreign countries. On the other hand, a malefic debilitated Sun in 11th house can trouble the native with unwanted expenses and debts usually passed on to him; spends much time, money and resources towards the needs and issues of his family; develops egoism and arrogance in him taking wrong decisions and can bring his downfall at one point in his life.

Sun in 12th Houses:

(i) Exalted Sun in the 12th House: An exalted benefic Sun in twelfth house of a horoscope is considered bad to very bad according to many Vedic astrologers and they consider it a cause of concern for the native, an exalted Sun in 12th house of a horoscope can bless the native with one or more than one permanent residences in foreign countries and keeps on visiting foreign countries again and again for the purpose of business as well as for enjoyment; good for mental peace sound sleep most of the times; enjoyment with many kinds of luxuries during his lifetime and good spiritual or religious interests. On the other hand, a malefic exalted Sun in the 12th house can trouble the native with problems in his married life; living away from his spouse for a long time; spending excessive amount of money towards luxuries and comforts and facing financial crisis and debts; suffering from diseases and spending money towards the treatment of diseases.

(ii) Debilitated Sun in 12th House: A benefic debilitated Sun in 12th house of a horoscope can take him to foreign lands and settling in foreign countries on the basis professions; very good progress in the fields like spiritualism, paranormal and occult and choosing one of such lines or faiths as his profession; renders very good social skills and management skills. On the other hand, a malefic debilitated Sun in 12th house can make the native over ambitious and extravagant facing many kinds of problems in his life; poses professional problems due to poor financial management skills; problems related to his children and the problems related to his married life and some health related problems.

(2) Effects of Exalted & Debilitated Moon in 12 Houses:

According to Vedic Jyotish, Moon is exalted in a horoscope when it is placed in Taurus and gains maximum strength. The Moon is debilitated in a horoscope when it is placed in the sign of Scorpio which means that Moon becomes weakest when placed in Scorpio.

Moon in 1st Houses:

(i) Exalted Moon in 1st House: A positive (benefic) exalted Moon in first house can bless the native with very good professional success by native's own efforts; very good in communication; wealth and comforts; peace of mind; making a creative artists; very charming and very attractive personality. On the other hand, a malefic exalted Moon 1st house can make the native very doubtful; having no normal relations with people; problems in his married life; keeping on doubting wife for no solid reasons and breaking the marriage; and health related problems.

(ii) Debilitated Moon in 1st House: A benefic debilitated Moon placed in 1st house can give very good results regarding the financial prosperity and professional success for most of his life, name, fame and power through politics as this placement is also favourable for good results through politics; going to foreign lands and spending many years in foreign countries as this placement usually does not promise a permanent settlement in a foreign country. On the other hand, a malefic debilitated Moon in 1st house can trouble the native on account of physical or psychological health issues and permanent disease right from his birth or his early childhood; causes delays and setbacks in professional sphere starting his careers late or very late in his life and it can distort his physical appearance.

Moon in 2nd Houses:

(i) Exalted Moon in 2nd House: A positive exalted Moon in second house can bless the native with very good financial prosperity; too much wealth; making a Dhan Yoga; very good education; successful in businesses of luxury and comfort; good married life and houses, Vehicles and other luxuries and comforts. On the other hand, a malefic exalted Moon in 2nd house can problems in the married life; getting divorce or separation from his spouse; diseases due to bad eating habits; spending much money towards the treatment.

(ii) Debilitated Moon in 2nd House: A benefic debilitated Moon placed in 2nd house can give good to very good results regarding profession and financial prosperity earning very good amount of money from his professions; very good social skills and diplomacy and most socially connected and networked person, making and maintaining relations and diplomacy, and growing throughout his life and very popular among his social circle; very good marriage and married life and wealth increasing after marriage. On the other hand, a malefic debilitated Moon in 2nd house can make the native overambitious, cunning and over smart in a negative way and hindering his progress or bringing his downfall; developing extreme desires for physical and materialist pleasures and accordingly spending a big part of his earnings, time and resource in order to fulfil such desires; indulge in bad habits of drinking, smoking, gambling and making him to face financial crisis more than once his life.

Moon in 3rd Houses:

(i)   Exalted Moon in 3rd House: A positive exalted Moon in third house can bless the native with very good creative abilities and recognition on international levels; good education and teaching as their profession; very good children and support from the children and name and fame by virtue of their children; spiritual progress and spiritual teachers or spiritual healers. On the other hand, a negative exalted Moon in 3rd third house can trouble the native by virtue of bad friends and colleagues and defamation; health problems and facing problems with bosses, seniors and the people in authorities; dispute, argument, bad reputation and financial losses.

(ii)   Debilitated Moon in 3rd House: A benefic debilitated Moon in 3rd house can render good or very good creative abilities and make successful professional as writers, singers, musicians, actors and other people of similar kind; gives good financial security gaining from his profession as well as from other sources like gifts or financial help provided by his brothers and sisters; gains from investment in stock markets or lotteries; renders very good business and trade sense and his own business. On the other hand, a malefic debilitated Moon in 3rd house can cause professional delays, financial setback and negativity in the workspace suffering from backbiting, conspiracies and biased behaviour in his workplaces; change of jobs many times; spoils his relations with his brothers, sisters and friends on account of misunderstandings created by a 3rd party in order to separate him from his brothers or friends.

Moon in 4th Houses:

(i)   Exalted Moon in 4th House: A positive exalted Moon in fourth house can bless the native with good health throughout the life; luxuries and comforts; and many properties during his lifetime and good conveyance. On the other hand, a malefic Moon in 4th house can bring serious problems in the married life; long separation or a divorce from spouse; financial problems and debts; legal litigation related to property matters and spending huge amount of money, time and resources in litigation and failure in judgment of litigation and bringing problems in the professional life.

(ii) Debilitated Moon in 4th House: A benefic debilitated Moon in 4th house can bless with his going to foreign countries and settling abroad on permanent basis; luxuries and pleasures; a position of status and power with government as officers. On the other hand, a malefic debilitated Moon in 4th house can cause serious problems in married life undergoing very painful experiences like delay the marriage, serious understanding, lack of physical as well as mental compatibility between the partners, severe arguments and even divorces; staying away from home and going homes very late as the atmosphere of his house is very disturbing and torturing for him; very bad financial management and spending much more than his incomes landing into financial crisis; long lasting disease to him troubling the native for a very long period of time.

Moon in 5th Houses:

(i) Exalted Moon in 5th House: A positive exalted Moon in fifth house can bless the native with love marriage and happy married life; professional success and good public image and relations; good dealing with people and connection to a large network of people and achieving success in life. On the other hand, the malefic exalted Moon in 5th house can trouble the native with difficulties in producing children; bad health related to their reproductive organs; bad in the love life and many break ups and much drama in love life; and financial losses.

(ii) Debilitated Moon in 5th House: A benefic debilitated Moon placed in 5th house can give financial prosperity; living easy, comfortable and luxurious life; very good skills of mathematic or calculations, a diplomatic nature, manipulative abilities, ambitions to be at the top and shining like successful politicians and diplomats, mathematicians, scientists, engineers, astrologers, numerologists and professions related to medical or food or dairy products; going to foreign countries and settling permanently as a very successful person. On the other hand, a malefic debilitated Moon in 5th house can give permanent health problems or diseases; very suspicious nature and handling relationships in very difficult way; mental incompatibility between him breaking one marriages; and very bad results for love life.

Moon in 6th Houses:

(i) **Exalted Moon in 6th House:** A positive exalted Moon placed in the sixth house can bless the native with a long life; and financial gains from inheritance or legal degree in suing some company for something bad they may have done to the native. On the other hand, a malefic exalted Moon in 6th house can bring serious problems in married life; legal litigations by filing a case against him by spouse; losses in the form of property, wealth or other assets as a result of penalty imposed by law; giving a big percentage of their wealth to his spouse as a settlement of divorce and very bad results related to the litigation or court cases; serious decease and spending huge amount of money towards the treatment and disease may prove fatal in some cases.

(ii) **Debilitated Moon in 6th House:** A benefic debilitated Moon placed in 6th house can make him successful as a lawyer, judge, police officer, army officer, doctor, physician, astrologer, psychic, and numerologist and so getting benefit from the troubles, problems and disputes of the other people; a position of power and authority with government or private organization being in charge and in control of many people. On the other hand, a malefic debilitated Moon in 6th house can cause serious problems in the marriage and married life and facing arguments, disputes and litigations in his married life or parting with his parental family on account of serious differences between him and his parents or living away from his parents for a very long period of his life; develops criminal interests in him making a notorious criminals, gangsters, mafia people, smugglers and other people of such kind spending a long period of time in jail due to his criminal activities.

Moon in 7th Houses:

(i) **Exalted Moon in 7th House:** A positive exalted Moon placed in the seventh house can bless the native with a very good married life with a beautiful and loving wife; rise of fortune after marriage as wives brings success, prosperity and happiness; getting wealth from his ancestors; position and power and a powerful leader or politician having a great number of followers; and residence in foreign countries or travel to many foreign countries. On the other hand, a malefic exalted Moon in 7th house can trouble the native with a disturbed marriage; facing serious issues with his wife and mother of the native can become the root cause of problems in his married life; waste of a great amount of time, money and resources in pursuit of materialistic pleasures and wasting much money and resources on account of his physical relations with many women and facing bad reputation.

(ii) Debilitated Moon in 7th House: A benefic debilitated Moon placed in 7th house can give good profession and financial stability and achieving good success and financial situation usually remaining balanced throughout his lives and not having any financial crisis; going to foreign countries on permanent basis and getting settled in a foreign country on the basis of his marriage which means that he can marry a foreigner and then settle in the country of his wife or husband. On the other hand, a malefic debilitated Moon can cause problems in the married life like mutual distrust or broken marriages; brings disgrace or defame due to allegations imposed on him by his seniors, colleagues, his wife or his in-laws and living with this disgrace for a very long period of time; health related problems like some disease claiming the life of the native.

Moon in 8th Houses:

(i) Exalted Moon in 8th House: A benefic exalted Moon placed in the eighth house is considered malefic, bad and associated with many bad results by many astrologers. But, a positive exalted Moon in 8th house of a horoscope can bless the native with very good professional benefits; practicing in the fields like astrology, palmistry, numerology and other similar fields; instant and easy gains many times and inheritance of a well established business from his father or forefather; good results in the fields like meditation, spiritualism and other such fields and some natives can achieve great spiritual or supernatural powers. On the other hand, a malefic exalted Moon in 8th house can trouble the native with disturbed peace of mind, restlessness, difficulty in sleeping; problems in the professional sphere; facing many difficulties and delays in his professional life; and bad health affecting the married life.

(ii) Debilitated Moon in 8th House: A benefic debilitated Moon placed in 8th house can bless the native with wealth and money by means of inheritance or gaining from his father or from his mother's family or his mother is the only child of her parents and so inheriting his mother's family properties or wealth; very good progress in spiritual, occult, paranormal and financial gains keeps on coming to him from time to time; keep on profiting on account of gifts, inheritance, financial assistance from his nears and dears even if he is not doing well in his professions and hence has no problems regarding financial issues. On the other hand, a malefic debilitated Moon in 8th house are disturbed throughout his life on account of problems coming from his marriage, his health, his poor financial status and facing financial crisis due to increased expenses and poor management of finances; diseases, particularly in the early and the last years of his life.

Moon in 9th Houses:

(i) Exalted Moon in 9th House: A positive exalted Moon placed in the ninth house can bless the native with money, wealth, prosperity and too much richness; getting birth in rich family holding a very respectful and special status in the society like a powerful and respected political or royal family or a powerful and respected religious family; inheritance of great amount of money and wealth from father or grandfather; very good spiritual or religious qualities; becoming the heads of some religious or spiritual organization and acquiring great wealth by virtue of their followers donating great amounts of money and wealth to him; trip to foreign lands for the purpose of business and great profits and prosperity in foreign lands; and good public reputation. On the other hand, a malefic Moon in 9th house can form a Pitra Dosha and suffer from the bad effects of Pitra Dosh; bad luck and bad fortunes and trying hard to achieve some success in his life.

(ii) Debilitated Moon in 9th House: A benefic debilitated Moon in 9th house can give very good results making very auspicious and beneficial yoga blessing him with countless beneficial things; good professional success, reputation, status, name, fame and taking him to foreign lands for profession and settle in foreign country on permanent basis or keep on travelling to many foreign countries for business or for pleasure; very good spiritual progress. On the other hand, a malefic debilitated Moon in 9th house can make him so naïve that most of the people around him take advantage of him making him prone to financial problems, cheatings, frauds, betrayals and even litigations in some cases; problems related to child birth or facing delays, treatments or other problems in child birth, but no complete denial of child on its own.

Moon in 10th Houses:

(i) Exalted Moon in 10th House: A positive exalted Moon in tenth house can bless the native with excellent professional life practicing in creative fields; getting settled abroad for job or business and spending most of his life in foreign country; good jobs and professions with less physical work and very well paid and no problems in professional life; and successful actors, producers, directors, musicians or lyricists in the fields like movies. On the other hand, a malefic exalted Moon in 10th house can trouble the native with problems in professional life; facing defamation or disgracing due to mistake done by him or false allegation; bad health and in case of females delaying in child birth and malfunctioning of reproductive organs or defect in some reproductive organ; making him a victim of wrong decisions, wrong judgments or discrimination; court decision going against him even when he deserves a right decision or the people in authority may misjudge his abilities or someone in authority may harm him due to some discrimination.

(ii)   Debilitated Moon in 10th House: A benefic debilitated Moon placed in 10th house can give very good profession working with big organizations connecting with and understanding the people, conveying his thoughts and plans for the people and his succeeding with the help of the people around them like a powerful politicians or as a lawyers, astrologers, engineers, teachers, preachers; take him to foreign lands spending many years of his life in foreign countries for professions or can settle in foreign countries on permanent basis; rising with the help friends, brothers and a network of people. On the other hand, a malefic debilitated Moon in 10th house can render criminal tendencies to him and working as gangsters, drug paddlers or drug mafia, smugglers, arms dealers and many other kinds of people dealing in illegal activities and some of these can operate on international level giving good financial results in the beginning but it finally takes away everything and ending up in jail and losses; gives diseases may be proving to be fatal for him.

Moon in 11th Houses:

(i)   Exalted Moon in 11th House: A positive exalted Moon placed in eleventh house can bring very good financial gains; rising to a financial success through his own efforts and luck; a very shrewd business sense having the ability and luck to rise from ashes to making fortunes by virtue of their professional skills and their luck; running his own businesses instead of working for someone else; very good health and a very good peace of mind and facing no major disease or any major financial problem. On the other hand, a malefic Moon in 11th house can bring problems in the married life; facing a long dispute, separation or divorce from his wife; affecting the mental health and suffering from mental problems or psychological disorders; tendency to doubt everyone including his wife and children; a feeling of strong insecurity to hide and protect himself from most of the people around him and behaving like an insane.

(ii)   Debilitated Moon in 11th House: A benefic debilitated Moon in 11th house can bless him with very good financial prosperity from the profession or other sources like sudden gains through stock markets, lotteries, gifts; give very good results regarding the marriage and married life and lead a happy married life and his fortune and financial prosperity increases after his marriage. The spouse plays an important role in his professional success; take him to foreign lands and the prosperity increases many folds in foreign countries. On the other hand, a malefic debilitated Moon in 11th house can cause problems in the marriage and married life like marrying late or very late; delays and setback in his professional life or delays or losses in his business, particularly from the business which is running in partnership with someone else.

Moon in 12th Houses:

(i) Exalted Moon in 12th House: A benefic exalted Moon placed in the twelfth house is considered bad by many astrologers, but actually it gives very good results in many horoscopes. A benefic exalted Moon in 12th house can bless the native with great amount of money by professions related to foreign countries, travel business, new innovations, new discoveries, entrepreneurship and similar fields; with a luxurious life and travels across the globe; spending a good amount of time and money towards his eating pleasures. On the other hand, a malefic exalted Moon in 12th house can trouble the native with financial problems and financial debts; bad results and suffer from the married life; separation from his parental family after marriage on account of misunderstandings between his parents and his wives; spending a good amount of money towards the treatment of diseases of native or his family members.

Debilitated Moon in 12th House: A benefic debilitated Moon in 12th house is a very good placement when it comes to going abroad and settling there permanently; good position of power and authority in government office or private organization and becoming top class officers; gives political interests and power becoming powerful political leaders or diplomats. On the other hand, a malefic debilitated Moon in 12th house can create serious flaws in his personality making him egoistic, arrogant, unrealistic, cruel or aggressive and these negative traits of his personalities land them in troubles, wasting of time in years of his life; bring his downfall by virtue of his arrogance, ego and other such negative qualities; permanent disease; may die due to a fatal disease or an accident or some other kind of unnatural death at a comparatively early age.

(3) Effects of Exalted & Debilitated Mars in 12 Houses:

According to Vedic Jyotish, Mars is exalted in a horoscope when it is placed in the sign of Capricorn which means that Mars gains maximum strength. Mars is debilitated in a horoscope when it is placed in the sign of Cancer which means that Mars becomes weakest in Cancer.

Mars in 1st Houses:

(i) Exalted Mar in 1st House: The placement of exalted Mars in first house is considered bad for the marriage by many astrologers as Mars in 1st house of a horoscope forms a Manglik Dosha which creates problems in married life of the native. However in actual practice, a benefic exalted Mars in 1st house can also give very good results to the native. A benefic exalted Mars in 1st house forms Mangal Yoga in stead of Manglik Dosha and bless the native with a very good marriage; very good professional success; big financial gains from his profession; luxury and comfort leading a prosperous life; very good progress in his profession gaining wealth and buying new houses after his marriage which brings good luck; strong body and free of any major disease; placement in the fields like Police, Army and other similar fields by virtue of their physical fitness. On the other hand, a negative exalted Mars in 1st house of a horoscope forms Manglik Dosha and causes serious troubles in married life; serious damages to the marriage; breaking of marriage or more than once divorces; problems in professional life; serious health problems and suffering on account of his bad relations with his brothers or friends.

(ii) Debilitated Mars in 1st House: A benefic debilitated Mars placed in 1st can give very good results regarding the financial prosperity and professional success and he is likely to be self made man instead of owning his business or profession from someone else; creative ideas to start successfully something new and different; forms a Manglik Yoga in the horoscope blessing him with very good results in his marriage and married life. On the other hand a malefic debilitated Mars in 1st house can trouble him in profession like he may not be able to start his professional career until a late or very late age, or losses and setback in his professions due to his wrong decisions; forms a Manglik Dosha in the horoscope causing problems in his marriage and married life and health related issues bothering him for a very long period of time.

Mars in 2nd Houses:

(i) Exalted Mars in 2nd House: A positive exalted Mars in second house can bless the native with a very good marriage; very good financial results becoming very rich with name and fame by virtue of his education, creative skills or his professional achievements; healthy and lucky children bringing money and fame to him by virtue of their success and good deeds; good progress in spiritual and religious fields. On the other hand, a negative exalted Mars in 2nd house can bring bad to very bad results and can be one of the worst placements of exalted Mars; breaking one or more marriages of the native; delay the marriage by many years; may not be able to marry at all throughout his life; bad results for the financial aspect; facing many unwanted expenses and waste of money; facing financial crisis and financial debts; health related issues right from the childhood of the native and troubling the native throughout his life.

(ii) Debilitated Mars in 2nd House: A benefic debilitated Mars in 2nd house can bless him with very good amount of wealth and prosperity and accordingly he will be rich or very rich due to his own efforts or through some family member or friends; render very good skills of financial management to him; forms a Manglik Yoga in the horoscope blessing him with very good results in his marriage and married life. On the other hand, a malefic debilitated Mars in 2nd house may be one of the most difficult placements of debilitated Mars and can trouble him on various fronts; forms a Manglik Dosha in the horoscope causing very serious problems in his marriage and married life; create major financial problems facing financial crisis in his life for very long periods of time; give permanent disease at very early in his life or he is born with health problem, disease or defect which stays with him throughout his life.

Mars in 3rd Houses:

(i) Exalted Mars in 3rd House: A positive exalted Mars in third house of a horoscope can bless the native with a strong and healthy body; great courage and determination; true warriors in many spheres; never admitting defeat and keeping on fighting with his problems, sufferings and enemies until he achieve victory over them; providing great strength to his friends; having a very big friend circle and having friends. On the other hand, a malefic exalted Mars in 3rd house can trouble the native with serious problems and losses caused by his enemies and competitors; problems, betrayal and big loss or even a fatal loss from his friends and brothers; choosing wrong paths to achieve his goals and fulfil his ambitions which can put the native in trouble many times in his life; problems like arguments and disputes in his married life and facing a long separation or a divorce.

(ii) Debilitated Mars in 3rd House: A benefic debilitated Mars in 3rd house can bless him with good profession or business getting help and profit through his in-Laws or working with his in Laws or getting financial help from his in-laws or by inheritance which is passed onto him by his in-laws. On the other hand, a malefic debilitated Mars can bring many kinds of problems in the marriage and married life like marrying late or very late or very painful marriages or may not be able to marry at all or he may suffer from more than one broken marriages; health related problems spoiling his relations with his friends, brothers and colleagues.

Mars in 4th Houses:

(i) Exalted Mars in 4th House: Many astrologers believe that the presence of an exalted Mars in the fourth house of a horoscope is very bad for the native as this placement forms a Manglik Dosha whereas in actual practice such a placement of exalted Mars in 4th house also from a Manglik Yoga instead of a Manglik Dosha and it can give very good results to the native. A positive exalted Mars placed in fourth house can bless native with happy married life with a beautiful and rich wife; love and support leading a happy marriage; many kinds of luxuries and comforts; possessing more than one luxury vehicles and houses; gains through the dealings in property, real estate and luxury vehicles; very good social skills; very big social circle and social network to achieve success in many spheres; very good creative skills; beauty and charm. On the other hand, a negative exalted Mars in 4th house can cause problems in the married life; Malefic exalted Mars in 4th house is forming a Manglik Dosha and so suffering from many problems in his married life; disturbed the peace of mind most of the times; his mother going through health problems and spending much money, resources and time towards the treatment of disease of his mother.

(ii) Debilitated Mars in 4th House: A benefic debilitated Mars in 4th house forms a Manglik yoga in the horoscope and gives good results regarding his marriage and married life by getting married at proper ages and leading happy married lives; very good results in profession requiring not much hard work and enjoying luxuries and comforts of life. On the other hand, a malefic debilitated Mars in 4th house can cause serious problems in the marriage and married life troubling him as per Manglik Dosha in 4th house; trouble him in his professional sphere like not becoming professionally successful for a very long period or facing many job changes, setbacks and losses through profession; can distort his logics who may not be able to take right decisions at proper times.

Mars in 5th Houses:

(i) Exalted Mars in 5th House: A positive exalted Mars placed in fifth house can bless the native with very good creative abilities and very good professional skills and financial prosperity; success in his profession, supernatural, paranormal or spiritual interests; practicing in the fields like astrology, palmistry, numerology, vaastu and other similar kinds of fields; very good, strong, healthy and courageous children especially male children and achieving great success, name and fame with the support of his children; raising his younger brothers or sisters and getting great support and respect from them throughout their life. On the other hand, a malefic exalted Mars placed in 5th house can cause problems in his education; discontinuation of his education; serious concerns in marriage and love life; getting romantically involved in love affair with his relative or a married woman bringing disgrace and troubles for him as well as his family members; bad relations with his brothers and children; strong opposition from his brothers and children; health related problems and undergoing surgeries or operations for treatment of these problems.

(ii) Debilitated Mars in 5th House: A benefic debilitated Mars in 5th house can give good results regarding the financial prosperity by becoming very successful police officers, army officers, lawyers and particularly criminal lawyers, doctors and particularly dentists; very good higher education and successful on account of the higher education achieved by him; name, fame, status, a life of luxuries and comforts and many other good things. On the other hand, a malefic debilitated Mars in 5th house can cause serious concerns regarding the financial prosperity due to his poor financial management; various kinds of unwanted expenses and taking his considerable amount earned and saved money leaving him with little money many times in his life; creates problems related to child birth and he may have to undergo medical treatment in order to produce child or can altogether deny a child to him though this fact needs to be supported from other dosha present in the horoscopes.

Mars in 6th Houses:

(i) Exalted Mars in 6th House: A positive exalted Mars placed in sixth house can bless the native with very good health and a strong body; professional success; achieving a post of power and authority as government officer in government department or by means of politics; making him a powerful, popular and revolutionary public leader and doing many revolutionary things; leading a big section of society and blessed with strong and dedicated followers going to any extent to support him; very brave and courageous and capable of winning over his enemies and completing difficult goals and accomplishing great heights in his life; very good leadership as well as administrative qualities having lead and rule a big section of society for a very long time and with great success. On the other hand, a negative exalted Mars in 6th house can cause serious problems like suffering from disputes and litigations related to wealth and properties; spending much time, money and resource to solve the problems troubling his father like litigations, financial losses or diseases; problems from his enemies and so losing his position of power and facing social disgrace due to his enemies; problems in the married life and health problems or diseases.

(ii)   Debilitated Mars in 6th House: A benefic debilitated Mars in 6th house can give good result in his professional sphere like successful doctors, chemists and all other people dealing in medicine, lawyers and especially criminal lawyers, people dealing in fast food or restaurant business and many other kinds of professionals dealing in various kinds of professions; makes him profit from his brothers and friends or profit through some court decision or litigation. On the other hand, a malefic debilitated Mars in 6th house can develop criminal tendencies in him engaging him in various criminal professions such as arms dealers, mafia people, terrorists or people associated with terrorist organizations, kidnapping mafia, contract killers and many other kinds of people dealing in similar kinds of professions; takes him to Jail for a very long time or serious injury received through some gang war or some crossfire with police or he may even die in a gang war or police crossfire or he may receive a death sentence from a court of law, though such extreme results need to supported by other Dosha present in the horoscope; troubles him on account of health problems undergoing surgeries, more than once in his life.

Mars in 7th Houses:

(i)   Exalted Mars in 7th House: Many astrologers believe that the presence of an exalted Mars in the seventh house of a horoscope is very bad for the native as this placement forms a Manglik Dosha whereas in actual practice such a placement of exalted Mars in 7th house also from a Manglik Yoga instead of a Manglik Dosha and it can give very good results to the native. A positive exalted Mars in 7th house give happy married life with a supportive, wise and courageous wife and acting more like a friend; very good communication social skills to succeed in the life; very good professional achievements and financial gains and reaching new heights; profession and residences in a foreign country and settling abroad and making good money there. On the other hand, a negative exalted Mars in 7th house forms a Manglik Dosha in the horoscope and as a result the he may have to suffer from serious problems in his married life like serous arguments, long separations, court cases, litigations and more than one divorces and in some extreme cases, one or more than one wives dieing due to some disease or accident; social disgrace by virtue of some love affair or some professional problems.

(ii)   Debilitated Mars in 7th House: A benefic debilitated Mars in 7th house forms Manglik Yoga in the horoscope and can bless him with a very good marriage and married life; gets married to rich and understanding partners who provides financial strength him and is very supportive; take him to foreign countries for profession or travel Tour or to settle in foreign country on permanent basis after his marriage; gives good results and financial prosperity, good reputation and status in the society. On the other hand, a malefic debilitated Mars in 7th house forms a Manglik Dosha in the horoscope and can cause many problems for him as per Manglik dosha in 7th house; very bad results regarding his comfort, convenience and peace of mind making him bothered, restless or desperate most of the times; brings disgrace and defame through some allegation imposed on him and troubles him with health related issues being psychological in nature.

Mars in 8th Houses:

(i)   Exalted Mars in 8th House: A positive exalted Mars placed in eighth house can bless the native with happy married life with supportive wife; very good results in his professional sphere; achieving great professional success and financial gains from his profession; long life. On the other hand, a negative exalted in 8th house of a horoscope can cause very serious problems in the married life; failing in his marriages and divorces; much drama, litigation and financial losses; serious diseases and spending much money, resources and time towards the treatment of the diseases; dieing due to the diseases after going through a long treatment, surgeries and much physical pain; making him a victim of cheats, frauds, false allegations, and financial failures and even getting murdered by one of his known enemies or one of his friend turned enemy.

(ii)   Debilitated Mars in 8th House: A benefic debilitated Mars in 8th house can bless him with a post of position or power with government or some private organization like successful police officers, army officers, bank officers, engineers, politicians, administrators; gives very good results regarding the spiritual and paranormal progress like becoming successful psychics, spiritual mediums, black magicians, tantrics achieving great success and power through these faiths and practices under the strong influence of this benefic placement of debilitated Mars; forms a Manglik Yoga in the horoscope giving very good results in the marriage and married life. On the other hand, a malefic debilitated Mars in 8th house forms a Manglik Dosha and can give very bad results regarding his marriage and married life as per Manglik Dosha in 8th house; connects him to the dark forces of paranormal and occult worlds may be gaining from these forces in the beginning, but they land into big troubles in the end and in some extreme cases may get hurt by them which can result in permanent mental instability or in some cases it can also cause serious harm to the body; try to engage in good faiths and practices only and should stay away from black magic, evil spirits.

Mars in 9th Houses:

(i) Exalted Mars in 9th House: A positive exalted Mars placed in ninth house of a horoscope can taking him to foreign lands and establishing in a foreign country on the basis of profession or marriage; happy married life with a foreigner and then settling abroad after his marriage; good understanding between him and his wife; getting support and complement from each other; very good professional success, name and fame and a very good public image and getting respected by the people around him. On the other hand, a negative exalted Mars in 9th house can cause problems related to the professional sphere; facing many delays, failures and losses in professional sphere; disgrace and defamation to him as well as his family; facing social disgrace or a legal litigation due to misconduct done by him; many problems and troubles in his married life.

(ii) Debilitated Mars in 9th House: A benefic debilitated Mars in 9th house of a horoscope can render very good results to him in profession, finances, wealth, comforts and many other departments; profit from an inheritance or other similar things and he can own a post of authority in his family business or a well established family business by means of inheritance; tours to foreign countries for the purpose of business and entertainment spending many years in a foreign country though he is not likely to settle there on permanent basis. On the other hand, a malefic debilitated Mars in 9th house forms a Pitra Dosha in the horoscope and can trouble him on account of debts, litigations and liabilities which may be transferred to him through his father or forefather; very bad results regarding the financial prosperity and wealth spending more and has financial crisis even though he may have been born in rich families; serious health issue or disease being genetically passed on to him.

Mars in 10th Houses:

(i) Exalted Mars in 10th House: A positive exalted Mars placed in tenth house of a horoscope can bless the native with a great professional success and very good professional skills and achieving success in his profession; strong and healthy body; leading a healthy life; getting name, fame and status in profession due to his qualities; leading a happy married life; getting successful astrologers, spiritual healers, palmists, doctors, consultants, physicians, child specialists and other similar kind of professionals. On the other hand, a negative exalted Mars in 10th house can create problems in the professional sphere; delays, problems and losses in his professional spheres; facing financial losses by cheating, facing a legal litigation and a social disgrace due to their alleged involvement in fraud or cheating activity; health problems and diseases and remaining jobless for a long time due to a serious and long lasting disease.

(ii) Debilitated Mars in 10th House: A benefic debilitated Mars in 10th house can give good results regarding the professional success; name, fame and a position of power with government which can come through politics, very good management and social skills and networking. On the other hand, a malefic debilitated Mars in 10th house can trouble him by serious financial problems and though he may earn well through his professions, but he is most of the times bothered on account of unwanted expenses, sudden losses and poor financial management; gives suffering from setbacks in professional spheres and suffering serious losses through profession; affects his decision making ability in an adverse way and accordingly facing losses and problems on account of wrong decisions or no decisions taken by him.

Mars in 11th Houses:

(i) Exalted Mars in 11th House: A positive exalted Mars placed in 11th house of a horoscope can give excellent financial prosperity; having Dhan Yoga; very rich and having more than one sources of income; huge financial gains and financial fortunes; building a big business empire if supported by the other factors of his horoscope; courage, adventure and having a very good understanding of professional matters; great financial success in their lifetime; good married life; good religious and spiritual interests; and achieving success in the fields of spiritualism, astrology, paranormal and other such fields. On the other hand, a negative exalted Mars in 11th house can cause problems with financial losses and delays in professional sphere; health problems and disease right from his childhood; problems in the married life like separation from his wife for a long period of time or serious arguments with his wife from time to time.

(ii) Debilitated Mars in 11th House: A benefic debilitated Mars in 11th house can bless him with qualities like hard work, perseverance and succeeding by virtue of these qualities; has keen interests towards the hidden and occult knowledge and faiths and rise in his life by practicing as astrologers, palmists or numerologists and other professions. On the other hand, a malefic debilitated Mars in 11th house can distort the sense of separating the right from wrong in him and making unrealistic decisions and chasing unrealistic goals and not paying attention to the good advice given by his nears and dears, as he thinks that he is the best judges of every situation and due to over confidence, over indulgence; make him suffer on accounts of delays and setbacks related to his professional sphere and in some cases affecting his marriage or married life.

Mars in 12th Houses:

(i) Exalted Mars in 12th House: A positive exalted Mars placed in twelfth house of a horoscope can take the native abroad by his profession; settling abroad; form a Mangal Yoga in the horoscope which can bless the native with a very good wife and a happy married life and settling in some foreign country on the basis of his marriage; good support from his wife; making him brave and courageous; very good at adjusting to new circumstances, new countries and to new people. On the other hand, a negative exalted Mars in 12th house of a horoscope can cause serious problems in the married life; a malefic Mars forming a Mangalik Dosha in the troubling him with many kinds of problems in his married life; living away from wife for a long period; limiting his marital pleasures; divorce; doubting most of the people around him including his own wife, brothers and sisters; serious relationship problems due to his habit; huge financial losses and debts; serious health and afflicted with a long lasting disease troubling him for a very big part of their life.

(ii) Debilitated Mars in 12th House: A benefic debilitated Mars in 12th house is a very good placement for going abroad and settling in foreign by owning his own more than one residence in foreign countries; gives very good luxuries and comforts enjoying in his lifetime; renders very good spiritual interests; forms a Manglik Yoga in the horoscope and gives very good results regarding his marriage and married life. On the other hand, a malefic debilitated Mars in 12th house forms a Manglik Dosha in the horoscope and causes serious problems in the marriage and married life as per effects of Manglik dosha in 12th house; gives health problems, professional delays, setbacks, financial problems caused due to bad spending habits and excessive unwanted expenses, and makes him suffer on account of injuries obtained from accidents.

(4) Effects of Exalted & Debilitated Mercury in 12 Houses:

According to Vedic Jyotish, Mercury is exalted in a horoscope when it is placed in the sign of Virgo which means that Mercury gains maximum strength when placed in Virgo compared to its placement in all the other signs. Mercury is debilitated in a horoscope when it is placed in the sign of Pisces which means that Mercury becomes weakest when placed in Pisces.

Mercury in 1st Houses:

(i) Exalted Mercury in 1st House: Positive exalted Mercury placed in first house gives a very good health; possess lean physique, active body; very good communication skills; profession requiring good communications skills; wisdom, analytical ability, good business and trading skills and very successful in his profession; great success; big money from his profession and extraordinary achievement in professional fields; happy married life; live longer. On the other hand, negative exalted Mercury in 1st house can trouble the native with problems in his married life due to his very much money oriented nature; disturbed the peace of mind; facing disgrace and defame through some court case or litigation.

(ii) Debilitated Mercury in 1st House: Benefic debilitated Mercury in 1st house can bless him with money, wealth and professional success; gives very good results regarding the marriage and married life generally getting married to religious, wise, benevolent and caring life partners under the strong influence of this benefic placement of debilitated Mercury; gives very good status, name, fame, good respect in the society, luxuries and comforts throughout his life. On the other hand, a malefic debilitated Mercury in 1st house indicates problems in the marriage and married life like marrying late or very late and facing many kinds of problems in married life; troubles with health problems or diseases showing up very early in his life or right from his birth but not likely to be fatal but troubling for a very long period of time or throughout his life; causes financial problems, problems related to peace of mind and some other kinds of problems.

Mercury in 2nd Houses:

(i) Exalted Mercury in 2nd House: Positive exalted Mercury placed in second house form a powerful Dhan Yoga blessing him with wealth, luxury, comforts, success and fame; very good speech abilities; good leadership; penetrating insight into things qualities and becoming very good public leaders or administrators; very good financial prosperity; very successful in his profession by virtue of his own efforts and qualities; good health, good education and a good family life and spending his childhood in very enjoyable circumstances. On the other hand, negative exalted Mercury in 2nd house can cause problems and many difficulties in the married life like arguments, clashes, separations and divorce; problems between him and his family members, especially his elder brothers and sisters; making him lazy and selfish; not achieving good professional success due to his laziness.

(ii) Debilitated Mercury in 2nd House: A benefic debilitated Mercury in 2nd house can bless him with very good financial prosperity, professional success, wealth and many other such things; render very good interest and progress in the field of spiritualism and paranormal and practicing as astrologers, numerologists, psychics, spiritual mediums, spiritual healers, tantrics, aghoris, black magician; make him profit from wills, inheritance, gifts and other kinds of sudden gains and good financial position throughout his life. On the other hand, malefic debilitated Mercury in 2nd house can cause problems in the marriage and going through difficult times in married life or undergoing one or more than one broken marriages or divorces; give problems related to child birth and suffering delays or undergoing prolonged medical treatment in order to give birth to his child; unwanted expenses which can keep on bothering him throughout his professional life and which may eat up a considerable part of his income.

Mercury in 3rd Houses:

(i) Exalted Mercury in 3rd House: Positive exalted Mercury placed in third house can bless the native with great creative abilities; achieving success in professions due to his creative abilities; very good writing abilities; good speech abilities and successful in profession requiring these abilities; great courage, confidence and determination and very good friends, supportive brothers and sisters helping and supporting to him. On the other hand, negative exalted Mercury in 3rd house of a horoscope can trouble with serious issues in relationships by his younger brothers and sisters, friends and colleagues; financial losses due to his brothers and friends; disgrace or defame due to something illegal or immoral done by his brother or friend; brothers or friends turning completely hostile and enemy towards him; facing problems form his bosses, seniors and other people in authority; facing false allegations or litigations initiated by his bosses and seniors; problems in the married life; separation from his wife for a long time.

(ii) Debitated Mercury in 3rd House: A benefic debilitated Mercury in 3rd house can bless him with professional success as police officers, army officers, doctors, lawyers, bankers and financial consultants, coaches and trainers, astrologers, reporters and other people related to print media and electronic media and many other kinds of people dealing in many other kinds of fields; going to foreign lands for short durations related to professions like reporters travelling to foreign countries for media converges, doctors travelling to foreign countries to attend some seminar, astrologers travelling to foreign countries on small business trips; very good skills of communication and writing becoming successful writers, authors, speakers under the strong influence of this benefic placement of debilitated Mercury. On the other hand, a malefic debilitated Mercury in 3rd house can trouble him in his professional sphere leading to conspiracies, back biting and other such things or changing his job, many times in his life; or facing disgrace and a court case on account of his involvement in some scandal or scam; health problems, problems in married life, spoiled relations with many of his relatives and many other kinds of problems.

Mercury in 4th Houses:

(i) Exalted Mercury in 4th House: Positive exalted Mercury placed in fourth house can give very good results regarding money, luxuries, houses and vehicles and accordingly leading a life of comfort and luxury right from his childhood; are born in rich families and hence enjoying a luxurious and comfortable life right from his childhood; very good skills of creation, manipulation and imagination and accordingly practicing in the fields requiring these qualities; becoming very good politicians achieving good position, power and success by his good communication, manipulation and imagination skills; good health and longevity; healthy and disease free life; live long. On the other hand, a negative exalted Mercury in 4th house can cause problems and troubles in their married life; health related problems suffering from long lasting diseases and spending a big amount of money; much physical and mental pain; problems related to the residential houses; facing litigations or debts related to his residential houses.

(ii) Debilitated Mercury in 4th House: A benefic debilitated Mercury in 4th house can give very good result regarding the wealth and prosperity possessing his own residences, vehicles and other things of comforts and luxuries; very good results in professional success and income like achieving a position of power and authority in government as an officer or a powerful post through politics; very good interests and progress in spiritual and paranormal worlds. On the other hand, malefic debilitated Mercury in 4th house can cause problems in the marriage and married life like late marriage, a disturbed married life or broken marriage; problems in professional sphere like changing many jobs in his life or changing many cities, states and countries throughout his life primarily due to his professions; relationship problems undergoing spoiled relations with his relatives, brothers, sister or wives.

Mercury in 5th Houses:

(i) Exalted Mercury in 5th House: A positive exalted Mercury placed in fifth house can bless the native with very good creative skills, wisdom, social skills, a tendency to work hard, a charming personality and very good trade skills and accordingly achieving success in various spheres of his life; very good, religious, spiritual, kind, considerate and idol children and achieving great professional success in their lives; children having great respect for him; getting many luxuries and comforts through his children; very good professional success, religious and spiritual advancement and visits to foreign countries for the purpose of enjoyment or trade. On the other hand, negative exalted Mercury in 5th house can cause financial problems; facing financial crisis frequently; diseases and at least one long lasting disease keeps on troubling him from time to time; spending much money, effort and time towards his parental family members and he may have to take care of one or more than one of his family members for a long period of time.

(ii) Debilitated Mercury in 5th House: A benefic debilitated Mercury in 5th house can bless him with good creative abilities achieving great things in professional spheres by virtue of his creativity; good interest and understanding of the hidden faiths and practices and practicing as astrologers, psychics, spiritual mediums, spiritual gurus or spiritual healers; good financial results and gain through stock markets, lotteries and other such concepts under the strong influence of this benefic placement of debilitated Mercury. On the other hand, malefic debilitated Mercury in 5th house can cause problems related to his education having to discontinue his higher education or may not getting any higher education at all; problems related to child birth or having a delayed conception, an abortion, birth of a physically or mentally challenged child or the birth of a dead child; problems in the married life due to extramarital affairs of one or both of the partners.

Mercury in 6th Houses:

(i) Exalted Mercury in 6th House: Positive exalted Mercury placed in sixth house can give very good results for financial prosperity; achieving good professional success as policemen, army people, doctors, consultants, lawyers, judges, and astrologers; very good overall health protected from any major disease or health problem. On the other hand, negative exalted Mercury in 6th house can trouble the native with some serious problems and constant troubled by his enemies and competitors; facing big losses due to his enemies and competitors; turn his brother or friend into an enemy and danger of facing a strong opposition from his own brothers, sisters and friends; complete professional failure, a punishment by law or some other kind of loss due to the betrayal of his brother, sisters or friend; facing a fatal attack and getting murdered due to betrayal of his brothers or friends; his friends joining his enemies to kill him.

(ii)   Debilitated Mercury in 6th House: Benefic debilitated Mercury in 6th house can give good results regarding the professional success becoming successful lawyers, doctors, engineers, computer programmers, astrologers; make profit from some will or inheritance though he may have to go through a legal case or litigation in order to benefit from such will or inheritance; good spiritual and religious interests and good social skills. On the other hand, a malefic debilitated Mercury in 6th house can trouble him with serious health or diseases or going to hospitals or undergoing surgeries many times in their his life or die with the disease; very bad financial status facing serious financial crisis more than once in life; bad results regarding his marriage and married life.

Mercury in 74th Houses:

(i)   Exalted Mercury in 7th House: Positive exalted Mercury placed in seventh house can give very happy married life and enjoying the marriage; well behaved, socially skilled, wise and practical and supporting wife in reaching his goals; rich and financially helping wife form their in-laws; good professional success; settling in a foreign country for a profession; travels in foreign countries for luxuries and comforts; having more than one house properties, many vehicles and many other kinds of luxuries at their disposal. On the other hand, a negative exalted Mercury in 7th house can cause facing serious issues in his married life due to mental incompatibility with his wife; the equation between the mother and his wife playing a major role in spoiling his marriage; his mother and his wife may not going along well with each other; suffering from long lasting disease; bad results regarding his professional life and suffering from some psychological disorders.

(ii)   Debilitated Mercury in 7th House: Benefic debilitated Mercury in 7th house can take him to foreign lands to settle permanently in some foreign country on the basis of profession or marriage; good results regarding the marriage and married life and living good and happy married life or better status and income after marriage; render good name, fame and status and good public image. On the other hand, malefic debilitated Mercury in 7th house can cause various kinds of problems in the marriage and married life undergoing much pain, emotional drama and many bitter experiences in his marriage; develops an escapist tendency he may leave and forsake his wife, family and all his responsibilities at one point of time and goes away to some asylum or some spiritual journey and may not ever come back again; health related problems troubling him for a very long period of time.

Mercury in 8th Houses:

(i) Exalted Mercury in 8th House: Positive exalted Mercury placed in eighth house can bless the native with a long life; very good interest and achievements in the fields like paranormal and supernatural and becoming astrologers, psychics, magicians; learning of hidden knowledge and making big amount of money and wealth as well as fame from such practices; happy married life with good understanding. On the other hand, a negative exalted Mercury in 8th house can witness many hurdles, problems and losses in his professions; bad for the education; discontinuing his studies at an early age; no happiness from his children; facing problems with children and separation from his children; suffering of professional failure of his children, failing marriage of his children or some other problems of children; disturbed peace of mind.

(ii) Debilitated Mercury in 8th House: A benefic debilitated Mercury in 8th house can bless him with great financial gains through many different ways or gains a considerable amount of wealth or business through inheritance of will or profit on account of gifts, help, stock market gains, lotteries; position of great power and authority in some government or private organization; very good results in the fields of paranormal and spiritualism becoming successful, wealthy and famous as astrologers, numerologists, psychics, magicians, spiritual healers, spiritual gurus and other people of similar kind. On the other hand, a malefic debilitated Mercury in 8th house can cause problems in the married life separating him from parental family after marriage on account of serious misunderstandings between his wife and his family or faces deception or betrayal in his marriage and many other kinds of problems; develop criminal tendencies in him and become criminals, murderers, arms dealers, terrorists and other such kinds of people; problems in professional life or health related problems to him.

Mercury in 9th Houses:

(i) Exalted Mercury in 9th House: A positive exalted Mercury placed in ninth house can bless the native with wealth, named and fame and good support passed on to him by his father or grandfather throughout his life; birth in rich families; his fathers wealthy, successful and well reputed in the society; going to foreign lands for the purpose of permanent settlement and settle in some foreign country for making fortunes; possessing a spiritual and religious nature and getting a position of power and authority in spiritual or religious organization. On the other hand, negative exalted Mercury in 9th house can trouble him with serious problems; facing many delays and losses in his professional careers; very strong enemies keeping on bothering ad troubling and making him unsuccessful throughout his life; suffering from permanent or long lasting serious disease giving much physical pain and proving fatal for him; financial losses and financial crisis and debts many times in his lives.

(ii) Debilitated Mercury in 9th House: Benefic debilitated Mercury in 9th house can give very good results regarding the professional success to him and tend to do jobs than businesses; good creative skills; and gains through court decisions or wills. On the other hand, malefic debilitated Mercury in 9th house forms a Pitra Dosha in the horoscope and can cause serious problems for him like remaining jobless many times in life, facing disgrace due to allegations imposed by his bosses, facing court cases filed by his employer or facing many other problems related to his professional sphere; face serious financial crisis, loans, debts and many other such things for a very long period of his life.

Mercury in 10th Houses:

(i) Exalted Mercury in 10th House: Positive exalted Mercury placed in 10th house can give excellent and great professional skills and professional success and setting new standards and milestones in his profession by his extraordinary achievements; life of luxury and comfort and great name and fame to him; getting very skilled, well behaved, social and understanding wife providing great support and becoming a pillar of strength for him; position of great power and authority to him in private or in the government through a job or through politics. On the other hand, negative exalted Mercury in 10th house can cause professional failures or delays or professional declines many times in his life; defamation or social disgrace due to some misconduct done by him leading to a complete professional failure or a resignation from a position of power; spoiling the relations with his wife and relatives and suffering from mental tension.

(ii) Debitated Mercury in 10th House: A benefic debilitated Mercury in 10th house can give very good results regarding the professional success as reporters, media persons, speakers, lawyers, judges, astrologers, spiritual healers, computer programmers, web designers, travel agents, engineers, researchers, analysts, police officers, bankers, financial advisors, stock market agents or dealers and people associated with travel trades, a life full of luxuries and comforts and purchase of houses and vehicles with his own income; name, fame and status like become ministers with the government under the strong influence of this benefic placement of Mercury. On the other hand, malefic debilitated Mercury in 10th house can cause problems in the professional sphere like delays, setbacks and losses in his professions; brings bad name or disgrace by virtue of his involvement in a scandal and suffering a lot due to his falsely implicated in many such cases; cause problems in various kinds of issues in his married life; and financial problems as well as health related problems.

Mercury in 11th Houses:

(i) Exalted Mercury in 11th House: Positive exalted Mercury placed in eleventh house form a powerful Dhan Yoga in his horoscope and bless the native with great financial gains and great financial benefits, money and wealth by means of trade and business throughout his life; financial gains through lottery, gambling, stock markets and other such practices; very good social circle and great success with the help of his friends and relatives and especially his brothers for the overall lifespan; living longer; happy married life and increase in financial fortune after his marriage. On the other hand, negative exalted Mercury in 11th house can cause delays and losses in professional matters; facing big financial losses either through his professions or through some kind of gambling or stock market; problems in married life related to his children; choosing illegal professions to make money and suffering defame or a jail sentence due to such activities; injuries due to accidents; having to lose a limb in an accident.

(ii) Debilitated Mercury in 11th House: A benefic debilitated Mercury in 11th house can bless him with very good financial gains in his profession or through other sources; or earns money through stock markets, lotteries, game shows, gifts and other such ways of making instant or fast money; leads a financially secured life and no major financial crisis throughout his life; renders name, fame and social status or becomes successful and famous as TV reporters, TV show hosts, movies actors, producers and other similar kind of people. On the other hand, malefic debilitated Mercury in 11th house can cause serious problems in the married life by his argumentative and suspicious nature and keeping on doubting his husband or wife for no solid reasons and thus creating problems in his married life leading to divorces or long separations; problems in his profession or child birth or delays or setbacks related to professional sphere or child birth.

Mercury in 12th Houses:

(i) Exalted Mercury in 12th House: Positive exalted Mercury placed in 12th house can take him to foreign land and settling in foreign country for his profession; some going and settling in foreign land during his childhood on the basis of his father being a permanent resident of a foreign country; very good spiritual interest and progress and achieving great progress due to his good social skills and having a big network of friends. On the other hand, a negative exalted Mercury in 12th house can cause serious financial problems and short of money most of the times; developing a tendency to spend much more than income; financial crisis; unwanted expenses and loss of money through diseases, accidents, court decisions and other such forms; debts that his fathers have passed onto him or towards treatment of disease which afflicts his father; problems in married life facing a long separation from his wife or a divorce.

(ii) Debilitated Mercury in 12th House: A debilitated Mercury in 12th house can take him to foreign lands and settling there permanently on professional basis and this aspect is further enhanced if Moon is also debilitated along with Mercury in the horoscopes; renders creative abilities to get success and recognition on international level by virtue of his creative abilities and such a fame and recognition is further supported if Venus is present in Taurus or Libra or if Jupiter is present in Sagittarius sign in such horoscope; very good progress in spiritual as well as paranormal fields and makes renowned astrologers, spiritual healers and other people of this kind. On the other hand, malefic debilitated Mercury in 12th house can cause problems in the married life like living away from his partners for a long period of time or faces serious disputes and litigations in his married life accompanied by physical violence and hospitalization; causes serious concerns with financial aspect causing serious losses and heavy financial debts to him; the most serious health problems or diseases causing much physical as well as mental pain to him and causing the death to him in

(5) Effects of Exalted & Debilitated Jupiter in 12 Houses:

In Vedic astrology, Jupiter is said to be exalted in a horoscope when it is placed in the sign of Cancer and He gains maximum strength. Jupiter is said to be debilitated in a horoscope when it is placed in the sign of Capricorn which means that Jupiter becomes weakest in Capricorn compared to its placement in all the other signs.

Jupiter in 1st Houses:

(i) Exalted Jupiter in 1st House: A benefic exalted Jupiter in 1st house can bless the native with very good financial prosperity; support from his father or grandfather and receives wealth in inheritance; very good name, fame, and power; spiritual and religious interests; very good professional skills and success; and successfulness in business. On the other hand a malefic Jupiter in 1st house can trouble the native with problems in his married life; suffering from serious diseases causing death of the native; having many enemies and physical injuries or even death by planned attack of his enemies.

1. Debilitated Jupiter in 1st House: A benefic debilitated Jupiter in 1st house can render a very strong character and personality and successful in profession like army, police force, sports persons, fitness trainers, coaches by virtue of his own hard work and efforts and he does not like to get success as inheritance or gift; make a successful businessmen by virtue of his own skills, calculative nature, perseverance, discipline and other qualities instead of owning a well settled business by means of inheritance. On the other hand, a malefic debilitated Jupiter can cause serious problems in the married life arising due to serious compatibility issues between the two partners; affect his religious and spiritual faiths and so not believing in religion, spirituality or even God; can afflict him with serious physical disease troubling for a very long period of time and damaging a limb or organ of his body.

Jupiter in 2nd Houses:

(i) Exalted Jupiter in 2nd House: A benefic exalted Jupiter in second house can bless the native with wealth; huge money and wealth through easy means; very good married life and getting married to very rich females who bring a great amount of financial fortune; very well behaved and good natured wives; very good house property; cars; luxuries and other luxurious things. On the other hand, a malefic exalted Jupiter in second house can bring serious problems in the married life; getting wife with very suspicious and troubling nature by virtue of their suspicious nature; suffering from a serious disease for a very long part of his life and death due to these diseases.

(ii) Debilitated Jupiter in 2nd House: A benefic debilitated Jupiter in 2nd house can bless the native with professional success, wealth and a life full of luxuries; is born in rich and influential families and enjoying the luxuries of his life right from his birth; a good marriage with an understanding life partners providing more strength to him and raising his social status after marriage; and name, fame and a position of power with some government body or some private organization. On the other hand, a malefic debilitated Jupiter in 2nd house can trouble the native on account of health problems and diseases; born with some health problem or disease which usually keeps on disturbing him throughout his life; developing permanent disease in the middle years of his life; problems in married life; getting addicted to excessive drinking, drugs, womanizing and other such vices which can make him suffer losses on account of finances, relationships and health.

Jupiter in 3rd Houses:

(i) Exalted Jupiter in 3rd House: A positive exalted Jupiter in third house can bless the native with very good financial gains by virtue of his own hard work; very good warrior skills who may become a renowned warrior or war expert; going to foreign countries and settling there for profession; and good creative abilities and success. On the other hand, a malefic exalted Jupiter in 3rd house can trouble the native with many kinds of problems, losses, waste and cheating and becoming enemies by friends, relatives and colleagues; financial losses by friend borrowing money and refusing to return it; giving guarantee for a loan which is not paid by his friends and relatives and in turn they do not stand by this native when he is in real need of them; losses due to professionally failure frequently; death in a war or fight and getting murdered by some unknown persons.

(ii) Debilitated Jupiter in 3rd House: A benefic debilitated Jupiter in 3rd house are seen successful in various kinds of professions like teachers, bankers, people working with financial institutions, advisors, managers; give very good understanding of supernatural and paranormal becoming astrologers, numerologists, psychics, magicians, spiritual healers; gives a position of power and authority with government and becomes officers with government, ministers, members of parliament, members of assemblies. On the other hand, a malefic debilitated Jupiter in 3rd house can cause problems or painful experiences in married life; causes differences between him and his brothers, sisters or friends and may have to part with his brother or sister; cause financial concerns due to his bad spending habits and health related issues.

Jupiter in 4th Houses:

(i) Exalted Jupiter in 4th House: An benefic exalted Jupiter in fourth house can bless the native with a luxurious life; getting birth in very rich families enjoying luxurious life right from his childhood; comforts and easy life; keep on getting money and other resources continuously throughout his lives; connection to foreign lands and receiving regular incomes from a trade established in foreign countries or some kind of import export business; living happily throughout his life and easy and peaceful death without any chronic disease or any other kind of painful death. On the other hand, a negative exalted Jupiter in 4th house can cause serious problems in the married life; problems in marriage and divorces; losses on account of disputes or litigations related to his residential property; serious health concerns and mental unstablity and may have to spend a long time in mental asylum for treatment of such disorders.

(ii) Debilitated Jupiter in 4th House: A benefic debilitated Jupiter in 4th house can give good professional success and financial prosperity when working in groups rather than working alone like networking and group work and social skills. On the other hand, a malefic debilitated Jupiter in 4th house can cause serious, very disturbing and shocking experiences in married life like serious arguments, exchange of bad words, physical violence, litigations, divorce; gives bad results regarding the profession like frequent job changes, remaining jobless for a long period of time; losses through business; causes loss of wealth and property through thefts, accidents, litigations and court decisions and disputes in residential houses and loss of property through accidents.

Jupiter in 5th Houses:

(i) Exalted Jupiter in 5th House: A positive exalted Jupiter in fifth house can bless the native with very good creative skills, higher education and shining by virtue of the same; very good financial prosperity; very good recognition for his works; making new discoveries and setting new standards; very good spiritual progress and connecting to the supernatural and paranormal worlds and practicing as spiritual gurus, spiritual healers, astrologers; possession of very good business skills and making big money from business; very good, creative and spiritually advanced children. On the other hand, a negative Jupiter in 5th house can bring problems; many broken love relationships; having not any love life at all and remaining single for a big part of life; problems in the education; discontinuation in his studies; leaving his family and going to Jungle or asylum due to his professional failures and family problems.

(ii) Debilitated Jupiter in 5th House: A benefic debilitated Jupiter in 5th house can give very good results regarding the professional success and financial prosperity; becomes successful as writers, actors, designers, architects, interior decorators dealing in profession requiring creative abilities; has a beautiful and loving wife by a love marriage or a marriage of his own choice instead of going for an arranged marriage; renders name, fame and fortune to him by his creative abilities or due to his professional success. On the other hand, a malefic debilitated Jupiter in 5th house can cause problems in the marriage and married life; performs an abnormal marriage against the wish of his parents or leaves the house of his parents in order to perform a marriage; faces serious problems and dramas in his married life, but does not break the marriage; causes financial and health problems psychologically.

Jupiter in 6th Houses:

(i) Exalted Jupiter in 6th House: A benefic exalted Jupiter in sixth house is considered bad by many astrologers. But, in fact, Jupiter can bless the native with great professional success; achieving great milestones in professional life; large amounts of money gain, luxuries and comforts and living very luxurious and comfortable life; very good creative abilities like a successful professional dancers, musicians, actors, stage artists, poets and writers and in the fields like astrology, palmistry, psychic readings and other such fields. On the other hand, a malefic exalted Jupiter placed in 6th house can bring serious financial problems; bad financial circumstances for a big part of his life; very heavy financial debts throughout his life; serious diseases; spending lots of money towards the treatment of diseases and even death; court cases and litigations and the native may even have to go to jail due to some court case or litigation.

(ii) Debilitated Jupiter in 6th House: A benefic debilitated Jupiter in 6th house can give good results regarding the profession as police officers, army officers, doctors, lawyers, engineers, managers or people associated with managerial jobs; blesses him with interest, knowledge and progress in the field of paranormal and occult as astrologers, psychics, black magicians, mediums. On the other hand, a malefic debilitated Jupiter in 6th house can engage him in illegal or immoral activities facing litigations, defamation and loss of reputation and spends a few years in jail; causes problems in married life like facing long court cases and going to jail due to complaints filed by his spouse; problems in his love life with one or two broken love affairs or betrayal by his lover

Jupiter in 7th Houses:

(i) Exalted Jupiter in 7th House: A benefic and exalted Jupiter in seventh house of a horoscope is considered very beneficial for marriage and married life; leading an undisturbed and happy married life; very good professional success and doing better in jobs compared to business; getting favours form his seniors or bosses and success in his job by virtue of these favours. On the other hand a malefic exalted Jupiter in 7th house can cause serious problems in the married life; getting married to altogether opposite personality and suffering great misunderstanding and problems in the earlier years of his married life; increased expenditures; very bad financial situations; separation with his wife and family due to serious misunderstandings; going to some asylum and leading the life of an ascetic; social disgrace and health problems.

(ii) Debilitated Jupiter in 7th House: A benefic debilitated Jupiter in 7th house can bless the native with a good marriage at proper ages and happily married life and rise in fortune in profession after marriage; very good financial gains; good results in professions and big profit in own business. On the other hand, a very strong malefic debilitated Jupiter in 7th house can cause the most serious problems in the married life like going through prolonged court cases, social drama, division of his property as a result of settlement towards marriage, physical violence, social disgrace and many other problems; breaks one, two, three or more of his marriages and such a malefic placement almost denies his marital pleasures and so some persons under strong influence of such a malefic placement of debilitated Jupiter choose to remain single after a couple of very badly broken marriages; also gives at least one permanent disease troubling him for a big part of his life and which can claim his life in the end and also bring very bad results for his financial condition.

Jupiter in 8th Houses:

(i) Exalted Jupiter in 8th House: An exalted Jupiter in 8th house of a horoscope is considered very bad by many Vedic astrologers and it is believed that such a placement always brings problems for the native. But in actual practice, a positive exalted Jupiter in eighth house can bless the native with good knowledge; wisdom and deep interest in paranormal and supernatural; great status and great supernatural powers and possessing a chair of power or authority by virtue of his profession. On the other hand, a malefic exalted Jupiter in

8th house can bring serious problems in married life; delay of marriage; marrying very late in his life; shorter lifespan; unnatural death; professional delays and addiction to alcohol, drugs and suffering from some serious disease.

(ii)   Debilitated Jupiter in 8th House: A benefic debilitated Jupiter placed in 8th house can give very good results regarding his income and keeping on profiting from other sources for their entire life; easy income from a comparatively comfortable jobs; post of power and authority with government when Jupiter is supported by the overall horoscope. On the other hand, a malefic debilitated Jupiter in 8th house can cause altogether different kinds of serious problems in the married life like getting married to a mentally abnormal person or a physically disable person or his spouse may become physically disable soon after marriage, he may start experiencing psychological disorders after marriage. This indicates Karmic debt of his wife towards him which he is supposed to pay in this life by serving a physically or mentally disabled wife. A malefic placement of debilitated Jupiter also gives at least one permanent disease to him after thirty years of age staying permanently with native.

Jupiter in 9th Houses:

(i)   Exalted Jupiter in 9th House: A positive exalted Jupiter in ninth house can bless the native with huge money and wealth; benefit from his parental wealth; great professional success; huge financial gain through his profession; good reputation in the society; very good spiritual and religious interests and becoming famous and successful in spiritual fields; very favourable for political gains; great positions of power through politics; and becoming very popular political or religious leaders having a huge number of followers. On the other hand, a malefic exalted Jupiter in 9th house can bless with downfall in his life due to bad characteristics or bad habits; becoming overconfident or proud by virtue of knowledge and power and committing serious mistakes; big loss in terms of money, reputation, power and position; facing a very bad reputation and public disgrace by involvement in a scandal being exposed and losing his reputation, power and success forever; and health related problems.

(ii)   Debilitated Jupiter in 9th House: A benefic debilitated Jupiter in 9th house can render very good financial gains and good permanent professional success making very good money from some sidewise business; very good results in the spiritual progress working for some religious of spiritual organization; gains from stock markets or gambling. On the other hand, a malefic debilitated Jupiter in 9th house can form a Pitra Dosha and give very bad results for the professional sphere seeing delays, losses and setbacks in his professions; give very bad results in married life facing many problems in his married life.

Jupiter in 10th Houses:

(i) Exalted Jupiter in 10th House: A positive exalted Jupiter placed in tenth house can bless the native with very good professional success; earning huge amount of money and wealth with his own efforts and skills; going to foreign lands and achieving success and money and wealth in foreign lands; benefit from professions associated with travels in foreign countries; success in profession of legal fields, medical fields, consultancy; great mental and speech abilities. On the other hand, a malefic exalted Jupiter in 10th house can cause suffering from loss of wealth, property and reputation by means of court cases and litigations; spending some time in jail due to a court decision going against him; serious concerns related to the peace of mind; anxiety, restlessness, insomnia or difficulty in sleeping or nervous system; and financial debts in his life.

(ii) Debilitated Jupiter in 10th House: A benefic debilitated Jupiter placed in 10th house can give very good results in the professional success; runs his father business or choose the same profession as his father used to engage in; can take him to foreign and settling in foreign countries on the basis of his professions, and though he is not likely to settle there permanently but he can spends many years of his life in foreign countries; can also render name and fame to him; bless him with a position of power and authority with some government organization or it can render a post of authority and status in some big private organization under its strong influence. On the other hand, a malefic debilitated Jupiter in 10th house can afflict him with criminal tendencies engaging him in criminal and illegal activities for a very long period of his lives and going to jail for engaging in such activities under its strong influence; can also trouble him with delays and setbacks in professional sphere bringing disgrace and defame to him and the later part of his life is usually very bad in terms of finances, health and reputation.

Jupiter in 11th Houses:

(i) Exalted Jupiter in 11th House: A positive exalted Jupiter placed in eleventh house can give great financial gains; good luck, very good professional skills and good convincing ability; very good married life; wife belonging to a more rich family; financial success and fortune by wife; achieving more after marriage; very supportive wives having good understanding of business; wife giving good advices; very good public reputation and great success. On the other hand, a malefic exalted Jupiter in 11th house can trouble the native by ill health of his mother and spending a big amount of money, resources and time on mother's treatment; unwanted and unexpected problems and trying again and again in order to achieve success; affecting the married life through very unsupportive wife having very harsh and doubtful nature; and health problems.

(ii)   Debilitated Jupiter in 11th House: A benefic debilitated Jupiter in 11th house can give very good results regarding the professional success and financial prosperity rising in his life due to his own efforts and luck; are self made man having the ability to start from a scratch and go to the top of a mountain; can render qualities like, name and fame, wisdom, keen business sense, sound judgment, benevolent nature and many other good qualities and accordingly he can rise in his life by virtue of these qualities; can get a position of status and authority with government and most of them are seen practicing as Judges, administrators or using his sense of Judgment in day to day work. On the other hand, a malefic debilitated Jupiter in 11th house can hinder the professional progress experiencing delays, setbacks and frequent changes in his professions or starting his careers late in his life; problems in married life; permanent disease keeping on bothering him for a very long period of his life.

Jupiter in 12th Houses:

(i)   Exalted Jupiter in 12th House: A positive exalted Jupiter placed in twelfth house can bless the native with good education and very good creative abilities and possessing multiple talents; going abroad for higher education and settling there after completing his studies; very good spiritual achievements and possessing supernatural powers. On the other hand, a malefic exalted Jupiter in 12th house can make the native suffer to his married life and profession and remaining jobless or without a job despite of possessing very good professional qualifications and skills; not producing healthy and lucky children and the children may become a cause of concern by virtue of failures in their married life, professional life, bad company or health problems; resigning from his family life and taking up the life of an ascetic and practice their spiritual interests forsaking his families forever and devote his life to spiritualism particularly in the later parts of his lives and making him wandering here and there for no solid reason and spending much money and much time on travelling from one place to another without any constructive and solid results.

(ii)   Debilitated Jupiter in 12th House: A benefic debilitated Jupiter in 12th house can take him to foreign lands for settlement on professional basis at early age and settling there on professional basis or not going to foreign countries but earning from the professions related to foreign countries or benefiting from foreign countries in one way or the other; supernatural and paranormal knowledge and becoming very highly advanced but hiding himself. On the other hand, a malefic debilitated Jupiter in 12th house can cause serious financial problems like poor financial management and spending much more than his income leading him into financial crisis many times in life; may be earning in millions in a very short period of time due to some other good Yoga in his horoscope but he is not able to save anything out of this multi millions earnings and soon after he has to face financial crisis due to poor financial management; has a strong tendency to eat up all his income under the strong influence of this benefic placement; trouble him with problems related to married life, family and family affairs, reputation and status and permanent diseases to him under its influence.

(6) Effects of Exalted & Debilitated Venus in 12 Houses:

Venus is exalted in the sign of Pisces, which means that Venus gains maximum strength in Pisces compared to its placement in all the other signs. Venus is debilitated in a horoscope when it is placed in the sign of Virgo which means that Venus becomes weakest.

Venus in 1st Houses:

(i) Exalted Venus in 1st House: A positive exalted Venus in first house of a horoscope can bless the native with very good facial features; having very beautiful and attractive face; very good creative and artistic skills and shining in fine arts; long lifespan; very good spiritual interests in spiritualism, paranormal and occult. On the other hand, a malefic exalted Venus in 1st house can trouble the native on account of physical problems and health right from his childhood having weak bodies; taking excessive interest in the pleasures of senses which can bring serious consequences; very strong sexual urges and coming in contact with many women in order to satisfy his excessive physical desires and in doing wasting a big amount of money and having disease sexually transmitted diseases; very bad reputation or defamation due to sexual scandal; and problems in married life.

(ii) Debilitated Venus in 1st House: A benefic debilitated Venus in 1st house can render abilities like intellect, charm, magnetism, good communications skills, diplomatic nature, calculative brain and other abilities achieving success by virtue of these qualities and practicing as businessmen, trade oriented professionals, research and analysis specialist, scientists, engineers, models, architects, interior decorators, dress designers and many other professionals of similar kind; ability to convince people and climbing the ladder of success and very good at attracting and convincing the people from opposite sex and as a results many such natives are seen more successful dealing with people of opposite sex, compared to people of the same sex. On the other hand, a malefic debilitated Venus in 1st house can develop over ambitiousness and extreme materialistic tendencies in him wasting a considerable amount of his time, money and resources in pursuit of physical pleasures, sexual pleasures and going to any limits to fulfil his sexual needs resulting in losses, problems in business, marriage, health and reputation and going to jail; problems in the married life and break his marriage in some cases due to extra marital affairs.

Venus in 2nd Houses:

(i)   Exalted Venus in 2nd House: A positive exalted Venus in second house can make a strong Dhan Yoga with big amount of money and wealth and established business of father or his paternal family; very good mother and the wealth passed on to him by his mother or mother's family; born in very rich families and gain from his father as well as his mother's family; very loving father and mother; very good professional success in singing and becoming very famous and popular among the people; placement of benefic exalted Venus in second house forms a powerful Raj Yoga; and getting position of power and authority. On the other hand, a malefic exalted Venus in 2nd house can cause many problems in profession; professional difficulties; complete professional failure some time; professional failure; involvement in some scandal; financial losses and increased expenditures and problems in his married life.

(ii)   Debilitated Venus in 2nd House: A benefic debilitated Venus in 2nd house can give very good results regarding financial prosperity, professional success and success in various kinds of professions; name, fame, recognition and status and position of power and authority like government officers, powerful political leaders and other people of such kind; good creative, management and administrative abilities helping him in becoming successful in various professional spheres. On the other hand, a malefic debilitated Venus in 2nd house can cause problems in married life due to altogether different natures of the partners; give concerns related to professional sphere; serious financial concerns many times in his life; suffers from his relations going bad with his wife, brothers, sisters, friends and some other relative and so has strong tendency to stay away from social relations and networking and he prefers to be alone ultimately leading him into depression.

Venus in 3rd Houses:

(i) Exalted Venus in 3rd House: A positive exalted Venus in third house can bless the native with very good professional success in writing, speech and creative abilities; settling abroad permanently on professional basis; professions related to foreign countries; travelling to foreign countries for professional purposes; doing better in business than in service or job; making great amounts of wealth and money from business related to foreign country; very capable, hard working and understanding children and such children support the native through thick and thin. On the other hand, a malefic exalted Venus in 3rd house can trouble the native on account of his bad relations with his brothers or relatives; facing serious problems due to some bad done to him by his relatives; serious problems in the married life by virtue of serious misunderstandings with his wife; a divorce; problems in the professional sphere; professional failure at one point in his life; and change of profession or relocating to a different location.

(ii) Debilitated Venus in 3rd House: A benefic debilitated Venus in 3rd house can bless him with professional success and financial prosperity and becoming successful in businesses of various kinds or shining in creative fields like, acting, modelling, dress designing, fashion industry and other such kinds of field related to creativity as well as glamour; and the ability to attract and charm people, particularly the people from opposite sex. On the other hand, a malefic debilitated Venus in 3rd house can cause problems in the married life like marrying late or very late; or marrying at proper age or even an early age but ending up in a mess or divorce; are very self centred and does not have much time and concern for people around him and at the same time he wants everyone around him to focus on him which lands him into many kinds of problems; and has bad health which generally show up in the 2nd half of his life.

Venus in 4th Houses:

(i) Exalted Venus in 4th House: A benefic exalted Venus placed in 4th house can bless the native with great financial prosperity; great fortunes; creative abilities, professional skills, patience and perseverance; good support from his mother; beautiful and caring wife; happy married life; shining in professional fields of creative talents as actors, musicians, writers, models. On the other hand, a malefic exalted Venus placed in 4th house can bring serious problems since childhood; spending a big part of his childhood away from his mother; serious problems in married life and living far away from his wife for a very long time; disturbed peace of mind; forsaking his wife and family and seeking refuge to asylum or some Jungle; suffering from serious disease related to nervous system or psychological disorder and staying in hospital for a long time for treatment of such disease.

(ii)   Debilitated Venus in 4th House: A benefic debilitated Venus in 4th house can bless him with good creative abilities, charming personality and successful in various professional spheres like lawyers, judges, doctors, models, astrologers, people dealing in food products selling his product with a pleasant smile instead of spending so many hours or days to convince his clients. On the other hand, a malefic debilitated Venus in 4th house of a horoscope can cause serious problems in married life due to mental incompatibility between the partners resulting like exchange of very bad words, physical violence, police stations, court rooms and much more; painful and permanent disease around 35 years of his life like surgeries, hospitalization claiming the his life at a comparatively early age where this malefic influence of debilitated Venus in 4th house is very strong.

Venus in 5th Houses:

(i)   Exalted Venus in 5th House: A positive exalted Venus in fifth house can bless the native with a very good married life with beautiful, wise and religious wife; love marriage; very good at social and professional skills; which becoming very good salesmen related to marketing field achieving good professional heights due to marketing abilities; very good spiritual interest visiting holy places and doing pilgrimages throughout his life; spending the last years of his life in some holy place of religious or spiritual importance. On the other hand, a malefic exalted Venus in 5th house can bring serious problems in love life; facing social disgrace or litigation or court case due to one or more of his love affairs; performing his marriage with his lover after running away from his families and facing legal litigations after marriage or staying away from his families for a very long period of time; delay and health related problems of birth of children; effects on his education or discontinuing his studies; suffering from a serious and permanent disease which cannot be cured.

(ii)   Debilitated Venus in 5th House: A benefic debilitated Venus in 5th house can bless him with beauty, charms, creativity, charisma and magnetism and accordingly succeeding in professional spheres like, modelling, acting, movies, theatre and other such fields requiring glamour and beauty; give many love affairs to him and benefiting or rising on account of his love affairs with people who help him achieve success in his professions and hence this placement is the best placements for having love life, enjoying the company of beautiful and influential people and then benefiting from that company. On the other hand, a malefic debilitated Venus in 5th house can cause problems in love life as well as married life; suffering financial losses, setbacks related to his careers, disgrace or defame and bitter experiences which keep on bothering him for a very long period of time on account of marriage and love affairs; cause problems and concern regarding child birth; professional delays and setbacks and health problems.

Venus in 6th Houses:

(i) Exalted Venus in 6th House: A positive exalted Venus in sixth house can bless the native with an attractive personality attracting the opposite sex very easily; very good social skills and a big social circle; successful in the fields like modelling, acting, singing reaching great professional heights; becoming a very successful, famous and popular actor or actress and he or she can make big amount of money and wealth by virtue of his or her profession; success in fields like medical profession, cosmetic industry, and hotel industry. On the other hand, a malefic exalted Venus in 6th house can trouble the native on account of legal litigations, court cases and defame due to some kind of scandal; affecting the married life or marrying very late or serious problems in his married life; more than one divorces and going through long legal litigations and drama to get his divorces finalized; affliction to diseases spending much time, money and many resources for treatment of such diseases and diseases may prove fatal for him; going to jail on account of some litigation or court cases on his professional cause.

(ii) Debilitated Venus in 6th House: A benefic debilitated Venus in 6th house can make him a police officer, an army officer, a lawyer, a Judge, a doctor, especially a gynaecologist, and many other kinds of professionals related to medicine and therapies. On the other hand, a malefic debilitated Venus in 6th house can suffer a lot in his married life and problems related to finances and health related issues.

Venus in 7th Houses:

(i) Exalted Venus in 7th House: A positive exalted Venus in seventh house of a horoscope can bless native marrying a rich and supportive wife; going to foreign countries and settling abroad on permanent basis after his marriage; or marrying someone from a foreign country; professional success; good creative ability and imagination power like dress designers, architects or interior decorators or jewellery and gemstones designers; getting good support from his father. On the other hand, a malefic exalted Venus in 7th house can cause loss of money and health; wasting a big part of money and wealth for fulfilment of his physical and materialistic desires; problems in married life due to baseless doubts on his wife.

(ii)   Debilitated Venus in 7th House: A benefic debilitated Venus placed in 7th house of a horoscope can give very good results in the professional success like big business rather than jobs and having more than one businesses; many luxuries, comforts, easy money; enjoying travels to foreign countries for the purpose of leisure and entertainment or establishing overseas business which can bless him with good profits. On the other hand, a malefic debilitated Venus in 7th house can make the native indulge in excessive physical pleasures and seeking the company of paid sex workers on regular basis to fulfil his sexual desires; cause serious problems in his marriage and married life like marrying late, breaks in marriage or his love relation under the strong influence of this malefic placement of debilitated Venus.

Venus in 8th Houses:

(i)   Exalted Venus in 8th House: A positive exalted Venus placed in eighth house of a horoscope can make the native succeed in professional fields like astrology, palmistry, numerology, yoga and meditation teaching; a good profession or well settled business passed on to him by his father or grandfather; inheritance of big business empires from his paternal family; long life and attracting the opposite sex; financial gains by attachment or affairs with a very rich and powerful females; physical pleasures; and very luxurious and comfortable life. On the other hand, a negative exalted Venus placed in 8th house can cause serious problems in married life; breaking of marriage by his extramarital affairs being exposed to his wife; involvement in excessive sexual pursuits and financial losses; facing a bad reputation due to his involvement in some kind of sex scandal; suffering from serious diseases and death of the native; and death in a fight or war with a weapon.

(ii)   Debilitated Venus in 8th House: A benefic debilitated Venus in 8th house can bless him with personality full of charm and magnetism and making very popular among people around him and attracting people from the opposite sex; very beautiful wife with a love marriage after a short period of courtship; very good professional success like becoming very successful businessmen expanding his business to foreign countries. On the other hand, a malefic debilitated Venus in 8th house of a horoscope can cause problems in marriage and married life like suffering on account of infidelity of spouse or facing a big financial deception or fraud from his spouse and permanent disease at the age around 30 years bothering him till the end of his life.

Venus in 9th Houses:

(i) Exalted Venus in 9th House: A positive exalted Venus in ninth house can bless the native with great amount of wealth and position of power passed on to him by his father as the father is likely to be very rich and powerful person; good support from mother; great success in business, position of power or authority; happy married life; great wealth in the form of big and luxurious houses, top class cars and vehicles and many other kinds of luxuries and travelling to foreign countries primarily for the purpose of enjoyment; owning permanent residences and luxurious vehicles in more than one countries of the world; good health, sound peace of mind and religious nature. On the other hand, a malefic exalted Venus in 9th house can bring serious problems in married life like clash in family; spending much time and money to towards the treatment of diseases which his mother or father may be suffering from; and bad reputation in cases and false allegations.

(ii) Debilitated Venus in 9th House: A benefic debilitated Venus in 9th house can bring very good all around results in profession, personal life, love life and many other spheres of his lives; take him to foreign land or some settling permanently in foreign lands blessing him with name, fame, status, authority and a position of power with government or private organization. On the other hand, a malefic debilitated Venus in 9th house can form a Pitra dosh in the horoscope troubling him on account of professional setbacks and delays and spoiling his relations on account of misunderstandings.

Venus in 10th Houses:

(i) Exalted Venus in 10th House: A positive exalted Venus placed in tenth house can give excellent results related to the profession and achieving great milestones; going to foreign lands for higher studies and permanently settling abroad after completing studies; very good academic education as well as creative skills; in case of females, she is capable of conceiving and delivering very healthy children; shining in astrology, graphic designing, interior decoration, textile business; successful models, actors, air hostesses, TV anchors and other similar kind of professionals. On the other hand, a negative exalted Venus in 10th house can bring problems in professional life encountering many obstacles; financial losses; serious problems in married life; suffering long separation or divorces due to such natives are so busy with his professional life that he is not able to give any time to his marriage and as a result their marriage can fail; bad children becoming his opponents.

(ii)   Debilitated Venus in 10th House: A benefic debilitated Venus in 10th house can bless him with very good professional results in professions related to medicines, hotels, law, food products, textiles, trading of some kind; gains from brothers, friends and sisters and name and fame to him. On the other hand, a malefic debilitated Venus in 10th house can create problems in professional spheres facing losses in his professions primarily due to wrong decisions taken by him or some suffering from remaining jobless for a good period of time, many times in his life; make him engaged in criminal professions and bringing disgrace and jail for him.

Venus in 11th Houses:

(i)   Exalted Venus in 11th House: A positive exalted Venus in eleventh house can bless the native with great financial prosperity; big gains from his profession; very charming and attractive to climb the ladder of success; getting very good results from people belonging to opposite sex; getting married to a very rich lady and as a result such a native can become very rich in a very short time; getting attention from very rich and powerful members of opposite sex; clever and charismatic benefited even from his enemies; gains through legal cases and litigations and becoming very rich by virtue of a court decision going in his favour. On the other hand, a malefic exalted Venus in 11th house can cause serious problems in married life; a long separation from his wife, a court case filed against him by his wife, a divorce and many other problems related to marriage and married life; spending some time in Jail by virtue of his wife filing the criminal case against him and giving a big amount of wealth to wife due to a court decision going against him; legal litigations in profession; suffering from a fatal disease spending much money and resources and mental agony.

(ii)   Debilitated Venus in 11th House: A benefic debilitated Venus in 11th house can take him to foreign lands or settling abroad and spending a very big part of his life in foreign countries; can marry someone from a foreign country and then migrating to that country where he may spend the rest of his life under the strong influence of benefic placement of debilitated Venus; bless him with wealth and prosperity, many after his marriage. On the other hand, a malefic debilitated Venus in 11th house can delay his marriage or marrying late or very late in his life or some facing many issues and problems in his marriage due to his aggressive nature and or due to difference of opinion between the partners; sudden financial losses through his profession or through other means facing financial crisis at least once in his life.

Venus in 12th Houses:

(i) Exalted Venus in 12th House: A positive exalted Venus in twelfth house of a horoscope can bless the native with a beautiful wife of a foreign origin and his settling to foreign country on permanent basis and becoming financially much stronger and more capable; happy married life; very good professional results and settling in foreign land on permanent basis; very good results in spiritualism and paranormal and choosing this fields as his profession or lifetime practice. On the other hand, a negative exalted Venus in 12th house can cause problems in the married life; getting married to a sexually much more or much less active woman than him; getting married to a women who is not loyal to him and cheating him and wife may running away with other person; social disgrace and mental pain; expenses much more than the overall income; spending big amount of money on disease or problem related to his wife or other family member of his wife.

(ii) Debilitated Venus in 12th House: A benefic debilitated Venus in 12th house can take him to foreign lands on the basis of permanent settlement or some setting office in foreign countries under the strong influence of such a benefic placement of debilitated Venus; give good results regarding the religious interests and social skills engaging in professions which are based on social networking and relations. On the other hand, a malefic debilitated Venus in 12th house can delay or disturb his marriage and married life by either marrying late or facing many problems in the first years of his marriages; gives disease to him suffering a lot on account of health or diseases throughout his life around age of 25 or some dieing due to disease after suffering much pain and spending a big part of his money towards the treatment of these diseases under the strong influence of this malefic placement of debilitated Venus.

(7) Effects of Exalted & Debilitated Saturn in 12 Houses:

According to Vedic Jyotish, Saturn is said to be exalted in a horoscope when it is placed in the sign of Libra which means that Saturn gains maximum strength in Libra. Saturn can work positively as well as negatively in a horoscope depending upon his benefic or malefic nature and the other deciding factors and overall tone of the horoscope. Saturn is said to be debilitated in a horoscope when it is placed in the sign of Aries which means that Saturn becomes weakest when placed in Aries. Debilitated Saturn can work positively as well as negatively in a horoscope depending upon the other deciding factors and overall tone of the horoscope. The word debilitated simply tells us about the weakness of Saturn in a horoscope and it does not relate to the functional nature of the Saturn in most of horoscopes and accordingly a debilitated Saturn placed in different houses of a horoscope can give benefic as well as malefic results depending upon the functional nature of the Jupiter as well as the overall tone of the horoscope.

Saturn in 1st Houses:

(i) Exalted Saturn in 1st House: A positive exalted Saturn placed in first house can render great creative potentials; shining in some kind of creative field like acting, music, dancing, drama, painting and other such fields; very good power of imagination, wisdom, manipulative abilities, analytical nature and diplomatic behaviour; success in political profession and achieving a position of power and authority through politics as he is good in communicating and influencing the masses or the public; rising the social status as his age increases; getting great support from his children leading to a great rise in his status and authority; becoming popular among people by virtue of his creative abilities, professional skills or his leadership qualities. On the other hand, a negative exalted Saturn in 1st house can cause problems in the marriage and married life; marrying late or very late; facing many difficulties in married life like long separation, divorce or even widowhood; indulge in bad habits of drinking and using other kinds of drugs and pursuing many women in order to satisfy his sexual desires causing financial problems, professional downfall, social disgrace and some diseases to him; fall many times in his lives.

(ii) Debilitated Saturn in 1st House: A benefic debilitated Saturn in 1st house can give very good results in professional income and status building up and rising slowly or sudden rise in his life; good amount of money, wealth and properties coming slowly but keeping on building year after year; good at communicating with people, connecting to people and attracting people and by virtue of these qualities, and he is able to succeed in the fields like politics or other such fields where the above mentioned qualities are required. On the other hand, a malefic debilitated Saturn in 1st house can cause serious problems for his marriage like easily break of marriage after a long time in court cases and financial divisions before the divorce is officially finalized; causes problems in the professional sphere including delays, long periods of inactivity and losses which are generally caused slowly and over a long period of time.

Saturn in 2nd Houses:

(i) Exalted Saturn in 2nd House: A positive exalted Saturn placed in second house of a horoscope can give very good financial prosperity; becoming rich by virtue of his creative abilities and business skills; very big financial gains from professions or from other practices like purchase and sell of house properties, stock investment or stock trading, gambling and betting and other likewise practices; happiness that he derives from his children; blessed with obedient and successful children and getting big financial gains from his children; name and fame; becoming very popular by virtue of his professional achievements . On the other hand, a negative exalted Saturn in 2nd house can cause serious problems in the marriage and married life; delayed marriage or not getting married throughout his life; disturbance in married life and divorces in his life; pursuing extramarital affairs breaking his marriage or marriages; facing serious financial crisis many times; huge financial debts which he may not be able to clear throughout his life.

(ii) Debilitated Saturn in 2nd House: A benefic debilitated Saturn in 2nd house makes a Dhan Yoga and blesses him with very good finance, wealth and prosperity and accordingly the natives are likely to possess good amount of wealth and money early in his life or born in rich or very rich families or start enjoying money and wealth right from the start of his life under the strong influence of such a benefic influence; makes him profit from professions or businesses related to foreign lands and is successful in businesses like import export, travel business and other such kinds of business or income from foreign countries, but does not promise a settlement in foreign country on its own. On the other hand, a malefic debilitated Saturn in 2nd house can cause major problems in his marriage and married life by not allowing him to marry until very late or a long separation; spoil his eating and drinking habits by getting get addicted to excessive consumption of alcohol, drugs and other similar kinds of addictions and suffering seriously on account of health loss and wealth loss or die due to overdose of some drug or alcohol, though such a death needs to be supported from the rest of his horoscope.

Saturn in 3rd Houses:

(i) Exalted Saturn in 3rd House: A positive exalted Saturn placed in third house of horoscope can bless the native with a happy married life; great courage, determination, leadership abilities, the ability to influence masses, sound judgment, bravery and some other similar kind of qualities and becoming very successful in his lives and shining in top class government services like IAS, IPS and other such services or becoming ministers, chief ministers or other head of states who have supreme power and authority; visiting foreign countries for the sake of his profession and benefiting from his foreign trips bringing him financial gains and professional success. On the other hand, a negative exalted Saturn in 3rd house can trouble the native with problems in his married life and suffering much pain and drama through his marriages and may not even get divorces with ease and fighting long legal battles in order to finalize his divorces; trouble on account of his friends and colleagues and facing many losses and spending a very big part of his earned income due to his friends and still may not getting respect and exploiting him; travelling here and there for most of his life without any gains or logic and dieing in an accident while travelling to some place.

(ii) Debilitated Saturn in 3rd House: A benefic debilitated Saturn in 3rd house can render creative abilities and become successful as musicians, writers and other professionals of this kind, very versatile personality, hidden knowledge and becoming a doctors, therapists, consultants, actors, astrologers, psychics, spiritual mediums, herbal medicine specialists and many other kinds of professionals; and takes him to foreign lands for short durations instead of being permanent. On the other hand, a malefic debilitated Saturn in 3rd house can create different kinds of problems in the professional sphere like remaining unrecognized and unsuccessful despite being very talented or despite possessing immense talents in some fields; produces a cursed talents like having got great talents but he is simply not able to convert his talent to success and fame due to lack of opportunity or due to some other reasons; causes spoiled relations and suffering from relationship problems with his wife, brothers, sisters, friends and colleagues; and financial problems and health problems for him.

Saturn in 4th Houses:

(i) Exalted Saturn in 4th House: A positive exalted Saturn placed in 4th house can bless the native with a rich and beautiful wife; rising financial status considerably after his marriage; good financial support from his wife or his in-laws; bigger professional successes with the help of his wives or in-laws; rendering analytical abilities, penetrating insight, diplomacy, manipulative abilities and other such qualities and climbing the ladder of success by virtue of these qualities possessed by him; developing interest for supernatural and paranormal phenomenon and shining in astrology, spiritualism, magic or black magic; residential houses and some houses gifted by his in-laws;

purchasing residential houses in foreign countries settling in foreign country. On the other hand, a negative exalted Saturn in 4th house can prove very bad for the mental health; suffering from problems like anxiety, restlessness, sleeping disorders and psychological disorder and spending a long time in a hospital or a mental asylum for the treatment of such disorders; serious problems in the married life and divorces in his life; bad results in profession; suffering financial losses, changing his place of residence many times life due to their profession.

(ii) Debilitated Saturn in 4th House: A benefic debilitated Saturn in 4th house can bless him with a life full of money, wealth, comforts and luxuries and have easy access to money, wealth, comforts and luxuries for most of his life starting right from his birth; have a well established business, a high paying job or a post of power, authority and status in government or private organization having him money, power and authority; beautiful residential houses and travel to many places for the sake of entertainment and leisure under the strong influence of this benefic influence of debilitated Saturn in their horoscopes. On the other hand, a malefic debilitated Saturn in 4th house can cause serious problems in the married life like suffering much tension, many arguments and disputes in his married life or divorces; distortion of his logic and vision like doubting each and everyone, lack of trust, stubborn and egoistic and making problems in marriage, spoiling the relations with his wife as well as with other people around him; indulging in materialist and physical pleasures and wasting a considerable amount of his wealth and earned money in order to fulfil his desires and facing serious financial crisis or selling of his residential houses and other immovable properties in order to get out of this financial crisis under the strong influence of this malefic placement of debilitated Saturn.

Saturn in 5th Houses:

(i) Exalted Saturn in 5th House: A positive exalted Saturn placed in 5th house can give great financial prosperity, wealth and a position of power, name or fame passed on to him through his father; father playing an important role in his life being very rich, famous and successful; influenced greatly by his father; good professional achievements and travel to many foreign countries for the purpose of business or higher education; good higher education; spiritual advancement being intellectual or spiritual philosophers. On the other hand, a negative exalted Saturn in 5th house can cause problems in the family life; problems due to his children; getting his children after long delays and much medical treatment; suffering on account of diseases keeping on troubling his children right from their childhood; giving birth to dead children or his children may die soon after conception; suffering on his education and financial prosperity; having to discontinue his higher studies due to some problems and suffering from losses in their profession from time to time; getting him involved in some unusual kind of love affair which can inflict financial losses bringing disgrace to him as well as his family members.

(ii) Debilitated Saturn in 5th House: A benefic debilitated Saturn in 5th house can bless him with creative abilities of many kinds and becoming successful as actors, movies stars, TV stars, stage artists, film directors, musicians, writers, architects dealing with designs and many other kinds of professional related to creativity taking his job very seriously and having a touch of strangeness and intensity added to most of his works under this influence of a benefic debilitated Saturn in 5th house of a horoscope; great wealth, professional success, spiritual progress and paranormal experiences and beneficial placement in his life. On the other hand, a malefic debilitated Saturn in 5th house can create problems in his marriage or love life like delays and setback in married life as well as love life or stay alone for a very long period of time after going through bitter experiences in marriage and love; makes him high on ego, arrogant, unrealistic and stubborn and accordingly suffering from failing relations, professional failures and other kinds of setbacks and downfalls due to these habits; gives problems related to child birth or delay in the birth of his children.

Saturn in 6th Houses:

(i) Exalted Saturn in 6th House: A positive exalted Saturn placed in 6th house can give very good results regarding the professional success as doctors, lawyers, judges, engineers, real estate dealers; with qualities like sound judgment, analytical nature, influencing people, wisdom, perseverance and rise; overall health and longevity; gain through court decisions and lucky to claim compensations. On the other hand, a negative exalted Saturn in 6th house can spoil the relations of the native with his father and having serious understanding problems and facing a complete opposition from his father or spending much money toward a disease or court cases bothering his father; causing many problems, delays, setbacks and legal litigations in his professional sphere; having strong enemies and opponents troubling him time to time and causing serious concerns for him in some cases.

(ii) Debilitated Saturn in 6th House: A benefic debilitated Saturn in 6th house can bless him with bravery and become successful as police officers, army officers and other people dealing in defence related professions and many other kinds of professionals which requires courage and bravery; victory over most of his enemies, rivals and competitors and proves to be very tough enemy or rival as it is often very difficult to win over him; name, fame and respect in society by virtue of his bravery and courage. On the other hand, a malefic debilitated Saturn in 6th house can trouble him by facing many problems from his enemies, rivals and competitors and facing serious failures, losses and defame due to conspiracies of his enemies or rivals; makes him suffer physical injuries in war, battle, shoot out, cross fire or in a planned attack by his enemies; problems related to his married life and immovable properties; facing court cases or litigations related to marriage or properties or hospitalization on account of health problems.

Saturn in 7th Houses:

(i) Exalted Saturn in 7th House: A positive exalted Saturn placed in seventh house can render great success in profession based in foreign countries or related to foreign countries and settling in s foreign country on the basis of his profession and doing better in foreign countries than in his own countries; great financial gains, prosperity and luxury after marrying a rich wife strengthening his financial position; very good communication skills, social skills and intelligence and rising in his life. On the other hand, a negative exalted Saturn in 7th house can give very bad results regarding married life; suffering from broken marriage and divorces; high on ego, stubborn and taking his decisions rashly causing problems and losses for a very long time; long lasting disease being.

(ii) Debilitated Saturn in 7th House: A benefic debilitated Saturn placed in 7th house can give very good results regarding the financial prosperity and success in various professional fields; renders good creative abilities, good social skills, good management skills, good administration abilities, good communication skills and many other such qualities making him good at management jobs or the jobs dealing with public; makes very good sales person, marketing people, advertising people, management professionals, popular and successful political leader by having abilities to connect to masses and influence them. On the other hand, a malefic debilitated Saturn in 7th house can create problems related to the marriage and married life suffering from a delayed or very late marriage, break up in a long term relationship up to 3 to 5 years, or lack of attention from his life partner or feeling ignored; financial crisis, debts, mortgages and litigations or law suits related to the non payment loans and debts or selling his business or residential property in order to clears such loans and debts.

Saturn in 8th Houses:

(i) Exalted Saturn in 8th House: A positive exalted Saturn placed in 8th house can bless the native with very good financial gains, success in profession and witnessing unplanned and unexpected financial gains throughout his life; practicing as professionals like, astrologers, psychics, black magicians and shining in supernatural and paranormal fields and making big name and fame along with money and wealth from such a profession; making him profit from parental property by means of a will or a court decision. On the other hand, a negative exalted Saturn in 8th house can seriously affect his lifespan shorter than the average life and dieing at a young age due to an accident or due to a fatal disease; affliction with a long lasting and fatal disease and going through much physical and mental pain due and finally to death; very bad results regarding the married life breaking his marriage causing divorces or one wife dieing due to some serious disease.

(ii) Debilitated Saturn in 8th House: A benefic debilitated Saturn in 8th house can give very good results and progress in the fields of spiritualism, paranormal and practicing as psychics, tantrcis, aghoris, black magicians, Voo doo magicians, spiritual mediums, spiritual healers, astrologers; or a successful doctor (particularly a surgeon or a child specialist), lawyer (especially a criminal lawyer), banker (especially dealing in the loan section of the bank), a physical fitness trainer like aerobics teacher or yoga teacher, police officers who deal with murder cases or crimes related to children, officers of anti terrorist squad and many other kinds of professionals; sudden financial gains. On the other hand, a malefic debilitated Saturn in 8th house can cause most serious problems related to the marriage, married life and long term relations like suffering from break ups, betrayals in love life, delayed or a very delayed marriage, broken marriages and divorces finalizing after a long history of court room drama and social drama; short lifespan or unnatural death or dieing in an accident, through a disease, through an attack by unknown enemies or another type of unnatural death; very bad effect in financial prosperity or professional progress or his overall health.

Saturn in 9th Houses:

(i) Exalted Saturn in 9th House: A positive exalted Saturn placed in 9th house of a horoscope can make a powerful yoga which can bless the native with money, wealth, name and fame; inheriting wealth and prosperity from his father or grandfather and being born in rich family enabling him to enjoy life right from his childhood; position of power and authority and achieving great power and authority with the help of his father or father's family; very popular among the public by virtue of his good and moral conducts; and gaining respect of his family members, relatives, colleagues and society. On the other hand, a negative exalted Saturn in 9th house can cause serious problems through a Pitra Dosha by afflicting the ninth house and getting very bad results and troubles in his married life finally breaking after much drama; delays and losses in profession and wasting a big part of earned money; bringing defamation due to serious allegations in false cases seriously affecting the public image and resigning from a post of power and authority due to these allegations.

(ii) Debilitated Saturn in 9th House: A benefic debilitated Saturn in 9th house can bless him with a post of power and authority in government becoming officers or top class officers with government, administrators, judges, chairmen of various government departments, members of parliaments, members of assemblies, members of corporations, mayors of cities, ministers or many other kinds of people holding a post of power and authority in government organizations; very and happy marriage and married life and major positive changes after his marriage like marrying a foreigner and settling in foreign country after their marriage, getting married to a rich, powerful and resourceful family and changing the status and opportunities quickly after his marriages or help provided by his in laws. On the other hand, a malefic debilitated Saturn in 9th house forms a Pitra Dosha in the horoscope and causes very serious problems like, delays in profession or no profession at all until very late in life, professional failures and setbacks, excessive expenses; brings disgrace and defame to him by virtue of involvement in some scandal or criminal case or develops criminal tendencies in him, domestic violence or serious physical injuries or possible murder or murders during domestic violence.

Saturn in 10th Houses:

(i) Exalted Saturn in 10th House: A positive exalted Saturn placed in 10th house can bless the native with great incomes from his profession and many other sources of income; very good for professional success; name and fame as the most recognized people on international level in presence of other good yoga in the horoscope; very good results regarding the marriage having a religious, spiritual and understanding wife caring him and his families and people around him. On the other hand, a negative exalted Saturn in 10th house can cause problems, delays, setback and losses in the professional sphere and long lasting disease and facing many delays and crisis in his professions due to health problems; false litigation or court case causing financial loss and spoiling his reputation of the native among his people.

(ii) Debilitated Saturn in 10th House: A benefic debilitated Saturn in 10th house can bless him with very good and successful in professional sphere practicing as doctors, surgeons, psychologists, dentists, management professionals, engineers, teachers, preachers, businessmen; power and status through politics becoming powerful politicians and getting a post of power and authority in government; takes him to foreign countries for professional affairs and stay in foreign countries for a long period of times but not a permanent settlement in foreign countries; progress in spiritual field and becoming top class professionals in the professions related to spiritual and paranormal faiths. On the other hand, a malefic debilitated Saturn in 10th house can trouble him with problems in his professional sphere like delays or extreme delays, setbacks, losses, failures, disgrace and many other bad things in his professions; very painful and bitter experience in marriage and the married life.

Saturn in 11th Houses:

(i) Exalted Saturn in 11th House: A positive exalted Saturn placed in 11th house of a horoscope form a powerful Dhan Yoga in the horoscope and accordingly bless the native with great wealth, more than one sources of income and financial fortunes; making his financial fortunes and business empires with his own efforts and qualities; very good leadership qualities, courage, bravery and support from his brothers and relatives achieving success; name, fame and a position of power and authority. On the other hand, a negative exalted Saturn in 11th house can cause problems in married life; breaking marriage due to mental incompatibility between him and partner due to they do not give proper time and attention to their life partners as they keep themselves very busy in their professional spheres; facing serious relationship problems and losing many friends and relatives due to his stubborn and aggressive nature and egoistic personality; health related issues keeps on troubling him for a long period of time.

(ii) Debilitated Saturn in 11th House: A benefic debilitated Saturn in 11th house can bless him with success in various professional spheres but change of his professions for his benefit; makes him successful as lawyers, judges, engineers, doctors, astrologers, spiritual healers, psychics, politicians, officers with government; gives very good results in the fields of spiritualism and paranormal. On the other hand, a malefic debilitated Saturn in 11th house can makes him suffer from sudden financial losses through stock markets, lotteries, treatment of suddenly appearing diseases, accidents, thefts; makes him extremely greedy engaging in fraud, cheating, criminal activities and other such things; suffers from relationship problems due to his greedy nature; he uses every person for his own benefit; serious health problem or painful disease.

Saturn in 12th Houses:

(i) Exalted Saturn in 12th House: A positive exalted Saturn placed in twelfth house can bring great financial prosperity and a life full of luxuries and comforts; going to foreign countries for professions and settling permanently in foreign country and benefiting from foreign countries; permanent residences in foreign country and visiting many foreign countries for the purpose of enjoyment and vacation. On the other hand, a negative exalted Saturn in 12th house can cause serious problems in the married life like physical or mental incompatibility with his wife, getting married to physically weak wife or wife suffering from some serious disease reducing the pleasure of the marriage to great extent, having to live away from wife for a long period of time due to professional reasons, arguments, separations, divorces and probable death of partners bad spending habits and unwanted expenses; facing financial crisis; suffering from permanent disease proving fatal; problems related to the immovable assets facing legal litigations and disputes related to his properties and losing one or more of his properties due to these litigations and disputes.

(ii) Debilitated Saturn in 12th House: A benefic debilitated Saturn in 12th house can take him abroad and settling in foreign country for profession; gives very good results in the fields of spiritualism and paranormal and becoming successful as astrologers, psychics, black magicians, magicians, tantrics, saints, spiritual healers, spiritual mediums; name and fame on international level by virtue of hisr expertise in one of the above mentioned practices. On the other hand, a malefic debilitated Saturn in 12th house can cause problems in the married life by marrying late or very late or divorces; serious and very long lasting financial crisis in his life; health problems, problems in relations, and problems related to professional sphere.

(8) Effects of Exalted & Debilitated Rahu in 12 Houses:

According to Vedic Jyotish, Rahu is exalted when it is placed in the sign of Taurus in a horoscope. It means that Rahu gains maximum strength when placed in Taurus. An exalted Rahu can work positively as well as negatively in a horoscope depending upon the other deciding factors and overall tone of the horoscope. The word exalted simply relates to the strength of Rahu in horoscope or kundali and it does not relate to the nature of Rahu. Rahu placed in different houses of a kundali can give benefic as well as malefic results depending upon the nature of Rahu as well as the overall tone of the horoscope.

Rahu is debilitated in a horoscope when it is placed in the sign of Scorpio which means that Rahu becomes weakest when placed in Scorpio. A debilitated Rahu placed in different houses of a horoscope can give benefic as well as malefic results depending upon the functional nature of the Rahu as well as the overall tone of the horoscope.

Rahu in 1st Houses:

(i) Exalted Rahu in 1st House: A positive exalted Rahu placed in 1st house can do wonderful things and can bless the native with a distinct personality; having and recognized by his own style in whichever field he goes; making him very independent, philosophical, adventurous; name and fame in his profession and producing very good and at top class scientists, explorers, philosophers, astrologers and other people of such kind. On the other hand, a negative exalted Rahu in 1st house can make the personality like hindering his progress in every sphere of life; troubling him with physical as well as mental disorders and undergoing treatment for psychological disorders.

(ii) Debilitated Rahu in 1st House: A benefic debilitated Rahu in 1st house can give very good results to him with some extraordinary qualities like quick wit, wisdom, very good speech abilities, fast processing brain, very good power of observation, analytical nature and goes to foreign countries and settles permanently in foreign country or he keeps on travelling to and from many foreign countries; or he becomes successful scientist, philosopher, astrologer, analyst, explorer under the strong benefic influence of debilitated Rahu. On the other hand, malefic debilitated Rahu in 1st house form Anant Kaal Sarp Yoga and can trouble him with many kinds of problems him like suffering from some permanent disease appearing right from the early years of life or from improper physical appearance or polio effects; makes him engaged in some criminal professions indulging in practices like black magic or other practices to harm other people and suffering a lot in the end by virtue of the bad karmas done by him; various kinds of problems as per effects of Anant Kaal Sarp Yoga.

Rahu in 2nd Houses:

(i) Exalted Rahu in 2nd House: A positive exalted Rahu placed in 2nd house can bless the native with attractive personality; doing very good in business, finances, speech and other fields; producing top class businessmen, explorers, astrologers, mathematicians, doctors and name and fame in his profession; wealth and prosperity; very good attitude towards living and enjoying the life. On the other hand, a negative exalted Rahu in 2nd house can afflict him with bad tastes of eating and drinking and addicted to alcohol, smoking, drugs and suffering on front of health, profession and finances among other things; engaging him in immoral or illegal professions; becoming smugglers, drug dealers or other such practices; forms a Kulika Kaal Sarp Yoga in the horoscope trouble him in various spheres of Kulika Kaal Sarp Yoga.

(ii) Debilitated Rahu in 2nd House: A benefic debilitated Rahu in 2nd house can bless him with very good results in financial spheres and has very good income, wealth and finances; gets sudden financial gains many times in his life; keeps on benefiting from many people and situations; very good speech abilities and does wonderful things by virtue of his speech power and ability to convince and influence other people through his communication skills and is successful in the professional requiring speech abilities and communication skills; very good in his studies and becomes toppers in his class or group of studies and such a placement of debilitated Rahu can produce scholars and highly educated and sharp people; has very good social skills and is the most strongly networked and social people and succeeds in his life by virtue of strong networking and social popularity. On the other hand, malefic debilitated Rahu in 2nd house can make him indulge in bad habits of eating and drinking and addicted to some kinds of drugs or alcohols which can cause serious loss of health and wealth or gets infected with some serious and life threatening disease due to excessive and regular consumption of drugs and alcohols; has very bad financial prosperity and social status suffering a lot on financial front or facing heavy and long lasting financial debts for a long period of time; form Kulik Kaal Sarp Yoga in the horoscope troubling as per Kulik Kaal Sarp Yoga to him.

Rahu in 3rd Houses:

(i) Exalted Rahu in 3rd House: A benefic exalted Rahu placed in 3rd house can render very good skills of writing, speech and other kinds of creative making him successful as a writer, author, musician, singer and other professionals of this kind; great courage and bravery and achieving success in his life and as a result; shining in the fields of army, sports, police requiring physical ability and bravery to succeed.

On the other hand, a malefic exalted Rahu placed in 3rd house of a horoscope or kundali can make him too ambitious and stubborn and suffering from failures in profession as well as personal life; serious differences between him and his friends, colleagues, brothers and sisters and siblings; failures in the 2nd half of his life mainly due to his overambitious attitude and due to not listening to his well wishers; form a Vasuki kaal sarp yoga troubling with bad effects of Vasuki Kaal Sarp Yoga.

(ii) Debilitated Rahu in 3rd House: A benefic debilitated Rahu in 3rd house can bless him with very good understanding of the hidden nature and becomes successful in the profession requiring the core understanding of hidden aspects of something like jewellers, architects, astrologers, palmists, doctors, therapists dealing with some kind of secret knowledge; gives very good appearance and health having lean and fit bodies and keeps fit throughout his life; has great courage and bravery; makes him travel to many foreign countries; becomes successful as sportspersons or police officers travelling to many foreign countries on regular basis; has very good communications skills and is very good at connecting to people and influencing people in order to gain benefits from them. On the other hand, a malefic debilitated Rahu in 3rd house can make him very selfish and cheats other people, ignores other people or sacrifices other people in order to have his own benefits and is generally disrespected by the people and society as he is not trustworthy; he can even sacrifice his own brothers and sisters in order to gain profits; has bad relations with his friends and siblings due to his selfishness or loses his friends or brothers permanently on account of this selfishness; he engages in some kind of criminal activities or criminal professions in order to get quick financial gains and such tendency lands him in big trouble going to prison by virtue of his involvement in criminal activities; forms Vasuki Kaal Sarp Yoga in the horoscope and trouble him on account of various kinds of problems as per Vasuki Kaal Sarp Yoga.

Rahu in 4th Houses:

(i) Exalted Rahu in 4th House: A benefic exalted Rahu in 4th house can bless the native with a very good family, luxuries and comforts; leading an easy, comfortable and secured life even if there professional success is not very good; support from relatives, friends or other people. On the other hand, a malefic exalted Rahu in 4th house can cause emotional disturbance right from his childhood and a sense of insecurity throughout his life; very bad health and bad in professional life; very bad results in married life; form a Shankhpal

Kaal Sarp Yoga and trouble him with bad effects of Shankhpal Kaal Sarp Yoga.

(ii) Debilitated Rahu in 4th House: A benefic debilitated Rahu in 4th house can bless him with very good results in professional spheres, becomes successful professionals with power, status, name and fame as a police officer, politician, minister, administrator, famous doctor and many other such professionals dealing in public, power and status; very good inclinations and progress in the fields like religion and spiritualism; a life of luxuries and comforts; goes and settles in foreign country on permanent basis. On the other hand, a malefic debilitated Rahu in 4th house makes him selfish, self cantered, egoistic and very aggressive losing many friends and relatives by virtue of these negative personality traits n his lifetime; develops a tendency to pick fight with every person causing him serious problems, setbacks and losses; gives very badly suffering from health problems and problems in his married life; forms Shankhpal Kaal Sarp Yoga in the horoscope troubling him on account of various bad effects of Shankhpal Kaal Sarp Yoga.

Rahu in 5th Houses:

(i) Exalted Rahu in 5th House: A benefic exalted Rahu in 5th house can bless the native with top class creativity and a successful professional requiring creativity making him successful or very successful as an actor, photographer, musician or other professional related to creative fields; bringing prosperity, name and fame of national or international level. On the other hand, a malefic Rahu in 5th house can cause problems related to the higher education; discontinuation of studies; problems related to child birth like delays or medical issues to produce a child and this issue can intensify in the case of female natives; form a Padam kaal sarp yoga troubling him with the effects Padama Kaal Sarp Yoga.

(ii) Debilitated Rahu in 5th House: Benefic debilitated Rahu in 5th house can give him very good results regarding the financial prosperity and making money through stock markets, trading, betting and even gambling to some extent; very good results in his higher education, very good character and conduct and very caring and understanding towards the needs of the other people; earns good amount of respect in the society by virtue of his noble characters; debilitated Rahu is one of the best placements in the horoscope by giving him spiritual progress in spiritual and paranormal worlds engaging him in spiritual and paranormal practices like astrologers, spiritual healers, spiritual mediums, magicians, black magicians, saints, tantrics. On the other hand, malefic debilitated Rahu in 5th house can make him lose considerable part of his earned money and wealth through practices like gambling, stock market trading and other such practices; very bad health suffering from permanent health problems or disease or claiming his life; gives problems related to child birth and professional sphere forms Padam Kaal Sarp Yoga in the horoscope troubling him on account of various kinds of problems as per Padam Kaal Sarp Yoga.

Rahu in 6th Houses:

(i) Exalted Rahu in 6th House: A benefic exalted Rahu in 6th house can bless the native with hard working nature, courage and special abilities and becoming successful as doctors, surgeons, advisors, lawyers, judges, people dealing in travel business or people dealing in import-export; rendering qualities like diplomacy, manipulation and communication skills; becoming successful as politicians, diplomats, mediators and other such kind of people. On the other hand, a malefic exalted Rahu in 6th house can cause serious troubles on account of diseases, enemies and financial losses; suffering from serious diseases for a very long period of time and he may even dieing due to such disease; defamation, financial losses and other problems like imprisonment due to court decision going against him; forming Mahapadam kaal sara yoga in the horoscope troubling him the effects of Mahapadam Kaal Sarpa Yoga.

(ii) Debilitated Rahu in 6th House: Benefic debilitated Rahu in 6th house can give good to very good results regarding the professional sphere and going to foreign countries and settling there on permanent basis or on long term basis; blessed with good speech abilities, analytical abilities, wisdom, sharp insight and the ability to make quick decisions and gives financial gains related to wealth coming to him as a decision of some court cases as he has a strong tendency to win court cases and litigations due to benefic influence of Rahu in his horoscope. On the other hand, malefic debilitated Rahu in 6th house can develop criminal tendencies in him and engaging him in criminal professions becoming notorious on international level by virtue of his engagement in top class criminal professions like drug supplying, smuggling, kidnapping for ransom and other such kinds of professions and he can operate his illegal business from foreign country due to this strong influence of such a malefic placement of debilitated Rahu in 4th house of horoscopes and is prone to get caught by the law at one point of his life or spending long period of time in jail or prison due to being convicted with some criminal offence; gives very bad results regarding the financial aspect, health and his well being; forms Maha

Padam Kaal Sarp Yoga in the horoscope troubling him on account of various kinds of problems as per Mahapadam Kaal Sarp Yoga.

Rahu in 7th Houses:

(i) Exalted Rahu in 7th House: A benefic exalted Rahu placed in 7th house can render a good social and public reputation, good public image; very good social and management skills; tendency to start his own business instead of working for someone; a post of power and authority in politics rendering name and fame in jobs related to foreign countries. On the other hand, a malefic exalted Rahu in 7th house can cause serious problems in married life witnessing much drama, arguments, separation and divorces; excessively indulgent in physical pleasures, particularly in sexual pleasures and wasting much of his time and money in order to fulfil his sexual desires; serious or even fatal sexually transmitted disease to him; form a Takshak Kaal Sarp Yoga troubling him with the effects of Takshak Kaal Sarp Yoga.

(ii) Debilitated Rahu in 7th House: Benefic debilitated Rahu in 7th house can give very good results regarding his marriage and married life marrying at proper ages or early ages and living a happy married life; benefit from his wife as well as his in laws and marrying a wife who is the only daughter of his parents who possess good wealth and property and such wealth and property is transferred to his wife and in laws also keep on helping him in his professional sphere; or marries to a person of foreign nationality and permanently settles in foreign country on the basis of their marriage; very good results in profession like business than in service or becomes very successful in his business by virtue of some different and new idea invented by him. On the other hand, malefic debilitated Rahu can give bad or very bad results regarding his marriage and married life like marrying very late or undergoing broken marriages which generally fail due to lack of mental compatibility between him and his wife; troubles him on account of bad reputation due to false allegation or his misconduct or due to his involvement in scandal; troubles him on account of health or losses through business or excessive indulgence in extra marital affairs destroying his marriage and married life; forms Takshak Kaal Sarp Yoga in the horoscope which can trouble him on account of various kinds of problems as per Takshak Kaal Sarp Yoga effects.

Rahu in 8th Houses:

(i) Exalted Rahu in 8th House: A benefic exalted Rahu placed in 8th house can render a long lifespan; good financial gains and support getting from his spouse; making him successful as astrologer, palmist, scientist, secret service agents and other such professionals which are dealing with something hidden. On the other hand, a malefic exalted Rahu in 8th house can cause serious problems in married life facing much drama and completely depending on his spouse and having to tolerate many bad words and bad behaviour of his spouse; creating criminal tendencies and engaging in criminal professions and going to jail due to criminal professions; form a Karkotak Kaal Sarp Yoga in the horoscope and troubling with the effects of Karkotak Kaal Sarp Yoga.

(ii) Debilitated Rahu in 8th House: Benefic debilitated Rahu in 8th house can give very good results regarding his professional success as doctors, police officers, army officers, bankers, people dealing in finance management firms, sports persons; takes him to foreign lands and becomes successful as astrology, palmistry, magic, black magic; blesses him with a good health and a strong ability to recover from diseases and health living healthy and long life. On the other hand, malefic debilitated Rahu in 8th house can cause serious problems related to his marriage and married life like marrying late or very late, having bad or very bad marriages keep on troubling him, getting very argumentative and quarrelsome wife in nature who keeps on harassing and troubling the native on account of small issues under the strong influence of such a malefic placement; suffers from injuries and surgeries on account of health problems, diseases and accidents; gets afflicted with serious and life taking disease or die in an accident or is killed in unplanned and sudden fight or attack; forms Karkotak Kaal Sarp Yoga in the horoscope troubling him on account of various kinds of problems as per Karkotak Kaal Sarp Yoga.

Rahu in 9th Houses:

(i) Exalted Rahu in 9th House: A benefic exalted Rahu placed in 9th house can take the native to foreign lands for the purpose of permanent settlement in abroad; making him to succeed in businesses connected to foreign countries; spiritual advancement of the native and as a result, some natives under a particular and successful spiritual leaders or spiritual philosophers; very broad minded and adapting to almost every kind of people and circumstances. On the other hand, a malefic exalted Rahu in 9th house trouble the native with serious problems related to his professional life and remain unemployed until very late in his life; form a Pitra Dosha in the horoscope posing various problems due to formation of Shankhchood kaal sarp yoga and its effects.

(ii) Debilitated Rahu in 9th House: Benefic debilitated Rahu in 9th house can give very good results regarding his spiritual progress or the gains coming from spiritual fields; gets a position of authority, respect and profit in spiritual or religious organization; takes him to foreign lands on permanent basis or travelling to many foreign countries on account of his profession and likely to get benefit from such travels under the strong influence of such benefic influence of debilitated Rahu; gets success in politics and becomes successful as politicians, ministers, administrators, police officers, judges and other professionals of similar kind. On the other hand, malefic debilitated Rahu in 9th house can trouble him on various fronts like exposing him to cheats and frauds and accordingly he suffers on account of people deceiving and cheating him many times in his life or suffers serious financial losses or bad reputation by virtue of misconducts planted and done by someone else; suffers from various kinds of diseases or facing difficulties in his professional life; forms Pitra dosha in the horoscope troubling him as per Pitra Dosha effects; forms Shank Chood Kaal Sarp Yoga in the horoscope troubling him as per Shank Chood Kaal Sarp Yoga effects.

Rahu in 10th Houses:

(i) Exalted Rahu in 10th House: A benefic exalted Rahu in 10th house can bring great success in his professional life; special abilities making him an expert in specific fields and becoming pioneers in his professional spheres of invention, discovery or creating many new things which were formerly not known to the world and earning name and fame by virtue of such innovations; not believing in following the trends but in setting trends. On the other hand, a malefic exalted Rahu in 10th house can trouble him with many problems like big losses in professions, disgrace or defame by virtue of false allegations, litigations or scandals which arise from their professional spheres; form a Ghatak kaal sarp yoga giving with respect to Ghatak Kaal Sarp Yoga.

(ii) Debilitated Rahu in 10th House: Benefic debilitated Rahu in 10th house can bless him with success in his professional sphere as doctors, engineers, analysts, discoverers, inventors, explorers, computer programmers or practices as astrologers, spiritual healers, spiritual mediums, numerologists, palmists, magicians, Vastu experts, psychics or good results in the profession requiring group activities or is blessed with very good social and networking skills and getting engaged in group oriented works with his colleagues and friends. On the other hand, malefic debilitated Rahu in 10th house can engage him in illegal professions and get involved in smuggling, drug dealing, gambling, kidnapping and other criminal activities which make him earn bad name for himself and his families and suffer from punishment given by court of law; or is tough criminals proving a big headache for the law enforcing agencies of his country and many other countries; suffers on account of professional setbacks and delays or undergoing in prison; faces serious financial crisis; forms Ghatak Kaal Sarp Yoga in the horoscope troubling as per Ghatak Kaal Sarp Yoga effects.

Rahu in 11th Houses:

(i) Exalted Rahu in 11th House: A benefic exalted Rahu placed in 11th house can bless the native with very good financial gains; achieving huge financial fortunes; very good social skills, adventurous spirit, keen business sense, risk taking ability and very good friends and he rising in his life due to qualities and possessions; very ambitious of dong anything. On the other hand, a malefic exalted Rahu in 11th house can impart gambling tendencies; criminal tendencies; engage in gambling or criminal activities and finally witnessing a huge losses and most of money and wealth earned through illegal or immoral means is lost; form Vishdhar Kaal Sarp Yoga in the horoscope causing many problems due to bad effects of Vishdhar Kaal Sarp Yoga.

(ii) Debilitated Rahu in 11th House: Benefic debilitated Rahu in 11th can give very good results regarding his professional success, financial gains and overall status coming from big organizations of people formed by him and succeeds as the leaders of such organizations; gets position of authority and power in some government organization or some private organization and travels to many foreign countries for his profession; has good sense of business, good judgment abilities and is well balanced and disciplined in his approach and all of these qualities help him in achieving success in his professional sphere. On the other hand, a malefic debilitated Rahu in 11th house can cause problems in his married life like staying away from his wife for a very long period of time on account of his professional demands, or has infidelity or loyalty issues from his wife side which means that his wife may engage in some extramarital affair or she may either bring disgrace to him or she may leave him due to such extramarital affair; suffers on account of losses coming from his profession, unwanted and unavoidable expenses leading to financial crisis, or health related issues; forms Vishdhar Kaal Sarp Yoga in the horoscope troubling him as per Vishdhar Kaal Sarp Yoga effects.

Rahu in 12th Houses:

(i) Exalted Rahu in 12th House: A benefic exalted Rahu in 12th house can take the native to some foreign lands very early in his life and he is settled or even study abroad and many settle in foreign countries on permanent basis; making him successful as a doctor, lawyer, astrologer, psychic and very good spiritual progress and the connection with other worlds. On the other hand, a malefic exalted Rahu in 12th house can trouble the native with many kinds of problems related to his health, wealth, marriage and professions; addicting to drugs, alcohol or other such intoxicants and choosing a life of total seclusion; unhappy in married life but not usually break the marriage, but making such a marriage very miserable and torturous for him; form a Sheshnag kaal sarp yoga in the horoscope causing problems related to Sheshnag Kaal Sarp Yoga.

(ii) Debilitated Rahu in 12th House: Benefic debilitated Rahu in 12th house can bless him with very good professional success getting a position of power and authority in some government organization; going to foreign countries and settling there on permanent basis; has good qualities like wisdom, optimism, perseverance, leadership ability, the ability to properly assess every situation and many other good qualities and climbs to the ladder of success by use of these qualities. On the other hand, malefic debilitated Rahu in 12th house can cause serious problems in various spheres like professional setbacks and downfalls, suffering in his marriage or relationship issues with their own parents or siblings, or makes him egoistic, adamant and selfish to some extent and witnessing various kinds of problems in different spheres of his life due to these negative qualities; or prone to see downfall from a top position due to not care about the opinion of his well wishers and the ethics of good and bad once he has decided to do something; forms Sheshnag Kaal Sarp Yoga in the horoscope troubling him as per Sheshnag Kaal Sarp Yoga effects.

(9) Effects of Exalted & Debilitated Ketu in 12 Houses:

According to Vedic Jyotish, Ketu is exalted when it is placed in the sign of Scorpio and so it means that Ketu gains maximum strength when placed in Scorpio. Ketu placed in different houses of a kundali can give benefic as well as malefic results depending upon the nature of Ketu as well as the overall tone of the horoscope. The Ketu is debilitated in a horoscope when it is placed in the sign of Taurus which means that Ketu becomes weakest when placed in Taurus. The word debilitated simply tells us about the weakness of Ketu in a horoscope and it does not relate to the functional nature of the Ketu in most of horoscopes and accordingly a debilitated Ketu can give benefic as well as malefic results in a horoscope.

Ketu in 1st Houses:

(i) Exalted Ketu in 1st House: A benefic exalted Ketu placed in 1st house can bless the native with great professional success related to the hidden knowledge like astrology, tantra like becoming renowned psychics, spiritual healers, spiritual teachers; top class position of power in government or in some big organization. On the other hand, a malefic exalted Ketu in 1st house can cause suffering from a serious physical or mental disease for a very long period of time; bad results in profession and problems in married life; having flawed personalities; form a Takshak Kaal Sarp Yoga causing more serious problems for him.

(ii) Debilitated Ketu in 1st House: Benefic debilitated Ketu in 1st house can render some distinct qualities, characteristics and interests to him succeeding in by virtue of these qualities and interests; blessed with a charismatic personality attracting and influencing people around him and achieving success in professional; possess qualities like bravery, courage and aggression and accordingly is seen successful in professional due to use of these qualities; gives a very good power of intuition capable of sensing future events in advance to help him to achieve success in life in fields like police force, politics, administrative services rendering him a position of power and authority. On the other hand, a malefic debilitated Ketu in 1st house can cause serious defects in the personality making him to suffer in various spheres like mistrust and doubt everyone around him and blaming or accusing even the closest people around them and so breaking the relations for a long time or becomes very arrogant, aggressive and rude giving very bad results regarding the married life facing serious issues like broken marriage; has health problems and diseases; forms Takshak Kaal Sarp Yoga in the horoscope troubling him as perTakshak Kaal Sarp Yoga effects.

Ketu in 2nd Houses:

(i) Exalted Ketu in 2nd House: A benefic exalted Ketu in 2nd house are born in rich families and hence they enjoy a good childhood; bless him with creative skills, social skills and climbs the ladder of success; special abilities and succeed as an architects, jewellers, interior designers. On the other hand a malefic exalted Ketu in 2nd house can cause problems and difficulties right form his childhood related to health, problems with parents or living away from parents, study related problems; bad results in married life like a late marriage, a broken marriage, a very troublesome marriage; serious financial problems and form a Karkotak Kaal Sarp Yoga in the horoscope posing various kinds of problems.

(ii) Debilitated Ketu in 2nd House: Benefic debilitated Ketu in 2nd house can make him very courageous and brave and accordingly he can excel in professional spheres like police force, army, sports activities and other spheres or makes him successful in professions like doctors, lawyers, surgeons, consultants, physicians, teachers; has good health and a long life; very good progress in education and does very well in his academic and school education. On the other hand, a malefic debilitated Ketu in 2nd house can face serious financial crisis in his or face long separations from his parental family on account of some false allegations or misunderstandings between him and his family members; or serious problems in married life like marrying very late or marrying two, three or even four times in his life as his marriages keep on failing one after the other; prone to have long lasting or permanent disease and suffering due to pain and financial losses through these diseases; forms Karkotak Kaal Sarp Yoga in the horoscope troubling him as per Karkotak Kaal Sarp Yoga effects.

Ketu in 3rd Houses:

(i) Exalted Ketu in 3rd House: A benefic exalted Ketu placed in 3rd house can render very good skills like writing, painting, sculpting; very good abilities like courage and bravery and he working like police force, army and other such fields; good religious and spiritual interests. On the other hand, a malefic exalted Ketu in 3rd house can make a native very adamant and egoistic failing many times in his life due to his bad personality traits; bad results in married life facing divorce, legal litigation, loss of reputation and much drama but no delay in marriage; form a Shankhchood Kaal Sarp Yoga causing various kinds of problems.

(ii) Debilitated Ketu in 3rd House: Benefic debilitated Ketu in 3rd house can bless him with very good results in professional spheres and goes to foreign lands and settling there on professional basis or travels to various foreign countries; renders qualities like bravery, courage and aggression excelling him in professional spheres like army or police force and become recognized warheads or commanders; renders good qualities like wisdom, the ability to separate right from wrong, the ability to take right decision at proper time, good financial management, the ability to judge and decide fairly while acting as a mediator and many other such good qualities and accordingly he is benefited by virtue of these good qualities possessed by them. On the other hand, malefic debilitated Ketu in 3rd house can make the native aggressive, rash and egoistic landing him into trouble many times in his life by indulging him in fights, wars and other violent activities of high risk or getting injured or dieing in one of such acts of violence by virtue of these negative traits; renders qualities like being impractical, living in dream worlds and pursuing irrational and unrealistic goals and accordingly wasting a long part of life pursuing and chasing unrealistic goals instead of working on something realistic and rational; forms Shank Chood Kaal Sarp Yoga in the horoscope troubling him as per Shank Chood Kaal Sarp Yoga effects.

Ketu in 4th Houses:

(i) Exalted Ketu in 4th House: A benefic exalted Ketu placed in 4th house can bless the native with a position of power and authority like becoming officers with government bodies or in some other organization; give a magnificent and royal personality; penetrating insight, great analytical abilities and a good sense of judgment and all these qualities can help the native in achieving success in his life. On the other hand, a malefic exalted Ketu in 4th house can cause serious distortions in the personality; getting afflicted with some serious psychological disorder bothering him for a very long period of time or afflict him throughout his lives starting from their childhood; very painful experiences and bad results in the married life form a Ghatak Kaal Sarp Yoga in a horoscope giving very bad effects.

(ii) Debilitated Ketu in 4th House: A benefic debilitated Ketu in 4th house can bless him with very good psychic abilities and spiritual interests practicing in various kinds of spiritual and paranormal faiths and practices as successful astrologers, psychics, spiritual healers, spiritual mediums, spiritual gurus, religious heads, magicians, black magicians, aghoris, tantric; strong powers of concentration and the ability to look through mysteries and secrets or some kind of hidden knowledge or faith; life of luxuries and comforts and position of authority, name and fame to him under its strong influence. On the other hand, a malefic debilitated Ketu in 4th house can cause serious problems in marriage and married life like broken marriages; suffer from physical injuries through accidents, serious diseases, loss of valuables through theft and many other kinds of problems; trouble on account of court cases and litigations related to his residential house or houses and change of place of his residence many times in life; forms Ghatak Kaal Sarp Yoga in the horoscope troubling him as per Ghatak Kaal Sarp Yoga effects.

Ketu in 5th Houses:

(i) Exalted Ketu in 5th House: A benefic exalted Ketu in 5th house can give a very pleasant and caring personality; interested in serving the weaker section of the society and society as a whole; very good financial prosperity; very good creative abilities; being successful in creative fields or business spheres. On the other hand, a malefic exalted Ketu in 5th house can trouble him in his education, particularly higher education; discontinuing his education at an early age; adverse results in love life and witness cheating, betrayals and heart breaks more than once in his life; trouble in the birth of a child and undergoing much treatment to them and delays in birth of a child; form a Vishdhar Kaal Sarp Yog in the horoscope troubling with various kinds of problems.

(ii) Debilitated Ketu in 5th House: A benefic debilitated Ketu in 5th house can give very good results regarding his education and academic progress; renders very good skills of business and a good understanding of professional matters and is seen successful as businessmen, very good salesman or marketing professionals as he has very good judging sense and the interest level of his buyers convincing them with right kind of approach; makes him an army officer, police officer, lawyer, judge and other such kinds of professionals as this placement can also render qualities like bravery, sound judgment, good speech abilities helping him succeed in professionals spheres requiring these qualities. On the other hand, a malefic debilitated Ketu in 5th house can cause serious problems in his married life going through broken marriages before finally settling in some marriage; breaking of his love affairs or betrayals, drama and heartbreaks in his love life or is cheated or betrayed by his love partners or lovers; suffer from delays, losses and setbacks in his profession; gives very bad results in child birth and health issues; forms Vishdhar Kaal Sarp Yoga in the horoscope troubling him as per Vishdhar Kaal Sarp Yoga effects.

Ketu in 6th Houses:

(i) Exalted Ketu in 6th House: A benefic exalted Ketu in 6th house can render great professional success as lawyers, judges, doctors, surgeons, astrologers, police officer; very good health and longevity; benefiting from court decisions and disputes; good spiritual advancement. On the other hand, a malefic exalted Ketu in 6th house can trouble him with the most serious and mysterious health related problems and visiting hospitals many times or undergoing surgeries multiple times; can corrupt him towards criminal activities engaging him in illegal and immoral professions which can ultimately land him in jail for a very long period of time; form a Sheshnag kaal sarp yoga with various kinds of problems.

(ii) Debilitated Ketu in 6th House: A benefic debilitated Ketu in 6th house can bless him with good or very good success in achieving positions of profit, status, authority and power in some government organization or in some private organization and working as police officers, judges, top class lawyers, administrative officers and people working on high level posts in government and private organizations; renders very good understanding of the root of every matter, ability to take calculated risks and initiative at proper times, the ability to connect to people and draw benefits from them and accordingly he is seen successful by virtue of these qualities; is successful and powerful politicians, diplomats and ministers with government. On the other hand, a malefic debilitated Ketu in 6th house can cause serious problems in his professional sphere and suffer from setbacks, delays and losses; develops criminal tendencies in him engaging in criminal professions and becoming gangsters, smugglers, arms dealers, mafia, terrorists or people working for terrorist organizations and suffering from serious physical injuries due to diseases or accidents or gang wars or encounters with law enforcing agencies; suffers losses on the hands of enemies and competitors having financial losses, physical injuries or other kinds of losses through their enemies and rivals; forms Sheshnag Kaal Sarp Yoga in the horoscope troubling him as per Sheshnag Kaal Sarp Yoga effects.

Ketu in 7th Houses:

(i) Exalted Ketu in 7th House: A benefic exalted Ketu placed in 7th house can render achievement of great professional success along with name and fame; going to foreign lands and settling there; good interest towards supernatural and paranormal. On the other hand, a malefic exalted Ketu in 7th house can cause serious problems in the marriage and married life like a late marriage, a very late marriage or no marriage in extreme cases, disturbance, drama, litigation, separation, divorces and many other kinds of problems; troubling him with permanent diseases and losing limbs or organs of his body; form Anant Kaal Sarp Yoga in the horoscope troubling him. The horoscopes should be very watchfully checked when it comes his marriage and married life.

(ii) Debilitated Ketu in 7th House: Benefic debilitated Ketu in 7th house can bless him with many good qualities like bravery, courage, determination, aggression, hidden faiths and practices achieving success in many spheres of his life practicing as astrologers, psychics, palmists, numerologists; name, fame and recognition in the society or fame on international level. On the other hand, a malefic debilitated Ketu in 7th house can give very bad results in his marriage and married life like marrying late or very late or not marrying at all or broken marriages before finally settling into marriage; suffers from bad name and loss of reputation in profession more than once in life; health problems or diseases or get afflicted with life threatening disease; forms Anant Kaal Sarp Yoga in the horoscope troubling him as per Anant Kaal Sarp Yoga effects.

Ketu in 8th Houses:

(i) Exalted Ketu in 8th House: A benefic exalted Ketu placed in 8th house of a horoscope or kundali can bless the native with knowledge of occult, paranormal or supernatural practicing as astrologers, psychics, palmists, magicians and other people dealing in similar kinds of faiths; do not have to work hard to gain his knowledge; wealth and fortune and inheriting wealth, fortune or some occult knowledge through inheritance. On the other hand, a malefic exalted Ketu in 8th house can shorten the lifespan and living a life shorter than the average life; cause his death in an accident or due to a fatal disease in early age; many kinds of problems in married life like separation and divorces or widowhood; form a Kulik kaal sarp yoga in the horoscope causing various kinds of problems.

(ii) Debilitated Ketu in 8th House: Benefic debilitated Ketu in 8th house can bless the native with long life and inherits business, wealth or some kind of knowledge; gain through inheritance and tradition is seen engaging in his family business or some traditional work; substantial financial gains, name, fame and recognition on international level through his professions and he leaves a legacy behind him which stays alive for many years after his death. On the other hand, malefic debilitated Ketu in 8th house can create serious problems in his marriage and married life or has to suffer with the death of his spouse, broken marriage or marriages and physical or mental disability of his spouse; or he may die an unnatural death like getting killed in some accident, surgery, getting shot by enemy, falling victim to natural calamity or something else of the same kind at a young age; suffers a lot in his life due to weak financial position, setbacks and delays in profession or sudden and unavoidable expenses; forms Kulik Kaal Sarp Yoga in the horoscope troubling him as per Kulik Kaal Sarp Yoga effects.

Ketu in 9th Houses:

(i) Exalted Ketu in 9th House: A benefic exalted Ketu placed in 9th house can give very good results regarding the birth of male children and producing more than one male child or producing only male children if this fact is supported by overall horoscope; becoming very successful in professions but late in his life; developing very good religious and spiritual interests, name and fame in the society. On the other hand, a malefic exalted Ketu in 9th house can cause serious problems in the professional sphere changing his profession many times due to failing in his profession or defaming or great defaming to him more than once in his lifetime; form a Pitra Dosha in the horoscope and also form a Vasuki Kaal Sarp Yoga in the horoscope and both of these dosha trouble him with various kinds of problems.

(ii) Debilitated Ketu in 9th House: Benefic debilitated Ketu in 9th house can bless him with good health, good physique and a long lifespan; gives very good understanding of business and professional matters and is successful in that; inherits wealth or business from his paternal family and getting help from paternal family members in many spheres of life; name and fame and gains a position of power and authority in some religious or spiritual organization. On the other hand, a malefic debilitated Ketu in 9th house can cause various kinds of problems for him due to Pitra dosha in 9th house damaging him on potential scales; gets bad name and loses reputation caused due to his bad karmas or due to bad deeds done by his family members, especially his children; suffers from serious financial crisis many times in his life and spoils his relations with his wife, siblings, family members and friends; forms Vasuki Kaal Sarp Yoga in the horoscope troubling him as per Vasuki Kaal Sarp Yoga effects.

Ketu in 10th Houses:

(i) Exalted Ketu in 10th House: A benefic exalted Ketu placed in 10th house can bless the native with great professional success rendering a political post of power, a post with government as an officer, a post in police force or army; having a tendency and ability to be at the top in every profession he goes to; good support from his children and his success and status rises with the help of his children. On the other hand, a malefic exalted Ketu in 10th house of a horoscope can cause problems related to his professional sphere seeing failures, losses and delays in his professional promotions; witnessing complete failure and becoming bankrupt at one point of time or he may have to start from a scratch again; starting his careers late in his life due to some reasons or he may have to remain jobless for a very long period of time; form Shankhpal kaal sarp yoga in the horoscope afflicting many kinds of problems.

(ii) Debilitated Ketu in 10th House: Benefic debilitated Ketu in 10th house can bless him with a top class position of power and authority in some government organization and is successful as police officers, administrators, judges and other people working on government posts of power and authority; becomes successful and powerful politicians, diplomats, ministers or chief ministers with the government, which is very much certain with this kind of benefic influence of debilitated

Ketu in a horoscope and is working in government or private organizations on the posts which execute power, authority and control over a big group of people; blesses him with a good physique, impressive body, bravery, courage, aggression, sound judgment, the ability to separate right from the wrong, leadership qualities and many other such qualities which help him attaining a position of power and authority and earns respect in the society. On the other hand, a malefic debilitated Ketu in 10th house can render foolishness, rashness, egoism, anger and arrogance to him and suffers by virtue of these bad characteristics possessed by him; goes through lots of pain, drama and unwanted events in his marriage and married life or divorces being finalized after a long history of litigations and court room drama; prone to be engaged in disputes, arguments and fights due to his aggressive and rash nature and faces serious consequences; forms Shankhpal Kaal Sarp Yoga in the horoscope troubling him as per Shankhpal Kaal Sarp Yoga effects.

Ketu in 11th Houses:

(i) Exalted Ketu in 11th House: A benefic exalted Ketu placed in 11th house can bless the native with huge financial gains; becoming successful with the help of his and some of them providing great help to him; help from his brothers and sons, particularly elder brothers; successful as top class police officers, top class army officers, doctors, surgeons, sports persons rendering name and fame to him mainly due to some deed of bravery done by him. On the other hand, a malefic exalted Ketu in 11th house can bring serious financial problems; face many unwanted losses and expenditures throughout his life; losing a big amount of money through gambling, diseases, accidents, thefts and other kinds of sudden losses; having heavy debts and remaining under debt almost for his whole life; suffering on account of betrayals of his friends or relatives; form a Padam Kaal Sarp Yoga in the horoscope troubling him with various kinds of problems.

(ii) Debilitated Ketu in 11th House: Benefic debilitated Ketu in 11th house can bless him with a financially secured life and gives good financial gains through professions to lead a happy life; renders great wealth, name and fame to him and fortunes of wealth and status in his; gets a position of power and authority in government or private organization; good results in advanced spiritual and paranormal experiences and becomes very advanced spiritual persons in his lifetimes. On the other hand, a malefic debilitated Ketu in 11th house can give very bad results in his married life facing issues related to child birth primarily due to incapability of his spouse to produce children, or infidelity of his spouse or he has to raise a child who is not his own child or produced due to extra marital sexual relations of his wife with someone else; financial problems and unstable profession like, setbacks, delays and losses through profession; health related problems; forms Padam Kaal Sarp Yoga in the horoscope troubling him as per Padam Kaal Sarp Yoga effects.

Ketu in 12th Houses:

(i) Exalted Ketu in 12th House: A benefic exalted Ketu placed in 12th house can make the native succeed as a top class politician; becoming successful as powerful politicians; gives a keen interest and understanding of the occult, hidden, paranormal or supernatural; produce top class, doctors, physicians, spiritual healers, therapists, astrologers giving top class results like becoming top class spiritual gurus or teachers. On the other hand, a malefic exalted Ketu in 12th house can make the native addict to drugs, alcohol, and other such kinds of things and suffering from a serious disease, a prolonged medical treatment, loss of a limb or organ; inclined towards criminal professions making him spend many years of his life in prison or jail due to his involvement in criminal activities; very bad result for the married life suffering a lot in his marriage; form a Mahapadam Kaal Sarp Yoga in the horoscope causing various kinds of problems.

(ii) Debilitated Ketu in 12th House: A benefic debilitated Ketu in 12th house can take him to foreign countries and settling permanently on the basis of his professions or marriage; gives very good results regarding professional success as doctors, lawyers, judges, engineers, police officers, officers working in foreign country situated high commissions or embassies, astrologers, mathematicians, businessmen, financial experts, bankers, consultants, charted accountants, financial advisors under the strong benefic influence of this placement; achieves great heights in the worlds of spiritualism and paranormal or reaches to very advanced stages of spiritual experiences, particularly in the later parts of his life. On the other hand, a malefic debilitated Ketu in 12th house can trouble him in his marriage and married life like lack of physical compatibility between husband and wife, serious arguments, long separations, or divorces or possible death of spouse; faces serious financial tightness or financial crisis many times in life which may be caused due to insufficient income or due to excessive and unwanted expenditure of the earned money or has debts; forms Mahapadam Kaal Sarp Yoga in the horoscope troubling him on account of various kinds of problems as per Mahapadam Kaal Sarp Yoga effects.

9.7 Predictions by Combust Lords & planets in 12 Houses:

Combust / Moudhya / Vikala / Ashtangata / Dagdha is a phenomenon which occurs when a planet comes very close to the Sun and gets burnt or loses its lustre and does not shine properly due to the closeness with Sun. When a Graha approaches too close to the burning rays of Surya, Graha becomes Combust or Vikala or Dagdha or Moudhya or Ashtangata. Combustion is caused due to a defined and certain longitudinal distances of the planet from Sun in the same sign. Combustion is deemed to be bad at it would rob away the effects of the planet. The combustion distance varies from planet to planet. Mars, Mercury, Jupiter, Venus and Saturn come at distance of 17°, 14° (12° when retrograde), 11°, 10° (8° when retrograde) and 15° on either side of Sun, they are subjected to combustion. On the other hand, when the planets come close to Moon, it is called Samagama. Nodes (Rahu/Ketu) do not have combustion as they are not radiating planets but merely shadow. Combust planet loses its power and becomes weak by combustion and it becomes unable to produce its results like other planets. This does not apply to Rahu & Ketu. Whenever a planet is combust, the native suffers with regard to the general significations of that combust planet and also suffers with regard to the significations of the house ruled by that combust planet in such a manner that the suffering becomes memorable. It is like a tragedy or a serious setback, particularly during the operational Mahadasha or sub-period of such a combust planet. In case the functional malefic planets are conjunct with or aspect the house, whose lord is combust; the significations of the said house(s) are almost totally destroyed. The only saving factor is the close aspect/conjunction of the strong functional benefic planet.

Benefic planets, if combust, need to be strengthened in order to get proper results for the house significations ruled by them. In case of positive and benefic combust planets in a horoscope, the best remedy is to strengthen these planets by wearing the suitable gemstones for these combust planets. Wearing gemstones for the positive combust planets provides them extra strength and power so that they become capable of protecting their significances and producing proper results of the house of their lordship. In case of negative or malefic combust planets, gemstones can not be prescribed to strengthen them, or gemstones should not be worn for negative or malefic planets, even if, they are combust. Strengthening negative combust planets with the help of gemstones can cause very bad results. In case of negative or malefic combust planets, the strength is provided with the help of Mantras and Poojas for them and not by using gemstones. Example: If Mercury is negative and combust in a horoscope, the best way to strengthen Mercury is to chant the Beej Mantra of Mercury for as many times a day as is prescribed by the astrologer so that Mercury may produce good results of its signification.

Combust planets lose much of their positive power and give only superficial benefits.

(1) Combust Lord Predictions:

(i) Combust lord of 1st house: He may have a life of a prisoner. He/She may have to face fear, disease and tension very often and this may result in lowered lifestyle, money crisis and obstacles in the way of success. One suffers from health problems and loss of status. If the combust lord of the first house is severely afflicted by Rahu/Ketu or the most malefic planet, it makes one vulnerable to addictions, serious health problems and loss of status.

(ii) Combust lord of 2nd house: He may behave improperly. She may even do some inconceivable acts. This combination can also cause eye pain, stammering and give a squandering nature. He/She may have a problematic life and this may result in lowered lifestyle, money crisis and obstacles in the way of success. One loses wealth, status, suffers on account of inharmonious marital relationship and male progeny. If it is severely afflicted by Rahu/Ketu or the most malefic planet, it makes one vulnerable to the diseases of eyes, thyroid, problems to teeth, problem of speech and results in sorrows on account of tragedy to male relations.

(iii) Combust lord of 3rd house: If the lord of the 3rd house is combust, this is harmful for the native's brothers. He/She might get a bad advisor and enemy in guise. He/She may feel mental stress and her self-esteem may be at a strike. One faces isolation and one's younger brothers face troubles. If it is severely afflicted by Rahu/Ketu or the most malefic planet, it makes one vulnerable to the breathing problems, mental retardation and financial losses through business ventures.

(iv) Combust lord of 4th house: Combust lord of the 4th house means the native's mother may be in trouble and pain. The native's land asset, house properties, conveyance, pleasures and happiness or pet may be in danger and he/she may be at risk from water and vehicle. Native's personal happiness is hampered in this situation. One loses mental peace and properties. One's mother and education suffer. If it is severely afflicted by Rahu/Ketu or the most malefic planet, it makes one vulnerable to the loss of assets, loss of parents, cardiac disorders and unhappiness through marital discord.

(v) **Combust lord of 5th house:** This may cause problems for children, failure of plans, problems in legs and unnecessary loss of money, knowledge and may break their love affairs. One's concentration is lost and one suffers unhappiness on the performance of children. If it is severely afflicted by Rahu/Ketu or the most malefic planet, it makes one vulnerable to the loss of children, losses through speculative investments, gambling, liver cirrhosis, stomach disorders, spinal inflammations and gall bladder problem.

(vi) **Combust lord of 6th house:** She or he may face defeat, conspiracy and may have to work under others in unfavourable +conditions. He may be hurt, afflicted by theft and may not be able to take intelligent decisions. One suffers from ill-health arising out of acidity, inflammation of intestine, renal disorders, etc. If it is severely afflicted by Rahu/Ketu or the most malefic planet, it makes one vulnerable to losses through thefts, fires, litigation and cheatings.

(vii) **Combust lord of 7th house:** It indicates separation from life partner. He may get in trouble from the opposite sex and develop illegitimate relationships. He might suffer from hidden diseases. One suffers from problems in marriage, partnerships and in foreign lands. If it is severely afflicted by Rahu/Ketu or the most malefic planet, it makes one vulnerable to marital discord, cheating in partnership business and suffers from severe difficulties in foreign lands.

(viii) **Combust lord of 8th house:** Weakness, failure, despair, hunger, disease and death are created by combust lord in the 8th house. One suffers from obstructions and delays and one's life is threatened. The combust lord of the eighth house when severely afflicted by Rahu/Ketu or the most malefic planet, it makes one vulnerable to fatal accidents and one's father may suffer financial setbacks and severe health problems. One also faces divorce in marital relationship.

(ix) **Combust lord of 9th house:** Ill fate, poverty and foolish behaviour are caused by combust lord in this house. The natives may be cursed by elderly people and teachers. One's father suffers from problem and the person gets setbacks in life. If it is severely afflicted by Rahu/Ketu or the most malefic planet, it makes one vulnerable to physical handicap and prolonged health problems.

(x) Combust lord of 10th house: He may fail at work, get demotion at the job, and may lead a difficult life where unpleasant incidents may happen very frequently. One gets persistent professional setbacks and has to move away from his native place. If it is severely afflicted by Rahu/Ketu or the most malefic planet, it makes one vulnerable to arthritis, government action, loss of reputation and sufferings on account of one's children.

(xi) Combust lord of 11th house: It means danger for elder brother of the native, bad news, loss of wealth, ear related diseases and separation from friends. One faces loss of income and trouble to friends and elder brother(s). If it is severely afflicted by Rahu/Ketu or the most malefic planet, it makes one vulnerable to blood Cancer, isolation and penury. But, if the 11th lord is exalted, though in combustion, there will be many gains.

(xii) Combust lord of 12th house: It may causes to suffer with various types of diseases, and may face a situation in which they are either imprisoned or in threat of imprisonment. One faces separation in marital relationships and loss of comforts. If it is severely afflicted by Rahu/Ketu or the most malefic planet, it makes one vulnerable to drug addiction, imprisonment and prolonged hospitalisation.

(2) Combust Planets' Predictions:

If a planet is in the state of combustion, the significations ruled by the planets are inordinately delayed, denied or damaged significations as mentioned below:

(i) If Moon is combust, the mental peace of the person is lost. One's mother or wife may not enjoy good health. There are problems in acquisition and maintenance of property. The people try to avoid the person. If such a combust Moon is under the close influence of Rahu or the lords of the eighth and twelfth houses, the person gets involved in many vices and loses mental peace. One is vulnerable to epileptic fits, lungs and mental disorders.

(ii) If Mars is combust, one is angry and suffers from injuries due to rashness and muscular disorders. One suffers from blood impurities, hypertension and multiple boils. If such a combust Mars is under the exact or close affliction of Rahu one becomes vulnerable to Cancer.

(iii) If Mercury is combust, one is confused, sensitive and slow in decision making. One suffers from over confidence or lack of confidence and nervous pressures. The peace of mind is disturbed. One becomes vulnerable to convulsions, paralysis, body-aches, breathing problems and skin diseases if the combust Mercury is under the influence of Rahu, especially from the malefic houses.

(iv) If Jupiter is combust, one suffers from liver disorders and fevers. One cannot concentrate on studies. There are obstructions in one's spiritual pursuits. One becomes selfish. One suffers on account of problems to one's male progeny and husband. If such combust Jupiter is badly placed and is under the severe affliction of Rahu/Ketu or the lords of the eighth or twelfth houses, one becomes vulnerable to liver cirrhosis, jaundice and diabetes.

(v) If Venus is combust, one faces loss of comforts and losses in business. One's wife suffers from ill-health. In the case of females they suffer from urinary infections, malfunctioning of uterus and fallopian tubes. One becomes vulnerable to skin problems, eye-diseases and renal problems if combust Venus is severely afflicted by Rahu/Ketu or the most malefic planet.

(vi) If Saturn is combust, one has to struggle more for success. One may suffer from bone fractures, pain in legs or spinal problems. One may get involved in leading the low class people or labourers. One does not enjoy good social respect.

10

Lords of Houses

10.1 Predictions by Lords in 12 House:

Lagnesh (Natal Lagna or Moon Lagna) is the performing agent in the Chart. Lagnesh represents body, appearance, features, wealth, profession, first marriage, social personality, social placement or social identity. Any planet in Lagna is rising planet and gains strength and dignity by occupying the lagna. Lagnesh become exceptionally strong when located in lagna. Chandra Rashi Lagnesh in Chandra Lagna will give his/her mother a leadership, public responsibility, and dignity.

Lagna Lord (Lagnesh) in 12 House:

Lagna lord is always stronger and benefic. If Ascendant Lord is well placed or un-afflicted, he/she will rise to a high position, will be happy, having good health and will attain gradual prosperity. There will be physical well-being, gains, and happiness. There will be good environments, good mood, fame, gain of maternal grandfathers' wealth, residence abroad, and pride during Maha Dasha of Lord of the Ascendant. If Ascendant Lord is badly placed or afflicted, the health of native will suffer during his Lagna Lord Maha Dasha. He/She will suffers with mental anxiety, accuse of personality, general retardation of success and failure in attempts leading to frustration. Followings are the specific effects of Lagna Lord Position in different house:

1. Lagna Lord in Thanu: L-1 is ruled by Swabhava of Mars. If Lagnesha is strong, full of vitality and well disposed, his identification in the society is done by one of the Pancha Maha Ppurusha Yoga (the characteristics of Graha concerned) and feels most natural and comfortable as an iconic image, physical athletic competitions, physical performance, in sporting, combat or hunting, in the center of attention receiving recognition for his expertise, working as an icon of the athletic in sports industry; or as a warrior and has high vitality, distinctive appearance, warrior-energy, high movement in body, self-determination, self-interest, enthusiasm for life. He has a strong self-image, focus, and drive toward achieving recognition for the characteristics of the graham (Lagnesha) occupying Lagna. His physical body appearance receives much attention -- positive or negative as the Graha determine. He is often photographed and his image becomes iconic for a movement, a product, an era, or an idea. He is naturally suited to an environment that emphasizes the personality, personal appearance, physical movement, and athletic prowess as "athlete" originally meant "mobility". He/She will have physical happiness and prowess. He/She will be intelligent, fickle-minded, will have two wives and will unite with other females. He becomes the focus in his/her life. He is socially identified strongly with the characteristics of the graha (Saturn) occupying Lagna.

L - 1 (Surya) in Bhava-1: If Surya is strong, full of vitality and well disposed in bhava-1 in his own sign in Kendra from Ascendant for Leo Lagna, he is excellent proprietor or entrepreneur, the creative director, the politician, a gifted actor; a brilliant theatrical director; emperor or monarch, self-employed, hunter, dancer, athlete, warrior, center-stage performer, iconic representative of a social movement, exhilarated in politics and political theatre. A very strong Surya in the radical lagna indicates a magnificent stage performer or military general. He marries an elder, experienced, practical and socially networked spouse. He may be admired for creative leadership in his profession and has prosperity in fields where independent decision-making and self-confidence are valued. If Kuja is favourable, pioneering, he gets success in politics and military professions. He is invincible in his approach to life, mathematical genius and a superb philosopher and famous.

L - 1 (Chandra) + Guru in Bhava-1: If Moon is strong, full of vitality and well disposed in bhava-1 in his own sign in Kendra from Ascendant for Cancer Lagna, he is a gifted actor, a brilliant theatrical director, gregarious; cunning, short-statured, valorous, happy, likes milk-based foods, interested in astrology, famous in the later part of life, subject to perils, and can suffer from somnambulism, or sleepwalking or daydreaming in life, and is handsome but fickle minded, scared of water. He can travel much in his 15th year.

L - 1 (Mangala) in Bhava-1: If mangal is strong, full of vitality and well disposed in bhava-1 in his own sign in Kendra from Ascendant for Aries and Scorpio Lagna, it forms a Ruchaka Yoga and he will have social identification with a strong physique, famous, king or an equal to a King, a ruddy complexion, attractive body, charitable disposition, wealthy, long-lived and leader of an army. He is martial, a leader, a great Commander, and an aggressive but patriotic ruler or an equal. He is vitally athletic, super-competitive, warrior & success in battle, endurance walker, and independence-winner, innovative, energetic, athletic, and in-charge and succeeds in a physical zed, muscular-movement environment and is active. Mars-1 is unfortunate for father, typically accidents to him. He will have good health, wealth and get higher position in defense or police. The person shall be brave, arrogant, and the victor. He earns lands and houses, millions of worth riches and money. His wife and children prosper well and enjoy good health, happiness and wealth. His brothers prosper and help him and cause him happiness. He gains huge profits by agriculture, horticulture too. He feels most natural and comfortable in youthful environments; wins in physical athletic events, in sporting and in combat or hunting competitions like Mr. Federar, and is in the center of attention for one's expertise or specialty,

L -1 (Budha) in Bhava-1: If Budha is strong, full of vitality and well disposed in bhava-1 in his own sign in Kendra from Ascendant for Bemini and Virgo Lagna, it forms a <u>Bhadra Yoga</u> and he is excellent, strong, intellectual, learned and rich, and becomes good in commerce and communication and will help relatives and will live up to a good old age. He is identified for his communication skills, talking, gesturing, reporting, writing, painting, tool-using, and technology-applying, arguing, criticizing, explaining, instructing, interacting with others, competition, and winning as political figure. He will have dispute with brother in the teen years. He is expert communicator, easy and friendly, personal styler. He is socially identified with mental abilities, science, adolescence, planning groups and associations. He prefers the company of younger people and has advertising and explaining skills, and travels far and wide; interested in the higher branches of learning, polite, kind and conciliatory and pilgrimage and great gain in 27th year. He is blessed with spouse and children; wealth, and truthful.

L - 1 (Guru) in Bhava-1: If Guru is strong, full of vitality and well disposed in bhava-1 in his own sign in Kendra from Ascendant for Sagittarius and Pisces Lagna, it formed a <u>Hamsa Yoga</u> his legs will have the markings of a conch, lotus, fish and ankusha. He will possess a handsome body; will be liked by others; righteous in disposition and pure in mind, idealistic, spiritual, broad-minded and selfless, fortunate, well built and having the voice of a swan. He gets a beautiful wife and possesses all comfort. He is religiously inclined and favourably disposed towards spiritual studies. He may be a writer, religious doctrine, dramatist model, a prosperous and high-profile career in modeling or drama, a mystical fiction writer or a multi-cultural teacher or an anthropologist-teacher. He has a natural facility with human languages, both spoken and literary and yields stupendous accomplishments in language, literature, linguistics and general humanistic understanding of civilization and culture. He is well liked by the public. He has good fortune; longevity, power and dignity to lead the social and commercial world; internal power through moral and religious self. He is optimistic, jovial, generous, prudent, sincere, courteous and amiable. He may hold a high position; such as, bankers, judges, doctors, lawyers, professors, preachers, government officers, shipping or large scale wholesaler. He is recognized by people for his proficiency, affluent, graceful.

L -1 (Shukra) in Bhava-1: If Sukra is strong, full of vitality and well disposed in bhava-1 in his own sign in Kendra from Ascendant for Taurus and Libra Lagna, it forms Malavya Yoga (Diplomacy) and he is strong-minded, fortunate, renowned, learned resolute, and has a well-developed physique, fame and name, immensely richness and wealth, happy children and wife, conveyances, sensual pleasures, music, dancing, fine arts, luxury and material comforts and pleasures. He is millionaire or a beauty queen, wealthy, good marriage and strong sense of justice and has a great deal of sexual enjoyment having good comforts in life. He always involves himselfself in good deeds, pleasant speech and skilled work. He is skilled in all trade, adviser, preceptor, home bound, powerful, fond of sour and salty food, mathematician, fond of beautiful garments, meritorious, social, artistic and has love for art, music, drama, poetry, singing-all that is beautiful, fair, generous and refined; admired by the opposite sex. He is very attractive, and has an important position in politics.

L- 1 (Shani) in Bhava-1: If Shani is strong, full of vitality and well disposed in bhava-1 in his own sign in Kendra from Ascendant for Capricorn and Aquarius Lagna, it forms Sashya Yoga and makes a philosopher of political economy and may be a president of a country along with Guru and he will command many good servants, but his character will be questionable. He has high vitality, distinctive appearance, warrior energy, movement, self-determination, self-interest, enthusiasm, a unique personality, competitive attitude, good life, wealth, intelligence and physical health at all times. He feels most natural and comfortable in youthful environments, physical athletic competitions, physical performance, in sporting, combat or hunting, in the center of attention receiving recognition for his expertise or specialty and surrounded by followers. He will command many good servants. His character will be questionable. He will be head of a village or a town or even a King, will covet other's riches and will be wicked in disposition. He no doubt becomes famous and happy but his sexual outlook would be perverse. He would be sporting with other men's wives and he would employ every unscrupulous means to gain other's money. Most corrupt and powerful or Mafia person would perhaps be having Sasa Yoga; otherwise they could not have minted millions at the cost of the poor man. In interpreting the Sasa Yoga, due consideration should be bestowed on the disposition of the Moon. If this luminary is free from affliction, the person having 'Sasa Yoga' will not covet other's wealth nor will he be unscrupulous. Where the Moon is not afflicted, the evil results attributed to Sasa Yoga can have an only 'restricted play.

2. Lagna Lord in Dhan: L-1 is ruled by Swabhava of Venus. If Lagnesha is strong, full of vitality and well disposed, he/she is socially identified with musical and artistic abilities, dramatist, industrialist, journalist, good memory, much stored wealth, active in biological and cultural conservation, treasuries of memory, wealth, knowledge, face, voice, speech, song, collections, record-keeping, beauty, knowledge of ancient languages and in particular the pronounce speech or singing. He/ She will focus on his family and accumulating wealth and is coming from a strong family background. He/She will be gainful, scholarly, happy, religious, and honourable, and has good qualities, and has many wives. He has treasuries of wealth, storage of bank savings, scholarship, expertise and physical suitability for curator and collector, professor, song-singers; speech-makers on topics of historic values; writer; philosopher; mathematician; song-singers; speech-makers on topics of historic values; writer; philosopher; mathematician; and owner of values, art, songs, traditions, memories, and family legacy behaviors. The attachment to family is strong. This is a common position for actors who look attractive, have nice teeth, strong memorization skills, and favorable speech or song. He/ She is physically attractive as Lagnesh, the lord of the body is in Shukra natural Bhava-2.

3. Lagna Lord is in Sahaj: L-1 is ruled by Swabhava of Budha. If Lagnesha is strong, full of vitality and well disposed, he is socially identified with teamwork, office work, business, government administration, commerce, local events management, writing of all kinds, editing the works of others, announcements, short-term travel, planning, scheduling, meetings, and small group process, favouring neighbour, commercial, and regional communications, pragmatic writing and media work, and sibling/team-mate/close-friend relationships, frequent business travel, messaging, business activities which generate self-made wealth such as writers, printers, publishers, announcers, reporters, evangelists, schedulers, planners, newspaperman, magazines, books, websites. He is most natural and comfortable with siblings. He is connected with his younger brothers and sisters, will acquire college degrees and spiritual initiations. He/She will be equal to a lion in valour, be endowed with all kinds of wealth, be honourable, and be intelligent and happy. If lagnesha is strong such as uchcha Kuja, he is powerfully identified with corporate travel.

4. Lagna Lord in Bandhu: L-1 is ruled by Swabhava of Chandra. If Lagnesha is strong, full of vitality and well disposed, he is socially identified by deeply patriotic, childhood education, patriotic values, parenting, shelters, ownership of the land and house properties, home and home culture, real estate, building and property management, vehicles, schooling, very parental, great teacher, a natural educator, school-teaching, fundamental education, greatly concerned for stable and productive education, most comfortable at home, and has a good education and financial security and provides better security and lifestyle for his people. He/She will be concerned with mother, home, heart and happiness. He/She will spend a lot of time at home, is concerned with the career of the spouse and will work for his/her spouse or work with her in his/her own home in perhaps a home-based business. He/She will be endowed with paternal and maternal happiness; will have many brothers, be lustful, virtuous and charming. He does not prefer to advertise or promote himself, but prefers to be sought out. He is associated with landed property. If the Graha is inauspicious, there may be disruption of ties to the land.

5. Lagna Lord in Putra: L-1 is ruled by Swabhava of Surya. If Lagnesha is strong, full of vitality and well disposed, he is socially identified as a natural celebrity with a special creative performance, intelligence, luck, speculation, gambling, authoring, performance arts, politics, literature, game and prefers a creative environment such as a theatrical stage, art gallery, music amphitheater, university lecture hall, or political campaign. He has good life in general, wealth, intelligence, physical health at all times, politics, well-known for theatrical celebrity (5), charismatic job, remarkably "lucky" in life, a natural celebrity with a special creative performance in the musical or theatrical stage and is dear to king. Children are fortunate with good appearance and good intelligence. He has too much romance in life, infused with a love of life and the physical body, fortune through speculative risk in politics, body-based performance such as acting, modeling, dancer, may be a brilliant, performance arts, speculation, gambling, creative environment Mesha lagna, Simha lagna, Thula lagna, Vrischika lagna, Dhanau lagna, Makara lagna, Meena lagna have the best results. He receives opportunities, much praise and applause for performances and privileges in creative activity, giving audiences, directing, acting, orating, playing games, being on stage. L- 1 (Surya) in Bhava-5 makes a great religious Guru like Guru Paramahamsa Yogananda, Saint Josemaria and is karaka for authorship (writer), esp. biography (writing about one's own self) like Rudyard Kipling.

6. Lagna Lord in Ari Bhava: L-1 is ruled by Swabhava of Budha. If Lagnesha is strong, full of vitality and well disposed, he is physically and socially identified with domains of medical clinics, battlefields, police and military, criminal attorney's office, money-lender and debt collection, poorhouse, bankruptcy court, social welfare agency. He is identified with medicines, drugs, poisons, illness, poverty, finance + Loan, social work and family conflict, accusations and blaming, criminals and crime, victimization, Low class, exploitation and animosity. He may be a professional in conflict management such as physician, attorney, police officer, or judge. He is a gifted advocate for victims in the practice of law and medicine, a conflict management professional such as physician, criminal or divorce attorney, police officer, social worker, or judge. He may offer a ministry of service in environments that manifest experiences of poverty, conflict, unemployment, war, enemies, toxicity, pollution, debt, and disease. He may have a praiseworthy social success in one of the conflict-management professions such as military, law, finance, social-work, or medicine. If Lagnesha is not strong, and not well disposed,he/she is identified with illness, poverty, conflict, unemployment, war, enemies, toxicity, pollution, debt, disease, exploitation, medicines, drugs, poisons, finance & loan-making, family conflict, accusations and blaming, criminals and crime, victimization, outsiders & low class, exploitation and animosity, and inability to achieve agreement. He/She will be devoid of physical happiness and will be troubled by enemies, if there is no benefic Drishti to 6th house. The affects of deaths, diseases and enemies overwhelm his/her life.

7. Lagna Lord in Yuvati Bhava: If Lagnesha is strong, full of vitality and well disposed, he/she is identified with negotiations, brokering, deals, partnerships, agreements, trades, match-making, terms of the contract, an agent of making deals within the law courts, in professional advising and counseling activities, executing barters, broker, match-maker, trader and negotiations of all kinds, such as, justice, equity, fairness, arrangement. He/She is identified with negotiations, brokering, deals, partnerships, agreements, trades, match-making, terms of the contract, an agent of making deals within the law courts, in professional advising and counseling activities, executing barters, broker, match-maker, trader and negotiations of all kinds, such as, justice, equity, fairness, arrangement. His/Her wife will not live long. It is a benefic; he/she will wander aimlessly, face penury and be dejected. He/She will alternatively become a king (if that Graha is strong). He/She is heavily focused on his/her partner, spouse, in life. If Lagnesh is Budha, he is a natural advocate. Attorneys with this placement are professionally successful in law courts and legal offices, mediation and arbitration tables, counseling rooms, and discussion with partners including spouses and is typical wealthy due to commercial acumen (unless the Rashi is unsuitable for either Budha or Shukra). If it is Mangala (Karaka for two unions), he is best match-maker for quick-in, quick-out of partnerships. It is favouable for a stable and respectable alliance and a strong factor for marital longevity and one is motivated to persist in the union even when unstable conditions exist. He will reside in foreign countries or in the house of his father-in-law. He may have multiple intimate relations and will be given to pleasures and will adorn himself, self with scents, and other beauty care products. If L-2 is in Yuvati Bhava, family interests and concerns are expressed through the marriage. L-3 in Yuvati Bhava, the spouse is administratively competent and inclined toward communicative aspects of business. L-4 in Yuvati Bhava, it is favorable for a stable and respectable alliance and a strong factor for marital longevity.

8. Lagna Lord in Randhra Bhava: If Lagnesha is strong, full of vitality and well disposed, he/she will be an accomplished scholar, be sickly, thievish, be given too much anger, be a gambler and will join others' wives. He/She will face serious physical harm, unforeseen difficulties and serious problems in lifetime. There will be sudden changes, shock, surgery, evolution, transformation, mysteries, and discovery of treasures, rebirth, secret dealings, treasures and piracy. Effects of any Graha in 8 are eruptive, secretive, occult, mysterious, hidden, magical, and changing-changing-changing. Lagnesh in Randhra or join a malefic is considered a "Deha-Kashta Yoga" meaning" bad , ill , evil, wrong, painful, grievous, severe, miserable, difficult, troublesome for the body". He will be devoid of bodily comforts. Most of the manual workers belong to this category. The Yoga is said to become defunct or Deha-Kashta yoga does not apply, if a benefic aspects or conjoins the ascendant lord in Randhra. Lagnesh in Randhra carries strong circumstances of sudden death. If the lord of the 1, who represents the body in the chart, conjoins a malefic planet or is in 8, and is not aspect or conjoins by a benefic, the simple pleasures of life, such as peace and physical comfort, are difficult to achieve throughout the life. If Lagnesh is a strong malefic or associated with a malefic (esp Rahu), one may be employed in emergency services, such as first-responder in fire, police, military, or medical intervention services and is at home in secret meetings, handling confidential or volatile materials, in tantric practice, in secret discovery modes and wants to be in domains where hidden matters are revealed, where healings occur as secrets are brought to the surface like Tantric, the psychiatrist's office, psychic reading room, stock market trading desk, oil drilling and mining, treasure hunter diving ship, police detective, archeological dig are all appropriate to his identification with concealed matters. Lagnesh in 8 is identified with secret transactions, hidden sources of wealth, including private investments, confidential therapeutic privilege, and classified

9. Lagna Lord in Dharma Bhava: If Lagnesha is strong, full of vitality and well disposed, he is socially identified with wisdom, preacher & teaching, fortunate, dear to people, devotee of Shri Vishnu, skillful, eloquent in speech, endowed with wife, sons, and wealth, sacred texts-knowledge, world travel, life philosophy, worship, and a fortunate person who can speak well, who has a good spouse and children, acquisition of wealth will be through her father and elders and who has a good future life as well. He/She will be fortunate, dear to people, be a devotee of Sri Vishnu, be skilful, eloquent in speech and be endowed with wife, sons and wealth. He/She is an overall fortunate and has all general good things in lifetime. L- 1 (Shani) in Bhava-9 produced Swami Vivekananda, Rush Limbaugh and Elizabeth Kubler Ross as a preacher. He is a preacher-teacher with philosophical beliefs, religious doctrine, ritual priesthood, the culture of liturgy and worship, or dharma teachings who feels most natural and comfortable in the university, the temple, delivering sermons (in person, by writing, or by communications media) . The perceived life-defining task is transmission of higher knowledge.

10. Lagna Lord in Karma Bhava: If Lagnesha is strong, full of vitality and well disposed, he is strongly identified with leadership, lawful environments, symbolic social roles, public responsibilities and institutional governance roles, positioned near the top of some hierarchy, well-known for dignified leadership roles such as parenting, business director, and governance of organizations, kinetic energy of the physical body movement and fleshly appearance and dynamically into high positions for public viewing and socially respected regardless of the job function. There will be easy success from positive recognition. He/She will be endowed with paternal happiness, royal honour, fame among men and will doubtlessly have self-earned wealth. He/She is heavily focused on the attainment of career, status position and success. If Lagnesh is fallen, there will be self-esteem issues can sabotage career performance. He is a dignified parent with the family's social standing, often going back several generations living long, devoted to his well-being, help to raise his children, and are persons of respect and reputation themselves. He is most natural and comfortable in the corporate or government in executive office, leading mass movements, and being a public icon. He is an exceptionally public person, quite famous and identified with prominent social position and recognized for lifetime public service, hierarchical leadership roles, making decisions which affect others, setting cultural standards for their group, attracted to domains of world government and policy, dignified public assemblies, regal ceremonies of state, the throne-chair in the corporate board room; enjoys roles such as village headman, president of trade guild, and has fame among men and will doubtlessly have self-earned wealth like USA Pres-41 George H. W. Bush "Sr." (with Surya), UK-Prince Charles & Meryl Streep (with Chandra), Mother Teresa of Kolkata & UK-Princess Diana Spencer (with Mangal), Albert Einstein (with Budha), Catherine Zeta-Jones & Larry King (with Guru), UK-Princess Anne Laurence & Tom Cruise (with Sukra) and poet and patriot Jiddu Krishnamurti & William Butler Yeats (with Sani & Rahu).

11. Lagna Lord in Labh Bhava: If Lagnesha is strong, full of vitality and well disposed, he is identified with environments that feature active networking and gain via links of profitable association in commercial and community organizations, in large social groups and assemblies, as "fund-raisers", and in the marketplace of goods and ideas, the principled and well-connected social progressive activist, social net worker, economies, distributive and associative networks, social movements, marketplace income, large group assemblies, gains and profits and is happy when moving from large event to large event. He/She will always be endowed with gains, good qualities, and fame. He/She is heavily focused upon the achievement of desires and certain gainful things in life and attached to friend and oldest siblings. If L-1 in 11 is Surya, he is engaged in political or entertainment career focused upon economics (11) or community development (11), global politics, theatre of philosophical explication. If L-1 in 11 is Mangala, he is engaged in muscular movement, career focused upon active pursuit of social and economic gain. If L-1 in 11 is Budha, he is engaged in large-assembly communications, rallies, meetings, planning, and advertising. If L-1 in 11 is Shukra, he is engaged in large-scale negotiation, deals, bargaining, brokering; also, can be decorating, or matching-making profession. If L-1 in 11 is Shani, he is engaged in extensive social networking necessary to establish effective governance. He/She is recognized for ability to coordinate and distribute resources on a large and well-ordered scale, like Elvis Presley who has the huge fan-base assemblies which crowded his concerts and the "mela" scene in Las Vegas and USA President-39, Jimmy Carter who concerned to distribute system-participatory opportunity via networks of charitable and political association just as easily mixing with the crowd.

12. Lagna Lord in Vyaya Bhava: If Lagnesha is strong, full of vitality and well disposed, he is socially identified with his "bed pleasures", visionary imagination, prayer and spiritual guidance, private bedroom, hospital and hotel, meditative spiritual retreat, divination, an internationalized individual who feels equally comfortable in the homeland and abroad, trait emerging most vividly in foreign lands during the Vimshottari Mahadasha period of the Lagnesh. He/She will be busy in charity, donations, and may live for some other purpose other than own life. He wants to be in an enclosed, sanctuary space like, bedroom, dormitory, attic, a spa, ashram, hospital, monastery, prison, peaceful courtyard, the ordered space of the meditation hall, the interior compound, the super-private meeting room, concentration camp, survivalist compound and any walled enclosure with very high levels of privacy in all undertakings, both personal and professional. If associated with other Graha, he will have high profile public life; creative ideas, decision-making skills, core relationships in a very private space and his appearance on the world stage. If Lagnesh is strong along with Surya, he can be a dramatic articulator of the collective imagination, like Madonna and Jack Welch. If Shukra is involved, may be a seductive tempter/temptress into a world of forbidden, private delights like Madonna and wealthy.

10.2 Predictions by Ascendant Lord-Yoga:

Sareera Soukhya Yoga: If the Lord of Lagna, Jupiter or Venus, are in Kendra, the Yoga is highly powerful and native is endowed with long life, wealth and political favours.

Deha-pushti Yoga: If the Ascendant lord is in a movable sign and aspect by a benefic gives rise to this Yoga. The native will be happy, will possess a well-developed body, will become rich and will enjoy life.

Deha-kashta Yoga: If the Ascendant lord join a malefic or occupy the 8th house, he will be devoid of bodily comforts. Most of the manual workers belong to this category. The Yoga is said to become defunct if a benefic aspects the ascendant lord.

Roga-grastha Yoga: (a) If the Lord of Lagna occupies the ascendant in conjunction with the lord of the 6th, 8th or 12th; or (b) if the weak Lord of Lagna (i. e. Shadbala Pinda (sum total of strength) falls short of the requisite quantity) joins a Trine or a Kendra, he will not have a healthy constitution. It will lack the requisite power of resistance so that he falls an easy prey to disease. He will possess a weak constitution and is sickly and will suffer from disease, infirmity, sickness, possessed by a demon; tormented; eclipsed; slurred; and has inarticulate pronunciation of the vowels

Krisanga Yoga: If the Ascendant lords occupies a dry sign (Aries, Taurus, Gemini, Leo, Virgo and Sagittarius) or the sign owned by a 'dry' planet (Sun, Mars, Mercury and Saturn), or the Navamsha Lagna is owned by a 'dry' planet and malefic joins the Lagna, he will be lean, emaciated, thin, spare, weak, feeble, small, insignificant and will have an emaciated or lean body and will suffer from bodily pains. The same result can be judged if the Lagna and the lord have acquired a large number of dry Varga or Divisions, such as Rashi, Hora, etc.

Deha-Sthoulya Yoga: If Lord of Lagna and the planet, in who's Navamsa the Lord of Lagna is placed, occupy watery sign (Cancer, Aquarius, Capricorn, Pisces, Scorpio and Libra) or the Lagna is occupied or and aspect by Jupiter from a watery sign or the ascendant must fall in a watery sign in conjunction with benefic or the ascendant lord must be a watery planet (Moon, Mercury and Jupiter and Venus), the native will have corpulent appearance, stoutness of the body and has no well-built or strong physical appearance. He will have bulky, big, huge, coarse, rough; dull, stolid, doltish, stupid, grown fat body.

Sada Sanchara Yoga: If the lord of either Lagna or the sign occupied by Lagna lord is in a movable sign or both the Lagna and the Navamsha Lagna are in movable signs, the native will always or most of the time, walking about, wandering, roaming, driving or riding, and he will almost always be a wanderer. This combination is very common in the horoscopes of travelling agents, diplomats and globe-trotters.

Bahudra-vyarjana Yoga: If the Lord of the Lagna in the 2nd, lord of the 2nd in the 11th and the lord of the 11th in Lagna or if a point of contact is established between the Lagna, 2nd and 11th houses and the lords of these three houses interchanging their respective positions, the native will earn lot of money and will amass a good fortune. Here again, the real value of the Yoga depends upon the strength of the lords concerned and how they are disposed in regard to the general scheme of the horoscope.

Swa-veerya-ddhana Yoga: If Lord of Lagna, being the strongest planet, occupies a Kendra in conjunction with Jupiter and the 2nd lord should join Vaiseshikamsa (a Graha which occupies a favorable sign in 13 out of 16 varga) or the lord of the sign, in which the lord of the Navamsha occupied by the Ascendant lord is, is strong, and join a Kendra or a trine from the 2nd lord or occupies his own or exaltation sign, or the 2nd lord occupies a Kendra or trine from the 1st lord or the 2nd lord being a benefic is either in deep exaltation or in conjunction with an exalted planet, then the native will earn money by his own efforts and exertions as a self-earned money and will have inherited wealth.

Daridra Yoga: If the lords of the 12th and Lagna exchange their positions and is conjoined; or is aspect by the lord of the 7th; or the lords of the 6th and Lagna interchange their positions and the Moon is aspect by the 2nd or 7th lord; or Ketu and the Moon are in Lagna; or the Lord of Lagna is in the 8th aspect by or in conjunction with the 2nd or 7th lord; or the Lord of Lagna joins the 6th, 8th or 12th with a malefic aspect by or combined with the 2nd or 7th lord; or Lord of Lagna is associated with the 6th, 8th or 12th lord and subjected to malefic aspects; or the lord of the 5th joins the lord of the 6th, 8th or 12th without beneficial aspects or conjunctions; or the lord of the 5th is in the 6th or 10th aspect by lords of the 2nd, 6th, 7th, 8th or 12th; or Natural malefic, who do not own the 9th or 10th, occupies Lagna and associate with or is aspect by the maraka lords; or the lords of the Lagna and Navamsha Lagna occupy the 6th, 8th or 12th and have the aspect or conjunction of the lords of the 2nd and 7th; then, this yoga indicates poverty, financial straits, wretchedness and miseries and the native is Daridra and roving, strolling, poor, needy, deprived of and a beggar, living a life of a horrible specter more grim like than even death, wretched life. He is compelled to live in squalid and unhealthy houses. His children have no facilities for education, health and decent existence and are shunned everywhere.

Predictions by Dhan Lord in Various Bhava:

If second Lord is well placed or un-afflicted, he/she will be blessed with wealth, additions to the family, success due to speech and good food. There will be overall happy financial position, inflow of finance, wealth possessions, education, grains, family, happy domestic life, Sanyas, inheritance, law suits and domestic comforts during Maha Dasha of Lord of the Second. If second Lord is afflicted or badly placed or afflicted, he/she has to incur heavy expenses on family affairs, eye trouble and fear from government due to financial irregularities. There will be bad food, suffering due to his/her bad speech or loose talking. There will be distress and possibility of death during Maha Dasha of Lord of the Second. As it is the lord of Maraka Dasha, so, if second Lord is associated with 6, 7, 8th house, the troubles will multiply. Followings are the specific effects of Lagna Lord in different house:

1. Dhan Lord is in Thanu Bhava, he/she will be endowed with sons and wealth, be inimical to his/her family, lustful, hard-hearted and will do others' jobs.
2. Dhan Lord is in Dhan Bhava; he/she will be wealthy, proud, will have two or more wives and be bereft of progeny.
3. Dhan Lord is in Sahaj Bhava, he/she will be valorous, wise, virtuous, lustful and miserly; all these, when related to a benefic. If related to a malefic, he/she will be a heterodox.
4. Dhan Lord is in Bandhu Bhava, he/she will acquire all kinds of wealth. If Dhan Lord is exalted and is yuti with Guru, he/she will be equal to a king.
5. Dhan Lord is in Putra Bhava, he/she will be wealthy. Not only he/she, but also his/her sons will be intent on earning wealth.
6. Dhan Lord is in Ari Bhava along with a benefic, he/she will gain wealth through his enemies; if Dhan Lord is yuti with a malefic, there will be loss through enemies apart from mutilation of shanks.
7. Dhan Lord is in Yuvati Bhava, he/she will be addicted to others' wives and he/she will be a doctor. If a malefic is related to the said placement by yuti with Dhan Lord, or by Drishti, his/her spouse will be of questionable character.
8. Dhan Lord is in Randhra Bhava, he/she will be endowed with abundant land and wealth. But he/she will have limited marital felicity and be bereft of happiness from his/her elder brother.
9. Dhan Lord is in Dharma Bhava, he/she will be wealthy, diligent, skilful, and sick during childhood and will later on be happy and will

visit shrines, observing religious code. He/She will earn finance or wealth by fortune and will earn money in the lottery and spend in religious work.

10. Dhan Lord is in Karma Bhava, he/she will be libidinous, honourable and learned; He/She will have many wives and much wealth, but he/she will be bereft of filial happiness.

11. Dhan Lord is in Labh Bhava, he/she will have all kinds of wealth, be ever diligent, honourable and famous.

12. Dhan Lord is in Vyaya Bhava; he/she will be adventurous, be devoid of wealth and be interested in other's wealth, while his/her eldest child will not keep him happy.

Predictions by Sahaj Lord in Various Bhava:

If third Lord is well placed or un-afflicted, there will be cooperation and help from co-born, rise in career, rise in self-earning, profitable short journeys and success in educational attainments during Maha Dasha of the planet. If Third Lord is afflicted or badly placed or afflicted, there will be quarrel and misunderstanding with brother/cousin/neighbour, mind will be in tension, activities will be performed at a low spirit; troublesome journeys, mental anxiety and may have to suffer due to his wettings/agreements. There will be unfavourable effects. There will be breaking of relation with younger siblings or younger brothers & sisters; and loss of vitality, personality and power to face the struggles, loss in business, and journey (short trips), parent's death, and diseases like cough, respiratory system during Maha Dasha of the planet. Followings are the specific effects of Lagna Lord Position in different house:

1. Sahaj Lord is in Thanu Bhava, he/she will have self-made wealth, be disposed to worship, be valorous and be intelligent, although devoid of learning.

2. Sahaj Lord is in Dhan Bhava; he/she will be corpulent, devoid of valour, will not make many efforts, be not happy and will have an eye on others' wives and others' wealth. The younger siblings will take the wealth his/her wealth. It also indicates loss to the younger siblings, but of course, all these lies in the nature of the planet concerned.

3. Sahaj Lord is in Sahaj, he/she will be endowed with happiness through co-born and have wealth and sons, be cheerful and extremely happy. He/She will have company of brothers and sisters.

4. Sahaj Lord is in Bandhu Bhava, he/she will be happy, wealthy and intelligent, but will acquire a wicked spouse.

5. Sahaj Lord is in Putra Bhava; he/she will have sons and be virtuous. If in the process Sahaj Lord be yuti with, or receives a Drishti from a malefic, he/she will have a formidable spouse.

6. Sahaj Lord is in Ari Bhava, he/she will be inimical to his/her co-born, be affluent, will not be well disposed to his maternal uncle and be dear to his/her maternal aunt.

7. Sahaj Lord is in Yuvati Bhava, he/she will be interested in serving the king. He will not be happy during boyhood, but the end of his/her life he/she will be happy.

8. Sahaj Lord is in Randhra Bhava; he/she will be a thief, will derive his/her livelihood serving others and will die at the gate of the royal palace.

9. Sahaj Lord is in Dharma Bhava; he/she will lack paternal bliss, will make fortunes through wife and will enjoy progeny and other pleasures.

10. Sahaj Lord is in Karma Bhava; he/she will have all lands of happiness and self-made wealth and be interested in nurturing wicked females.

11. Sahaj Lord is in Labh Bhava; he/she will always gain in trading, be intelligent, although not literate, be adventurous and will serve others.

12. Sahaj Lord is in Vyaya; he/she will spend on evil deeds, will have a wicked father and will be fortunate through a female.

Predictions by Bandhu Lord in Various Bhava:

If fourth Lord is well placed or un-afflicted, he/she will be helpful to family; will gain good relations from mother; acquire land, house, property, agricultural land; acquire vehicle and overall a happy period. There will be acquisition of house and land, family affairs, happiness, possession of vehicles, higher education such as master's degrees, or doctorates, short journeys and pleasure trips, gain of fame and wealth, pleasures, luxury, increase in savings, reputation/popularity, and pets during Maha Dasha of the planet. If fourth Lord is afflicted or badly placed or afflicted, t there will be Distress due to mother; Disputes in family leading to loss of property; litigation; discomfort of conveyance; danger from water; no peace of mind during Maha Dasha of the planet. Followings are the specific effects of Lagna Lord Position in different house:

1. If Bandhu Lord is in Thanu Bhava, he/she will be endowed with learning, virtues, ornaments, lands, conveyances and maternal happiness.

2. If Bandhu Lord is in Dhan Bhava, he/she will enjoy pleasures, all kinds of wealth, family life and honour and be adventurous. He/She will be cunning in disposition.

3. If Bandhu Lord is in Sahaj Bhava, he/she will be valorous, will have servants, be liberal, virtuous and charitable and will possess self-earned wealth. He/She will be free from diseases.

4. If Bandhu Lord is in Bandhu, he/she will be a minister and will possess all kinds of wealth. He/She will be skilful, virtuous, honourable, learned, and happy and be well disposed to his spouse.

5. If Bandhu Lord is in Putra Bhava, he/she will be happy and be liked by all. He will be devoted to Sri Vishnu, be virtuous, honourable and will have self-earned wealth.

6. If Bandhu Lord is in Ari Bhava, he/she will be devoid of maternal happiness, be given to anger, be a thief and a conjurer, be independent in action and be indisposed.

7. If Bandhu Lord is in Yuvati Bhava, he/she will be endowed with a high degree of education, will sacrifice his/her patrimony and be akin to the dumb in an assembly.

8. If Bandhu Lord is in Randhra Bhava, he/she will be devoid of domestic and other comforts, will not enjoy much parental happiness and be equal to a neuter.

9. If Bandhu Lord is in Dharma Bhava, he/she will be dear to one and all, be devoted to God, be virtuous, honourable and endowed with every land of happiness.

10. If Bandhu Lord is in Karma Bhava, he/she will enjoy royal honours, is an alchemist, be extremely pleased, will enjoy pleasures and will conquer his/her five senses.

11. If Bandhu Lord is in Labh Bhava, he/she will have fear of secret disease; he will be liberal, virtuous, charitable and helpful to others.

12. If Bandhu Lord is in Vyaya Bhava he/she will be devoid of domestic and other comforts, will have vices and be foolish and indolent. The mother will be strong in his/her religion and it will cause a great impact him. So, he/she will have strong faith, belief in spirituality or religion.

Predictions by Putra Lord in Various Bhava:

If fifth Lord is well placed or un-afflicted, there will be attainment of education; passing a competitive examination; birth of a child (according to age); gain and promotion in career; political success; sudden gain of money; sudden rise in attraction to women. There will be progress in education, child birth and happiness from the children. There will be love affairs & relationships, gain by speculation, lotteries, luck, and pregnancy during Maha Dasha of the planet. If fifth Lord is afflicted or badly placed or afflicted, there will be sickness of child; worst affliction may lead to death; worries on account of children like failure in examination; poor performance in studies; victim of fraud and deceit and a life full of distress during Maha Dasha of the planet. Followings are the specific effects of Lagna Lord Position in different house:

1. Putra Lord is in Lagna; he/she will be scholarly, be endowed with progeny happiness, be a miser, be crooked and will steal others' wealth.

2. Putra Lord is in Dhan Bhava, he/she will have many sons and wealth, be patter families, be honourable, be attached to his/he spouse and be famous in the world.

3. Putra Lord is in Sahaj, he/she will be attached to his/her co-born, be a tale bearer and a miser and be always interested in his/her own work.

4. Putra Lord is in Bandhu, he/she will be happy, endowed with maternal happiness, wealth and intelligence and be a king, or a minister, or a preceptor. He/She will be money earner right from boyhood. He/She will live in a luxurious building with beautiful surroundings for his/her children to enjoy. His/Her mother will have good longevity. He/She may have more daughters than sons. He will become rich due to investments and speculations.

5. Putra Lord is in Putra Bhava, he/she will have progeny, if related to a benefic; there will be no issues, if malefic is related to Putra Lord, placed in Putra Bhava. Putra Lord in Putra Bhava will, however, make one virtuous and dear to friends.

6. Putra Lord is in Ari Bhava, he/she will obtain such sons, who will be equal to his/her enemies, or will lose them, or will acquire an adopted or purchased son.

7. Putra Lord is in Yuvati Bhava; he/she will be honourable, very religious, endowed with progeny happiness and be helpful to others.

8. Putra Lord is in Randhra Bhava; he/she will not have much progeny happiness, be troubled by cough and pulmonary disorders, be given to anger and be devoid of happiness.

9. Putra Lord is in Dharma Bhava, he/she will be a prince, or equal to him, will author treatises, be famous and will shine in his/her race.

10. Putra Lord is in Karma Bhava; he/she will enjoy a Raja Yoga and various pleasures and be very famous.

11. Putra Lord is in Labh Bhava, he/she will be learned, dear to people, be an author of treatises, be very skilful and be endowed with many sons and wealth.

12. Putra Lord is in Vyaya Bhava, he/she will be bereft of happiness from his/her own sons, will have an adopted, or purchased son.

Predictions by Ari Lord in Various Bhava

If sixth Lord is well placed or un-afflicted, then he/she will have victory over enemies - success all of a sudden; rise in personal efforts, good health; redemption of debt during Maha Dasha of the planet. If sixth Lord is afflicted or badly placed or afflicted, he/she will suffer loss through enemies/colleagues/business opponents, reversals in life; constant troubles due to labour unrest, shortage of raw material; sudden troubles from income Tax/Excise. If Sun and Jupiter, being the lord of Sixth house, are badly afflicted, there ill be humiliation in family circles; loss of vehicles; loss of job due to negligence. There will be danger from enemies and ill health. There will be debts, law suites, challenges from competitors or thieves and robberies, accidents, acute illnesses, obstacles in life, mental worries, calamity, fighting, imprisonment, misery, success over enemies, loss of moneys, cheatings and calamities (troubles) through women during Maha Dasha of the planet. Followings are the specific effects of Lagna Lord Position in different house:

1. Ari Lord is in Thanu Bhava; he/she will be sickly, famous, and inimical to his/her own men, rich, honourable, adventurous and virtuous.

2. Ari Lord is in Dhan Bhava; he/she will be adventurous, famous among his/her people, will live in alien countries, be happy, be a skilful speaker and be always interested in his/her own work.

3. Ari Lord is in Sahaj; he/she will be given to anger, be bereft of courage, inimical to all of his co-born and will have disobedient servants. He/She has to face tremendous enemies and non co-operation from neighbours. It makes him/her angry and loses confidence. Enmity to brother is also predicted.

4. Ari Lord is in Bandhu Bhava, he/she will be devoid of maternal happiness, be intelligent, be a tale bearer, and be jealous, evil-minded and very rich.

5. Ari Lord is in Putra Bhava, he/she will have fluctuating finances. He/She will incur enmity with his/her sons and friends. He/She will be happy, selfish and kind. There is a draining away of the negative things in life such as death, diseases and enemies.

6. Ari Lord is in Ari Bhava; he/she will have enmity with the group of his/her kinsmen, but be friendly to others and will enjoy mediocre happiness in matters, like wealth.

7. Ari Lord is in Yuvati Bhava, he/she will enjoy happiness through wedlock. He/She will be famous, virtuous, honourable, adventurous

and wealthy. He/She will be lucky with respect to 12 house of final emancipation. His/Her expenditure will be curtailed and there will not be any incarnation.

8. Ari Lord is in Randhra Bhava; he/she will be sickly, inimical, will desire others' wealth, be interested in others' wives and be impure.

9. Ari Lord is in Dharma Bhava; he/she will trade in wood and stones, 'Pashan' also mean poison, and will have fluctuating professional fortunes.

10. Ari Lord is in Karma Bhava; he/she will be well known among his/her men, will not be respectfully disposed to his/her father and will be happy in foreign countries. He/She will be a gifted speaker.

11. Ari Lord is in Labh Bhava, he/she will gain wealth through his/her enemies, be virtuous, adventurous and will be somewhat bereft of progeny happiness.

12. Ari Lord is in Vyaya Bhava; he/she will always spend on vices, be hostile to learned people and will torture living beings.

Predictions by Yuvati Lord in Various Bhava:

If seventh Lord is well placed or un-afflicted, he/she will marry during this period; can successfully enter into partnerships; chances of foreign tours; if already married, then happiness, good relations with wife, pleasure and enjoyment from wife or opposite sex; acquisition of luxury household items, and marriage of self or someone in family. There will be travel, trade, marriage, partnership in business, conjugal happiness, honour and reputation in foreign country, residence in foreign lands, and journeys to distant places during Maha Dasha of the planet. If Seventh Lord is afflicted or badly placed or afflicted, he/she may suffer due to wife and partnership, and extent of sufferings depends on the association of malefic planets with 7th Lord. There will be bad relations with in-laws; separation from wife, contacts with ill-reputed women during Maha Dasha of the planet. As it is Marakasthan, the Lord of 7th house, if it is in 2nd, 8th and 12th house or associated with 2nd, 8th and 12th lord, will give physical disturbance/ailments. There will be distress to spouse and the possibility of the death, litigations and courts cases, divorce, and increase in expenditure. Followings are the specific effects of Lagna Lord Position in different house:

1. Yuvati Lord is in Thanu Bhava, he/she will go to others' wives, be wicked, skilful, devoid of courage and afflicted by windy diseases.

2. Yuvati Lord is in Dhan Bhava; he/she will have many wives, will gain wealth through his/her wife and be procrastinating in nature.

3. Yuvati Lord is in Sahaj, he/she will face loss of children and scarcely there will be a living son. He/She may have a daughter. He/She will be devoted to his/her wife. His/Her in-laws will help him/her whenever he/she is in crisis. He/She will have immense literary talent and will enjoy fine arts, ornaments and dress. He/She will have good public relations which advance his/her development.

4. Yuvati Lord is in Bandhu Bhava, his/her wife will not be under his/her control. He/She will be fond of truth, intelligent and religious. He/She will suffer from dental diseases.

5. Yuvati Lord is in Putra Bhava, he/she will be honourable, endowed with all (i.e. seven principal) virtues, always delighted and endowed with all kinds of wealth.

6. Yuvati Lord is in Ari Bhava, he/she will beget a sickly wife and he/she will be inimical to him/her. He/She will be given to anger and will be devoid of happiness.

7. Yuvati Lord is in Yuvati Bhava, he/she will be endowed with happiness through spouse, be courageous, skilful and intelligent, but only afflicted by windy diseases.

8. Yuvati Lord is in Randhra Bhava, he/she will be deprived of marital happiness. His/Her spouse will be troubled by diseases, be devoid of good disposition and will not obey him/her.

9. Yuvati Lord is in Dharma Bhava; he/she will have union with many women, be well disposed to his/her own wife and will have many undertakings. He/She and his spouse will have faith in religion.

10. Yuvati Lord is in Karma Bhava; he/she will beget a disobedient spouse, will be religious and endowed with wealth, sons. He/She works with his/her spouse.

11. Yuvati Lord is in Labh Bhava; he/she will gain wealth through his/her spouse, be endowed with less happiness from sons etc. and will have daughters.

12. Yuvati Lord is in Vyaya Bhava, he/she will incur penury, be a miser and his/her livelihood will be related to clothes. His/Her spouse will be a spendthrift.

Predictions by Randhra Lord in Various Bhava:

If eighth Lord is well placed or un-afflicted, but eighth house lord should not be stronger than ascendant lord, there will be elevation in rank, benefit through court judgments, many quarrels will be settled; and acquisition of immovable or permanent assets, inheritances or money from others such as wills and insurance policies, monetary gains from partner, during Maha Dasha of the planet. If eighth Lord is afflicted or badly placed or afflicted, he/she will have ill health, distress and sorrows. There will be rivalry and opposition in many fields. The Lord of 8th house, the strength will be more during its Maha Dasha, if associated with 2nd and 7th lord, is a potential killer and so there will be death of self or in family and loss of reputation during Maha Dasha of the planet. There will be the possibility of death and financial losses. There will be accidents, worries, inheritance, mental pain, obstacles, and gain from in laws, imprisonment, investigation, scandals, bankruptcy, and trouble from enemies, lost property and unions with others. Followings are the specific effects of Lagna Lord Position in different house:

1. Randhra Lord is in Thanu Bhava; he/she will be devoid of physical felicity and will suffer from wounds. He/She will be hostile to gods and Brahmins.
2. Randhra Lord is in Dhan Bhava; he/she will be devoid of bodily vigour, will enjoy a little wealth and will not regain lost wealth.
3. Randhra Lord is in Sahaj Bhava, he/she will be devoid of fraternal happiness, be indolent and devoid of servants and strength.
4. Randhra Lord is in Bandhu Bhava, he/she will be deprived of its mother. He/She will be devoid of a house, lands and happiness and will doubtlessly betray his friends.
5. Randhra Lord is in Putra Bhava, he/she will be dull witted, will have limited number of children, be long-lived and wealthy. He/She will be subject to dire attitude. His good actions and altruistic behaviour may go unnoticed which gives him/her the creeps.
6. Randhra Lord is in Ari Bhava; he/she will win over his/her enemies, be afflicted by diseases and during childhood will incur danger through snakes and water.
7. Randhra Lord is in Yuvati Bhava, he/she will have two wives. If Randhra Lord is yuti with a malefic in Yuvati Bhava, there will surely be downfall in his business.
8. Randhra Lord is in Randhra Bhava, he/she will be long-lived. If the said Graha is weak, being in Randhra Bhava, the longevity will be medium, while he/she will be a thief, be blameworthy and will blame others as well.
9. Randhra Lord is in Dharma Bhava, he/she will betray his/her religion, is a heterodox, will beget a wicked spouse and will steal others' wealth.
10. Randhra Lord is in Karma Bhava; he/she will be devoid of paternal bliss, is a talebearer and be bereft of livelihood. If there is a Drishti in the process from a benefic, then these evils will not mature.

11. Randhra Lord along with a malefic is in Labh Bhava; he/she will be devoid of wealth and will be miserable in boyhood, but happy later on. If Randhra Lord is yuti with a benefic and be in Labh Bhava; he/she will be long-lived.

12. Randhra Lord is in Vyaya Bhava; he/she will spend on evil deeds and will incur a short life. More so, if there is additionally a malefic in the said Bhava.

Predictions by Dharma Lord in Various Bhava

If ninth Lord is well placed or un-afflicted, he/she will have prosperous time with family; gain in wealth and rank; success in higher education/may go to other place for higher education; doing religious and charitable acts; pilgrimage and fruitful journey; develop power of intuition; blessings of preceptors. There will be educational gain, religious gain and unexpected gains of wealth. There will be fortunes, foreign- travels, higher education, writing books, prosperity, pilgrimages or journeys to gain spiritual knowledge, gain from lottery and Diksha (spiritual initiation) during Maha Dasha of the planet. If Ninth Lord is afflicted or badly placed or afflicted, there will be decline in prosperity; domestic unhappiness; weak financial status; fall in respect and reputation during Maha Dasha of the planet. Followings are the specific effects of Lagna Lord Position in different house:

1. Dharma Lord is in Lagna, he/she will be fortunate, will be honoured by the king, be virtuous, charming, learned and honoured by the public.

2. Dharma Lord is in Dhan Bhava, he/she will be a scholar, be dear to all, wealthy, sensuous and endowed with happiness from wife, sons etc.

3. Dharma Lord is in Sahaj Bhava, he/she will be endowed with fraternal bliss, be wealthy, virtuous and charming.

4. Dharma Lord is in Bandhu Bhava; he/she will enjoy houses, conveyances and happiness, will have all kinds of wealth and be devoted to his/her mother. He/She will be fortunate to have a luxurious house and be healthy.

5. Dharma Lord is in Putra Bhava, he/she will be endowed with sons and prosperity, devoted to elders, bold, charitable and learned.

6. Dharma Lord is in Ari; he will enjoy meagre prosperity, be devoid of happiness from maternal relatives and be always troubled by enemies. His/Her father's health may be affected and may have clashes with his/her mother. Wealth will be inherited from his/her father.

7. Dharma Lord is in Yuvati Bhava, he/she beget happiness after marriage, be virtuous and famous.

8. If Dharma Lord is in Randhra Bhava, he/she will not be prosperous and will not enjoy happiness from his/her elder brother. He/She will lose his/her father early in life and will be raised without a father.

9. Dharma Lord is in Dharma Bhava; he/she will be endowed with abundant fortunes, virtues and beauty and will enjoy much happiness from co-born.

10. Dharma Lord is in Karma Bhava, he/she will be a king, or equal to him/here, or be a minister, or an Army chief, be virtuous and dear to all.

11. Dharma Lord is in Labh Bhava, he/she will enjoy financial gains day by day, be devoted to elders, virtuous and meritorious in acts.

12. Dharma Lord is in Vyaya Bhava; he/she will incur loss of fortunes, will always spend on auspicious acts, and will become poor on account of entertaining guests.

Predictions by Karma Lord in Various Bhava

If tenth Lord is well placed or un-afflicted, he/she will be successful in his/her attempt to improve his sources of income; his/ her promotion or professional elevation; gain of authority and status; good financial gain and political success. There will be recognition and awards from the Government. There will be gain in career, profession, business, honour & respect, popularity, reputation, status in life, power and authority, government favour, livelihood, living in foreign lands and Government employment or authority. There will be gain of wealth of the father, earned money, position, and place of settle in life, self-fulfilment in carrier and promotion during Maha Dasha of the planet. If tenth Lord is afflicted or badly placed or afflicted, there will be fall in position; financial losses; dispute and disgrace in profession; set-back in career; involved in misdeeds during Maha Dasha of the planet. Followings are the specific effects of Lagna Lord Position in different house:

1. Karma Lord is in Thanu Bhava, he/she will be scholarly, famous, be a poet, will incur diseases in boyhood, and be happy later on. His/Her wealth will increase day by day.

2. Karma Lord is in Dhan Bhava, he/she will be wealthy, virtuous, honoured by the king, charitable, and will enjoy happiness from father and others.

3. Karma Lord is in Sahaj Bhava, he/she will enjoy happiness from brothers and servants, be valorous, virtuous, eloquent and truthful.

4. Karma Lord is in Bandhu Bhava, he/she will be happy, be always interested in his/her mother's welfare, and will Lord over conveyances, lands and houses, be virtuous and wealthy.

5. Karma Lord is in Putra Bhava, he/she will be endowed with all kinds of learning, and he/she will be always delighted, and will be wealthy and endowed with sons.

6. Karma Lord is in Ari Bhava, he/she will be bereft of paternal bliss. Although he may be skilful, he/she will be bereft of wealth and be troubled by enemies.

7. Karma Lord is in Yuvati Bhava, he/she will be endowed with happiness through wife, be intelligent, virtuous, eloquent, truthful, and religious.

8. Karma Lord is in Randhra Bhava, he/she will be devoid of acts, long-lived and intent on blaming others. His/Her career is not good and has great difficulty in establishing a career or has some sudden or serious troubles, which greatly harm the progress in his/her career.

9. Karma Lord is in Dharma Bhava, he/she born of royal scion will become a king, whereas he/she being an ordinary person will be equal to a king. This placement will confer wealth and progeny happiness.

10. Karma Lord is in Karma Bhava, he/she will be skilful in all jobs, be valorous, truthful and devoted to elders.

11. Karma Lord is in Labh Bhava, he/she will be endowed with wealth, happiness, and sons, virtuous, truthful and always delighted. He/She can earn merit and reputation along with money. He/She will display a happy exterior always and show bonhomie and geniality. This earns him/her good reputation and goodwill among public. He/She will have fame and reputation.

12. Karma Lord is in Vyaya Bhava; he/she will spend through royal abodes, will have fear from enemies and will be worried in spite of being skilful.

Predictions by Labh Lord in Various Bhava:

The eleventh lord has much to do with gains and achievement of one's desires and it's placement in the various twelve houses tells us a lot about achievement of desires in life and where his/her desires lay. For example, if the 11th lord is in the ninth house, then fortune flows to the achievement of his/he desires. If the eleventh lord is in the eighth house, then great trouble comes in the achievement of his/her desires. In this way one of the twelve houses exerts itself upon the eleventh lord and he controls the things.

If eleventh Lord is well placed or un-afflicted, there will be continuous flow of money, happiness and prosperity; in family due to improved financial status; increase in bank balance; assets; good position in society and most sought after. There will be gains of wealth and the possibility of diseases. There will be gain in accumulated wealth, fulfilment of desire, children marriage, gains from profession, elder sibling, love affairs and girl friends, envision of new and different possibilities during Maha Dasha of the planet. If Tenth Lord is afflicted or badly placed or afflicted, there will be problem with brothers; loss and disrespects, ear trouble; chances of being cheated financially and problem of money invested in risky ventures during Maha Dasha of the planet. Followings are the specific effects of Lagna Lord Position in different house:

1. Labh Lord is in Thanu Bhava, he/she will be genuine in disposition, be rich, happy, and even-sighted, is a poet, be eloquent in speech and be always endowed with gains.
2. Labh Lord is in Dhan Bhava, he/she will be endowed with all kinds of wealth and all kinds of accomplishments, charitable, religious and always happy.
3. Labh Lord is in Sahaj Bhava; he/she will be skilful in all jobs, wealthy, endowed with fraternal bliss and may sometimes incur gout pains.
4. Labh Lord is in Bandhu Bhava; he/she will gain from maternal relatives, will undertake visits to shrines and will possess happiness of house and lands.
5. Labh Lord is in Putra Bhava, he/she will be happy, educated and virtuous. He will be religious and happy.
6. Labh Lord is in Ari Bhava, he/she will be afflicted by diseases, be cruel, living in foreign places and troubled by enemies.
7. Labh Lord is in Yuvati Bhava; he/she will always gain through his/her wife's relatives, be liberal, virtuous, and sensuous and will remain at the command of his spouse.
8. Labh Lord is in Randhra Bhava; he/she will incur reversals in his/her undertakings and will live long, while his/her spouse will predecease him/her.
9. If Labh Lord is in Dharma Bhava, he/she will be fortunate, skilful, and truthful, honoured by the king and be affluent.
10. Labh Lord is in Karma Bhava, he/she will be honoured by the king, be virtuous, attached to his/her religion, intelligent, truthful and will subdue his senses.
11. Labh Lord is in Labh, he/she will gain in all undertakings, while his/her learning and happiness will be on the increase day by day.

He/She will have a comfortable life with his/her partner, children and riches. It makes a powerful Dharma yoga. Immense gains will accrue to him/her.

12. Labh Lord is in Vyaya Bhava; he/she will always depend on good deeds, be sensuous, will have many wives and will befriend barbarians.

Predictions by Vyaya Lord in Various Bhava

If twelfth Lord is well placed or un-afflicted, he/she will spend money in good causes; have a pleasurable life, good deeds and respect during Maha Dasha of the planet. If twelfth Lord is afflicted or badly placed or afflicted, then there will be ill health; expenditure on treatment; increase in secret enemies; chances of imprisonment and if 6th lord is also weak, then there will be rise of debts during Maha Dasha of the planet. There will be distress and danger from diseases. There will be loss or expenses, death, bed comforts and couch pleasure, miseries, sufferings, troubles, betrayals, law suits, imprisonments, hospitalisation, conjugal relations with opposite sex other than wife, sorrows, debts and loss of goods. There will be mental agony, disputes, the foreign trips, number of the foreign trips, benefits or loss due to the foreign trips and settlement in the foreign lands. Followings are the specific effects of Lagna Lord Position in different house:

1. Vyaya Lord is in Thanu Bhava; he/she will be a spendthrift, be weak in constitution, will suffer from phlegmatic disorders and be devoid of wealth and learning.

2. Vyaya Lord is in Dhan Bhava; he/she will always spend on inauspicious deeds, be religious, will speak sweetly and will be endowed with virtues and happiness.

3. Vyaya Lord is in Sahaj Bhava; he/she will be devoid of fraternal bliss, will hate others and will promote self-nourishment.

4. Vyaya Lord is in Bandhu Bhava; he/she will be devoid of maternal happiness and will day by day accrue losses with respect to lands, conveyances and houses.

5. Vyaya Lord is in Putra Bhava, he/she will be bereft of sons and learning. He will spend, as well as visit shrines in order to beget a son.

6. Vyaya Lord is in Ari Bhava; he/she will incur enmity with his/her own men, be given to anger, be sinful, miserable and will go to others' wives. The 12th lord is in the 6th is a Vipareeta Raja Yoga which gives wealth, fame and all sorts of comforts.

7. Vyaya Lord is in Yuvati Bhava; he/she will incur expenditure on account of his/her spouse, will not enjoy conjugal bliss and will be bereft of learning and strength.

8. Vyaya Lord is in Randhra Bhava; he/she will always gain, will speak affably, will enjoy a medium span of life and be endowed with all good qualities.

9. Vyaya Lord is in Dharma Bhava; he/she will dishonour his elders, be inimical even to his/her friends and be always intent on achieving his/her own ends. The religion, spirituality, dharma and god and misfortune (bad luck) will have a great affect upon the losses in his/her life. It means that losses will come to him/her because of his/her bad luck and his/her faith, spirituality, or acts of god.

10. Vyaya Lord is in Karma Bhava; he/she will incur expenditure through royal persons and will enjoy only moderate paternal bliss.

11. Vyaya Lord is in Labh Bhava; he/she will incur losses, be brought up by others and will sometimes gain through others.

12. Vyaya Lord is in Vyaya Bhava; he/she will only face heavy expenditure, will not have physical felicity, and be irritable and spiteful.

11

Lagna Wise Horoscope (Predictions)

11.1 Aries (Mesha) Lagna:

Even though Mars is the Lord of 8th House for Aries Lagna, he will be benefic and helpful to other auspicious planets like Moon, Jupiter and Sun. Saturn, Mercury and Venus are malefic and adverse. The mere Conjunct of Saturn with Jupiter will not produce auspicious effects although they own a Kona and a Kendra. If Jupiter is at the disposal of a malefic, he will surely give inauspicious results. Venus is a direct (independent) killer. Saturn will also inflict death, if associated with an adverse Planet (Venus).

Sun: Surya is L – 5 measured from Mesha lagna, Mesha Chandra, Mesha Navamsha and so is very benefic planet for Mesha Lagna and is a friend of Mars (Lagnesha). If Surya is well-disposed and strong, Surya brings good luck, highly intelligent performance, and a royal position to enjoy all aspects in life, winning by speculation, good children and great pride by children. Surya will provide public roles, leadership responsibilities in politics and contests for power in elections, theatrical performance, games, entertainments, royalty, courtly life, literary arts, creativity of children and avocation with youth group activities. **Remedial Ratna:** If Surya is weakly disposed, then native will profit considerably by wearing a Ruby. He will be blessed with children, honours and favours by Government. The beneficial effects of the gem will be more pronounced in the major period of "Sun." It will give more beneficial results if worn with Red Coral the gem stone of Mars, the lord of the Ascendant.

Moon: Chandra is L-4 measured from Mesha lagna, Mesha Chandra, Mesha Navamsha and so is neutral planet for Mesha Lagna and is a friend of Mars (the lord of the Ascendant). If Chandra is well-disposed and strong, Chandra will provide secure foundations, schooling, and all kind of pleasures, good childhood home, shelters and house properties, vehicles of all kinds, patriotism and love of the homeland.

Remedial Ratna: If Moon is weakly disposed in the horoscope, native will profit considerably by wearing the Moti (pearl) and will developed in the area of the education of disadvantaged children (Bhava-4). The native will get happiness from mother, success in education, gain of landed and house property. The results will be more pronounced in the major period and sub-periods of Moon. It will give more beneficial results if worn with Red Coral the gem stone of Mars the lord of the Ascendant.

Mars: Kuja (Mangal) is L-1 & L-8 measured from Mesha lagna, Mesha Chandra, Mesha Navamsha and so is very benefic planet as Lagnesha. If Mangal is well-disposed and strong, it makes the native very physically strong and competitive for winning competitions through physical skill, recognition in physical appearance, a competitive businessman and social rewards to him/her for his strong physical body engaged in healing service, such as, massage therapy, psychotherapy, hatha yoga instructors unless Kuja is oppressed by Shani. Mangala will make a skilled and active sports enthusiast, a hunter, an athlete, dancer, surgeon or dentists and police officers. If Kuja occupies Makara in Vikrama (10th) bhava, he/she marks excellence in business and a propensity to create self-made wealth. He will be exalted in the field of sales, marketing, administration, meetings, teamwork, colleagues, siblings and mental tasking. If there is a Parivartana Yoga with Shani rising in Scorpio (once in every 29 years a batch of these people are born), the native has tremendous self-made wealth through determined hard work, dominating aggressiveness, and sharp business judgment. **Remedial Ratna Munga (Coral):** Mars is lord of the Ascendant. The person should wear Coral all his life. Red Coral will bestow good health, longevity, courage, name, fame and happiness. It will neutralize the evil effects if Mars is ill disposed or is afflicted. "

Mercury: Mercury is L-3 & L-6, lord of two bad houses the 3rd and 6th, measured from Mesha lagna, Mesha Chandra, Mesha Navamsha. Mercury is the most malefic planet for Mesha lagna and is unfriendly with Mars (the lord of the Ascendant). Budha and Kuja is adversarial planet to each other. Mercury is problematizer for the native, particularly for siblings (due to natural karaka of 3rd house), planner, and accountant, organizer and media worker and so they feel victimized and disturbed. Budha is especially a pernicious planet for Mesha lagna. Mesha natives feel frustration when Budha occupies Mesha. Budha can even signify a divorce. **Remedial Ratna:** Mercury is lord of two bad houses, the 3rd and 6th. It is not advisable for the native to wear Emerald. However, Emerald may be used in the major period of Mercury if he is in his own sign in the 3rd or the 6th house.

Jupiter: Guru is L-9 & L-12 measured from Mesha lagna, Mesha Chandra, Mesha Navamsha. On account of his being lord of the 9th (a Trine), Jupiter is auspicious for Aries Ascendant and is friendly to Lagnesha, Mars. Guru supports both public and private religious activities. Guru casts his auspicious aspect upon three bhava and thus four of the twelve sthana gets benefit from the graha-drishti of the all round good planet (L-9). Wise and expansive Guru controls good bhava-9 and bhava-12 (the house of loss of identity and loss of body). Remedial Ratna: Yellow sapphire will make the native intellectual, charitable, religiously inclined, and respectful towards elders and bestow on him good fortune, wealth, honours, name and fame. These results will be more pronounced in the major and sub-periods of Jupiter.

Venus: Shukra is L-2 & L-7 measured from Mesha lagna, Mesha Chandra, Mesha Navamsha. Venus is the malefic planet for Mesha lagna and is unfriendly with Mars (the lord of the Ascendant). The male native will resist divorce and avoid breaking business partnerships even when conditions are very inhospitable. **Remedial Ratna:** Venus is lord of the two death inflicting houses. Moreover Mars, the lord of this Ascendant is an enemy of Venus. It is, therefore, not advisable for the native s of this Ascendant to wear a Diamond. However, Diamond may be worn in the major period of Venus, if necessary, if Venus is in his own sign in the 2nd or 7th house.

Saturn: Shani is L-10 & L-11 measured from Mesha lagna, Mesha Chandra, Mesha Navamsha. Shani is the malefic planet due to Badhakesha (L-11) for Mesha lagna. If Sani occupies Lagna, His debilitation Sign (Aries), the native will suffer isolation and his identity is oppressed by herd mentality or crowd movements and the native has fear of entering into new connections with people and ideas. Saturn in 2nd house (Taurus), the house of his very intimate friend, Venus, will give position and income as Shani is lord of 10 and 11, more especially if that lord Venus is strong or if there is the aspect of Jupiter. Saturn will certainly cause some wasteful expenditure and some trouble in the family like loss of children. Saturn is in the 10th in his own sign, Capricorn, will give rise to Sasa Yoga and so native will command good servants, but his character will be questionable. He will be head of a village or a town or even a king, will covet others' riches and will be wicked in disposition. Remedial Ratna: Saturn is not considered a very auspicious planet for Aries Ascendant. But, if Saturn is posited in the 2nd, 4th, 5th, 7th (in his sign of exaltation),

9th, 10th (in his own sign) and 11th (in his own sign), Blue Sapphire can be worn in the major period of Saturn. The wearing of Blue Sapphire will enhance the good effects of the yoga."

Rahu: Rahu is co-lord of Kumbha, L-11, "gains", measured from Mesha lagna, Mesha Chandra, Mesha Navamsha. Rahu's effects for Mesha lagna depend significantly on Rahu's Bhava, Rashi, and Drishti. Rahu amplifies the effect of any graha who are sharing Rahu's house. Rahu magnifies the effects of the lord of His occupied Rashi. Rahu and Ketu are said to give positive results always in the 3rd, 6th, and 11th bhava from the lagna or from Chandra. Rahu-Ketu are exalted in Vrishabha-Vrischika. Rahu and Ketu will give good results when Mangala and Shukra are well-disposed. The uchcha Rahu for Mesha lagna becomes a virulent maraka in dhana bhava. Vimshottari periods of Rahu are associated with malignant diseases of Shukra (STD, sugar-related, drug-addictions). **Remedial Ratna:** Rahu-ratna (gomedha) may become a beneficial gem for matters of professional networking, earnings in the marketplace, and acquisition of gains and achievements (not only monies but ideological goals as well). However Rahu in either rashi of maraka Shukra is dangerous for Mesha lagna native. Rahu-ratna should be considered only if the lord of bhava-11 and its occupants are highly auspicious.

Ketu: Ketu is co-lord of Vrischika L-8. Ketu effects for Mesha lagna depend significantly on Ketu Bhava, Rashi, and Drishti. Ketu amplifies the effect of any graha who is sharing Ketu house. **Remedial Ratna:** Ketu-ratna is highly harmful gem for the native, *except* for matters of surgery and transformative healing, emergency medical and disaster response, and management of confidential financial information. However even for these purposes, apply Ketu-ratna with caution and only if the lord of bhava-8 and its occupants are auspiciously placed.

Planet Effects in Twelve Houses for Aries (Mesh) Lagna

1st house:

Sun: Sun in Aries ascendant (Lagna) is exalted. This makes him/her wealthy, lucky, honourable and good position like leader in life. It bestows power to create conditions of dignity and honourable life. He/She is wealthy and benefited from children. He/She is intelligent and learned. He/She has changeable residence. He will overpower the opponents. In middle of life, there will be some loss and reversals in respect of wealth, profession or rank. It is helpful to others. He/She may suffer from bad blood circulation, palpitation and bile. If afflicted, benefit of power, authority and fluctuating finances. One faces ups and downs in life.

Moon: Moon in Aries ascendant (Lagna) makes him/her very lucky, wealthy and of honourable positions. He/She is learned, benefited from children. He/She has more daughters than sons. In middle part of life, there will be some reversals. He/She has foreign travel, afflicted eyes, loss through relations and mother's family. He/She has influence of opposite sex on life.

Mars: Mars in Aries ascendant (Lagna) makes him/her a good administrator, wealthy, landlord and successful in life. He/She is respected by Govt. and people. He/She is advisor and a good orator. There will be some scar or wound in the body. He/She is rash temper, independent views, bold and courageous.

Mercury: Mercury in Aries ascendant (Lagna) gives him/her afflicted health, enemies, stone hearted and full of vices, a glutton, liar, quarrelsome but will be interested in occult sciences, astrology etc. His/Her spouse will not be so beautiful and will be of questionable character. He/She is destroyer of ancestral property.

Jupiter: Jupiter in Aries ascendant (Lagna) makes him/her respectful to preceptors and religious. Wealthy, learned, respected, intelligent, holds a good position in Govt., faithful, character will be above board. He/She is dedicated to public and generous.

Venus: Venus in Aries ascendant (Lagna) blesses him/her with conveyance, wealth and sweet spoken relations with opposite sex, more daughters, travelling and foreign travels. In middle life, he/she has some reversals, legal litigations in financial matters and troubles in eyes.

Saturn: Saturn in Aries ascendant (Lagna) is debilitated and will make his/her life miserable, and he/she is unsuccessful, strained relations with wife, destroyer of ancestral property. He/She has journeys, envious and will have strained relations with relatives. Saturn makes him/her proud, unfaithful and gives opposition to mother.

Rahu and Ketu: When Rahu or Ketu are in the Ascendant Aries and are not associated with any malefic planet, he/she is blessed with respect, wealth, rank in life and army. He/She is brave and of commanding personality, conveyance and good luck will follow. But if conjoined or aspect by malefic, reverse results will be experienced.

Second House:

Sun: Sun indicates that he/she will gain through investments, has pleasure, children, and gain in possessions. If Sun is afflicted, his possessions are lost in speculations, pleasure and through young people.

Moon: Moon indicates that he/she will be successful in business and other fine arts, sweet and soft spoken, gains by deals in property and land and will have knowledge of foreign languages.

Mars: Mars is lord of Ascendant and 8th house. When Mars posited in second house indicates severe illness or death in childhood, if survives then long life, disrespect and ill fame, benefits through industrious efforts, gain through the dead and if afflicted, losses and troubles in life.

Mercury: Mercury indicates that he/she will gain through educations, short journeys, writings music, gain through service, employees and small animals. When Mercury is afflicted, it indicates that he/she will lose through Govt. and enemies in middle of Life, troubles with kindred over money matters and losses through sickness, servants and animals.

Jupiter: Jupiter indicates that he/she will gain through foreign trade, science, learning, publication, travel, invention or banking, will be social with all, respected, religious, and devoted to one's country and gain through secret mission and occult. If afflicted, it indicates loss of money through enemies and generally unfortunate conditions.

Venus: Venus indicates that he/she will gain through personal ingenuity and industry and by marriage. He/She will be a good orator. If afflicted, it indicates loss through partnership, contracts, law suits, women, death of wife or partner.

Saturn: Saturn indicates that he/she will gain through industry, trade, profession of Govt. service in offices, friends and acquaintances, through legislative interests or development through his own creations. If afflicted, it indicates losses through above factors.

Rahu: Rahu indicates that he/she will be dubious and insincere in speech; tender hearted, gain through sovereign, wrathful and happy and disease in mouth or face.

Ketu: Ketu indicates that he/she will have vile speech, sinister look, devoid of learning and riches. He/She will normally be eating at other's tables.

Third House:

Sun: He/She will have interest for travels, sports, drama and adventure, pleasure through kindred and gain through journeys, with children or young people. He/She will be blessed with brothers and relations, respected, lucky, authority in Govt. service and strained relations with father in middle of life. If afflicted, there will be reverse results.

Moon: He/She will gain through brothers, blessed with wealth, respect and lucky, fond of music and amusements. If afflicted, there will be reverse results.

Mars: He/She will gain mental development, short journeys, gain through brothers and kindred, delayed opportunities, psychic and mysterious experiences. If afflicted, there will be death of brother, loss through short journeys.

Mercury: He/She will interest in studies and industrial economy, gain through education, writings, short journeys and brothers. If afflicted, there will be reverse results.

Jupiter: There will be learning gain and progress through research, travel, investigation, exploration or writings, occult learning. If afflicted, there will be loss, troubles through relations, friends or neighbours loss through short journeys and writing.

Venus: There will be marriage in relations or neighbours, gain from wife, gain through educations writings, short journeys. When afflicted, there will be loss through above and legal or religious disputes and enmity with brothers or neighbours.

Saturn: He/She will gain through partner's relations, short journeys, questionable character, Profit from business, writing, Govt. and honours, friendship with neighbours and others through writings and journeys. When afflicted, there will be reverse results.

Rahu: He/She will be proudly, hostile to brothers, strong willed, long lived and wealthy.

Ketu: He/She will have long life, strength, wealth and fame, live happily with his wife and eat good food. When afflicted, he may lose a brother.

Fourth House:

Sun: There will be gain through property, conveyance, parents and love for home, command respect and authority till end of life, lucky mother and native children will gain through gifts. If afflicted, there will be reverse results.

Moon: There will be gain of property through old people and antiquities, assistance of parents, profit and gain at birth place, good conveyance, famous and wealthy. If weak or afflicted, there will be reverse results.

Mars: He/She will have medium age, gain through lands, mines, inheritance, and possession, home connections and affairs with the father, success late in life, occult investigations, and gain through dead and from abroad. If afflicted, there will be danger through falls, flood and storms, trouble over inheritance of land and property.

Mercury: He/She will have soft and sweet spoken, intelligent, learned and artful, respect, name and authority and blessed with conveyance, property and wealth. If afflicted, there will be reverse results.

Jupiter: He/She will be lucky like a king, wealthy, famous, and respected due to business, trade or other position, builder of charitable institutions, visit to religious place, and other works connected with public welfare. If afflicted, there will be secret sufferings, loss through father and above sources.

Venus: He/She will have property by marriage, chaste partner, happy wedded life, and gain through parents, land, mines and house hold goods. When afflicted, there will be loss through above pursuits and litigation over property and may there be two marriages.

Saturn: He/She will have gain and honours through parents, land and property, success at close of life, interested in reclamation, colonization, cooperative movements, horticulture, mining, architecture or Archaeology, fortunate in property and love for father. If afflicted, there will be reverse results.

Rahu: He/She will have friends, happy, medium life, not intelligent and will cause sorrows.

Ketu: He/She will lose his land, property, happiness and conveyance, residence in foreign land and strained relations with wife.

Fifth House:

Sun: He/She will be lucky, wealthy and much pleasure, love affairs and much adventures, speculations and investment prosperous, children and pleasure through them. He/She will get rank and respect in youth and wealth in old age. If Sun is afflicted, there will be reverse results are denoted.

Moon: He/She will have gain and pleasure through father's good fortune; possessions descend to children, self made man and will be diseased and worried in childhood, If Moon is afflicted or weak, there will be reverse results will be experienced.

Mars: He/She will have delight in pleasure, amusements, sports, speculations and children and success through them. When Mars is afflicted, the children will be unfortunate; there will be danger through excessive pleasure, speculation and children.

Mercury: He/She will have pleasurable journeys, mental pleasure through children, brothers, reading study, accomplishments and travels to pleasure resorts. If afflicted, there will be loss through the above.

Jupiter: He/She will have liberal or unconventional views in regard to union, free living, pleasure through journeys, and gain from travel, philosophy, air flights, sports, foreign investments and speculations. If afflicted, there will be secret sorrows through children, sports, love, speculations and journeys.

Venus: He/She will have gain by investment, speculation, pleasure, entertainment, pleasure through wife. There will be loss through speculations, enmity with children, if Venus is afflicted.

Saturn: He/She will have gain and honour through speculations; children will rise to honour but suffer from sickness, also gain through children, friends and beneficial circumstances. If afflicted, there will be reverse results.

Rahu: He/She will have troubles through travels, from Govt. He/She may be childless, hard hearted and suffers from belly ache.

Ketu: He/She will lose children through death or abortion, stomach diseases, will be evil minded and wicked.

Sixth House:

Sun: He/She will have gain through careful speculations or by children's earnings, pleasure by hygienic methods, pets and hobbies, grief from progeny, loss of respect in middle of life, otherwise wealthy, respected and blessed with property, may reside in foreign land due to some compelling circumstances.

Moon: He/She will have gain through service, servants and small animals will build his own house, may have grief in youth due to death of parents. If Moon is weak and afflicted, there will be less gain through them, loss through opponents.

Mars: He/She will have gain through servants, employees, fond of pets, small animals. If afflicted, there will be loss through above, operation and injury on head.

Mercury: He/She will have interest in study and methods of healing or social economy, good health, and success with employees, poultry, medicine, healing or social service. If afflicted, there will be severe illness or injury and difficulties through brother.

Jupiter: He/She is born with silver spoon in a respectable family, wealthy, blessed with property, good rank and service, death of elders in the family, many enemies, will be under debt in middle of life, last part of life will be comfortable. If afflicted, there will be sickness through travelling, difficulty in work in foreign land, or in connection with export.

Venus: He/She will marry to a lady below his status mentally or socially, has sickly and evilly disposed employees but gain through them. When afflicted, there will be reverse results and loss through sickness.

Saturn: He/She will have modest worldly position, gain and honour through service, and employment, gain from friends among navy, air force, army or working class, faithful servants and interest in social welfare. If afflicted, there will be reverse results.

Rahu: He/She will have troubles through enemies, servants and employees, wealthy and long lived. He/She will suffer from disease in the anus.

Ketu: He/She will have very magnanimous and have best qualities. He will attain everlasting fame, firmness and high authority, realised hopes and victory over enemies.

Seventh House:

Sun: He/She will have pleasure, close association or understanding with partner, but discord with children, loss by theft and partnership, wicked, arrogant, and fond of travel. If afflicted, there will be trouble and loss through love affairs. Debilitated Sun or if afflicted, it may deny or delay marriage and If married, there will be divorce, separation or bickering in married life.

Moon: He/She will gain through marriage, land and women generally, have no permanent profession, and may cause plural marriage. If Moon is weak or afflicted, there will be loss through above.

Mars: He/She will have a rich partner or one to whom money comes unexpectedly, partnership with others, fond of opposite sex. If afflicted, there will be troubles in financial matters, death of partner, danger of death by violence, suicide, accident or war, loss and trouble through law, union and open enemies and material unhappiness.

Mercury: He/She will have marriage as a result of journey, writing or with relation, skilled in many arts, fickle minded and will have no respect for his words. If afflicted, there will be law suits with servants, difficulties with disreputable women and sickness, unfortunate union, and danger of robbery while travelling.

Jupiter: He/She will have marriage to a stranger of education refinement whose relations will be opposed to the native, wise, tactful, learned and brave, gain through marriage, travels, partnership and contracts. If afflicted, there will be troubles through deceit and treachery concerning unions, law suit, marriage, partnership, contracts, and troubles through women generally and public enemies.

Venus: He/She will have gain by marriage, contracts, business and dealing with others especially with women, success in law suits, marriage in good family, and the partner may prove cold, untrue or hostile. If afflicted, there will be troubles in unions and through open enemies, competition and theft.

Saturn: He/She will have gain through law suits and dealing with public generally, honour and reputation assisted by honourable marriage and partnership in responsible concern, the partner will love with desirable friendship and social connections and success in law suits. There will be delayed marriage, if Saturn is exalted in 7th house.

Rahu: He/She will have loss of wealth through intrigues with women, suffers separation from his beloved, loses his manhood, influence of opposite sex and becomes self willed.

Ketu: He/She will have will suffer in self respect, company of bad women, afflicted with bowels disease, loss of wife and of power of virility, and delayed marriage.

Eighth House:

Sun: He/She will suffers through children, death of children before the native, loss through speculations, strained relations with father and loss of power and authority.

Moon: He/She will gain by legacy or estate when weak or afflicted, there will be death of father, danger to mother during confinement and loss of inherited property.

Mars: There is danger of death through assassination through enemies in youth, danger from fire, wound, reptile or animals and is skilled in many arts, no permanent residence and profession, occult experiences, gain by legacy, a comfortably fixed partner, spiritualistic experiences. If afflicted, there will be disappointment over legacy, troubles in financial matters, operations, wounds, danger to husband longevity. Mars is in own sign.

Mercury: There will be sickness and troubled life, journeys on account of troubles, false accusations or trouble on account of death or bequests, financial reward for faithful service and troubles in service.

Jupiter: He/She will gain by long journeys concerning legacies or dead, psychic experiences, secret enemies, many misfortunes, death of secret enemies, trouble over inheritance, unsatisfactory end, and persecution regarding religious, scientific or educational convictions also through publications.

Venus: He/She will gain by legacy and the partner, gain of money or property by marriage, death of partner and difficulty in inheritance. If afflicted, there will be loss of money through marriage and partnership and trouble in financial matters.

Saturn: He/She will gain by honours in handling the estate and money of others, also through law suits, legacies, inheritance, insurance, of the dead, gift of legacies from friends and relations and an easy end and death of friend.

Rahu: There will be short life, wicked deeds, disease of wind and limited issues and defective limb.

Ketu: There will be short life, sinful actions, quarrelsome, injury from weapon and disappointment and separation of dear ones.

Ninth House:

Sun: He/She will be blessed with wealth, rank and authority; kingly status in life, dutiful children who will be lucky, pleasure in foreign land, long travels, charitable and religious, gain from children, and the children may become preacher, scientists or explorer. If afflicted, there will be reverse results.

Moon: He/She will gain through science, religion, long journeys and wife's relatives, long gainful travels, good conveyance. If weak or afflicted, there will be loss in foreign land and reverse results than above.

Mars: He/She will long journeys, religious, or psychic experiences, linking for science, invention law, philosophy and all matters of higher mind, prophetic dreams or visions, gain through partners' relations and, of dead, spiritual experiences. If afflicted, there will be troubles with foreigners, legal or educational affairs, troubled or dangerous voyages.

Mercury: He/She will have long journey, gain through publication, science or religion, foreign journey and marriage, work in connection with foreign affairs or universities, intelligent and learned, gain through service and brothers, a comfortable and joyous life. If afflicted, there will be sickness abroad, bad wife's relations, and danger through over study and reverse results than above.

Jupiter: He/She will have travelling for education, scientific or for religious purposes in foreign country. Prophetic dreams, many fine qualities. Gain through culture and development, secret mission, investigation, publications or secluded research, sacrifice for science or religion. If afflicted, there will be shipwreck or air crash, difficulties with relations, providential help in case of difficulty if one is religious.

Venus: There will be marriage to stranger from far off place, gain by wife's relations and in-laws, partner's journey to foreign land, gain through books, sea trading, long journeys, philosophy, religion and science. If afflicted, there will be loss through above factors.

Saturn: There will be honourable voyages, professional journeys, and honour through learning, writing, publishing, research or philosophy. One will not care for any body, knowledge of foreign languages, regains all the lost property, last part of life will be lucky, friendship through travel and learning, friends through educated, ministers, writers, explorers or inventors. If afflicted, there will be, reverse results.

Rahu: He/She will speak against head of family, village or mayor and do wicked deeds.

Ketu: There will be sinful actions, deprived of his father, unlucky, long journeys, indigent and will slander the good.

Tenth House:
Sun: He/She will be renowned and wealthy, gain in speculation, honourable children, gain in business or in theatrical world, pleasure and honour through wife, father-in-law or own father. If afflicted, there will be then reverse results.
Moon: He/She will gain through possessions, trade, profession, public or Govt.; work or by fruits of the earth, kingly life, landed property and good conveyance, determined, brave and native of foresight, fond of opposite sex and favour through women. If afflicted, there will be loss and reverse results.
Mars: It is exalted here. He/She will have merit, honour and success, rise to high social and professional position, gain of inheritance, power and authority, well versed with working of machinery, ammunitions, fond of opposite sex and favour through women, learned and interested in poetry. When Mars is afflicted, he/she will earn displeasure of superiors or Govt.; dishonour, humiliations, loss of parents or trouble through them, danger of violent death through Govt. order or war.
Mercury: He/She will have professional or honourable journeys and gain through them, honour and renown through writings and other accomplishments, journeys with mother, honour through service or national activities. If afflicted, there will be dishonour, sorrows, and mental anguish through brothers and losses through above.
Jupiter: He/She will have honour, credit, high position, rank and authority, gain and honour through science, literature or travel, success in foreign land and affairs many friends and gain through them, success in hopes and wishes. If afflicted, there will be reverse results like disgrace, loss from superiors, troubles in long journeys.
Venus: He/She will have an honourable partner beneficial to the professional career, gain by occupation, profession, merchandising, Govt.; and in-laws. If afflicted, there will be financial loss through employer, rivals, employment, public disgrace or scandal through a union or enemy.
Saturn: He/She will have gain and honour through profession, and Govt.; successful hopes and efforts, friends among high positioned persons and gain through them who are of high standing in social circle and Govt. If afflicted, there will be reverse results.
Rahu: He/She will have famous, limited issues, fearless, wicked deeds and will engage himself in other's business.
Ketu: He/She will have obstacles in life, vile acts, energetic, bold and widely renowned.
Eleventh House:

Sun: He/She will have good position, respect and will be generous and charitable, promotions, authority and rank in life, gain from children, Govt.; and property, successful in hopes and wishes, profit through speculations, or pleasure among legislators, ambassadors and sportsmen. If afflicted, there will be reverse the results.

Moon: He/She will be blessed with comforts, conveyance and property, wealthy, respected, gain from children and business, successful hopes, gain through friends, old age will be comfortable. If weak or afflicted, there will be reverse the results.

Mars: He/She will have large circle of friends and gains through them, much pleasure in life, hopes and wishes often attained, gain through profession, Govt.; employer and well wishers, and gain in legacies and through friends. If afflicted, there will be reverse results.

Mercury: He/She will have gain through journeys and brothers, hopes and wishes are fulfilled, gain from studies, interest and gain from legislative activities, political and social welfare. When afflicted, there will be reverse results.

Jupiter: He/She will have fortunate friendships on voyages and gain through foreign lands and foreign friends, acquaintances among relatives, travellers, scientists and legislators. If afflicted, then he/she will face troubles, disappointments and obstacles, deceitful friends and losses through them.

Venus: He/She will have gain through friends, accidental fortunes, two wives or marriage with widows. If afflicted, there will be troubles with wife and friends and loss though them.

Saturn: He/She will have success in ambitions, eminent friends among legislators, persons of authority and professional positions, helpful to others, many friends among radical people and gains through them. If afflicted, there will be reverse results.

Rahu: He/She is prosperous, long lived, has children, and may suffer from ear disease.

Ketu: He/She will hoard money through illegal means, good qualities, will enjoy himself well, and will command all facilities and successful in hopes and wishes.

Twelfth House:

Sun: He/She will have pleasure and gain through investigation, of mysterious things and research nature, secret sorrows through children, speculation can cause ruin. If afflicted, there will be danger or imprisonment and loss through Government.

Moon: He/She will have afflicted health due to long journey by parents, sorrows, end of life in seclusion or devoted to the study of occult subjects.

Mars: He/She will have gain through occult affairs and secret missions, success in middle life, difficulty over inheritance, great sorrows, and anxiety concerning death or imprisonment. It afflicted, there will be death or imprisonment through enemies, secret unhappiness, suffering and misfortune.

Mercury: He/She will have occult learning, work of secret nature in C. I. D. or C. B. I., interest in archaeology or submarine life. If afflicted, there will be danger of imprisonment through travelling, secret sorrows, sickness, enmities, suffering and misfortune.

Jupiter: He/She will have occult abilities, powerless enemies, secret investigations, fond of animals, limitations, troubles and obstacles prove to be blessings in disguise by developing inner growth and understanding. If afflicted, there will be difficulty and sorrow through religion, science and journeys. If writer or inventor, then difficulty in the completion of job. In middle or latter part of life seeks, there will be seclusion for development, occult learning.

Venus: He/She will have unhappy marriage, secret sorrows, jealously, sickness, partner or opponents cause imprisonment or fear of it or danger of death at their hands, gain through secret mission, hospitals sanatorium, occult science or large animals. If afflicted, there will be denying of marriage.

Saturn: He/She will have loss of office or honour, dignity through business associates who becomes secret enemies, unfortunate conditions, deceitful friends and loss through them, difficulty in employment, pleasure in peaceful, quite harmonious or secluded places, and good friendship among occult people. There will be sorrowful friends and losses through the above, if Saturn is afflicted.

Rahu: He/She will do sinful acts secretly, spend much and suffer from a water-disease.

Ketu: He/She will spend money on vile acts, sinful deeds, will destroy wealth, of questionable character and will suffer from eye disease.

11.2 Taurus (Vrishabh) Lagna:

Jupiter, Venus and Moon are malefic. Saturn and Sun are auspicious. Saturn will cause Raja Yoga. Mercury is somewhat inauspicious. The Jupiter group (Jupiter, Moon and Venus) and Mars will inflict death.

Sun: Surya is L-4 measured from Vrishabha lagna, Vrishabha Chandra, Vrishabha Navamsha and therefore is a benefic planet due to Kendradhipati dosha for the Taurus ascendant. If Sun is strong in the kundali, there will be great pride to him in childhood in home, family care, shelters, vehicles, schools, access to social and emotional security, including education, vehicles and higher degree, license, exam pass, school-teaching and parenting. Remedial Ratna: Manika (Ruby): Sun is benefic as a lord of the 4th house, but Sun is an enemy of Venus, the lord of the Ascendant. Therefore, Ruby is to be worn in the major period of Sun, or if He is in own sign or in the 10th house where he gets directional strength or exalted. Wearing a Ruby will bestow mental peace, success in educational activities, gains of land and property, happiness from mother and comfort of conveyances.

Moon: Chandra is L-3 measured from Vrishabha lagna, Vrishabha Chandra, Vrishabha Navamsha and is a malefic planet for the Taurus ascendant. As L-3, lord of 8th from 8th, Chandra is a bit decisive and has dicey influence and extreme fluctuations caused by tsunami-quality churning of the wheel of death and rebirth. Sibling relationships are strengthened but mother conditions are prone to sudden upheaval, and extreme emotional upset. If Chandra is associated with Rahu or Ketu, the mother's life is certainly subject to considerable psycho-mental volatility. Mother will have her own business. Remedial Ratna for Chandra (Moti - Natural Pearl): Moon is lord of the third house for the Taurus Ascendant. Pearl should not normally be worn by the native. However, if the Moon is in his own sign in the 3rd house, the Pearl will do well in the major period of the Moon.

Mars: Kuja is L-7+ L-12 measured from Vrishabha lagna, Vrishabha Chandra, Vrishabha Navamsha and is a benefic planet due to Kendradhipati dosha for the Taurus ascendant. Presuming Vital and competitive Kuja is strong in the horoscope, it controls good alliances and agreements. The spouse is naturally competitive, active, but aggressive. Spouse will work hard and play sports hard. Sleep is less peaceful and he takes extreme measures to ensure his sleep due to mental restlessness and bed pleasures are sexually (Kuja) activated. Kuja dosha is less harmful for Vrishabha lagna because Kuja aspects to swakshetra Vrischika Rashi in 7th house, which gives the spouse more strong personality rather than harming marriage relations. Remedial Ratna for Mangala (munga/red coral): Mars is lord of 12th and 7th and is a benefic planet for the Taurus Ascendant. Therefore the native of this Ascendant should wear a Coral. However, if Mars is in his own sign in 7th or 12th, Coral must be worn in the major period of Mars.

Mercury: Budha is L-2 + L-5 measured from Vrishabha lagna, Vrishabha Chandra, Vrishabha Navamsha. Mercury is lord of the 2nd, the house of wealth and the 5th, a trine, and so is a very auspicious or benefic planet for the Taurus Ascendant. Presuming Budha is strong in the horoscope, it empowers him to talk, explain, analyze, articulate, announce, memory, money, historical knowledge, beautiful face, voice, language, speech and song, all matters of bhava-2 and children, politics, fashion, charisma, games and gambling, winning prizes, awards and trophies, fortune and luck, creative arts & literature, divine Intelligence, speculation, prestige, fame, fashion, theatre centre-stage roles, celebrations and self expression, all matters of bhava-5. There are four main wealth houses, such as, Bhava-2 is for banked savings; Bhava-5 is wealth through gambling or speculative wins; Bhava-9 is wealth in the form of assets provided for acts of faith; and Bhava-11 is earned income wealth. Budha becomes a highly auspicious as L-2 and L-5 for Taurus Ascendant and, if strong, gives stored wealth, conversational and sexual power to Vrishabha Lagna native. The sources of the wealth will be as per the bhava and rashi of Budha position. Example: USA Pres-29, Warren G. Harding of Vrishabha lagna had publications and announcements of Mithuna, Bhava-2. Budha is L-2 for Vrisha lagna and therefore Budha may function as "maraka" in the Vimshottari Dasha period of Budha. However, the other maraka for Vrishabha lagna is Mangala, a natural aggressor Maraka. Remedial Ratna for Budha (Panna - Emerald): The native of this ascendant can always use Emerald for good results. It will help in acquiring wealth, children, intelligence and good fortune. The beneficial results will become more pronounced in the major period and sub-periods of Mercury. Wearing an Emerald with Diamond, the gem stone of Venus, the lord of this Ascendant, will magnify the good results.

Jupiter: Guru is L-8+ L-11 measured from Vrishabha lagna, Vrishabha Chandra, Vrishabha Navamsha and is most malefic planet for the Taurus ascendant. Apart from that, the lord of the Ascendant Venus is not a friend of Jupiter. But, if Jupiter is in the Ascendant, 2nd, 4th, 5th, 9th, 10th or 11th (in his own sign) and strong, the native will have financial gains, abundant income, gains of wealth, secret or privileged information; generous and liberal friends and wisdom teachings. However, eighth lord (L-8) Guru in the karmasthana (10th) cause very bad trouble in career. Remedial Ratna for Guru (Yellow Sapphire - Pukhraj): If lord of the 11th Jupiter is in the Ascendant, 2nd, 4th, 5th, 9th, 10th or 11th (in his own sign), the native will have financial gains if he wears a yellow sapphire in the major period of Jupiter.

Venus: Shukra is L-6+ L-1 measured from Vrishabha lagna, Vrishabha Chandra, Vrishabha Navamsha and is an auspicious or benefic planet for the Taurus ascendant even though He is also lord of Ari Bhava (6th). If there are dignified graha in 6th, the native goes to the carrier of medicine. If L-6 Shukra occupies Ari bhava, one becomes an agent of addiction to sweet foods, candies, ladies, alcohol, and numbing drugs. The first wife may have health problems (6th), and has well-developed social personality (1), but typically argumentative mentality. It is said that a graha which rules two bhava of a nativity will give the strongest results for its Mulatrikona rashi. Shukra's mulatrikona Rashi is Thula (6th) for Vrishabha. If Shukra occupies bhava-6 in Thula rashi, there will be success in cosmetic or reconstructive medicine, import-export business and work in foreign lands, decorator and artistic designer. Remedial Ratna for Shukra (Flawless Diamond - Heera): Diamond can always be worn by the native of this Ascendant for longevity and for advancement in life. Combination of Diamond and Emerald will prove very advantageous for the native of this Ascendant.

Saturn: Shani is L-9 + L-10 measured from Vrishabha lagna, Vrishabha Chandra, Vrishabha Navamsha and Shani is benefic planet and Yogakaraka (Most Beneficial) for Vrishabha Lagna. One of the best characteristics of Vrishabha lagna is that three bhava will receive the auspicious drishti and one bhava will have position of Shani as dharma-karma-adhipati. Three houses thus get benefit from the drishti of the all good ninth lord and the pretty-darned-good 10th lord. Shani as L-10 brings plenty of public responsibilities, and performing as public leadership and pursue profitable personal business. Example: OSHO Rajneesh, yogakaraka (L-9+L-10) Shani is in Bhava-8 provides leadership via hidden knowledge or secret money. UK-Queen Victoria I, yogakaraka (L-9+L-10) Shani in 11 provides leadership via market places and social networks. USA Pres-29, USA Pres-31, Herbert Hoover, yogakaraka (L-9+L-10) Shani in 10th provides leadership via executive roles, ordering and regulating behaviors, imposition of policy and protocol in hierarchical organizations. Career: Shani is lord of both the karma bhava for leadership roles and the dharma bhava for loss-of-impetus (12th from10th) in leadership roles. Therefore, (L-10) Shani gives a series of gains and loses in positions. It is rare for Vrishabha native to obtain one leadership role and hold it for extended periods. Rather, he gets financially comfortable and respectable position holding basic decency and remarkably steady level of luxury so long Shukra is well-disposed. There is likely not too much ambition to "get to the top or highest post" but rather a pleasant, competent, lawful routine regularity of professional position that allows respect without too much pushing for recognition. The most productive period for public reputation and social recognition occurs during Shani mahadasha; but each Shani bhukti is also a step toward his longer-term credibility in management and executive roles. If the Shani mahadasha occurs during middle age then his career becomes quite dignified. There is no need to fight for competition or push against nature, but slowly and steadily, one rises into visible positions of social responsibility that are much appreciated and generally well-paid. Second husband: In a female nativity, Shani is significator of the second husband, who will be well-known or possessed of a high public dignity and responsibility in one's community (10) and may be involved in being a father, moral guide, and possibly an ethnic priest. Remedial Ratna for Shani (Blue Sapphire - Neelam): Saturn is a Yogakaraka planet being lord of the 9th and 10th houses for the Taurus Ascendant. Therefore, if a Blue Sapphire is worn by the native of this Ascendant he will always enjoy happiness, prosperity, name and fame, advancement in the professional sphere and favours from Government. The results will be more pronounced if Blue Sapphire is worn in the major and sub-periods of Saturn. It will prove more advantageous to wear Blue Sapphire with Diamond, gem stone of Venus and the lord of this Ascendant.

Rahu: Rahu is co-lord of Kumbha (bhava-10) measured from Vrishabha lagna, Vrishabha Chandra, Vrishabha Navamsha. Rahu's effects for Vrishabha lagna depend significantly on Rahu's bhava, rashi, and incoming drishti. Rahu amplifies the effect of any graha who share Rahu's house and also the lord of His occupied Rashi. Rahu and Ketu are said to give positive results in the 3rd, 6th, and 11th bhava counted from the lagna or counted from Chandra. Since R-K is exalted in Vrishabha-Vrischika, R-K will give good results when Mangala and Shukra are well-disposed. If Shani is strong, then Rahu may be empowered to aid matters of professional recognition and acquisition and development of respected social positions. If Rahu is in bhava-2 in Mithuna, Rahu Periods are notable for acquisitive uprising like a rising of cobra from the snake-charmers basket. Rahu-ratna (gomedha): The Rahu-ratna (gomedha) may become a beneficial gem for purposes of enhancing the career (10) and high-visibility public leadership roles (10), if Shani and occupants of 10 are highly auspicious. Otherwise, Rahu, without concern for impact on others, may create a public backlash toward the end of Rahu mahadasha.

Ketu: Ketu is co-lord of Vrischika (bhava-7) measured from Vrishabha lagna, Vrishabha Chandra, Vrishabha Navamsha. Ketu's effects for Vrishabha lagna depend significantly on Ketu's bhava, rashi, and drishti. Ketu amplifies the effect of any graha who share Ketu's house and also the effects of the lord of His occupied Rashi. Ketu is co-lord of Vrischika (bhava-7) measured from Vrishabha lagna, Vrishabha Chandra, Vrishabha Navamsha. Ketu's effects for Vrishabha lagna depend significantly on Ketu's bhava, rashi, and drishti. Ketu amplifies the effect of any graha who share Ketu's house and also the effects of the lord of His occupied Rashi. Ketu-ratna (Cat's Eye): Ketu-ratna can become a beneficial gem for the Vrishabha native, except if the native is positively involved in contract negotiation, advising professions, or relationship development. However even for these purposes, Ketu-ratna can be worn with caution and only if the lord of bhava-7 and its occupants are auspiciously placed.

Yoga for Vrishabha lagna: If Chandra is in lagna, Surya in Vrischika (full Moon uchcha in lagna) and Shukra in Thula and Budha in bandhu bhava, one will get exceptionally high social position.

Raja Yoga for Vrishabha lagna: (i) If Shani is in Vrishabha and Shukra in Kumbha, there is an exchange of Rashis between Karma's Lord and Lagna's Lord, and it will make the native associated with the king in a great manner (BPHS, Ch. 40, Sl. 13). (ii) If the lord of the Lagna (L-1) and the 10th (L-10) exchange houses with each other then they form a Raja Yoga. They will confer high position, reputation and power (BPHS Ch. 5, Sl. 41).

Vrishabha- Chandra Yoga: If Chandra is in Vrishabha Lagna, it will give treasures, fields and pastures, money, luxury, sensual pleasures, wines and foods, perfumes and oils, ancient knowledge, languages, history, records and archives, old songs and stories, capital accumulation, acquisition of goods, collections, banking, stored wealth, stored memories, value-hoards of every description.

Planet Effects in Twelve Houses for Taurus (Vrishabha) Lagna

First House:
Sun: When Sun is posited in Lagna it is in Taurus sign and in enemy's camp. He/She will be of rash temperament, fond of scents and luxuries; interested in cloth business like as ready made garments, general cloth, comforts from wife and conveyance is indicated. He/She should not enter into partnership with any body otherwise or he/she will suffer loss unless 7th house is aspect by Jupiter. He/She will be fond of dress and self make up, are tall and

gain through inheritance, land and property.

Moon: Moon is Taurus is strong and exalted, it indicates threat. He/She will be a wealthy man, of fair complexion and sanguine temperament, will respect elders, will be intelligent, lucky and attain high position in life, will be blessed with more sisters then brothers, will have influence of opposite sex on his life, fond of good food and dress and may suffer from cough or bronchitis in life.

Mars: Mars in Taurus is strong and in enemies Sign. He/She is found to be very cautious about their self respect, will be obstinate, or rash temperament, clever and unsocial, wife will be of afflicted health, and he/she can be subjected to poor blood circulation high blood pressure but even heart attack cannot be ruled out, dependant on his children and troubles in hopes, his health will not be robust, comforts from children and threats from enemies is indicated. He will be proud and will try to overpower his wife, relations with other women and gain through them, normally he will not stay at his home and will have more comforts outside and danger of imprisonments and wound.

Mercury: Mercury is in friend's camp. He/She will be intelligent and learned. He will plan his affairs in advance and will be blessed with success, wealthy, open minded, sexy and under the influence of opposite sex. He will have medium span of life. He will enjoy comforts of family life, wealth and children. His mother parents will also be quite wealthy.

Jupiter: No doubt Jupiter is in enemy's Rashi. It is lord of 8th house so malefic; but since here it is posited in Lagna, it is quite strong. It will confer rank, authority, respect, wealth and blesses with long life, self respected and comforts from wife and progeny, quite, social, and generous and will hold a good rank in society, very kind to lower rank persons. He will waste his property and may come under debt. He will be courageous and make donations. He will have gain from foreign trade of export and import. Legacies, overpower his enemies and obstacles.

Venus: Venus is in own house. He/She will be of beautiful, stout and of well proportioned body. He will enjoy long span of life, learned and respected by the Govt.; he will have a position of authority, wealth and comforts from conveyance, attracted towards opposite sex to the maximum, (one should restrict this as this may become a source of disrepute and troubles in life). Such a native will over power his enemies. He will be popular, happy and sexy and fond of scents and luxurious life. On the whole, he will be respected and lead a comfortable life, troubles through servants, will have beautiful and attractive wife. If afflicted, one may not be involved in sex scandals and disputes due to woman.

Saturn: Saturn is in friend's camp aspect 3rd, 7th and 10th houses and a Yoga Karka. He/She will be shy, immoral, fewer comforts from wife, relations with brothers will not remain cordial, education will be limited, not so witty and intelligent, reversal in profession, destroyer of ancestral property, and relations with Govt.; and officers will not be harmonious. If other aspects and yoga indicate, he/she may commit suicide and have long journeys, fortunate with strangers and foreigners.

Rahu: In case of Rahu, he/she will reside in foreign land. He will enjoy respect, regard, and high status in life and will be a man of authority. He will face some difficulties in middle part of his life.

Ketu: If Ketu is posited here, he/she will not have any permanent position and profession. He will suffer at the hands of enemies.

Second House:

Sun: Sun posited in second house is in a good position. It confers wealth and status in life, gain from agriculture, land, property and estates is indicated. He will have a forceful and effective speech. He/She will face litigation on account of property, trouble through eye disease, a man of word and authority.

Moon: Moon in this position will give limited progeny. Wealth and comforts will be a one's command. He/She will be inclined to non-religious activities of sweet speech but not clear and worries through opponents, money through educational affairs, short journeys and fond of music.

Mars: One will be in Govt.; service and will be surrounded by the opponents. He will be blessed with wealth and property but the same will be wasted. Last part of life will be troublesome. Others will be benefited from his assets. Gain from children is not indicated. Gain by marriage and secret affairs and methods.

Mercury: No doubt one is born in poor family but he will be learned, soft spoken, intelligent, beautiful and wealthy. Eyes will be afflicted. There will be gain from business and trade, from children, pleasure and investment.

Jupiter: He/She will be learned, intelligent, command over many languages and scholar. No accumulation of wealth. Middle part of life will be troublesome, losses and a troubled life. Realisation of debts, gain through deceased and friends.

Venus: One will have a good complexion, stout and well proportioned body, influenced by the opposite sex, sweet speech. One will have hazel eyes, losses through other women, wealthy, learned and a native of attractive manners. One will make efforts to gain money.

Gain by marriage, intelligent out spoken and fluent in speech. He may have two marriages (subject to other checks). If Venus is afflicted one will have illicit relations with opposite sex for the sake of money.

Saturn: He/She will be devoid of comforts from parents, greedy, will face financial difficulties in life, weak eyesight, and more expenditure than income. In middle part of life and during 14th, 36th, 48th and 60th years of life, one will face difficulties and gain through foreign merchants and persons and publications and gain in profession and from Govt.

Rahu and Ketu: Position of Rahu and Ketu in second house of Taurus Ascendant is not auspicious. This denotes poverty and troubled life. All property and wealth will be wasted. One will not be respected by the people and Govt. This position also indicates that one will work outside his place of residence. Danger of poisoning in the life is denoted.

Third House:

Sun: Sun in Cancer is in friend's house and deprives a native from the comforts of elder brother but brothers will be lucky. He will have low paid job. One will suffer from syphilis and diabetes. His wife will also suffer from blood defects. The eyes of native will be afflicted, may reside and travel with relatives. Father is put to sorrows.

Moon: Moon in own house, confers one with wealth, good rank and authority. One will be learned and intelligent. More sisters than brothers are indicated. Benefit from higher education. Good deeds and speech and inclination towards religious work.

Mars: He/She will be devoid from comforts of brothers. Younger brother will suffer a loss, of rash temperament, disease of ears, relation with widows are indicated. His wife will be a patient of blood circulation, not virtuous and will be of unhealthy thinking, legal or religious disputes, troubles on short journeys and occult learning.

Mercury: He/She will be coward, gain from brothers, sexy, immoral, relations with opposite sex, a good artist. He may suffer from secret diseases like syphilis, gain through education, writing, neighbours and short journeys, will marry in a respected and wealthy family. One will be endowed with authority and wealth. But the wealth and property will be wasted in disputes due to women, more daughters than sons. Belly trouble is indicated and psychic and mysterious expenses.

Jupiter: Jupiter in Cancer is exalted. This is not a happy position and indicates wealthy, and learned, but all will be destroyed and one will be under debt, early part of life miserable, death of brothers, property litigations. No gain from brother. Young age will be somewhat happy for a short duration.

Venus: He/She will have long life, happy, wealthy and well respected, happy married life, gain from in-laws and good position and respect through in-laws and sisters, journey to foreign land, fond of music. One will suffer from rheumatic pain, cough and bronchitis, marriage with some one of kin or to a neighbour, legal troubles or disrepute with a woman during journey or in foreign lands. One needs caution in choosing life partner.

Saturn: No happiness from brother, diabetic, defective eye sight, and brothers will be prosperous, strained relation with younger brother, sickly wife, and abortions and break in education, progress through research, and travels and gain through in-laws and honour through journeys and Government.

Rahu or Ketu: He/She will have lucky brothers, trouble in eye, fewer comforts from brothers, average profession, loss, wastage of time in property, litigations and break in education.

Fourth House

Sun: Sun is posited in own house. It denotes gain from ancestral property, in middle of life, litigations due to property. Good conveyance, property and wealth but still unsatisfied. Mother's parents will suffer loss, reversal in life. Last part of life uncomfortable, loss of property, strained relations with relatives, loss through storms and floods.

Moon: Moon posited in fourth indicates travelling and writing in connection with home affairs and property. If afflicted or weak, troubles through brothers over the above.

Mars: He/She will have diseased body, loss of land and property, weak and diseased mother, not well for uncles and maternal uncles, strained relations will relatives, property by marriage, chaste wife and happy married life, limitations at end of life.

Mercury: He/She will have happy life after 30th years, wealthy, well respected and trade in many items, gain from mother's parents, and death of mother in young age of the native, love of home, gain from parents, land or mines.

Jupiter: He/She will be religious and charitable, wealthy, learned, and well respected, blessed with happiness, property and land, honest, even tempered and man of good deeds, comforts from wife, gain of ancestral property and legacies.

Venus: He/She will have of fickle mind, learned respected, and man of authority, promotion and gain in middle part of life, happy, comforts in life, gain from brothers, and troubles through servants, sickness through worries or domestic affairs, wife will be beautiful and assist her husband, despite opposition from wife in domestic matters marital life will be happy. If afflicted, especially by a planet connected with 8th to 12th house, native may gets imprisonment due to women, faces a lot of defamation at the hands of women.

Saturn: Abundance of enemies, early death of mother, reverses in profession, ancestral property will be destroyed. He/She will have not much favour from father, low type profession, gain through in-laws, scientific inheritance.

Rahu or Ketu: He/She will have strained relations with relatives and brothers; comfortable life for a short period otherwise troubled and movable property will be destroyed.

Fifth House:

Sun: Sun in Virgo indicates a good family and property, early death of parents, obstinate, shortage of progeny, loss of property, troubles and turmoil due to death of elders.

Moon: He/She will be learned and intelligent but no gain and comforts there from, abundance of opponents, relations with opposite sex, less gain from children, more daughters, pleasure through journeys, children, brethren and travel.

Mars: He/She will have grief from children in middle of life, poor, ill famed and sorrowful life, and loss by speculation, secret sorrows and difficulty through love and speculation.

Mercury: He/She will be wealthy and man of authority, a good administrator, and no birth of son and no gain from children. There will remain no heir in families of uncle and grand uncles, gain through speculations, investment and entertainments in life.

Jupiter: He/She will be intelligent, learned, wealthy and famous, will lead a comfortable and prosperous life, and may go to foreign land due to some property disputes in the family, much happiness and pleasure through children and friends and loss through speculations.

Venus: He/She will be womanish nature and fond of music and dance, many professions, death of wife in youth, more daughters, sexy, influence of the opposite sex, weak constitution and may be deprived from the comforts of life and sickly children. It is its debilitated sign, cunning and clever to trap ladies, sexual relations with other women and of other's wives of sake of pleasure and entertainment, not real love for his wife, no problem in married life. If afflicted, there will be scandal due to woman in sex relations.

Saturn: He/She will be sickly, immoral, shortage and death of children. He will be devoid or progeny even after two marriages, no benefits from parents, loss of money, strained relations with elder brothers, liberal and free living and pleasure in journeys.

Rahu or Ketu: He/She will be immoral, no comforts from children, any limb of the body may be lost, troubled and worried life, will be a source of trouble to the parents.

Sixth House:

Sun: Sun is debilitated in Libra and is not favourable, may cause early death of mother in early age, troubled and sorrowful life. Ancestral property will be destroyed, strained relations with relatives, will be a source of loss and worries to mother's family and may build his house.

Moon: He/She will be arrested due to some blemishes, obstinate, and sorrowful, quarrelsome and ill reputed, disease of ears, difficulties through brothers and sickness or injury through journeys.

Mars: He/She will be worried from opponents, litigation of property and travel to foreign land, danger from fire and animals, after youth, life will be miserable due to many obstacles and troubles, disease of blood.

Mercury: He/She will be worried, no gain from education, poor, loss of wealth through employees, restless and no accumulation of property and Sickly wife.

Jupiter: He/She will be a good friend, sweet speech, worries from opponents, diseases, danger of ill fame in youth, unsuccessful in hopes and faithful servants.

Venus: He/She will have a diseased and worried wife, danger and loss through enemies, less comforts from relations, defective health and no sincere friends. If well aspect then he/she will be good health, gain in service, from poultry, medicine or social service, wife is normally of manly features and behaviour, beautiful voice but slightly harsh, such persons get good and beautiful wife but charms fades away after marriage, causing differences after some period.

Saturn: He/She will have victory over enemies, well wisher of all and good friends, wastage of money, less comforts from relations and loved ones, ill health fear from animal and thieves, benefit from Govt. rich, tactful and malicious, difficulty in foreign land.

Rahu or Ketu: He/She will have loss through enemies, troubled and worried life, most part of life will be through turmoil and losses and loss of wealth through women.

Seventh House:

Sun: He/She will be respected, a man of high rank and authority with powers of administrator, success and happiness due to relatives; diseased and worried wife; will be intelligent, generous and judicious and a good and powerful politician, will have step mother; after the age of 40th years he will attain more fame and authority. If afflicted there will be loss of property through marriage of partnership, bickering in married life.

Moon: He/She will have quarrelsome, wicked and obstinate; very tactful and mischievous. He will have influence of opposite sex, loss through wicked women; favour from Govt., and officers; marriage through journey or to one of kin.

Mars: He/She will be devoid from comforts in early part of life; worries and disturbed mind from wife; wise, clever, good orator and argumentative; cruel and plans fruitlessly for his advancement. In middle of life he will meet with success, respect and will be blessed with wealth, success is law suits.

Mercury: Up to 40th years of age, one will lead a life of worries, troubles turmoil. Early part of life will be more troublesome; disputes and losses in profession and worries get wealth and happiness from wife. Near the age of 50th years, one will be relieved from worries; favour from Govt.; and officers and rest part of the life will be happy, discord with children, love with partner; gain by marriage, contracts especially with women.

Jupiter: He/She will be intelligent, clever and learned but devoid of wealth; married in a respected and reputed family but in spite of all resources, influence and capability, there will be no success and comforts, many enemies, not good profession; under debt; will earn moderate money through many sources after struggle; a rich partner or to whom money comes unexpectedly and success in law suits.

Venus: He/She will have immoral deeds, fond of music, dance and opposite sex; attracted toward vices; bad and objectionable company; more than one wife is indicated. If afflicted, there will be marital unhappiness, loss and troubles through in-law and employee's trouble. His wife will be tall and lean, soft skin and pretty but slightly quarrelsome and selfish, not cooperative with family members; extravagant; wants to rule the family. Native will maintain balance in stress and strain and troubles. Native will not have much pleasure from his wife and go to others; marital unhappiness. If afflicted, native may get disrepute because of woman or girl involved in sex scandal; may suffer from diseases due to over indulgence in sex.

Saturn: He/She will have no comforts from marriage; two marriages (subject to other checks), indifferent, rich, devoid of respect and wealth unsuccessful in efforts and profession; delayed marriage and unhappy married life; marriage to stranger whose relatives will oppose; gain through law suits.

Rahu or Ketu: He/She will have rash temperament, obstinate and quarrelsome; many opponents and enemies, disputes in life, ill fame due to actions of his wife; disrespected and ill famed, people may hate him. He will be worried and diseased.

Eighth House:

Sun: Sun in Sagittarius is strong but here Sun is posited in 8th house which indicates that the native will be worried and under debt. Not so wealthy. He/She will have less comforts; disputes with everyone due to undue interference in all affairs. Any part of the body will be defective, quarrelsome, cruel and sorrowful; less comforts from parents.

Moon: He/She will have No ancestral property, early life troubled and miserable; less comforts from mother and brothers; false accusation or death of brethren.

Mars: Early part of life will be happy. Last past of life will by full of troubles and turmoil. He/She will be under debt; untimely unhappy death; loss of property, wealth and respect.

Mercury: He/She will be Dependent on others; loss of respect and prestige; worries and troubles during youth; loss of wealth, under debt, subordinate position in life. After 30th years of age, one will suffer series of miseries; will suffer from diseases connected with blood, gain through legacy or money of partner; suffers through children.

Jupiter: He/She will have a good crafts man but no gain from that; unhappy and under debt. Will engage him in many professions but will remain worried and financially tight; loss and troubles through friends.

Venus: He/She will have loss of wealth due to women; death at some auspicious place. He will engage himself in many professions; loss of wealth; danger of some secret disease; gain by the dead; spiritualistic experience; death of friends, gift or legacies from friends; troubles from wife and expenditure on account of her. Native is licentious, passionate and debauchee. One is involved in illicit relations with opposite sex.

Saturn: He/She will be long lived, punishment by the Govt., not well respected; thievish nature, unscrupulous and unhealthy; gain by long journeys, estates, law suit and insurance of deceased persons.

Rahu or Ketu: Early part of life will be troublesome. He/She will have danger of death from poison or enemies; troubles and loss through wicked persons; happy life from middle age.

Ninth House:

Sun: One will be famous, respected, Govt. Servant and a person of authority. Good deeds, enmity in family due to agriculture land or property; troubles, disputes with relations of wife due to property. Middle of life will be troublesome and full of turmoil, after that life will be happy and comfortable. If afflicted, Loss to foreign property and in above matters.

Moon: Gain of ancestral property; famous and good businessman; comforts from parents. Gain from trade and business; a comfortable life provided the Moon is strong in Pakashbala and well aspect; long journeys and gain through publishing and Science.

Mars: He/She will be lucky, famous and born in a respected family favour and authority from Govt. One will hold a magisterial post but he will be strict and cruel in behaviour and decisions; will be entangled in many cases but will be exonerated after efforts. His brothers will also be lucky and hold a good position; marriage to a stranger from after foreign travel. If afflicted shipwreck etc.

Mercury: He/She will have changeable profession; command over servants; connections with high ups; suspicious, a man of powers. The most suited profession for him is trade and business; gain by books, long journeys, philosophy and science; dutiful children and pleasure in foreign land.

Jupiter: He/She will have birth in a respected family; low deeds; malicious thoughts and actions; loss and un-fulfilment of hopes; journeys, loss and troubles through journeys; death in a distant land by drowning or voyages.

Venus: He/She will be a dutiful and faithful wife; early marriage. Gain from in-laws; learned, intelligent and respected; a man of good thoughts; an honest worker; long journeys, prophetic dreams or Vision. Such natives are normally lucky after marriage. In case of death of wife, good luck and comforts disappears. His wife will be tall, round face, proud and short tempered. She is selfish; helpful to husband. Wife will be more involved in worldly affairs rather than family members. If afflicted, marriage will be out of caste or to a widow of older age with illicit relations. Not good relations with parents; un-liked by women; may have relations with friend's wife etc.

Saturn: He/She will be learned, respected and intelligent, strong Willed, selfish, covetous, suspicious, prudent, loss of wealth, strained relations with elder brothers; inheritance from wife's parents; prophetic dreams, many fine qualities, gain through learning, writing, publishing and research.

Rahu and Ketu: He/She will have troubles and loss through opponents unpopular, fewer comforts from parents and one may not be an ardent follower of his religion and may change to other.

Tenth House:

Sun: Sun in Aquarius in 10th house incites the native to remain out to control of his parents, father will thus remain angry and it is a probable that one may be denied from the ancestral property. One is self conceited and obstinate, and will be respected by people; field of political career is suitable for such a native. He may join any service in latter part of his youth. There may be some troubles in service and one may earn displeasure and disfavour of Govt. and officers.

Moon: He/She will have a good rank and profession but danger to respect. One may be devoid from comforts of children and parents; troubles and grief from brothers. One will remain indebted to others. Gain through journeys, writing and professional pursuits.

Mars: He/She will have Death or separation from wife in youth; grief from children, loss of wealth and property. One may loose his savings even; a troubled and worried life. His profession will be somewhat odd like smuggling, murders etc. One can be exiled from the city or country.

Mercury: He/She will be A good businessman, teacher or professor; bad deeds; less respected; grief from a son, merchandising, Govt. and in-laws.

Jupiter: He/She will be learned, intelligent and well respected; generous, honest and good deeds; learned in Vedic, ancient literature and Aryuvedic science; a few sons will command respect in people and will hold a position of command and authority; death of native through war or violent death.

Venus: He/She will have Gain through wife, fond of music, poetry, drama from childhood; a lucky wife; blessed with good house and conveyance. All luxuries of life will be available to him; a good rank and position of authority. He will be owner of land, palatial buildings, respect and good top position in society. Gain from land, estates and property. Care should be taken that Venus is not being aspect by any malefic planet. Temperamental incompatibility exists between husband and wife. Two marriages can also be possible. Normally such persons are very calculative, and attach least importance of marriage, neither, have any relations with any woman and prefer to remain unmarried. They marry when after accumulating money and when well establishing in life.

Saturn: Saturn in 10th house is posited in own house. Contrary to others, the author is of definite views that it will make the native practical, able, diplomat, ingenious, well versed in machinery, residence in foreign land, technocrat; strained relations with father; honour, credit, and esteem through science, literature or travel.

Rahu or Ketu: He/She will have immoral acts, residence in foreign land, many journeys; will leave his place of birth; changeable and low profession; worried and unlucky; loss through opposition and back biting of opponents.

Eleventh House:
Sun: Sun in Pisces is strong and in his friend's housed. It indicates that the native will be greatly benefited through Govt. and Govt. service. He will be respected and have a post of authority and command; generous, good fame and comforts from relations. There will be some danger to respect and profession in the middle age. Elder brother will be well fixed up. The native will lead a comfortable life. Many hopes and wishes attained in old age.
Moon: He/She will be clever and obstinate; low profession, will generally be obstacles at the time of promotion and gain; diseased, troubled all ill fame. People will not respect him. He will interfere in the property affairs of his sister. Sometimes he will get respect but on the other hand will also be defamed and carry ill repute, friends through journeys, fruitful hopes and gains.
Mars: He/She will have abundance of agriculture land and gain there form; gain of money from disputes and litigations; frequent journeys in life; danger from enemies; confrontation with some big family in middle age.
Mercury: He/She will be low profession, under debt and poor; shortage of money; old age not happy and full of troubles; accidental fortune; fruitful hopes and love for children.
Jupiter: He/She will be learned and well educated; wealthy, but wealth will be destroyed through sons and relations, no sincere friend's and opposition from relations; loss and troubles through women; death among friends.
Venus: He/She will have gain of wealth through women; more than one marriage; respected and famous; favour from Govt. and officers; comforts from property and good conveyance; sexy and licentious; extravagant. The life will be comfortable and native will lead a luxurious life; wishes attained, large circle of gainful friends; interest in legislative and political activities. One is immoral, sexy and licentious; spends money on women for sexual pleasures; contacts with ill reputed ladies; entices and seduces the women for pleasures and gain of wealth through women.
Saturn: He/She will be diseased and of weak constitution, lean body, devoid of wealth, loss of money, scheming, will face difficulties; less comforts from brothers; fortunate friendship on voyages and with high ups; ambitions ideals.
Rahu or Ketu: He/She will be poor, under debt, ill famed, diseased and troubled; low deeds and actions; at times, one will face humiliation before elders.
Twelfth House:

Sun: Sun in 12th house is exalted. It will confer respect and authority to the person; long life. He will hold a position and authority like a minister; lead a very comfortable and luxurious life; will spend a good part of life in mountainous region or area; will command more respect and gain in foreign countries; afflicted health; end of life in seclusion or devoted to the study of occult subjects.

Moon: He/She will have voyage; danger from Govt. and officers; not so happy, extravagant, early part of life troublesome; good deeds and also charitable spending; sorrows through brothers; occult Learning; danger of enmities or imprisonment while travelling.

Mars: He/She will have wastage of money in property and litigation; service in Army; more than one wife. Old age will be miserable and loss of money; unhappy marriage, secret sorrows and death at the hands of enemies.

Mercury: He/She will be learned and skilled in many arts; gain from business and hard work; little happiness; more expenditure than income; moderate wealth; a good planner and man of strong WILL-power; capable of making inventions; inclined to good deeds and will act for the betterment of people; gain through secret methods; sorrows through children.

Jupiter: He/She will be under debt, a preceptor of religion, journey to foreign land, generous, good deeds but some time will act in an immoral way; difficulty over inheritance, fear of imprisonment and deceitful friends; good friendship with occults.

Venus: He/She will be skilled, poet and fond of music, dance and other women; licentious, changeable residences; a good author, gain of wealth through drama, dance and even from immoral trafficking of women; more pleasures of beds with opposite sex; becomes immoral; highly sexed; spends money on women for the gratification of sex desires; wife is normally quarrelsome; connections with organisations involved in illegal and immoral activities related to women.

Saturn: He/She will have loss of office and honour through business associates who will become enemies also and through religion, science and journeys; may seek seclusion in middle part of life; difficulties in employments.

Rahu: He/She will have sinful deeds and actions secretly; extravagant and watery disease.

Ketu: He/She will have wicked deeds and spending on the same, wealth will be destroyed, eye diseases and quarrelsome. There will be many enemies.

11.3 Gemini (Mithuna) Lagna:

Mars, Jupiter and Sun are malefic, while Venus is the only auspicious Planet. The Conjunct of Jupiter with Saturn is similar to that for Aries Lagna. Moon is the prime killer, but it is dependant on her association.

Sun: Surya is L - 3 measured from Mithuna lagna, Mithuna Chandra, Mithuna navamsha. Surya is malefic for Mithuna Lagna unless Surya is in his own sign or exalted in the horoscope. Surya is also an inimical of Mercury (the lord of the Ascendant). However, Surya is natural tanupati (Karaka of Ascendant). Surya in bhava-3 (Simha), in own sign, makes a magnificent writer or a local politician and places him in centre stage in business administration, writing skills, publishing, announcements; planning, scheduling & reporting; meetings and teamwork and favours communications with respect to sales, marketing, advertising, media production (films, internet visual arts, music,), short-journey services (holiday and business travel agent), seminars and conferences, editing, technical and skills working with the hands like painting, scribe or transcription, tool-using, type-setting, inks and literary press. However, Surya in the dushthamsha supports ego-identification. If Surya is not placed in a dushthamsha 6-8-12 from lagna, Mithuna is capable of making a central figure on the team work like movie script-writing Surya in 4 (Kanya), indicates a remarkable writer and a teacher. Surya in 10 (Meena), makes a capable writer and a social activist leader like the scientist, Einstein's leadership skills, that he was asked to accept the role of prime minister of Israel but he declined. Surya in bhava-6 makes one a big dramatist, actor or model, but provides break in contract from the industry due to mental disorder. Surya in bhava-12 makes one a dramatist, model, communicator but provides a severe dissolution of engagement from media industry productions due to mental breakdowns. Surya in 8-Makara, makes a splendid writer and publisher, but provides sudden upheaval or series of catastrophic mental upheavals leading to suicide due to suspected bi-polar disorder or similar like Mr. Monroe. Remedial Ratna for Surya: Manika (pure Ruby): Sun is the lord of the 3rd house. It is malefic and will not be useful for the native of this Ascendant unless Sun is in his own sign in the third house and then also it should be worn only in the major period of Sun when it will make the native courageous and successful in all his ventures. "

Moon: Chandra is L - 2 measured from Mithuna lagna, Mithuna Chandra, Mithuna navamsha and so Chandra is neutral and maraka for Mithuna Lagna unless he is in his own sign in the horoscope.

Chandra (L-2) is an inimical of Mercury (the lord of the Ascendant) and Shukra (natural dhana-pati) and is in "double trouble". Therefore, Moon is highly fluctuating and becomes a conservationist in His role as lord of the house of the Treasury. If Moon is in bhava-6 (Vrischika), one has to leave his country due to some force measures and can become ill. If Chandra is yuti with Ketu, the native's family may be perfectly stable and organized, but he will not connect emotionally with others and may lack empathy and provides a feeling to suicide action. However, the Mithuna native with a strong and healthy Moon may enjoy a vibrant and deeply supportive relationship to the family history along with luxury and (Laxmi) wealth. If Chandra is well disposed the native will eat and drink delicious nourishment. If Chandra is poorly disposed, matters of the mouth including tongue, teeth, and voice may become problematic. Chandra is a maraka for the Mithuna lagna; Guru-Chandra and Chandra-Guru periods are both candidate triggers for separation from the fleshly body, i. e. death. According to BPHS ch 43, the period of Chandra-Guru can be fatal. Remedial Ratna for Chandra (Moti - pure natural Pearl): Moon is lord of the 2nd house, the house of wealth and the house of death (maraka house). Therefore, Pearl may not prove very suitable. But when the Moon is ill placed and afflicted and the native is going through an adverse time in financial matters, wearing of Pearl is expected to avert financial difficulties. If Moon as lord of the second is exalted in the 12th, and is in his own sign or be in the 11th, 10th or 9th house, the wearing of a Pearl will prove beneficial, in the major period of Moon. On the whole, however, Pearl is not a very agreeable stone for this ascendant as Mercury the lord of the Ascendant is not friendly with Moon.

Mars: Kuja is L-6+ L-11 measured from Mithuna lagna, Mithuna Chandra, Mithuna navamsha and so Kuja is most malefic for Mithuna Lagna unless he is in his own sign in the horoscope. Kuja is also an inimical of Mercury (the lord of the Ascendant and 6th house) and Shani (natural labhpati-11th house) and so is in "triple trouble". As lord of 6th and 11th (6th-from-6th), Kuja is a highly adversarial (malefic) graha for Mithuna lagna. Kuja is by far the most difficult graha (malefic) for Mithuna natives. The placement of Mangala will determine the effects like debt, disease, divorce, gains of marketplace, income and friendship networks. Periods of Mangala are likely to encourage marital conflict, personal illness or injury, even though Kuja might be well-placed in horoscope. The best profession for Mithuna lagna native is an advocate or an attorney, and if Mangala is strong in Kendra, he will be very successful in handling even the most aggressive accusations against one's legal clients. L-6 Kuja in bhava-1, bhava-4, or bhava-12 provides very challenges for marriage due to the L-6 (a natural malefic) anddue to Kuja Dosha by drishti upon yuvati bhava. Example: Bill Gates is the most acutely sufferer of his health and harmony in life due to Kuja in bhava-4. Female, in particular, tends to female-illnesses such as breast and reproductive cancers due to Mars in 4th. Remedial Ratna (Munga - Red Coral): Mars is not an auspicious planet for the Gemini Ascendant. Avoid wearing Coral to the native of this ascendant. However, Red Coral can be used if Mars is in his own sign in the 11th or 6th in the Major period of Mars.

Mercury: Budha is L-1+ L- 4 measured from Mithuna lagna, Mithuna Chandra, Mithuna navamsha. Despite His kendradhipati dosha, lagnesha Budha is a highly auspicious benefic graha for Mithuna lagna. If Budha is well placed, the native is well indoctrinated and he easily completes examinations, obtains licenses, and earns diplomas, good capabilities in business, in writing, as well as skill in acquiring property and vehicles, good early childhood home, has good ability to talk, instruct, explain, analyze, articulate, announce and has luxurious life, personality, physical body appearance, competitions, innovation, muscular movement, vitality and birth lands and properties (esp. marine coastal), buildings, vehicles, property management, stewardship, schooling, diplomas, examinations, and licensing. Lagnesha Budha in bhava-10 provides a public figure. Lagnesha Budha in bhava-4, the uchcha state, provides property and vehicle ownership, the securing of shelters, and matters of global education. Remedial Ratna for Budha (Panna - pure Emerald): The native of this Ascendant should always wear Emerald. This will bestow him with good health, wealth, longevity, lands and property, domestic happiness, success in the field of education, a smooth and obstruction free fife. The beneficial results will be more apparent in the major period and sub periods of Mercury. The beneficial effects of any Yoga in Kundali will be enhanced by wearing an Emerald.

Jupiter: Guru is L-7+ L-10 measured from Mithuna lagna, Mithuna Chandra, Mithuna navamsha and so Guru is malefic for Mithuna Lagna due to Kendra-adhipati dosha unless he is in his own sign or exalted in the horoscope, such as Jupiter in the Ascendant, 2nd (sign of exaltation) or 4th, 5th. 7th, 9th, 10th (his own Sign), or 11th, he is considered benefic and will give good results. Brihaspati is the "bad boy" badhesha for Mithuna lagna and gives hindrance, harm, harassment. L-7 Brihaspati can generate troublesome experience in relationships. Even if Guru is well-disposed, it can provide fall from leadership power and loss of collegial relationships. Marriage and children are also affected by the badhesha function of Guru and falls below expectation. The marriage and children have not realized their growth potential. Guru-ratna (yellow sapphire - Pukhraj): Jupiter is lord of the 7th and 10th houses and is Maraka and suffers from Kendra-adhipati dosha. Those apprehending death on account of old age or some illness or there are indications of illness or short life, should not wear Yellow Sapphire, as Jupiter is a strong death inflicting planet for this Ascendant. It will be useful to wear a yellow sapphire in his major and sub-periods.

Venus: Shukra is L-5+ L-12 measured from Mithuna lagna, Mithuna Chandra, Mithuna navamsha. Being the lord of the 5th, a trine, Venus is an auspicious or benefic planet for this Ascendant. Moreover, Venus is a friend of Mercury, the lord of the Ascendant. Presuming Shukra strong indicates possessiveness, pleasures, satisfaction, treasures-and-pleasures, luxuries (Laxmi), possession of children and political power (5) and all types of sweets. Shukra rules bhava-5 (romance) and Guru rules the yuvati bhava, suggesting multiplicity of alliances. Shukra being L-5 and L-12 indicates getting indulge in clandestine, private sensual and extramarital romance. Shukra's agency promotes reproduction and pleasures from theatre-politics-fashion, as well as pleasures of the more private variety association with the bedroom and long-term foreign residence. If Shukra is well placed there is much happiness from children from the first wife living in a foreign land. Remedial Ratna (Flawless Diamond – Heera): Being the lord of the 5th, a trine, Venus is an auspicious planet for this Ascendant and is a friend of Mercury. Consequently wearing of a Diamond in the major and sub-periods of Venus will bless the native with children, happiness, intelligence, name, fame and good fortune. Wearing of Diamond with Emerald will always be of immense benefit to the native.

Saturn: Shani is L-8 and L-9 measured from Mithuna lagna, Mithuna Chandra, Mithuna navamsha. Saturn is lord of the 9th, a trine, and Mercury, the lord of this Ascendant, is a friend of Saturn too. So, Saturn is a very auspicious or benefic planet for Mithuna lagna. Presuming Shani strong, it gives wisdom after age of 60; the L-8 Shani begins to give remarkably valuable wisdom about the cycle of deaths and rebirths. Remedial Ratna for Shani (Blue Sapphire - Neelam): Saturn is lord of the 9th, a trine; Saturn is an auspicious planet for this Ascendant. It will, therefore, be quite beneficial for the native to wear Blue Sapphire in the major period of Saturn. Mercury, the lord of this Ascendant, is a friend of Saturn. Therefore, wearing of an Emerald, with Blue Sapphirewill prove more advantageous.

Rahu: Rahu is co-lord of Kumbha, L-9, bhagya bhava, measured from Mithuna lagna, Mithuna Chandra, Mithuna navamsha. Rahu becomes a beneficial and bhagya amplifier for Mithuna lagna. Rahu is said to give positive results in the 3rd, 6th, and 11th bhava from the lagna or from Chandra. R-K are exalted in Vrishabha-Vrischika as per BPHS, R-K will give good results when Mangala and Shukra are well-disposed. However, Rahu being in any Rashi of maraka, Guru is dangerous for Mithuna lagna. Remedial Ratna for Rahu (Gomedha): Rahu is co-lord of bhava-9 (Kumbha), therefore Rahu-ratna (gomedha) may become a beneficial amplifier gem for matters of fatherhood, religious priesthood, taking of important vows, performance of ritual ceremonies, achieving professorship, university culture, and world travel.

Ketu: Ketu is L- 6 + L – 8 measured from Mithuna lagna, Mithuna Chandra, Mithuna navamsha. Ketu is temporal co-lord of Ari bhava-6 (11th from 8th), Vrischika. Ketu is the natural co-regulator of bhava-8 (Makara). Ketu effects depend significantly on Ketu bhava, Rashi, and drishti. Ketu un-connects the effect of any graha sharing Ketu house and scatters the effects of His planetary lord. Even if Ketu is auspicious, the nature of litigation and lawsuits (6) shall certainly get "scattered". **Remedial** Ratna for Ketu (Cat's Eye - Vaidurya): Vaidurya could be a beneficial gem in the case that Shani and Mangala are both well disposed.

Planet Effects in Twelve Houses for Gemini (Mithuna) Lagna

First House

Sun: Lord of 3rd house posited in Lagna indicates rash temperament, comforts from brothers, gain or wealth through others; wealthy and learned; intelligent and blessed with medium life span; interested in Astrology and Mathematics.

Moon: He/She will be selfish, intelligent, medium age, interested in business, wealthy and generous; become rash on trifles, influenced by women. Youth may become troublesome. If Moon is weak and afflicted, cough and lung disease is indicated. Money comes readily.

Mars: He/She will have short life; many real friends; two wives (subject to other indications); brave, clever and knowledge of law; strained relations with friends and relatives; gain from land and watery profession; wealthy; less comforts from children; victory over enemies; fortunate actions and successful hopes.

Mercury: He/She will have comforts from father, intelligent and strong Will power. He will be learned, skilled, a good mathematician and interested in Astrology; sweet speech and of good conduct; a comfortable life and wealthy; remains away from family; fortunate inheritance; gain through land, property and parents.

Jupiter: He/She will be intelligent and learned; social, good thoughts and actions; gain and comforts from children and wife; generous and man of justice; a man of authority, wealthy and happy; first half part of life very happy and latter half average; love of women and benefit through them; honour through merits, success through hard work.

Venus: He/She will be wealthy, sweet tempered and beautiful; gain of wealth through women; intelligent and hard worker; conscious about his respect; fond of music and relation with high ups; promotion in life; well respected and famous; will lead a kingly life; good deeds; many Love affairs; secret sorrows, many troubles and limitations; wife is loving and affectionate, intelligent and educated. Such a native seduces other women and girls for the sake of fun and pleasures of sex. Sometimes wife is barren with inauspicious bearings but attractive for men.

Saturn: He/She will have good early life, immoral, diseased, early death of wife of separation (subject to other checks), miserable life in youth; strained relations with brothers; cordial relations with others; selfish, boastful, disrespect in old age, insincere, ingenious, taste for Chemical and Mechanical sciences and affliction by skin diseases; legacies and money through deceased; accumulation of wealth; long journeys, fortunate with foreigners and strangers.

Rahu or Ketu: He/She will be whitish colour, diseased and off middle stature; medium life span; quarrelsome, harsh in speech and downfall of father just after his birth. The wife will not be so young and so well educated and intelligent.

Second House:

Sun: Sun in Cancer makes a native of rash temperament, spend-thrift not so wealthy; one will have no gain from ancestral property. The brothers and sisters of the native will also lead an average life. Right eye will be afflicted and gain of money through educational affairs, writings, and music.

Moon: He/She will earn through different sources and hard labour; gain from trade and agricultural land, from parents and relatives. He will be respected, promoted and lead a comfortable life in youth; favour from maternal relations. He/She will be fond of dress and good diet; sweet speech and kind hearted; will face some difficulties in old age.

Mars: He/She will have a troubled life. Not favourable for brothers and sisters; poor, unhappy and destroyer of property; danger to respect; danger of punishment without any fault; benefits through industrious activity.

Mercury: He/She will be secretive, revengeful, average wealth, no permanent profession; inclined toward business; gain from parental property; after 35th years of life gain of wealth through a woman; gain through land, property and estate.

Jupiter: He/She will have a true and sincere friend, learned and knowledge of law; scholar of astrology; a man of authority and rank. Early life is troublesome and full of miseries. Old age will be comfortable; one will gain wealth and respect; gain, through marriage, trade, profession, industry and Govt. office.

Venus: He/She will have comforts from property and conveyance, many children, sudden gain of wealth; respect and favour from Govt. and officers; gain of ancestral property; gain through secrecy and occult investment and children. He/She will be emotional and lot of involvement with women other than his wife. Sometime two marriages are indicated in some cases.

Saturn: He/She will be unfaithful, unscrupulous, pleasure seeking and adulterous, a few children cunning, rich, disrespected, deceitful and malicious and devoid of motherly care. He/She will be unfaithful and of wicked nature; gain by realising debts and money of others.

Rahu of Ketu: He/She will have afflicted health, adulterous, jack of all but master of none; seeks help from others; learned and interested in Tantric science; loss of wealth and property due to bad company; a good business man but loss of wealth.

Third House:

Sun: He/She will have many brothers and sisters, a good rank in life, a brave and commanding; continuous journeys in life.

Moon: He/She will be kind hearted, coward and fearful. He will not have a command over any family member; comforts from brothers and sisters; respect of women in the family is to be safeguarded; gain through education, writing and short journeys.

Mars: He/She will have less comforts from brothers and sisters; diseased; strong willed, brave and of rash temperament; greedy for the worldly comforts, respect and favour from Govt. and officers, likelihood of mental disturbance when sick.

Mercury: He/She will be respected and blessed with a good rank; gain from maternal relations; many brothers and sisters; loss of parental property; loss of brevity and virility; master of Astrology; troubles and turmoil during youth.

Jupiter: He/She will have gain and respect from rules; kind hearted and sweet tempered; many brothers and sisters; blessed with some award and rank during middle part of the life; enmity with brethren and neighbours; marriage with relative or neighbour.

Venus: He/She will have not many brothers and sisters; more than one marriage; fond of other women, music, drama and dance, will hold a good rank; act as a solicitor in village or institutions; deaths in the family; disappointment, sorrows through friends; troublesome short journeys. Wife is attractive but proud.

Native will be sexual, licentious and fond of other women, forms relations during journey or in neighbourhood.

Saturn: He/She will have no comforts from brothers and sisters; difficulties in middle part of life, unfortunate, conflicting, and evil minded, earn disputes through relations of other women, diseased and may be sentenced; gain through research, travel, investigations and writing; psychic experiences.

Rahu or Ketu: He/She will have less progeny of brothers and sister; sexy; unhappy, disrespected, strained relations with relatives; may be sentenced or death imprisonment, subject to other checks of the birth chart.

Fourth House:

Sun: He/She will be fond of travelling, friendship with learned persons; less comforts from property and land and also from mother, full reward of efforts and hard work will not be enjoyed. In middle of life, there will be loss in maternal family. He should not stand guarantee for anybody; otherwise he will suffer a loss.

Moon: He/She will be intelligent and learned, fame due to some invention or some craftsmanship; respect and gain from rulers; increase in parental and self property; gain from mother or maternal side. He may be attached with some king or minister and gain from parents, land or mines.

Mars: He/She will have diseased and troubled mother, disputes with brothers and sisters, loss of wealth and of some body part; less comforts from relatives; unfruitful attempts and hopes; a troubled life; inheritance through friend's fortunate in property and love for father.

Mercury: He/She will have comforts from mother; a good mathematician, learned and intelligent; respect and regards during youth; owner of land and property; comforts from father and conveyance; changeable residence; maternal relations will not be so wealthy; loss through storms and floods; gain through lands, mines.

Jupiter: He/She will be learned, well respected and blessed with long life; comforts and owner of property and conveyance, sexy and adulterous; many relations with other women; gain from parents and maternal relations, property by marriage, and happy married life.

Venus: He/She will have more daughters, skilled, fond of music and dance. He will make new property and late gain of ancestral or other property; low deeds, malicious, less comforts from mother and her relations; unsuccessful in hopes; will become prominent after youth; love for home; secret sufferings, restrictions at end of life; low and mean acts. Venus being debilitated here may cause two marriages; secret love affairs not to the interest of native. One is more interested in music, drama but involves himself in low acts.

Saturn: He/She will be narrow minded, erratic, boastful, peevish, and unfavourable for parents or mother, quarrelsome, no immoveable property; early death of mother; no regard for religion; gain of ancestral property.

Rahu or Ketu: He/She will be poor and under debt; loss of ancestral property; may be a cause of loss to parents; knowledge of many languages.

Fifth House:

Sun: Sun is debilitated in this house. This will not confer a comfortable life, less comforts in life, early life will be happy. He/She will have loss and shortage of progeny; loss through maternal relations; poor and unhappy; very late birth of a child; gradual loss of respect and wealth; loss of ancestral property; mental pleasure through journeys and children.

Moon: He/She will be learned, intelligent, good orator and with good argumentative power; many children, gain from maternal side; lucky children through second wife; many friends; sharp rise in life and one will be blessed with respect and wealth; gain through speculations and investments.

Mars: Mars is in enemy's camp. This indicates derangements of brain; may become a cause of loss of his own or brother's children; last life full of difficulties and, turmoil; unrealised lending of money; loss of ancestral property in repayment of debt; disease of nerves is indicated; life full or troubles and turmoil.

Mercury: If Mercury is strong and not combust the native will be intelligent, learned and owner of land; god fearing, religious, respected and man of authority, less and delayed progeny is indicated; journey to foreign land; gain of wealth from maternal side. If Mercury is weak reverse are the results and gain and pleasure by father's good fortune.

Jupiter: He/She will have birth in good environments; comforts from children, respect from Govt., and officers; learned; travel to foreign country, waste of money on litigations; worried and troubled life; gain through speculations.

Venus: If Venus is strong then it will make the native intelligent and learned, respect from ruler. Gain from education; popular, jolly and of good deeds; high position, and respect in youth. He will be famous, wealthy and will enjoy all comforts of life, more daughters; secret sorrows and difficulty through love affairs and adventures; prosperous children and gain through them. Such persons are very passionate; do not get wife of his choice, so they are indifferent to married life and remain unhappy. In female's chart, such ladies remain sometime barren and suffer from menstrual troubles and problems.

Saturn: The Saturn is exalted here in this house. He/She will be famous, tactful, powerful, respected, founder of institution; will hold good position in life being learned and well respected. High and good ideas benefit from mother, devoid or children from first wife, death of first wife and comfort from children born from second wife; free living, pleasure through journeys.

Rahu or Ketu: Such a person will be religious, respected enjoy good fame and wealth; grieved from the death of elder brother or children; many brothers; gain from the help of others; relations with other women.

Sixth House:

Sun: Sun in 6th house is in Vrischika or Scorpio sign and the position indicates difficulties and troubles in life; disease of head or brain; disputes and enmity with parents and brothers; disturbed professions; danger from ruler and Officer; termination of service is indicated.

Moon: The native is clever, talkative, and brave and will suffer loss and difficulties on account of above habits; loss to maternal relations; diseased, journey to foreign land, gain through inferiors, poultry or hygiene.

Mars: He/She will have many enemies due to his own talks and loss there of; diseased, troubled, and untrustworthy; can be subjected to diabetes and other blood disease; some severe illness; faithful servants, interest in social welfare.

Mercury: He/She will be of kind temperament, a few friends, kind hearted and of cool mind; trouble and ill health in early age; loss through sickness and servants; gain through cloth, food and small animals.

Jupiter: He/She will have residence in foreign land; learned, famous and respected; two wives, under debt and without any ancestral property; troubles due to disputes and litigations; no permanent profession; loss of wealth and troubles due to enmity.

Venus: Venus is benefic for this Lagna, it makes the native highly sexual, licentious, profligate and of loose morals. He/She will be licentious, sexy and gain from women; voyage, victory over enemies and litigations in life; troubles through employees, gain through speculations, children and hobbies; may suffer from diseases due to sexual over indulgence; may marry second time. Wife will be normally of very temperament with ideas like man; gain from women.

Saturn: He/She will have danger from enemies and thieves, rash, leg disease, average life, termination of services due to displeasure of officers, skin disease, self conceited, unhappy, rash and violent; dangerous sickness; difficulty through work in Foreign Land.

Rahu or Ketu: He/She will have Victory ovary enemies, famous in foreign land, fickle minded disease of legs, ambitious of fame and respect; sexy and fond of opposite sex and danger from reptiles.

Seventh House:

Sun: Sun in Sagittarius is good. It makes the man learned in foreign languages and law. He/She will have danger from rulers; famous and well educated. Profession will always remain in danger; disfavour from superiors. He will be harmed by his officer in connection with case of a woman and will be successful and gain out of that; marriage as a result of journey, writing; disturbed married life.

Moon: He/She will be intelligent, learned and master of law. Well respected and famous; two beautiful wives, happy, lucky and gain of wealth from mother; gain through marriage, contacts especially with women.

Mars: He/She will have many enemies, grief from wife, and waste of ancestral property; more than two marriages; diseased and ill health; deaths in family of elder brother; troubles with employees.

Mercury: He/She will be intelligent, sweet, wise and lucky; comforts from conveyance; cautious for respect; troubles in early life; death of wife; respect, fame, wealth and comforts in old age; fond of opposite sex, martial unhappiness; gain by land and women.

Jupiter: He/She will be learned, intelligent and respected; residence in foreign countries, happy from wife and a native of good rank and authority; less comforts from parental property; brothers will be lucky and holding authority; gain through law suits.

Venus: He/She will be licentious and sexual, many conveyances, comforts from servants and wealth; immovable properties and palatial buildings; comforts of life and respect; troubles in youth from some cruel officer; will hold a good post of authority; troubles through deceit and treachery, contracts and law suits; discord with children, loss by theft. Wife will be tall and beautiful; maintains calm and patience even in case of adverse circumstances. She is not very passionate; good advisor to husband. Native will not truly love his wife; married life continues but not so happy. If Venus is afflicted, one may not marry or have two marriages. One is lazy, licentious and wants luxurious life. Numbers of romances are there. Shows lot of respect to girl in love and does not look down his wife. Late marriage is also indicated from 21-26 or 28-32 years.

Saturn: He/She will have disturbed married life, abundance of enemies; one will remain in debt, late marriage, sickly wife, licentious, adulterous, no comforts from progeny; unhappy life; syphilis disease; money comes to partner unexpectedly.

Rahu or Ketu: He/She will be diseased and troubled; less comfort from wife, Loss in profession, danger to respect and troubles through enemies; under debt, troubles and turmoil; danger from Govt.; and death of wife.

Eighth House:

Sun: Sun in Capricorn is in enemy's camp. It will give troubles and turmoil in the life. He/She will be under ancestral debt. He/She will be Devoid from parent's comforts from childhood; danger of poison or death from enemies and also from reptiles; short life; disease and troubles; disfavour and punishment from the ruler.

Moon: He/She will have gain in foreign land; early part of life is full of difficulties; poor and under debt; medium age, untimely death and a troubled life; may be a cause of loss of maternal relations; disease of bronchitis; gain by legacy and goods of the dead.

Mars: He/She will have loss due to disputes with brothers and enemies; troubles in early life; loss through fire; diseased and troubled. In middle part of life, one will be blessed with comforts and wealth; occult experience; financial reward for faithful service.

Mercury: He/She will be intelligent and learned but no gain from education; diseased and troubled and may be subject to some fatal disease; danger from learned persons and disputes with friends; after youth there will be troubles; some accident in middle age; maternal relations will be happy and wealthy in the early life; gain by legacy or estate.

Jupiter: He/She will be devoid of any gain in spite of the fact one in learned and intelligent; gain from ancestral property. He/She will have Journey to foreign country; grief from progeny in youth; not good behaviour with elders; gain from wife; gain through law suits and handing other's money.

Venus: In friendly sign, mostly benefic results are indicated but being in 8th house some evil effects are also to be experienced. There is lack of marital happiness; enjoys and gets pleasures from women other than this wife and he gains money from them.

Saturn: He/She will have displeasure from Govt. or superior, diseased, left leg weak, loss through enemies; intelligent and peevish; gain by long journeys.

Rahu or Ketu: He/She will be troubled and turmoil life, danger of death through poison from opponents and enemies. He/She will have diseased, loss of wealth; early life troublesome.

Ninth House:

Sun: He/She will have authority, respect and regards from Govt. and superiors; gain of wealth and comforts of life. He/She will have grief due to death of parents in early age; long journeys; gain through publishing and science.

Moon: He/She will have loss of money and property and may lead to extinction of family. Saturn in this house will give always a worried mind and troubled life; gain by books, trading at sea and long journeys.

Mars: He/She will not be religious, miser and unscrupulous. He/She will have enmity with officers and subordinates; opponent to relations and friends. Father will be diseased and troubled and enmity with brothers, sickness abroad; danger through over study; relations with high ups.

Mercury: He/She will be lucky, learned, and intelligent and birth in a wealthy family and relations with more women. Parents will be famous and intelligent. He/She will have fewer comforts from mother; long journeys, religious, psychic, inventions and law.

Jupiter: He/She will be a man of authority and rank; religious and learned; inclination towards religion and may be charitable and founder of religious institutions. He/She will have progeny from second wife. He/She will be generous and good famed person; famous due to religious activities and good orator; honourable voyages, gain from writing, learning and publishing.

Venus: He/She will have comforts from wife, wealthy, famous and man of powers; marriage in a wealthy family and gain there from; gain of wealth and respect from Govt.; and officers; gain from friends, public enterprises. He/She will be fond of pleasures, and social life and of opposite sex; sincere and faithful in love; sudden and unexpected experiences in love; dogmatic in views; popular, independent and fond of freedom and attached to other man's life.

Saturn: If Saturn is well aspect by Jupiter, then he/she will be founder of religious institutions, good deeds, diplomatic, intellectual and learned. If badly aspect, then he/she will be immoral, destroyer of parental property, licentious, adulterous, litigations and loss through thieves; death in a distant land.

Rahu or Ketu: He/She will have troubles and difficulties in life; waste of ancestral property and may bring ill fame to the family; many enemies and opponents, gain from parents at later age. He/She will have loss through journey; not religious and immoral deeds.

Tenth House:

Sun: He/She will have grief from death of father in early life. He/She will be destroyer of ancestral property; loss of wealth; respect and comforts in young age; average life; gain through journeys and writings.

Moon: He/She will be owner of agricultural land and an agriculturist; birth in a good family, wealthy, gain from mother's property. But in spite of all, less comforts in life are indicated; loss through grieves; a changeable life; waster of ancestral property; grief from mother in youth; gain by occupation, merchandising, Govt.; and in-laws.

Mars: He/She will have sickness, licentious, non-religious and a changeable life being a man of authority he will not lead a comfortable life; sorrows and dishonour; friendship with persons in good position; gain from Govt.; and officers.

Mercury: He/She will be religious, generous and God fearing; first half will be quite comfortable and in last part of life, waste of property is indicated. He/She will be devoid from comforts in spite of having rank and respect; may change his religion; merit, honour professional success to a high position; gain from trade, public or Govt.; work or by fruits of earth.

Jupiter: He/She will be wealthy, lucky and lead a comfortable life. He/She will be a man of authority and power; well respected and blessed with property; cannot remain subordinate to others, a man of independent views; noble and famous, more respect and rank after middle part of life, opposition from friends, helpful to others; ambitious ideals.

Venus: He/She will be adulterous and connections with many women in first half of life; extravagant; journeys to foreign land; happy and a man of powers; disrepute due to his wife; loss of wealth; fond of music; last part of life will be lucky and comfortable; deceitful friends and loss of children; successful hopes. Such natives do not bind the selves in marriage bonds and are inclined to remain unmarried. They only marry after having good money; afflicted married life; may causes two marriages; marital tendencies than marriage bonds, may incline not to marry.

Saturn: He/She will be disrespected, greedy, sudden rise and fall in life. He/She will be cheat, immoral and stubborn; low type profession of doubtful nature, non-religious and disrespected.

Rahu or Ketu: He/She will have wealth and comfortable; non-religious; traveller, may change his religion, may face punishment in youth; brave and skilled in many arts; disrespected, immoral acts and quarrelsome.

Eleventh House:

Sun: Sun is in exalted position in Aries of eleventh house for Gemini ascendant. This indicated birth in a good and respected family; respect, gain and regard from Govt. and officers. He/She will be wealthy, respected and famous; gain from property of brothers or relations. He/She will be blessed with long span of life; brothers will be prosperous.

Moon: He/She will have famous, wealthy and respected; comforts from conveyance; two wives, worries and troubles from second wife; more sisters; some fatal disease may grip you, avoid cold; maternal relations will be prosperous in the first instance but latter on will become poor; accidental fortune.

Mars: He/She will be wealthy and owner of property but loss of wealth and being under debt; loss to brothers; more journeys in life; a good doctor or surgeon of a good soldier, and may be an officer. He/She will have sickness among friends and family; interest in legislative, political and social activities.

Mercury: He/She will have change of country and troubles in early life. In youth he will be quite comfortable, learned, command over languages and skilled; a comfortable and happy life. He may become preceptors of high ranking persons; a scholar of Mathematics; gain from wealthy person and religious guide; successful hopes.

Jupiter: He/She will have average age but prosperity and fame as the age advances, knowledge of law and will have a good rank. He/She will be respected, independent and wealthy man. He/She will have a good progeny and comforts there from; comforts from conveyance; loss through friends.

Venus: He/She will be lucky; wealthy and will lead a comfortable life; well respected and famous; owner of land and property. He/She will be a man of authority and powers. He/She will have grief from the death of brother in youth; comforts and gain from conveyance and servants; good for marital gains; friends will be people of culture and taste; conservative feelings, aims and wishes are fulfilled; etiquette and courteous.

Saturn: He/She will have happy early life rash temperament, troubles from brothers; loss of money and of ancestral property; short life and unhappy in old age; help from old age persons, give advice that asked for.

Rahu or Ketu: He/She will have prosperous with wealth, respect and famous; elevation in life due to Govt. favour for some period; respect and regards from brothers and relations; good friends. He/She will have shortage of progeny; may suffer loss of wealth.

Twelfth House:

Sun: He/She will have malicious thoughts, gain of wealth through unfair means; respect and regards from the Govt. and officers; some defect in eyesight of right eye. He/She will have a happy life due to far sightedness; great sorrows through brethren Sufferings; interest in occultism.

Moon: He/She will have many professions, low type jobs dispute with mother and her relations; eyesight in left eye will be afflicted. He/She will be poor and under debt; happiness for a short period in early life; gain by affairs of secret nature, through hospitals, occult science and animals.

Mars: He/She will be a man of authority or a magistrate. He/She will be very cruel, adulterous, drunkard, passionate and loss of money there from; happy and go lucky life in youth; may face imprisonment and loss through servants; occult science; powerless enemies.

Mercury: He/She will be scholar of a languages and gain there from; favour from rulers; command over many languages; no happiness from ancestral property; troubles due to woman; a good religious speaker; on the whole will lead an average life; secret unhappiness; occult science; afflicted health; sorrows through long journey.

Jupiter: He/She will be learned, intelligent and scholar of law, skilled; well respected and lucky from childhood; two marriages; gain and happiness from second marriage; expenditure of money on good deeds; unhappy first marriage, secret sorrows, jealous and sickness; loss of honour and dignity.

Venus: He/She will have waste of ancestral property; kingly luck; acquisition of wealth and property; respect from ruler and comforts, and gain from conveyance; occult science and secret investigation; speculation may cause ruin.

Saturn: He/She will have movements and journeys, quarrelsome, deceitful friends, loss through enemies, change in profession, secretive enmity, unhappy and loss of wealth; difficulties over inheritance religious science or journey.

Rahu or Ketu: He/She will be quarrelsome, many enemies, under debt and disrupted; skilled with weapon and can be a magician;

journey to foreign land; less comforts from progeny; happy and comforts for a short period in last part of life.

1.4 Cancer (Karka) Lagna

Venus and Mercury are malefic; Mars, Jupiter and Moon are auspicious. Mars is capable of conferring a full-fledged Yoga and giving auspicious effects. Saturn and Sun are killers and give effects, according to their associations.

Sun: Surya is L-2 (the house of wealth) for Karka lagna measured from Cancer (Karaka) Ascendant, Karaka Chandra, Karaka navamsha and Sun is also a friend of Moon, the lord of Cancer Ascendant. Sun is a neutral planet but inclined to give benefic results to native. The second house is also a Maraka house (house of death), so Sun is strong Markesh or maraka planet for cancer Lagna. He is more beneficial, if he occupies his own house or is exalted in the horoscope. Presuming Sun strong, Karka native will have public roles and professional leadership responsibilities related to speech, song, language, story-telling, memory and historic conservation. Remedial Ratna for Karka lagna Sun (Manika - pure Ruby): If there is loss or lack of wealth or trouble in eyes on account of the unhappy disposition or affliction of Sun in the chart, wearing a Ruby will prove very helpful. The second house is also a Maraka house (house of death). Therefore, it will be better if Ruby is worn along with Pearl which is the gem stone for the lord of Cancer Ascendant. The beneficial results will be more pronounced in the major period and sub-periods of Sun."

Moon: Chandra is L-1 for Karka lagna measured from Cancer (Karaka) Ascendant, Karaka Chandra, and Karaka navamsha. Lagnesha Moon is very beneficial or benefic planet for the Cancer Ascendant. Lagnesha Chandra provides emotional sensitivity and deepens the natural connection to the parents. Native is comfortable with boating, fishing, and life near the sea shore. If Chandra is strong or exalted, in own sign, in parivartan with his house lord, or in Kendra, the Karka lagna native will enjoy worldly success. Guru, Shukra, and other benefic will contribute and amplify the Moon powers to enhance talent and capability. If Chandra occupies Vrishabha (Taurus) in labha Bhava, one has many friends and good business intuition. Female friends will provide many valuable opportunities. Uchcha Chandra in Bhava-11 gives connections, networks, goals, accomplishments, and the marketplace suggests that this native will be well-known in his community for traits of beauty, artistic sense, and financial stability, love of fine foods and drinks, and celebrations. This person has a wonderful team spirit and will be well-liked even if other planets cause distress. If Chandra occupies Vrischika (Scorpio), His sign of debilitation in Putra Bhava, the individual may find that his creativity and self-expression is in the dark side. If lagnesha Soma in the 8th house [2nd from 7th], the first spouse will bring the wealth for him. Remedial Ratna for Chandra (Moti = pure natural Pearl): Pearl will prove very beneficial for the Cancer Ascendant and he should wear it through out his life. The Pearl will ensure good health and long life. It will make life smooth, free from mental disturbances and will improve finances."

Mars: Kuja L-5+ L-10 measured from Cancer (Karaka) Ascendant, Karaka Chandra, and Karaka navamsha. Mars being lord of the 5th a trine (trikona) and 10th house quadrant (Kendra) is a very auspicious or benefic planet, Yoga Karaka and is friend of Moon, the lord of Ascendant. Presuming Mars strong, he will be blessed with children, intelligence, good fortune, name, fame, honours and success in his professional career. Yogakaraka Kuja has the power to bring fame, winnings, leadership roles, and public dignity for him. He is more beneficial, if he occupies his own house or is exalted in the horoscope. Neechcha Kuja residing in radical lagna usually gets neechcha bhanga and becomes very beneficial for public recognition (10) through vigorous, self-defending competition in politics and entertainment (5) and makes him a famous sea captain, industrialist, politician, entertainer, or adventurer. Mars in Bhava-2, may be he is a highly energized public (10) who can suit the facts of the world to one's own financial (2) purposes like India-PM Indira Gandhi, USA Pres-43, G. W. Bush or financier Donald Trump. Remedial Ratna for Mars (flawless Red coral – Munga): Mars being lord of the 5th a trine (trikona) and 10th house quadrant (Kendra) is a very auspicious (yoga Karaka) planet for the Cancer Ascendant. He should always wear flawless Red coral and he will be blessed with children, intelligence, good fortune, name, fame, honours and success in their professional career. If Red Coral is worn with Pearl, the gem stone of the lord of Cancer ascendant, results will prove more beneficial. The beneficial results will be more pronounced in the major period of Mars.

Mercury: Budha is L-3 + L-12 measured from Cancer (Karaka) Ascendant, Karaka Chandra, and Karaka navamsha. Mercury is lord of two inauspicious houses, the 3rd and the 12th, so he is malefic planet unless he is in his own sign in the 3rd and 12th. Moreover, Moon, the lord of this Ascendant, is an enemy of Mercury too. If Budha is strong, he tends toward shallow, short-term thinking, unless

Budha is quite well disposed or there are good occupants in Bhava-12. He is especially successful in publications business such as bookstore or media advertising. However the Karka native is not a philosopher but rather one's mentality is looking for short-term advantage (3) and opportunity for bedroom sanctuary (12) or indulgence of articulated fantasies (12).

Remedial Ratna for Karka lagna Budha (Panna - pure Emerald): Mercury is lord of two inauspicious houses, the 3rd and the 12th. It will, therefore, be not advisable for natives of this Ascendant to wear Emerald. Moreover, Moon, the lord of this Ascendant, is an enemy of Mercury. If Mercury is in his own sign in the 3rd and 12th. Emerald can be worn, if at all necessary, in the major period of Mercury.

Jupiter: Guru is L-6+ L-9 measured from Cancer (Karaka) Ascendant, Karaka Chandra, Karaka navamsha. Being lord of the Bhagya sthana (9th), a Trine, Jupiter is considered to be an auspicious or benefic planet for Cancer (Karaka) Ascendant. However Guru as L-6 can function as a divorce agent. If the Jupiter is in the Ascendant, he will be exalted and in 9th, he will be in his own sign and these dispositions give rise to powerful Raja Yoga. He is more beneficial, if he occupies his own house or is exalted in the horoscope.

If Guru is yuti Lagnesha Chandra, he is prone to mix sincere religion with deep corruption

Guru in Bhava-6 makes him a sincere humanist, a wonderful teacher especially of the underprivileged, but the circumstances of his school or temple will be full of conflict, illness, and debt. Vimshottari periods of will be an end of agreement phase in relationships and degradation of the professional status due to spiritual priorities. During a Guru bhukti, he will be nourished by attending religious services, practicing in your own home, travelling to holy places, enjoying the company of priests and wise persons, and also taking continuing education classes in your profession. Guru period will be excellent for global travel and he finds himself working less and travelling more. It is a splendid time for travelling abroad with those like-minded. Remedial Guru-Ratna (Yellow sapphire – Pukharaj): Being lord of the 9th, a Trine, Jupiter is considered to be an auspicious planet for this Ascendant. Consequently wearing a yellow sapphire will bless Karka-1 with good children, high education, charitable and religious inclinations, good fortune, wealth, name and fame. The results will be more pronounced in the major and sub-periods of Jupiter. Jupiter in the Ascendant (sign of exaltation) or in 9th (his own sign) will give rise to powerful Raja Yoga and a Yellow Sapphire will enhance the beneficial results of the yoga. Karka-1 of this Ascendant can expect more favourable results if he wears a Yellow Sapphire with Red Coral or Pearl as Mars is a Yoga karaka and Moon is lord of the Ascendant and both are the friends of Jupiter.

Venus: Shukra is L-4+ L-11 measured from Cancer (Karaka) Ascendant, Karaka Chandra, Karaka navamsha. Venus is lord of the 4th (Kendradhipati Dosha) and 11th houses (inauspicious house), and accordingly Venus is not an auspicious but a malefic planet for Cancer Ascendant. Moreover, Moon, the lord of this Ascendant, is an enemy of Venus. Still we feel that the 4th and the 11th are auspicious houses and so Shukra is by and large a beneficial graha for Karka lagna, if he occupies his own house in 4th and 11th or is exalted in the horoscope. Money and Finance: If Shukra (karaka for wealth as L-11) is well placed and strong, it provides a special opportunity to him for financial gain of wealth via market place gains and L-4 for the wealth via owned properties, shelters, vehicles in Vimshottari periods of Shukra. He will get well-educated or land-holding (4) female/business partner having socially and aesthetically well-connected. His partner will be a good housewife having strong interest in the children's education and home-loving and the couple may enjoy an active social network. Shukra brings him wealth and pleasure through large social networks and gainful market relationships. Seva via large organizations and providing financial support and alliances can empower wealth to him. Remedial Shukra Ratna (Flawless Diamond - Heera): "Venus is lord of the 4th and the 11th houses for the Cancer Ascendant. Accordingly Venus is a malefic planet for this Ascendant and Moreover, Moon, the lord of this Ascendant, is an enemy of Venus. Still it is considered that the 4th and the 11th are auspicious houses and wearing of a Diamond in the major period of Venus will prove fruitful, particularly if Venus is in his own sign in the 4th or in the 11th.

Saturn: Shani is L-7+ L-8 measured from Cancer (Karaka) Ascendant, Karaka Chandra, and Karka navamsha. The 7th is a death inflicting (maraka) house. Saturn is very malefic planet for Cancer (Karaka) Ascendant, still can give good results, if he occupies his own house or is exalted in the horoscope. Shani provides burdensome or difficult marriage-unions, such as, spouse tending to be older, or hailing from a lower class than him. The partner's character may be rigid and inflexible, with little room for growth or change in the marriage behaviours. Shani in Bhava-6 or Bhava-9 encourages religious sectarianism, great sense of duty in providing service to the sick and poor. Shani in 7th house gains Drik-bala and often transform the marriage into a celibate spiritual practice. **Remedial Shani-Ratna**

(Blue Sapphire – Neelam): Saturn is the lord of the 7th house, and 8th house. The 8th is very inauspicious houses and the 7th is a death inflicting (maraka) house. Moon, the lords of Ascendants, is also enemies of Saturn. Native should never wear a Blue Sapphire.

Rahu: Rahu is co-lord of the dusthamsha Bhava-8 (Kumbha) measured from Cancer (Karaka) Ascendant, Karaka Chandra, Karaka navamsha, therefore Rahu is quite inauspicious or malefic planet unless he occupies his own house or is exalted in the horoscope. Rahu and Ketu are said to give positive results in the 3rd, 6th, and 11th Bhava from the lagna or from Chandra. In addition, some authorities posit that since R-K are exalted in Vrishabha-Vrischika (per BPHS), R-K will give good results when Mangala and Shukra are well-disposed. Remedial Rahu-Ratna (Gomedha): Rahu in either rashi of maraka Shani is dangerous for Karkata lagna. For Karkata lagna native, a Gomedha is beyond consideration and should be completely avoided. Rahu-ratna (gomedha) is generally quite inauspicious unless Karkata-1 has a specific purpose for enhancing the role of emergencies and sudden, forced changes in one's life. Nevertheless, specialized and rare tantrik practitioners may benefit from wearing the Gomedha after close examination of their unique goals and intents, if the lord of bhava-8 and its occupants are exceptionally auspicious. Also under the proper circumstances Rahu can support careers in surgery and transformative healing, emergency medical and disaster response, and management of confidential financial information.

Ketu: Ketu is co-lord of Bhava-5 (Vrischikha) measured from Cancer (Karaka) Ascendant, Karaka Chandra, Karaka navamsha, therefore Ketu becomes a beneficial or benefic graha for the Karka native. Ketu effects for Karkata lagna depend significantly on Ketu bhava, rashi, and drishti. Ketu amplifies the effect of any graha who are sharing Ketu's house; and also Ketu magnifies the effects of the lord of His occupied Rashi. Remedial Ratna for Karaka Lagna Ketu- (Cat's Eye - Vaidurya): Ketu is co-lord of Bhava-5 Vrischikha, therefore **Vaidurya** may become a beneficial gem for the Karkata native. Ketu-ratna may be profitable if Karka-1 is positively involved in politics, theatre, or special services for children (such as musical training or other evocation of individual talent school). However even for these celebratory undertakings in the service of children, politicians, and thespians, the Ketu-ratna should be applied with caution and only if the lord of Bhava-5 and its occupants are auspiciously placed.

Planet Effects in Twelve Houses for Cancer (Karka) Lagna

First House:

Sun: He/She will have changeable temper, respected and tall in stature; complexion will be sanguine bright with reddish tinge on the face; comforts from conveyance; many friends; disputes with father; gain from trade and business; diseases of belly and eyes; money comes readily.

Moon: He/She will be beautiful, long life, gain from mother, respected and wealthy, gain from land and property. He/She will be intelligent and clever. He/She will have fond of scents; bestows power, victory over enemies; long fortunate life; honour through merits and hard work.

Mars: He/She will be reddish in complexion, changeable career. He/She will have fond of music and dance; strained relations with wife; many friends and children no permanent residence; troubles in 50th year, many love affairs.

Mercury: He/She will be short life, waste of ancestral property, liar, cheat and sexy; residence outside place of dwelling; fond of opposite sex, disputes with relation and friends; fear from enemies and friends; will earn by own hard efforts; secret sorrows, danger of punishment, journeys and removals.

Jupiter: He/She will be wealthy and learned, skilled in many arts, a comfortable and happy life. He/She will have blessed with palatial property and houses; long life; friends of high ups; a man of high rank and authority; will lead a kingly life; in the start and end builder of charitable institutions; long journeys, learning, wisdom, fortunate with strangers and foreigners, voyages and in laws.

Venus: He/She will have happy and comfortable life; wealthy, conveyance, property and gain from relations and real friends; respected, fond of dress, diet and opposite sex; two marriages; helpful to others and victory over enemies; fond of garden and planting trees; fortunate actions, successful hopes, gain from ancestral property or business of land and property.

Saturn: Saturn here is in enemy's camp of Moon and will make him/her tall, of whitish complexion, coupled with blackish tinge, Non religious, licentious, adulterous, pleasure seeking, plumy, can be subjected to syphilis and venereal diseases. He/She will have limited comforts and average intelligence; less comforts from mother, average financial position and destroyer of parent's property; victory over enemies, public enemies, love of women and gain through them.

Rahu or Ketu: He/She will have diseased, ugly face and of low type profession; troublesome and fretful life. If it is being aspect by benefic planets then one will be respected, famous and wealthy.

Second House:

Sun: He/She will be wealthy, respected and famous, owner of land and property. As Sun is in own house, he/she will have authority, respect and regards from Government of the day; a good orator and impressive; strong willed; in middle life maximum honour, wealth, comforts and happiness; well respected and famous; troubles in eyes; money through hard work and industry.

Moon: He/She will be poor, loss in family, waste of time and money in litigations; fond of travelling in jungles and mountains; diseases of eyes, belly and throat; benefit through industrious activity.

Mars: He/She will have loss of property in early life; a strong willed man; will earn wealth by hard labour but will be wasted; right eye will be affected at any time in life; gain through children and investments, industry, trade, profession and Government.

Mercury: He/She will be intelligent, learned and respected; command over many languages and true to his words; average wealth; in old age will get wealth and ancestral property; average life; will remain under debt; bronchitis disease is indicated in the family; gain through educational affairs, short journeys writings and music, secrecy, occult and large animals.

Jupiter: He/She will have gain of ancestral property but loss there of and will be under debt; many enemies; danger from opponents and friends; troubles in life and a desperate man, gain through foreign merchants, science, publication, travel and inventions; loss through sickness and servants.

Venus: He/She will have gain of ancestral property, sweet temper and kind; a man of authority, wealthy and famous from early life; more women in the family; gain through dealing with land, property and estate, many friends or legislative interests.

Saturn: He/She will be poor and under debt, early life full of struggles in litigations and domestic disputes, eye trouble, financial difficulties up to middle age and loss of money; gain through deceased and realisation of debts.

Rahu or Ketu: He/She will have gain of money from unauthorised sources and manners; many enemies. There will be bad smell from mouth and body. He/She will have arrest and troubles in youth and disfavour of Government officers; danger to respect; many difficulties, troubles and turmoil in life.

Third House:

Sun: Sun in Virgo sign of Cancer ascendant makes a person tall and slander, but very well proportionate body, good complexion, dark hair and cheerful. He/She will have fond of recreations; danger from disease of bronchitis or T. B., loss and difficulties due to brothers.

Moon: He/She will have shortage of brothers, comforts for a short period, and worries from enemies, in last part of life one will become happy, wealthy and lead a comfortable life.

Mars: He/She will have shortage of brothers. Brother will be patient in young age; more daughters; troubles and turmoil in youth and old age; a troubled and sorrowful life.

Mercury: He/She will be learned, intelligent and wealthy; comforts from brothers and sisters; medium length of life; troubles and difficulties in early age. In middle life, he will have wealth, comforts, respect and authority; journey to foreign lands.

Jupiter: He/She will have many brothers; fond of music and dance; scholar of occult sciences; grief due to failure and losses of brother, in youth.

Venus: He/She will have gain of wealth through sisters; will waste his money on other women; in youth, one will be blessed with respect and rank but they will be temporary; escape from danger and troubles many a times. He will face difficulties for a short of life.

Saturn: He/She will have devoid of parental property, the only brother, wife will be ugly and immoral, last phase of life will be full of worries and financial stringency; early death of parents; less comforts from children; marriage with kin or neighbour.

Rahu or Ketu: He/She will be devoid of brother or sister; low paid profession; journeys in life; change of religion; danger to respect; troubles in old age.

Fourth House:

Sun: Sun in debilitated sign will cause loss to family and property. He/She will have down fall in environments; loss of parents in early age; litigations, troubles and turmoil in old age; may be a cause of downfall of family; gain through parents, land or mines and through household goods.

Moon: He/She will have average property, a small house and land for carrying on. He/She will be skilled in arts; dangerous from maternal relations; gain through land, mines, inherited property; success late in life, occult investigations.

Mars: He/She will be brave and talkative. He/She will be submissive, no comforts from house, conveyance of land, knowledge of many languages; loss to parents through the native; not happy in early age; comforts in old age; troubled home or domestic affairs and through servants; gain and honour through parents, land and property.

Mercury: He/She will be ambitious, traveller, journey to foreign land, a good mathematician, less comforts and gain from brothers. He/She will be disrespected and unhappy; loss of property through business, trade and journeys; gain through writings.

Jupiter: He/She will have loss of property, under debt through enemies; no permanent profession; friends will be insincere and disrespected in the family; sickness of father, trouble in family and servants.

Venus: He/She will have comforts from conveyance, ancestral property, wealthy, respected and comforts from parents; command over official language; early life portion of life will be like a king; old age will be a little discomfort able; gain in property and through father; loss through storms and floods; fortunate in property.

Saturn: Saturn is exalted in this house and aspects 7th, 10th and Lagna houses, and will confer good lucky children, who will be a source of comforts from brothers and sisters; unsuccessful in many efforts and will depend upon others. He/She will be endowed with conveyance and property; early death of mother, changeable occupations and wife will not be much beautiful; property by marriage, chaste wife and happy domestic life.

Rahu or Ketu: He/She will have troubles in young age, may be cause of downfall of maternal relations; fall from conveyance; loss of wealth; loss of limb due to fall; devoid from the comforts of houses and land; dangerous for father.

Fifth House:

Sun: He/She will have a kingly life, wealthy, prosperous man of authority; more progeny; due to extravagance some financial difficulties in middle age. Part of life will be troublesome. Children will be lucky; comforts in old age due to children; gain by speculation, investments, pleasure and entertainment.

Moon: He/She will have gain from children, not so intelligent, immoral deeds. Children will be lucky for a short period; diseased mother.

Mars: He/She will have much progeny, love for children, comforts from land but this period will be short. Early life will have much pleasure, love affairs; gain through speculations or investment.

Mercury: He/She will have changeable profession, no gain from education, worried and poor man; no gain from children; medium length of life. The life will be full of worries and unlucky native; secret sorrows and difficulty through love, speculation and children of games; travel to pleasure resorts.

Jupiter: He/She will be learned, wealthy and respected and religious;

cautious for respect; much progeny; kingly status in life; intelligent, clever and a high ranking; sickly children, illness due to over indulgence in pleasure or sport; pleasure through journeys.

Venus: He/She will have face ups and downs in life; big family but not comforts from them; any ancestral property; loss or children, wealth through pleasurable pursuits, gain through children, friends and father.

Saturn: He/She will have ugly and licentious wife, troubles through children, possibility of two marriages, discomforts from relations. He will be full of worries; wealthy, children from second wife will be lucky and no parental property; loss by speculation; danger through excessive pleasure.

Rahu of Ketu: He/She will have life will be worried and full of turmoil. He/She will be grieved and without comforts; two marriages; in latter part to life, children from second wife are indicated.

Sixth House:

Sun: He/She will have quarrelsome, rash temperament and litigation minded; loss of wealth; under debt, troubles and turmoil; will bring disrespect to father; of strong will power; loss and troubles through enemies; gain through inferiors.

Moon: He/She will have short life, bronchitis, disputes and arguments with enemies; troubles in life; gain through food, clothing and employees.

Mars: He/She will have a daring, trouble in middle life, loss of land and ancestral property; money through speculation; a famous, brave and good military officer; loss and disputes with Govt.; and officers. There will be numerous troubles and difficulties in life; sickness of children; gain through service.

Mercury: He/She will be killed in many arts but no gain from that; diseased and inactive; journeys in foreign lands; enmity with people; unhappy and loss through enemies; troubles through employees; sickness or injury through journey.

Jupiter: Jupiter is no doubt is his own house but it will not confer good results being in a Dusthana house. So the person will be disrespected although he may be a religious and kind hearted man; many enemies; will lead an average life, troubles and worries in middle age. He/She will have shortage of wealth and under debt; friends and relations will act as enemies; worried on account of debt. If well aspect reverses are the results.

Venus: He/She will have loss of wealth in pleasurable pursuits; loss of rank and service, no comforts from conveyance, adulterous, litigation and troubles; death of wife; life full of worries and troubles;

faithful servants, interest in social welfare; may built own house.

Saturn: He/She will have a good position for Saturn. It will confer on the native fame, respect and wealth. He/She will be famous, intelligent and victory over enemies; well versed in many engineering branches particularly or Mechanical and Agriculture; successful in solving the difficult most problems; rheumatic pain in legs and neck; happy last life, long, well versed and famous mechanical engineer; sickly wife, marriage below social status; dangerous sickness, loss of money given to others.

Rahu or Ketu: He/She will be a good and capable scientist; skilled in many arts. He can also be a good surgeon; many journeys; victory over enemies, will face litigations and disputes in middle of life and will be disrespected by the people; an afflicted health; under debt.

Seventh House:

Sun: He/She will be intelligent, a good mathematician, gain through women. Not so sexy, downfall and troubles from rulers; disrespected. In last part of life may be under worries and disputes with the Govt.; gain by marriage, contracts and from women.

Moon: He/She will be respected and famous; gain from business, gain of enormous money through some contract in middle age. Happy married life; blessed with property; fond of opposite sex.

Mars: He/She will be wealthy, famous respected and brave; courageous and land owner; victory in disputes and arguments; a good orator; two marriages; no comforts from first wife; discord with children; loss by theft; gain through law suits and good marriage.

Mercury: He/She will have many enemies, loss and disputes in middle age; two marriages; no happiness even from second wife; loss of wealth through law suits; a troubled and worried life generally through women.

Jupiter: Jupiter is debilitated in Capricorn sign and will not be beneficial. He/She will be learned but no gain from that; under debt, clever and indwelling, worried, loss of wealth through quarrels and law suits, changeable profession, a troubled life. Disease of hearts, disrespected; loss through business; no comforts from wife and will be very popular with the people; public enemies, marriage with a stranger whose relations will oppose.

Venus; He/She will have no happiness and success in early life; clever and intelligent; relations with opposite sex; loss in business, a quarrelsome and wicked wife, a troubled life, rheumatic pain; comforts and relief after middle age; success in law suits; property by marriage, land and women.

Saturn: He/She will have two marriages, unhappy children, disturbed married life; wife will be of harsh temperament, quarrelsome, reverses in business and service, litigations and debts. Life will be full of difficulties and miseries. If well aspect, there will be gain, through law suits and profit and comforts in life.

Rahu or Ketu: He/She will have many relations with women and loss through them; loss in life by having confidence in people; rash temperament and worried; loss of property; danger to respect and ill fame; disputes and litigations in life.

Eighth House:

Sun: Sun in Aquarius in 8th house is not auspicious. He/She will have short life, rheumatic pain, under debt, quarrelsome and poor; will face financial difficulties and may become bankrupt; danger to respect; many enemies, loss of ancestral property; troubles and worries through Govt. If well aspect then, there will be gain by legacy, goods of dead and money of partner.

Moon: He/She will have danger from enemies and thieves, short life, may be a patient of T. B. or bronchitis; troubles from wife; lean body; lazy and spendthrift; will cause loss and dispute to maternal relations, occult experience; mediumistic.

Mars: He/She will be adulterous and licentious; troubles and worries after middle age; indigestion and weak stomach; disputes with people. Rheumatic pains and may die on this account; loss or injury to nay part of body due to fall from high place or conveyance; relations with other women; two marriages and loss of wealth; loss through speculations and children; gain through estate, law suits, legacies, inheritance, insurance etc. of deceased persons.

Mercury: He/She will have long life, poor and wicked friends, subordinate to others; no happiness in spite of the fact that one is well skilled but no gain from such skill; may be afflicted to cough, flu and asthma; false accusation; unsatisfactory end and troubles over inheritance.

Jupiter: He/She will have short life, early life will be comfortable; success and wealth in middle life; relation with other women; wealthy, blessed with property; less comforts from parents; downfall for a short period and may become ill famed; worried and in trouble due to some secret disease; under debt; disfavour and worries from officers and Govt.; financial reward for faithful service; gain by long journeys and psychic.

Venus: He/She will enjoy a fairly long life; good tempered and habits; worries from wife, fond of travelling, death at some religious place; loss of a limb due to fall or accident from conveyance; repayment of elder debt; gifts or legacies from friends; gain through legacies or

estate.
Saturn: He/She will have long life, licentious, adulterous, respected, many enemies, loss of health and wealth; may become victim of syphilis, bile and venereal diseases; money by marriage, death of partner.
Rahu or Ketu: He/She will have undesirable company; will talk ill about his elders; spendthrift and licentious; earning of wealth due to hard labour; many enemies; loss of wealth and will become bankrupt; disputes and litigations in life.
Ninth House:
Sun: Sun in Pisces indicates loss of ancestral property; gain of wealth since 2nd lord is in Trikona house. The native will be the cause of loss for maternal relations; worried and diseased father, disfavour from Govt. and officer; disturbed and worried old age.
Moon: He/She will be wealthy, happy, and famous and blessed with property; unhappy brothers; cautious for his respect; a lucky native; well respected and owner of other's property by fair or unfair means; very lucky and wealthy; will enjoy full comforts of life. In old age he will be more respected and wealthy; long journeys, religious or psychic, liking for science, inventions and law, prophetic dreams; gain through in-laws and through foreigners in foreign land.
Mars: He/She will have less comforts from brothers, not religious, worries and unsuccessful in hopes; a troubled and worried life; dutiful children and gain through them; pleasure and gain in foreign land; travelling for education; prophetic dreams.
Mercury: He/She will be skilled and learned; not much gain from ancestral property; command over many arts and languages; foreign travel, devoid from comforts of parents and elders in youth. If afflicted shipwreck; gain through science and publications.
Jupiter: Jupiter in own and trine house. He/She will be lucky, wealthy, learned, famous and respected; a man of authority, power and command. In youth, he will become an administrator and will lead life like a prince; religious, founder of charitable and religious institutions; last part of life will be troublesome for education, scientific or religious pursuits; prophetic dreams.
Venus; He/She will be wealthy, famous, respected and man of powers; favour and gain from Govt. officers and brothers; happy brothers; disputes and worries due to in-laws or maternal relations; fond of travelling; respect, authority and wealth will be gained through a court of law; more daughters; gain through science, religion, long journeys and in-laws; friendships with high ups.
Saturn: He/She will have diseased, no parental property and loss

from enemies; marriage to a stranger and gain from them; death in a distant land by drowning or while on voyage.

Rahu or Ketu: He/She will have disputes with brothers, no permanent property; loss of service or profession during youth; less gain from ancestral property. Non religious, mean and vice actions; a wicked life. The native will not stick to his promises.

Tenth House:

Sun: Sun in Aries in tenth house is exalted. It bestows respect, wealth, authority and favour from Govt.; and officers; gain and comforts from parents; will lead a life of authority and respect; comforts from property and conveyance.

Moon: He will be a king or ruler or will have an equivalent authority; blessed with land and authority and will rule different parts of the country. He will have many vice habits; last part of life will be troublesome and may be deprived of the property in any way. If weak and afflicted, there will be reverse results.

Mars: Early life will be comfortable; respected, favour, and gain from Govt. but in middle age, one will face difficulties, changeable profession, brothers will be average in finances, and he/she will support them also.

Mercury: He/She will be intelligent, respected and wealthy; cautions for his respect. He may have too many jobs at a time; in old age may not become defamed; honourable children, renown in speculations, gain through wife's father or his own mother.

Jupiter: He/She will be famous and very lucky, a man of authority and powers; a very lucky guy; in middle of age may be transferred to some foreign country in an administrative capacity; may teach others; in old age may not face loss of wealth; honour through literature of travel; success in foreign affairs.

Venus: He/She will have comforts from conveyance, intelligent, active and of sweet speech; trade in Gems and jewels; worried due to his wife and enemies. In old age he may face some troubles; gain through high ups, trade, profession, public or Govt.; work.

Saturn: He/She will have success in hopes, good position in life, bad company. Early part of the life will be in difficulties, quarrelsome, shameless, liar, pleasure seeking, disrespected, and well versed in arts or amusement affairs; honourable partner; danger of violent death; drain through matters of the dead persons.

Rahu or Ketu: He/She will be cruel and of rash temperament, victory over enemies in disputes; disrespected and ill famed. In middle of age, one will face difficulties and also short of finances.

Eleventh House:

Sun: He/She will have a permanent profession but average income in life; adulterous, gain from brothers and friends. He may take place of his brother and accrues gains; confident, proud and bold, victorious over enemies; accidental fortune.

Moon: He/She will be wealthy, owner of land and property; gain from business; many sisters; respect and rank; coward, comforts from conveyance, large circle of gainful friends.

Mars: He/She will be unhappy, loss of parental property in early age; a few brothers; a troubled life; happy and comfortable life in old age but through influential and persons in authority; great attachment between the native and his children. Successful hopes and gain through legislatures, ambassadors and sportsmen.

Mercury: He/She will be generous, kind hearted and God fearing; no ancestral property; gain of money through journeys; gain from brothers; successful hopes through progressive studies.

Jupiter: He/She will have cautious about respect; two marriages, not much respected; loss through brothers; may become bankrupt in life; may deny the repayment of debt as if has not taken. The wife will be fruitful; success will come after the birth or a child; interested in politics and gain; voyages and gain in foreign land.

Venus: He/She will have gain and happiness through friends, fond of society, favour from females and happy associations; well respected and comforts from conveyance; two marriages subjects to other checks; lucky sisters; gain from business and trade; many hopes and wishes attained in old age, and gain from friends.

Saturn: He/She will be wealthy and prosperous, early part of life average, lucky, owner of property, well respected and gain from conveyance; false and deceitful friends and loss through them; danger of losing the children.

Rahu or Ketu: He/She will have no permanent profession and source of income; disputes and litigations with brothers, loss of ancestral property. Wife will be of questionable character; old age will be troubled.

Twelfth House:

Sun: He/She will have no ancestral property, loss of wealth in early age; ill famed; gain through secret nature; under debt; life in far off lands, danger of exile or life apart from kindred; worries from enemies; will again after middle of life; gain through hospitals.

Moon: He/She will have a native of weak constitution, mind is self absorbed. Love of mystery fanciful fears, success in isolated positions and remote corners and voyages; will be helpful to others; fear of imprisonment, secret unhappiness and enmities.

Mars: He/She will be cruel, licentious, disrespected, enmity with other people; children cause secret sorrows; sickness of an inflammatory nature occurs to the partner; ruin through speculation; escape from bondage; loss of dignity and honour, difficulty in employment.

Mercury: He/She will be kind hearted, well wisher of all; taste for occult sciences and secret schemes and plots; gain from business; changeable professions; self absorbed mind and happy; occult power, powerless enemies, secret investigations and limitations.

Jupiter: He/She will be learned, wealthy, kind hearted and generous; God fearing; may seek seclusion in latter part of life; helpful to mothers. If well aspect then there will be success with large animals and in remote places; private enemies through servants, work of secret nature.

Venus: He/She will be intelligent, happy and wealthy; comforts from conveyance; deceitful friends; a good argumentative and orator, secret love affairs, leading to enmity of women; jealousy, peaceful seclusion to one's own taste; voyage, occult friends; gain and respect through wife; afflicted health, end of life in seclusion or study of occult science.

Saturn: He/She will have no permanent profession, bad repay master, journeys, and unhappy marriage; no gain from business, journey to foreign lands, quarrelsome and secretive; difficulty over inheritance.

Rahu of Ketu: He/She will have changeable professions, litigations and troubles in early life; happiness in last part of life; journey to foreign lands; a long outstanding litigation will end in your favour, of independent views and obstinate.

11.5 Leo (Simha) Lagna

Mercury, Venus and Saturn are malefic. Auspicious effects will be given by Mars, Jupiter and Sun. Jupiter's Conjunct with Venus (though, respectively, Kona and Kendra Lords) will not produce auspicious results. Saturn and Moon are killers, who will give effects, according to their associations.

Sun: Suryais L-1 measured from Cancer (Karaka) Ascendant, Karaka Chandra, Karaka navamsha and is an enormously positive graha with the potential to bring great power and fame in politics and theatre, great pride in one's physical appearance and the products of one's unique social personality. If Sun is strong, it is very good for Politics and makes leaders like, Emperor Napoleon Bonaparte (Sun in Simha/1); Richard Nixon, (Sun in Dhanusha/5); G. H.W. Bush 41 (Sun in Vrishabha/10); JFK, Jr. had he lived (Sun in Vrischika/4). Sun favors literary and dramatic arts, theatre.

Career: Surya is karaka for career visibility and public regard, and as L-1 for Simha lagna, it provides public roles and professional leadership responsibilities. Career for Simha lagna tends toward spectacular rise and fall. The Simha politician who starts high office on a "landslide" election win may end one's term of office only a few years later, as an object of public contempt.

Remedial Ratna for Simha Lagna Surya - Manika (pure Ruby): Sun is lord of the Ascendant. Ruby is, therefore, a 'must' for natives of this Ascendant. They should always wear a Ruby, more so when the Sun is not well placed, weak or in any way afflicted in the chart. Ruby will protect them from health and enemies give vitality to the body, long life, and success in administrative and other posts. It will also enhance the spiritual power of the native.

Moon: Chandra is L-12 measured from Cancer (Karaka) Ascendant, Karaka Chandra, Karaka navamsha. If Chandra is well, the native has well-being. If Chandra is disadvantaged, the native will need to compensate and cope in order to gain the essential levels of security and protection needed in this lifetime.

L-12 Chandra provides emotional need for residence in distant lands near the sea-coast of distant shores on foreign beaches and on long ocean-going journeys.

The marriage may be problematic one due to Shani as lord of 7). Simha typically seeks an elder spouse or one senior in business rank and one profitably connected in the marketplace. If Moon (L-12) is in yuvati bhava (Kumbha-Chandra for the Simha lagna),

He is irrepressibly lustful. He receives money from his spouse and gains renown in distant lands. Associations and partners are social outcasts or foreigners involved in shady deals."

Remedial Ratna for Simha Lagna Chandra - Moti (pure natural Pearl): Moon is the lord of the 12th house. Therefore, the natives of this Ascendant should avoid wearing a Pearl. The Pearl may, however, be worn in the major period of Moon only, if Moon is in the 12th in his own sign. The pearl is not recommended for Simha natives unless perhaps the native desires long sojourn in a foreign land, a deeper meditation practice, or more intimate communion with the spirits in preference to the vitality of connection with human beings. Pearl may stimulate loss of personality-identity by weakening attachments to the flesh body and its social attributes.

Mars: Kuja is L-4+ L-9 measured from Cancer (Karaka) Ascendant, Karaka Chandra, Karaka navamsha. Kuja is L-4+L-9 and so is Yoga

Karaka for Simha lagna. Kuja periods will bring high educational accomplishment. For the Simha lagna with Kumbha Chandra, Shukra and Mangala create a special pair which controls 3-4 and 9-10. When both Shukra and Mangala are strong, the native is much empowered according to their location. Mangala is Yogakaraka as L-4+L-9 from Simha lagna and L-3+L-10 from Chandra. Shukra is Yogakaraka L-4+L-9 from Chandra and L-3+L-10 from Simha lagna

Behari says that Mars in Leo is eminent in its own fiery element and destroys everything unwanted in life and actively seeks spiritual experiences. He has to sacrifice material pleasure for spiritual realization, either by denial of marital happiness or absence of social associations or loss of family inheritance or abandonment from one's siblings. He stands completely on his own. Accidents, surgery, problems with the blood or diseases connected with the genito-urinary tract are the worldly trials which impel you toward a higher understanding.

Remedial Ratna for Simha Lagna Kuja- Munga: Mars as lord of the 4th (quadrant) and 9th house (a trine) is very favourable and auspicious planet (Yoga karaka) and wearing of Red Coral will be followed by the same beneficial results as in the case of Cancer Ascendant. The native will acquire, lands and property, will have good fortune, will become virtuous and longevity of his father will be improved. If the Mars is posited in his own sign in the 4th or 9th the beneficial results will become more pronounced. Wearing of a Ruby and the gem stone of the lord of this Ascendant -- along with Red Coral -- will accelerate the beneficial results."

Mercury: Budha is L-2+ L-11 measured from Cancer (Karaka) Ascendant, Karaka Chandra, Karaka navamsha. If Budha is strong and well placed, native wants to talk, instruct, explain, and announce and has hoardings, accumulation, acquisition, money, collections of records, libraries, historical knowledge, beauty, face, voice, speech and singing ability. He gains income, profit by marketplace associations, networks, financial systems, connections, community and large assemblies.

Budha becomes a highly auspicious as L-2 / L-11 and becomes a wealth graha for Simha lagna. The sources of the wealth may be found in the bhava and rashi of Budha, such as, Paramahamsa Yogananda's material wealth arose primarily from sales of his extraordinary books due to Budha + lagnesha Surya in 5, creative literature and in Dhanusha for humanistic religion; George H. W. Bush -41 has material wealth derived mainly from his government positions and his ownership of natural resources (oil + gas, also cattle + land) due to Budha + lagnesha Surya in 10 for government positions; and in Vrishabha for natural resources.

For Simha lagna , presuming that Budha is fairly well disposed, periods of L-2+L-11 Budha generates considerable wealth via family-of-origin hoarded assets and knowledge-values, as well as earned income from the marketplace of goods and services.

Budha's materialistic behavior may increase wealth, solicit death, strengthen a second marriage, increase aptitude for languages, improve relations with the family history, intensify the articulator powers of speech and song, and improve the capacity to store and acquire virtually any type of items of value, from food to furniture to art collections to currency money. If well disposed+ L-2 + L-11 calculating Budha = an excellent graha for the financial professional.

Bhukti periods of maraka L-2 (Budha) can be fatal. Example: Simha-lagna USA Pres-37, Richard Nixon died of a debilitating stroke age 81, Ketu-Budha period. Simha-lagna JFK, Jr. (son of the USA Pres-35, JFK) died of airplane crash into the ocean age 39, Shani-Budha period. Simha-lagna India-PM 1984-89 Rajiv Gandhi died of injuries from a bomb explosion age 46, Rahu-Budha period.

Remedial Ratna for Simha Lagna Budha - Panna (pure Emerald): Mercury is the lord of the 2nd (house of wealth) and 11th (house of gains and fulfillment of ambitions).

If the native wears an Emerald in the major period of Mercury and his sub-periods, he will get harmony in his family, substantial financial gains, name and fame. The results will be more pronounced if Mercury be in his own sign in the 2nd or in the 11th."

Jupiter: Guru is L-5 + L-8 measured from Cancer (Karaka) Ascendant, Karaka Chandra, Karaka navamsha.

Remedial Ratna for Simha Lagna Guru- Yellow Sapphire (pukharaja): Jupiter is lord of the 5th and 8th houses. Being lord of 5th house, a trine, Jupiter is considered as an auspicious planet for this Ascendant. Consequently it will be beneficial for the native of this Ascendant to wear a yellow sapphire, particularly in the major and sub-period of Jupiter. It will prove more useful if it is worn with a Ruby, the gem stone of the lord of Ascendant. ."

Venus: Shukra is L-3 + L-10 measured from Cancer (Karaka) Ascendant, Karaka Chandra, Karaka navamsha. Shukra is mutual enemy of lagnesha Surya. Shukra is L-3 (self-owned-commercial business) + L-10 (governance and public-interest leadership) and provides natural capability to lead and govern, either by self-earned wealth (3) or by manual craft such as writing, drawing, planning behaviors. The native switches gracefully between the public sector and private business, earning wealth from both as one capitalizes on commercial skills when in governing roles and upon government contacts when in private business. Shani regulates 7 being 10th-from-10th and presuming Shani is reasonably well disposed, Simha are often capable bureaucrats with a pragmatic Shani instinct for step-wise careful climbing ups the ranks of the target hierarchy. He may be engaged in scientific systems consulting, computerized governance. Politicians = excellent for socially diverse partnerships + wife's help with large-scale community fundraising wives

Shukra is strength for the Simha lagna and enhances success in writing, publications, announcements; commercial verbiage of sales, marketing, advertizing; relationships with siblings and team-mates; and ensemble work in any venue. If uchcha Chandra or uchcha Rahu occupy Karma bhava, the career flourish in matters of women, fashion, luxury, and the social-emotional experience of partnering in human relationships.

For the Simha lagna with Kumbha Chandra, Shukra and Mangala create a special pair which controls 3-4 and 9-10. When both Shukra and Mangala are strong, the native is much empowered according to their location. Mangala = Yogakaraka L-4+L-9 from Simha lagna + L-3+L-10 from Chandra. Shukra = Yogakaraka L-4+L-9 from Chandra + L-3+L-10 from Simha lagna.

Remedial Ratna for the Leo Ascendant Shukra: Venus is lord of the 3rd and the 10th houses. Here again Venus is not considered an inauspicious planet for this Ascendant.

If, as lord of the 10th, Venus is posited in the 6th, 8th or 12th (where he will be exalted)

Diamond will help to offset the setbacks in profession. If Venus is in his own sign in the 10th, it gives rise to the yoga, namely Amala and Malavya (one of the Panchamahapurusha yoga). The effects of this yoga have been described as under:

(1) Amala yoga: This yoga is caused when the 10th from the Moon or Lagna is occupied by a benefic (natural) planet. As a result of Amala yoga the person will achieve lasting fame and reputation. His character will be spotless and he will lead a prosperous life. (2) Malavya yoga is caused if Venus should occupy a quadrant which should be his own or exaltation sign. As a result, the person will have a well developed physique, will be strong minded, wealthy, happy with children and wife, endowed with clean sense organs and renowned and learned. The good effects of this yoga will get an impetus if a Diamond is worn during the major periods and sub-periods of Venus.

Saturn: Shani L-6 + L-7 measured from Cancer (Karaka) Ascendant, Karaka Chandra, Karaka navamsha. Periods of Shani provides ill health and faltering of the will. Shani makes trouble for Simha native's marriage. L-7 Shani and Surya, lord of Simha, are bitter enemies. L-6 Shani is an especially difficult graha for Simha native as it is L-6 from both radix lagna (material enemies) and Chandra lagna (emotional resistance to agreement):

Death: Physical separation of the soul from the flesh is triggered mainly by the Vimshottari period of the "maraka" lords of bhava-2 or bhava-7, which for Simha lagna are Budha and Shani. Other maraka are Mangala and Rahu-Ketu giving the effect of a maraka graha ruler. The Shani drishti to Ayur-bhava (8) or even strong Shani in 8 will deliver a fairly long lifespan, otherwise the periods of Shani or Budha will be the death timing for Leo Lagna native. Examples: India-PM 1984-89 Rajiv Gandhi left his body in Rahu/Budha period. JFK, Jr. left his body in Shani/Budha period.

Remedial Ratna for Leo Ascendants Shani - Blue Sapphire (Nilam): Saturn is the lord of the 6th and 7th houses. The 6th and 8th are very inauspicious houses and the 7th is a death inflicting (maraka) house. The lords of these Ascendants are also enemies of Saturn. The native should never wear a Blue Sapphire.

Marriages: As a general rule, Simha lagna and Karkata lagna are both benefitted by marrying older, professionally established partners. For peaceful marriage, avoid Shani ratna_(blue topaz, blue sapphire, lapis lazuli, and amethyst).

Rahu: Rahu is co-lord of bhava-7, Kumbha measured from Cancer (Karaka) Ascendant, Karaka Chandra, Karaka navamsha. Yugasthana is not a trine angle from Simha lagna. Rahu's effects for Kumbha lagna depend significantly on Rahu's bhava, rashi, and incoming drishti. Rahu amplifies the effect of any graha who are sharing Rahu's house; and also Rahu magnifies the effects of the lord of His occupied Rashi. Rahu and Ketu are said to give positive results in the 3rd, 6th, and 11th bhava from the lagna or from Chandra. In addition, some authorities posit that since R-K are exalted in Vrishabha-Vrischika (per BPHS), R-K will give good results when Mangala and Shukra are well-disposed. Numerous other schemes for evaluating the elusive aprakasha graha also exist.

Remedial Ratna for Leo Ascendants Rahu - Gomed: Rahu is co-lord of bhava-7, Kumbha. Yugasthana is not a trine angle from Simha lagna. However, during the major period of Rahu, the Rahu-ratna (gomedha) may become a beneficial gem for purposes of enhancing the career (7 is 10th from 10) and deal-making roles such as diplomat or negotiator (7). Rahu is especially gifted in dealing with people from exotic or criminal or otherwise unusual backgrounds.

Ketu: Ketu is L-4, Scorpio measured from Cancer (Karaka) Ascendant, Karaka Chandra, Karaka navamsha. Ketu's effects depend significantly on Ketu's bhava, rashi, and drishti. chidakaraka Ketu un-connects the effect of any graha sharing Ketu's house. Ketu scatters the effects of His planetary lord. Ketu = temporal co-lord of bhava-4 = 9th-from-8th = Vrischika. Ketu is the natural co-regulator of bhava-8.

Remedial Ratna for Leo Ascendants Ketu – Vaidurya: Under certain carefully assessed circumstances, if the lord of bhava-4 and occupants of bandhu bhava are auspiciously placed, and a "scattering" non-discriminating effect could be perhaps helpful, the Cat's Eye might aid special cases of settlement and protection for spiritualized purposes, for example the founding of a local retreat-center or monastery (4). These are rare circumstances but they do occur. Under very special circumstances may provide exceptional results for the spiritual educator looking to establish a mystery school. Especially successful for this unique purpose of continuation of the fire-worshipping priesthood when the radical lagna or Chandra = Ketu-ruled Nakshatra of Magha.

Planet Effects in Twelve Houses for Leo (Simha) Lagna

First House:

Sun: He/She will have irritable, rash temperament, liberal and generous; wealthy, a native of authority, power and rank; strong willed and respected; very cautious for his respect; ambitious but he has to adjust with his wife; gain through hard work, long fortunate life and victory over enemies.

Moon: He/She will be a good conversationalist. He/She will have many journeys to mountainous and jungle regions; attracted towards vice habits in youth; a good orator, delicate health, helpful to others but he may get bad name for helping others and religious; secret sorrows, occult sciences or new thought studies, danger of punishment by law.

Mars: He/She will have rash temperament, quarrelsome and irritant; health will be average and may be subjected to any bloody disease; boils can also appears on body and face; gain of ancestral property; happy in early age but last part of life will be poor, worried and diseased, comforts from wife and children; a man of strong will and will admire good persons; gain through property, land and long journeys; fortunate with strangers foreigners, in-laws and voyages.

Mercury: He/She will be born in a good, wealthy and respected family but will be devoid of their comforts; worried in middle age; liar and may have even a thievish nature. He/She will remain worried on account of overwork; less progeny. He/She will have relations with other women, enmity with friends and relations; less comforts in life; money comes readily; real friends and supporters; victory over enemies, successful hopes and fortunate actions.

Jupiter: He/She will be a judicious, religious and pious native; respected, wealthy, and a man of authority and powers; blessed with property, respect from rulers, generous and charitable. He may grab the property of his enemies; many love affairs.

Venus: He/She will be healthy, learned and intelligent; long life, popular in friends, relation with other women, command over foreign languages licentious and adulterous; loss of health and wealth in pleasurable pursuits and brethren. Old age will be troublesome and worried; gain from neighbours and writings; honour through matters of merits and Government.

Saturn: Saturn is in enemy's camp. He/She will be dark complexion, adulterous, sad face, bad thoughts hatred and flatterer, journeys, no progeny, diseased, unhappy old age, bad company, quarrelsome, envious, disturbed family life, unhappy life but a good writer. He/She will have public opponents and troubles through servants.

Rahu or Ketu: He/She will be a liar, licentious, ill famed and revengeful; disputes and bickering with people; no happiness in life; short life; unsuccessful in hopes and failure in business in spite of efforts; disrupted in middle age.

Second House:

Sun: He/She will be a good orator and have strong and forceful power of arguments. He/She will have loss of ancestral property; a commanding voice, victory over enemies, as the age passes eyesight will be afflicted; happy and wealthy in old age; may be owner of agriculture land; a self made man; good health; a long fortunate life; harmony and triumph over difficulties.

Moon: He/She will have a big family but will remain outside the family; less gain from ancestral property; of sweet and soft speech; skilled in many arts; may create property by his own labour; secret sorrows; all difficulties can be lightened by occult or new thought studies and science.

Mars: He/She will have rash temperament, poor and worried; disfavour from Govt. in middle age; gain of ancestral property; loss in business and troubles in service; grief from children; gain through land, property; money also can be gained from foreign merchants, science, learning and publication including travel.

Mercury: He/She will be learned, intelligent and of sweet speech; afflicted eyesight in middle of life; a wealthy and comfortable native; brothers and sister will enjoy a comfortable life; respect and gain of wealth; gain through friends, legislative and self efforts.

Jupiter: He/She will have birth in a respected and good family; promotions, comforts and gain up to middle age; a man of words; clever and intelligent; gain through deceased; respected, owner of land, lucky up to middle age, after than there will be downfall for some period; a happy old age; gain through investment, and children.

Venus: He/She will be intelligent, religious and kind hearted, but no gain, will have impressive speech; devoid of money and will lead a miserable life; gain in money by industry, trade and Govt.; money through educational affairs, short journeys, writings, music.

Saturn: He/She will be liar, dangerous, loss of money in litigations, adulterous, believer of many religions, poor, without any permanent residence, loss in business and loss through theft and amusements. Last part of life is unhappy. When well aspect, there will be money through service, employees and small animals; gain by marriage.

Rahu or Ketu: He/She will be poor or troublesome; disfavour and troubles; disrespected and ill famed; loss of ancestral property; under debt; will have some secret disease; disputes and enmity with people.

Third House:

Sun: Sun is debilitated sign. He/She will have a worried and grateful if; less comforts from brothers and a few sisters; unsuccessful hopes; downfall and loss; many enemies; mar troubles; short journeys, opportunity delayed.

Moon: He/She will have comforts in foreign land; under debt, disease of syphilis may attack him; dependent on relations and brothers, sorrows and troubles through kens and friends, occult learning. If it is aspect by Jupiter, there will be good results than above and gain from occult sciences.

Mars: He/She will be devoid from brothers and sisters and worries and troubles on their account; brave and un-fearful; will earn through hard effort; diseases of chest, headache and lungs; half life will be full of worries and troubles; progress through research, travel and writings.

Mercury: He/She will be a learned, respected, famous and a scholar of astrology and occult science when aspect by Jupiter; blessed with power of intuition; less comforts from brothers; loss of ancestral property; gain through education, writings and publications; friendship among neighbours and kindred, friends through writing and journeys.

Jupiter: He/She will be licentious and adulterous; out of control from the parents. He/She will have disputes and enmity with relations; will be comfortable and lead a pleasurable life for a short period; loss in vice acts; troubles through brothers; fortunate in realising debts and gain through deceased; liking for travel, sports, drama and adventures.

Venus: He/She will be extravagant due to women, adulterous, many relations with women, sexy and licentious; fond of music; a good power of arguments; gain from learning, writings, short journeys and brothers; respectful, gain through in-laws and Govt.

Saturn: Saturn is exalted in Libra. He/She will have many brothers and sisters; diseased, unsuccessful in plans due to illness started in young age; inclined to other religious, gain of parental property and in the affairs of opposite sex, wealthy in last part of life. If afflicted, there will be enmity with brothers and relations or a neighbour.

Rahu or Ketu: He/She will have many brothers and sisters, elevation and comforts in middle age; successful plans and hopes; ear disease in youth; pious, hate the opposite sex and may enunciate the world.

Fourth House:

Sun: He/She will be in favour of parents, will enjoy the long life of parents. He/She will have success and happiness after middle age; gain of ancestral property but will not be much benefited out of that. He will gain in worldly affairs; gain through land and mines; success late in life.

Moon: He/She will have death of Mother in youth, may be adopted by someone; acquisition of property in middle of life; gain from foreign land, clever and gain from business and trade; secret sufferings.

Mars: He/She will have changeable profession, gain from parental property. In middle life either mother will die or be seriously ill; gain and wealthy due to help of brothers; loss through natural calamities.

Mercury: He/She will be not so educated; cautious for respect may be cause of loss for his maternal relations; gain of ancestral property; loss for father; gain of property and wealth in last part of life; gain through land and mines.

Jupiter: He/She will be learned and intelligent. He/She will have generous, knowledge and command over rules and law points; will have self created property; love of home, well respected, wealthy and happy in last phase of life; gain through children. If afflicted, reverse are the results.

Venus: He/She will have loss of wealth and property in adulterous actions; happy and wealthy, blessed with conveyance in youth;

residence in foreign land; success, gain, honour and land at close of life.

Saturn: Saturn is in enemy's house. He/She will be loss of agricultural land, proud, worried and may become selfish; high position in life by dint of hard labour, faces many difficulties in life, gain through marriage; unsuccessful in plans and schemes due to proudest; no gain from parental property; long life to mother.

Rahu or Ketu: He/She will have journey to distant lands, no gain from parents; will be a source of troubles for parents; a good rank but no comforts; loss through animals.

Fifth House:

Sun: He/She will be very wealthy, learned and respected; cautious about his respect, a native of fame, authority and power. In middle of life, one will be at the helm of the affairs; comforts from children; respect and favour from Govt.; journeys to foreign land; delight in pleasure, amusements, sports and children and success through these.

Moon: He/She will have birth of sons but fewer comforts from; more daughters; intelligent but average living; devoid from comforts of mother in youth; journeys in life; loss of ancestral property; gain from relations of mother; a kind hearted, and well wisher of all; secret sorrows and difficulty through love, speculation, children and gambling.

Mars: He/She will have rash temperament and quarrelsome; worried from children. Children will not remain within control; happy married life. Wife will be of good nature; gain through father, liberal or unconventional ideas in regards to union and free living.

Mercury: He/She will be intelligent and learned; worried in early age; respected and social; gain of wealth in old age, successful hopes and a native of high thought; gain through children, friends etc.

Jupiter: He/She will be generous, respected, learned and well versed in worldly affairs; love affairs, respect gain, authority and rank from Govt.; early life will be average; adventures; a wealthy and will lead a comfortable life; worried from progeny; may be under debt in life. From middle age onward, one will lead life of a prince and also gain from children; will lead a respectable life.

Venus: He/She will have enmity with father and children; fond of music, dance and of opposite sex; loss of wealth due to women; loss and worries from Govt. and entangles in litigations; accomplishment and travel to pleasure resorts; gain through speculations.

Saturn: He/She will be adulterous, no happiness from sons, rheumatism, more than one attachments, cruel, may be murderer;

disrespected being licentious and bad company; loss by speculation; sickly children, illness due to over indulgence in pleasure and sports.

Rahu or Ketu: He/She will be non-religious, unchaste, worried and ill famed; worried and may suffer loss through sons; changeable professions; loss or shortage of progeny.

Sixth House:

Sun: He/She will have a troubled and worried life, loss and disfavour from Govt.; litigations and disputes in life but victory over enemies; enmity with relations and many persons; diseases of eyes and intestines may attack the native; unhappy and uncomfortable life; sickness among children; pleasure through food, clothing and small animals.

Moon: He/She will be poor and under debt; loss and troubles through the foreign rulers but will be saved in the end; windy complaints and T. B. in old age; troubles through employees and animals; limitation on account of sickness.

Mars: He/She will be licentious and adulterous; disrespected; many disputes and one have to face many challenges in life; loss through Govt.; servants, opponents and may face many troubles. In the end, he will have victory over enemies and may be a bit comfortable in old age; loss through export.

Mercury: The native will lead a happy life in an average way; of delicate health, without any opportunities and enemies, will remain balanced in case of money but at times will feel difficulties and may be under debt; may be a case of loss to maternal relation. Middle life will be full of miseries and troubles; interest in social welfare, faithful servants and gain through them.

Jupiter: He/She will be diseased and worried; devoid from the comforts of elders and face disputes, disrespected, loss through enemies and opponents and of property in the litigations; sickness among children, gain through careful speculation.

Venus: He/She will be adulterous and licentious, loss of wealth through enemies, women and vices; fickle minded, under debt; a troubled life; may become victim of diseases like windy complaints, syphilis etc; modest position; difficulties through brothers; gain and honour through service, employment; injury or sickness through journey.

Saturn: He/She will have diseased, victory over enemies, loss through litigations and opponents, disrespected, disturbed from wife, loss of money but will become a bit comfortable after young age; sickly and bad wife.

Rahu or Ketu: He/She will have loss of wealth and property in litigations and disputes; short life, disrespected; a troubled and worried life.

Seventh house:

Sun: He/She will have strained relations with wife, association with opposite sex, fickle minded; loss through business or troubles in service; will face litigations and disputes especially in the middle age; marital unhappiness.

Moon: He/She will have voyage, relation with many women; changeable and unprofitable profession. Financial status will be average. There will be disputes, losses and difficulties in the early age; troubles through deceit and treachery, union, contracts and partnerships etc.

Mars: He/She will be licentious and adulterous, disputes and quarrels in early life and becomes ill famed; strained relations with wife; loss in business. But in middle of life, one will be blessed with comforts, respect and wealth; gain by marriage, land and women. He/She will have marriage with stranger of education and refinement whose relations will oppose.

Mercury: He/She will be unhappy due to relation with many women. He/She will have changeable professions, disrespected and loss in business. The financial condition will be good to average. Wife will love; desirable friendship and social connections; success in law suits; gain by marriage. If afflicted, there will be reserve results.

Jupiter: He/She will have marriage in a wealthy family and gain there from. The wife will be faithful and gentle; gain from his profession. After the middle age, one will face a serious accident (not bodily) due to which he will face many difficulties, loss of wealth by theft and may be under debt; pleasure or close association and understanding with partner but discord with children.

Venus: He/She will be adulterous and licentious; gain through opposite sex. One will oppose his elders and preceptor; lucky in young age; a good conversationalist; gain in profession; marriage as a result of journeys, writings or to a relation; gain through law suits and marriage.

Saturn: This is a Moolatrikona Rasi of Saturn. One will be deep thinker and slow worker; gain from residence in foreign land; selfish, shameless and quarrelsome. Wife will be of independent views but there will be bickering with wife and non-religious; success in law suits, marriage in good family but wife will be untrue or hostile.

Rahu or Ketu: He/She will have loss or wealth in disputes and litigations; danger to respect, loss in life due to opposition. Short life and may be ill famed.

Eighth house:
Sun: He/She will have loss of ancestral property. He/She will suffer from blood and intestine diseases; loss of wealth through failure and unsuccessful in hopes; peaceful death through irregularity; death of father during young age of the native; occult experience.
Moon: He/She will have death of mother and some other old man in early age, worried and troubled; under debt and poor; change of religion or non-religious, short life, loss through relations, friends and opponents; many enemies and unsatisfactory end, secret enemies die and trouble over inheritance.
Mars: He/She will have short life, death due to wound or fall; disrepute in middle age; poor and under debt; loss through employees if afflicted. Gain by legacy or estate and long journeys; psychic experience.
Mercury: He will be endowed with abundant land and wealth. But he will have limited marital felicity and be bereft of happiness from his elder brother; gain through his wife's relatives, be liberal, virtuous, sensuous, and will remain at the command of his spouse.
Jupiter: He/She will have birth in a good respected family; not much wealthy. He will be religious and preach others; average, normal and comfortable life, suffers through children and loss through speculations; spiritualistic experience.
Venus: He/She will have more than one marriage; gain through opposite sex; loss through relations with other women; headache, diseased in middle age, relations with friends and relatives will not be much cordial; gain through estates, law suits, legacies, insurance of other; false accusation, sickness or death of brother, journeys on account of troubles.
Saturn; He/She will have long life, fear from thieves, fire and enemies; diseased; disrespected profession, displeasure of Govt.; and superiors and bad company; dangerous illness, financial towards for faithful service; property by marriage; death of partner; some inheritance but difficulty over it.
Rahu or Ketu: He/She will have danger from enemies and thieves; loss of wealth; drunkard and remain under debt; loss of property due to enemies or litigations; danger to respect; troubles and turmoil in life.
Ninth House:
Sun: He/She will have good stature, strong and stout, not clear but good complexion; large eyes; noble, violent, courageous, gain of wealth, man of authority and rank; gain of ancestral property; victory over enemies; famous and a terror to enemies; delighting in war-like

actions and enterprise; long journeys, psychic dreams; gain through in-law; liking for science, inventions etc.

Moon: He/She will have loss in disputes and litigations; may be the cause of loss to maternal relations, worried and troubled; medium span of life; death of father in young age; no comforts from ancestral property in spite of the fact that the native will be blessed with such property, long journeys, secret missions, secluded research; sacrificed for science or religious.

Mars: He/She will be devoid in youth from the comforts of father; wealthy and will lead a comfortable life; respected; many journeys in life for educations, scientific or religious purposes; prophetic dreams; many fine qualities.

Mercury: He/She will be learned, intelligent, generous, religious, and skilled; fond of long travelling and wealthy; friends amongst high ups; like a king and helpful to people, gain by books, trading at sea, religion and science; gain and success through long journeys; friendship through travel and learning.

Jupiter: He/She will be lucky and learned. He/She will have long life and generous; pilgrimage, no dutiful children and good deeds; helpful to others; long life to father; gain from publication of own books, good and renowned author; loss and troubles in old age; pleasure in foreign land; afflicted with windy complaints; death in a distant land, partner has troubles with relatives over money matters; providential help at the time of difficulty.

Venus: He/She will be wealthy and learned. The native will be blessed with property, conveyance and comforts; early death of father; will become unhappy and face troubles at times in if many long journeys in life; gain through philosophy, publishing and science, brothers are likely to go for journey to foreign and marry; a religious and intellectual mother.

Saturn: Saturn is in enemy's camp. He/She will have worried, loss of parent's property, unlucky as the age advances specially in old age; tourneys without gain, break in education, disagreement with father; fond of occult science; sickness abroad while travelling; danger through over study; marriage to a stranger from afar and gain thereby; partner's journey to foreign land; family environments will not be congenial.

Rahu or Ketu: He/She will have change of profession, many journey, troubled and worried life, of wicked nature, actions and thoughts; under debt but in spite of all, he will be well respected; cautious for his respect; an intelligent native.

Tenth House:

Sun: He/She will be respected and wealthy; a man of power and authority; gain and respect from Govt.; and officers; success in disputes and litigations; a comfortable life; merit, honour, success, rise to high professional or social position. If afflicted reverse results.

Moon: He/She will have voyage, respected and respectable profession; victory in disputes and litigations; gain and respect in foreign land; fond of travelling. He may get ill fame at a time in life; victory over enemies and cautious for respect. If weak and afflicted then there will be liability to disgrace, discredit and persecution from superiors and Govt.; resulting in a long journey or retirement and seclusion; sorrows through mother or troubles with in-laws.

Mars: He/She will have victory over enemies, respected and native of authority and powers. He can be a magistrate; brave but very strict; wealthy; gain from trade, profession or Govt.; work or by fruits of earth; honour and credit through science, literature or travel; success in foreign affairs; reverse if afflicted.

Mercury: He/She will have a good profession well behaved and learned; a good mathematician and doctor of medicines; generous and charitable; comforts to parents from him; respect and regard in middle of life; friendship with high ups and honours through them.

Jupiter: He/She will be wealthy and learned; a man of powers and authority; a judicious and have comforts from children and wealth. He can be a judge or head of a village; favour and authority from Govt.; and officers; renown in speculation, business or theatrical work; gain through wife or in-laws and his own mother.

Venus: He/She will be wealthy, native of authority, respect and regards from Govt.; and officers; Owner of land, property and conveyances; learned, well behaved; gainful journeys; a lucky; loss after middle age, worries and troubles; fond of music and dance; honour or renown through writings or other accomplishments; journey with mother.

Saturn: He/She will be not obedient to father, liar, licentious, lucky after middle age, respected and blessed with immovable property and agricultural land, sickness through dishonour, sorrows on account of mother or in-laws. If it is aspect by Jupiter, there will be reverse results.

Rahu or Ketu: He/She will be skilled and well versed in jobs of machinery; respected, gain favour from Govt.; in middle of life, one may face disputes and may be under debt; a worried and troubled man and on such troubles and worries, one may go to foreign country where he will be benefited.

Eleventh House:

Sun: He/She will be intelligent, learned and respected; wealthy and man of powers and authority; respect and gain from Govt.; large circle of friends, comforts from conveyance. Elder brothers will also be well fixed in life. But in a middle age, one may face reversals, worries and troubled life. Service or profession may face serious troubles if afflicted. Pleasure in life, hopes and wishes often attained.

Moon: He/She will be sweet tempered, skilled and clever; victory over enemies; gain through friends; a good profession and respected native; blessed with property, respect and authority. He will rule others; successful hopes. If afflicted, there will be reverse results.

Mars: He/She will be wealthy and respected man; devoid from elder brother, if the same is, strained relations with him; brothers and sisters will suffer from loss of progeny. Till middle age, life will be miserable, troubled and worried from enemies; may be a cause of loss to maternal relations; voyages. He will be adversely rewarded for his good actions and behaviour with others. Many hopes will attain in old age; gain from foreigners and in foreign land.

Mercury: He/She will have comforts from conveyance, blessed with wealth and property; gain from friends, respected and generous; loved by friends and relations; more expenditure than income; well defined hopes and wishes; accidental fortune.

Jupiter: He/She will be respected, wealthy and comfortable; gain and comforts from conveyance; a native blessed with land and property, favour and power from Govt.; fond of children; will have a good rank and authority; generous and famous; builder of houses; successful hopes; pleasure among legislators and ambassadors etc.

Venus: In early and old age, he/she will face difficulties and worried; loss of wealth. In middle age, one will be happy, respected, and wealthy and gain through journeys; gain from brothers; hopes and wishes accomplished through progressive studies; eminent friends amongst high ups and legislators; ambitions ideal; helpful to his associates and others.

Saturn: He/She will be devoid of elder brother and if blessed with, no happiness from him; bad thoughts, bold and courageous; litigations and a liar; livelihood from agriculture land, Interest in politics and social life.

Rahu or Ketu: He/She will be skilled, respected, famous and intelligent; gain from foreign land; brave, courageous and respected; accumulated wealth will be wasted in old age; danger of respect; loss and troubles through enemies and opponents.

Twelfth House:
Sun: He/She will be a man of authority, powers, rank and wealth;

success in middle life; an honest, generous and cautious for his respect; connections with high ups; respected by the people, if well aspect; otherwise fear of imprisonment, secret unhappiness, occult science; sufferings and misfortune.

Moon: He/She will be respected and famous; cautious or his respect; average living; residence in foreign land; early life troublesome, after that life be happy; occult abilities, powerless enemies, secret investigations, limitations, afflictions and adversities prove to be blessing in disguise by development of inner power.

Mars: He/She will be adulterous and extravagant; loss of property, business, and wealth in middle age; adoption of another profession; difficulties through religion, science, journey; long journey; afflicted health, end of life in seclusion.

Mercury: He/She will be worried and loss through enemies; troubled and poor; good friend among occult persons; pleasure in peaceful, quite and secluded places. No gain from education; a troubled and worried life; gain by affairs of secret nature, hospital and animals; deceitful children, and sorrowful friends.

Jupiter: He/She will have extravagant, average profession; sorrows through children; loss of service; generous and respected; honest and will spend the money on good jobs and public welfare; speculations cause ruins; difficulty over inheritance or fear of imprisonment.

Venus: He/She will be respected, happy and wealthy native; comforts from wife and conveyance; owner of land and property if strong; after middle age there will be down fall resulting loss of wealth, respect and property; loss of office, honour, and dignity; difficulty in employment; occult science gainful; danger of enemies and imprisonment while in journeys.

Saturn: He/She will be under debt and unlucky; loss in middle age; danger of death at the hands of enemies; may be a cause for extinction of family; loss of money and ancestral property; fear of imprisonment; work of secret or mysterious nature; interest in archaeology and in submarine life; unhappy marriage.

Rahu or Ketu: He/She will have unfruitful journey to foreign land. In early life there will be troubles and worries, litigations and disrespect. After middle age, he/she will be happy, lucky, gain of wealth, respect and regard from Govt.

11.6 Virgo (Kanya) Lagna

Mars, Jupiter and Moon are malefic, while Mercury and Venus are auspicious. Venus Conjunct with Mercury will produce Yoga. Venus is a killer as well. Sun's role will depend on his association.

Sun: Surya is L-12 measured from Virgo (Kanya) Ascendant, Kanya Chandra, Kanya navamsha and so is malefic planet for this ascendant unless Sun is in his own sign in the 12th house or exalted in the horoscope. Because Ravi = L-12, the native may have a less dramatic personality or leadership responsibilities within private. Remedial Ratna for Kanya lagna Sun (Manika - pure Ruby): It is not advised the native of this ascendant to wear a Ruby. But when Sun is in own sign in the 12th house or exalted, Ruby can be worn if necessary but preferably along with Emerald (the gem stone of lord of Virgo Ascendant) during the major period of the Sun."

Moon: Chandra L-11 measured from Virgo (Kanya) Ascendant, Kanya Chandra, Kanya navamsha. Moon the lord of the 11th and so he is malefic Planet unless Moon is in his own sign in the 11th house or exalted in the horoscope.

Presuming Chandra is well placed and strong, he may gain from ocean and its products and elder sibling especially a sister, cousin, or a close-in-age aunt, can be most helpful as a mentor in this life. Remedial Ratna for Kanya lagna Moon (Moti - (pure natural Pearl): It is not advised the native of this ascendant to wear a Pearl. If Pearl is worn in the major period of Moon, the native gets financial gain, acquires wealth, achieves name and fame and is blessed with children. If the Moon is in Taurus, in the 9th, a Pearl can be worn always for success of ventures, peace of mind and acquisition of wealth and conveyances. If the Moon is afflicted in any way in the horoscope wearing of Pearl will neutralise the ill effects.

Mars: Kuja L-3 + L-8 measured from Virgo (Kanya) Ascendant, Kanya Chandra, Kanya navamsha and so is a great functional malefic or highly inauspicious for the Virgo Ascendant unless Mars is in his own sign in the 3rd or 8th house or exalted in the horoscope. He has an overtly sexual, animal instinctive, L-8 (Kuja), and has a tendency to get involved in impulsive sexual relationships with girlfriend, a team-mate or office-mate or neighbour and can get into trouble that may threaten his marriage. Mangala for Kanya lagna becomes a karaka for the "office marriage type relationship or sexualized connection" and the native spends more time in office getting involved with a team-mate than with his own spouse. His communication style (L-3) is overly aggressive to the point of being self-destructive (L-8) unless Mangala or Budha are very nicely disposed. He can have typically quite sudden death involving blood disease, severe injury, or results of conflict. From 1, 2, and 5 bhava, Mangala casts drishti upon randhra bhava and gives sudden and possibly violent death from causes given by Mangala's rashi. Example: Kanya lagna, Mangala in bhava-1 (Kanya) gives intestinal disease, argument, iatrogenic causes, particularly mixed drugs, poison, spoiled food, often from a hostelry or jail. Kanya lagna, Mangala in bhava-2 (Thula) gives kidney disease, adrenaline surge, disease of external genitals, sexually transmitted disease. Kanya lagna, Mangala in bhava-5 (Makara) gives bone disease, skeletal disorder, and multiple broken bones. Kanya lagna, Mangala in bhava-7, death is swift and direct for the first and second spouses. Mangala-Mesha-8: circumstances of death involve weapons, engines that propel moving vehicles (car crash), or both like JFK, in a moving car when killed by rifle bullet. **Remedial Ratna for Kanya lagna Mars (Munga - red coral):** Wearing of Red Coral by the natives of this ascendant should be avoided as far as possible. However, if absolutely necessary, Red Coral may be used in the major period of Mars, if Mars is in his own sign in the 3rd or the 8th.

Mercury: Budha is L-1+ L-10 measured from Virgo (Kanya) Ascendant, Kanya Chandra, Kanya navamsha and so is a great functional benefic for the Virgo Ascendant in the horoscope. Presuming Budha (L-1) is strong and well placed, he has very good personality & physical body, vitality, contests and competitions ability, innovation and fashion, professional or social dignity: leadership roles, position and attributes of distinction, rank and reputation, promotions, government bureaucrats, policies and procedures, legislators and laws matters. Budha (L-10), is a booster for a communicative career in administrative, information-intensive, and policy specialties. As L-10, Budha for Kanya lagna is a significator of career, public reputation, and social status. The native tends to favour professions in the verbal, media, explanatory, or communicative arts, with a special affinity for careers that involve writing. Budha as lord of 10th navamsha for a Kanya native indicates a great facility in sales, marketing, advertising, public relations, or any of the communicative-administrative activities of commercial business. The person may be an extraordinary sales professional if Budha is well disposed. Remedial Ratna for Kanya lagna Budha (Emerald - Panna): Emerald will always act as a protective charm or talisman for the natives of this Ascendant and will also bestow good health, longevity, success and advancement in profession and name, fame and honour. If Mercury is afflicted or not auspiciously placed in the birth chart, wearing of Emerald will neutralise the evil effects of such disposition. "

Jupiter: Guru is L-4 + L-7 measured from Virgo (Kanya) Ascendant, Kanya Chandra, Kanya navamsha. Counted from Mithuna lagna or Kanya lagna, Brihaspati is the "bad boy" badhesha who gives hindrance, harm, harassment. As lord of two Kendra, Guru acquires kendradhi-pati dosha and so he is a great functional malefic for the Virgo Ascendant unless Jupiter is in his own sign in the 4th or 7th house or exalted in the horoscope. Presuming Guru is strong and well placed, he will have good home life having compatible spouse (unless Guru is damaged). Otherwise L-7 Brihaspati can generate troublesome and failed expectations and fewer fulfillments in relationships to both personal and professional, marriage, Children, grandchildren and marriage and children are affected by the badhesha function of Guru for Kanya Lagna. There is a general life feeling overall for most Mithuna-lagna and Kanya-lagna natives that marriage and usually children too have not realized their growth potential. The spouse's career (10th from 7th = 4) is affected by badhesha Guru and during Guru Periods it too will experience unexpected setbacks, hindrance, and harm due to spousal misplaced confidence. However, Guru periods helps in significant educational completion, such as learning a new trade skill or a professional diploma, presuming Guru is healthy and there is no severe resistance in house-4 or house-7. **Remedial Ratna** (Pukharaja/yellow sapphire): Wearing of Red Coral by the natives of this ascendant should be avoided as far as possible. Still if the Jupiter is giving rise to Hamsa yoga, 5th, 9th, 10th or 11th, yellow sapphire should be worn in the major and sub-periods of Jupiter. He should not wear this gem stone, which is in apprehension of death on account of old age or ill nor if his horoscope indicates short life, as is Jupiter a strong death inflicting planet for this Ascendant.

Venus: Shukra is L-2 + L-9 measured from Virgo (Kanya) Ascendant, Kanya Chandra, Kanya navamsha. Venus, being lord of the 2nd house (house of wealth) and the 9th (Bhagya), a trine, is a very auspicious planet for the Virgo Ascendant. Shukra becomes a highly auspicious L-2 / L-9 wealth graha for Kanya lagna. Presuming Venus is strong and well placed, he will enjoy the company of women, wealth (banked savings assets-2, gambling or speculative win assets-5, or earned income assets-11), knowledge as well as world travel, religion and his wife will have financial skills and an international lifestyle. As L-9 Shukra favours artistic and musical talents for the grandchildren and accomplishments in arts and music for students of the guru. Shukra is also a maraka graha who will effect a saturation of pleasure, when the time has come. **Remedial Ratna for Kanya lagna Shukra (Diamond – Heera):** Wearing of a Diamond will ensure gain of abundant wealth, birth of children, good fortune, and honours. The results will be more pronounced during the major and sub-periods, particularly if Venus is in his own sign in the 2nd or the 9th. Wearing of Emerald, along with Diamond through out life will ensure long life and success in all ventures.

Saturn: Shani is L-5 + L-6 measured from Virgo (Kanya) Ascendant, Kanya Chandra, Kanya navamsha. Saturn is lord of the 5th (a trine) and the 6th houses for the Virgo Ascendant and so Saturn is an absolutely auspicious (benefic) planet for this Ascendant. However, Shani will bring conflict in marriage, losses from lawsuits, health issues, and financial strain from loan debt. Periods of L-5 Shani -or- Guru -or- L-9 Shukra will produce children, eventually - unless Shani occupies house-5 or house-9 in an unfavourable sign or house-5 is otherwise damaged. Shani = the L-6 can offer particular difficulties of chronic illness and exacerbation of resistance to keeping agreements in contractual relationships, during Shani periods. Remedial Ratna for Kanya Lagna Shani (Blue Sapphire - Nilam): Blue Sapphire can be worn by the native of this Ascendant in the major period of Saturn."

Rahu: Rahu is co-lord of maha-dusthamsha bhava-6 (Kumbha) measured from Virgo (Kanya) Ascendant, Kanya Chandra, Kanya navamsha and so Rahu is a malefic planet except Rahu is in his own sign in the 6th house (Kumbha) or exalted in the horoscope. Rahu is a malefic planet except for careers in medical clinical practice and human services ministries, e.g. social work, druggist, police, divorce law, bankruptcy, criminal attorney. Rahu effects for Kanya lagna depend significantly on Rahu bhava, rashi, and drishti. Rahu effects for Kumbha lagna depend significantly on Rahu bhava, rashi, and incoming drishti. Rahu amplifies the effect of any graha who are sharing Rahu house; and also Rahu magnifies the effects of the lord of His occupied Rashi. Rahu and Ketu are said to give positive results in the 3rd, 6th, and 11th bhava from the lagna or from Chandra. In addition, some authorities posit that since R-K are exalted in Vrishabha-Vrischika (per BPHS), R-K will give good results when Mangala and Shukra are well-disposed. Remedial Ratna for Kanya Lagna Rahu (Gomedha): Rahu-ratna(Gomedha) is harmful or highly unlikely to be a beneficial gem except for careers in medical clinical practice and human services ministries, e.g. social work, druggist, police, divorce law, bankruptcy, criminal attorney only in case, the lord of bhava-6 and its occupants are auspiciously placed and strong. If Rahu Himself occupies Kumbha-6, the results can possibly be superb. However conditions for a positive Rahu-6 result should include presence or drishti of all malefic toward Ari Bhava.

Ketu: Ketu is co-lord of bhava-3 (Vrischika) dusthamsha measured from Virgo (Kanya) Ascendant, Kanya Chandra, Kanya navamsha and so Ketu is a malefic planet except Ketu is in his own sign in the 3rd house (Kumbha) or exalted in the horoscope. Ketu effects depend significantly on Ketu bhava, rashi, and drishti. Chidakaraka Ketu un-connects the effect of any graha sharing Ketu house. Ketu scatters the effects of His planetary lord. Ketu is temporal co-lord of bhava-3 (8th-from-8th – Vrischika). He will study yogic literature, engage in penance, and establish a new school of thought. Ketu will bring cataclysmic changes to his life, resulting from the distrust, humiliation, and animosity generated by colleagues and associates. After attaining great proficiency in trade, he will be forced to alter your life pattern. After having confronted opposition from several quarters, he will gain equanimity and spend time more of less like a recluse. He will seldom open his inner feelings and ideas to others, and will experience a great metamorphosis in life. Remedial Ratna for Kanya Lagna Ketu (Cat's Eye – Vaidurya): Vaidurya is unlikely to be a beneficial gem except for matters of mystical-transformative writing and publication.

Planet Effects in Twelve Houses for Virgo (Kanya) Lagna

First House:
Sun: Sun, 12th lord in Virgo in Lagna, indicates that he/she will be fond of music, painting and poetry; respected by rulers, being learned and intelligent when strongly aspect by Jupiter; devoid from the comforts of father in early age; loss of ancestral property. If Saturn aspect Virgo Lagna, it will make one lazy and easy going, make changes in your environments and he will form new associations and friendships; secret sorrows; gain through occult.

Moon: He/She will have a joyous nature, short life subject to other checks; not so well respected, more daughters, unsuccessful hopes and desires; diseases of windy nature and cough are indicated; good and sincere friends and gain; victory over enemies, fortunate actions and successful hopes.

Mars: He/She will have comforts from wife and friends; a learned, intelligent, clever and brave guy; worried and troubles from opponents and enemies; will be subject to disrepute, disputes with officers and relations; selfish and self conceited nature; fond of music, dance and pleasure of life; gain respect in business; medium length of life; a good progeny and helpful to others; gain through writings and learning.

Mercury: He/She will be wealthy and respected; learned and blessed with long life; will attain rank and authority till middle age; a kind hearted life. Good health; honour through merits, success through industrious efforts; gain through mother and Govt.; reverse results if afflicted.

Jupiter: He/She will have gradual rise in life, Comforts from wife and progeny; respected, generous, skilled and endowed with wealth; a good hospitable and expenditure on ill health; fond of scents and pleasurable life; an average life; gain from land, property, inherited property, love of women and benefit through them, partnership; public enemies.

Venus: He/She will have wealth, comforts from ladies; popular everywhere, endowed with conveyance and property, but in spite of all worried and troubled; defects in eyesight, gradual downfall in life, unhappy old age, loss of money in youth will be recovered in old age; long journeys, learning, wisdom, fortune with strangers, foreigners, wife's relations and voyages; interest in science, law, inventions and philosophy.

Saturn: Saturn is in friend's house. He/She will be lucky and well respected, well wishers of all, gainful for brothers, worried in middle part of life, long lived; more careful for self respect; immovable property and land, well respected by Govt; and superiors. If Saturn is retrograde or weak, then reverse results will be enjoyed by the native. Many love affairs.

Rahu or Ketu: He/She will have average financial set up; many enemies, there be a loss of any limb, loss of property and diseased.

Second House:

Sun: Sun is in debilitated sign in second house of finance. It indicates, over spending, under debt, troubled and worried native Disputes in family; early life troublesome due to parents, defective eyesight; loss of wealth in vices; gain through secrecy, occult and animals, if aspect well.

Moon: He/She will be wealthy in middle age; troubled eye; a big family, more daughters, gain from business and trade; gain through friends, own hard work or legislative interest.

Mars: He/She will be clever and critics; under debt, loss of property due to vice habits, troubles and worries in life; money through educational affairs, Journeys and writings; fortunate. If afflicted reverse results.

Mercury: He/She will be learned and intelligent, soft spoken, gain from education and blessed with wealth and comfortable life; cautious for his respect; gain from his business; gain by industry, trade, profession of Govt., offices. If afflicted and combust losses and wants and reverse results.

Jupiter: He/She will be learned and well skilled but will be disrespected; relations with many women; as average financial status and will be under debt; gain by marriage, land, property and estate.

Venus: He/She will be lucky and wealthy; a comfortable life respected and gain through elders; two mothers; gain from business and trade from foreign merchants, learning, publications, travel or banking.

Saturn: He/She will be worried and unlucky, devoid of mother's comforts and affection; short life, unlucky in early part of life, but lucky and wealthy in last part of life; gain through children and investment; loss through servants.

Rahu or Ketu: He/She will be troubles and worries, litigations and loss, death of parents in middle of life, residence in foreign country; victory over enemies; loss of wealth in disputes and litigations.

Third House:

Sun: He/She will have gain in foreign land, lucky for brothers and victory over enemies. But he will suffer from his brothers and sorrows; troubled journeys; occult science.

Moon: Early life will be troublesome, less comforts from brothers. He/She will have gain due to sisters; a comfortable and pleasurable life; will face happiness and unhappiness alternatively; friendship through kindred, neighbours and gain.

Mars: He/She will be adulterous and licentious; loss of wealth and ancestral property through women and bad habits; brave and good relations with brothers; victory over enemies; affliction of chest and heart is indicated; gain through learning, writings, short journeys and brothers; psychic and mysterious experiences.

Mercury: He/She will have troubles and worries and middle age, no comforts from brothers, loss through litigations, in spite of the fact that native may be a lawyer or have a command over law yet he will suffer and bear loss in litigations, when afflicted. Otherwise, there will be respect, gain, honour and advancement through wife's relations, short journeys, writing etc.

Jupiter: He/She will have love and affection with brothers; religious, learned and wealthy. He will gain through women and brothers; marriage with relative or neighbour; legal or religious disputes.

Venus: He/She will be adulterous and licentious, two marriages or wives or keeps; influence of the opposite sex and gain through them; gain of wealth in middle age; victory over enemies; gain through travel, learning, research and writings.

Saturn: He/She will have a large family and many relations; fair minded, victory over enemies and wealth; will be benefited through other's wealth; difficulties from brothers; liking for travels, sport, drama and adventure; mental sickness.

Rahu or Ketu: He/She will have a troubled; loss of wealth due to displeasure of Govt.; and superiors; strained relations with brothers; loss or property due to cheating, so one should be very cautious.

Fourth House:

Sun: He/She will be well respected and wealthy; blessed with long life, famous, loss from ancestral property and all comforts of life; will own his self created property; death of mother in middle age; secret sufferings, restrictions and limitations at the end of life.

Moon: He/She will have gain from business and trade in unauthorised manner; death of mother or devoid of her comforts in early age; troubles and worries in middle life; gain in inheritance; fortunate in property; love to father.

Mars: He/She will be respect and gain in foreign country; journeys, gain and comforts from mother and her relations; death of mother in middle age; a comfortable and happy life; gain in property of the deceased.

Mercury: He/She will be learned and intelligent; blessed with long life, wealthy and comfortable life; owner of land and property; gain from mother; a comfortable residential house. But he will be a miser; gain from mother and land; gain through land, mines, and inheritance and from father; success late in life; occult investigation; interest in colonization, reclamation, cooperative, horticulture, mining, architecture etc.

Jupiter: He/She will be intelligent, learned and wealthy; long life, blessed with property and land but disrespected. In middle life, one will rise, successful hopes and efforts and will be blessed with comforts of life and wealth; property by marriage and chaste wife; happy married life, assistance of father; loss through storms and floods.

Venus: He/She will be a handsome native; comforts from conveyance, land property and mines; well respected in middle age and death of mother; scientific and gain through inheritance.

Saturn: He/She will be lucky, respected, loss of ancestral property in early part of life; may be source of loss for parents of mother Loss of money; in middle of life, worries, litigations, devoid from mother's affection and distressed; sickness of father through changes and worries; lucky in latter part of life; love of home.

Rahu or Ketu: In early age one will settle at one place. After that residence and gain in foreign land; source of worries for mother and loss to her relations; changeful life.

Fifth House:
Sun: He/She will be learned, intelligent and interested in Astrology; but a troubled and worried life; progeny will be considerably delayed, may be blessed with a son after middle age; unwanted and insincere friends; short period or comforts in life; secret sorrows and difficulty through love, speculation, children or gambling.
Moon: He/She will have gain of money in foreign country; blessed with children; relations with elder brothers will be good; gain from children; rheumatic or windy complaints; pleasure in life through friends and beneficial circumstances.
Mars: He/She will have many enemies; early life will be troublesome, bad company, thievish nature, danger through excessive pleasure, resorts; danger of death of a son in middle life. Old age will be comfortable; the native will have wealth, property and will lead a prosperous life; unfortunate children; long journeys.
Mercury: He/She will have early troubled and worried life; unhappy from children; after middle age, there will be gain of money, respect, rank and authority. The native will be sincere and intelligent; success through pleasure, sports, speculation and children.
Jupiter: He/She will have a troubled and worried life; loss of ancestral property; enmity with children; disrepute and disrespected; no comforts from progeny; death of son in youth; financial conditions will be below average; gain and pleasure from father's fortune.
Venus: He/She will be wealthy, learned, intelligent and a man of authority and rank; more daughters; gain through wife in middle, age a happy and prosperous life; blessed with conveyance, property but under debt; gain by speculation, investment, children, liberal or unconventional; free living, pleasures, journeys, voyages and foreign trade.
Saturn: He/She will be worried, distressed in early life, not very intelligent, loss of ancestral property, birth of children who will remain sickly; love affairs; journeys in foreign lands, disrespected and under debt. If Saturn is aspect by Jupiter, reverse are the results; also death of young son and gain from children; adventure, speculations or investments. When afflicted reverse results.
Rahu or Ketu: He/She will be disrespected and troubled by the Govt.; no comforts from progeny; worries and loss through enemies. In old age may get some comforts through children. There will be many troubles and worries in early age.
Sixth House:
Sun: Early life will be comfortable. After middle age, the life will be full worries and troubles; loss of wealth through relations and opponents;

diseased, ill famed, troubles through employees.

Moon: He/She will have residence in foreign land; no comforts from wife children of elder brother and loss through enemies and opponents; more journeys in middle age; may the health be afflicted due to T. B., cough, rheumatic, pain, faithful servants; interest in social service.

Mars: He/She will have punishment from the Govt.; loss of ancestral property; grief from brother and wife; many enemies but will have victory over them; poor and under debt; disrespected and ill famed; injury through journey. If aspect is well, reverse will be the results.

Mercury: He/She will have ill health and lean body, average life, troubled and worried life, loss through enemies; may be the cause of loss to maternal relations. There may be some reversal in profession. At times in life there will be gain and happiness; an average life; gain through food, clothing, employees; modest worldly position; gain and honours through service and employment.

Jupiter: He/She will have loss through opponents, a troubled and worried life; diseased mother, strained relations with wife; no gain from mother, under debt in early age; also disrespect and ill fame; stomach disease. In old age, he will be relived of all troubles, will gain money and leads a comfortable life; sickly wife, loss through servants. If well aspect, there will be reverse results.

Venus: He/She will have many enemies, adulterous and licentious, strained relations and loss through wife; loss of wealth, disrespected and ill famed; sickness through travelling; gain through inferiors when not afflicted.

Saturn: He/She will have victory over numerous enemies; worried, displeasure from Govt; at times, litigations, troubled, diseased, danger and loss of wealth and respect; gain and good relations with superiors and officers; blessed with conveyance, and good health. If Saturn is week or retrograde, evil effects will predominate. If it is well aspect by Jupiter, there will be reverse results.

Rahu or Ketu: He/She will have average financial status; diseased and troubled; many enemies and loss through them; disrespected and under debt; loss and worries from Govt and superiors.

Seventh House:

Sun: 12th lord posited in 7th house indicates strained relation with wife; diseased and unhappy; no comforts from ancestral property; litigations with the Govt; and disfavour from officers resulting loss of wealth, prestige, and reputation; under debt, victory over enemies; troubles from contracts, partnership and law suits; troubles through women.

Moon: He/She will be respected and famous gain from business and profession, lucky brothers; wealthy and will enjoy kingly status; a devoted, beautiful and virtuous wife; relations with opposite sex. In middle age, he will suffer loss and troubles due to his wife; success in law suits.

Mars: 8th and 3rd houses lord is in 7th house. He/She will have gain from ancestral property; but troubled married life; two marriages. Ladies will be quarrelsome; less comforts from first wife; gain through rich partner, money comes unexpectedly.

Mercury: Mercury, of course, lord of Ascendant aspects ascendant. He/She will be a learned and intelligent native; medium life span; wealthy but gain through some art and own labour; may there be two marriages, head of a family or village and able administrator. There will be some worries and troubles in youth; a lucky and happy life on the whole; gain through law suits; honourable marriage and respect and honour through it; partnership in responsible concern; fond of opposite sex. If afflicted and combust then reverse results are indicated.

Jupiter: Jupiter is in own house aspecting Lagna. He/She will be wealthy, respected and famous; blessed with high rank, authority and powers; gain from marriage in a good family, wife will be virtuous and respectful; gain from in-laws; will enjoy a kingly status; success in law suits, untrue partner.

Venus: He/She will be adulterous and licentious; fond of music, dance and opposite sex. His wife will be beautiful but out of control and of opposite temperament; voyage and native of authority and power; gain through law suits, contracts; honour and reputation through honourable marriage.

Saturn: He/She will be poor, little education, mean profession and bad company; under debt, wife not beautiful and diseased but faithful and loved by her husband. The native will be licentious and adulterous; quarrels or law suits with servants; difficulties with disreputable women and sickness; discord with children; loss by theft.

Rahu or Ketu: He/She will have weak health and lean constitution. Wife will be of opposite views; troubles in youth. But he will be intelligent and respected, comforts and happiness after middle age; residence in foreign land. In old age he will lead a comfortable and happy life.

Eighth House:

Sun: Sun is in his exalted sign but lord of 12th house in 8th house of friend Mars causes Vipreet Raj Yoga. It will give worries and troubles in the start of any work, but the same will meet with success in the end. In middle of life, there will be comforts and happiness. There will be enmity with opponents, troubles from superiors and Govt.; loss in profession and property, unsatisfactory, troubles over inheritance at the end.

Moon: Chandrashtama is termed always bad and indicates affliction of parts of body such as breast, stomach, uterus, ovaries and mind of mother and self in the middle of life. This position indicates short life, troubles and worries in life, separation from parents in early age, many brothers and sisters, death of friends, gifts or legacies from friends and easy demise.

Mars: First half life will be joyous and comfortable. He/She will have otherwise many enemies, under debt, bad company and loss of ancestral property. Blood diseased will afflict him/her. There will be false accusation, journeys on account of troubles, and gain by dead. He/She will be spiritual.

Mercury: He/She will have weak in constitution, diseased and troubled, loss of wealth, property and also litigation and disputes for ancestral property, short life, death of persons in family, gain by legacies and good of the dead, occult experiences and mediumistic.

Jupiter: He/She will have strained relations with wife, under debt and troubled, grief of children, early life will be happy but sudden downfall, troubles and worries will upset the schedule of life resulting in loss of wealth and respect. Last part of life will be happy. There will be death of wife, some inheritance but difficulty over it and gain by legacy or estate.

Venus: He/She will have up to middle age happy and comfortable life but after that one will face troubles, worries loss of wealth and of ancestral property. Health will also be afflicted, gain by long journeys concerning legacies or goods of deceased.

Saturn: He/She will be poor and lucky, youth life miserable, loss and grief in middle life, loss of ancestral property and children, reversal in service, afflicted health with bile. It is also called ashtama Sani. There will be financial reward for faithful service, loss through speculations and gambling.

Rahu or Ketu: He/She will be troubled and worried in middle life, danger from animals, under debt, short life, success in litigation. Old age will be happy and comfortable and will be blessed with respect and wealth.

Ninth House:

Sun: He/She will have birth in a respectable family, respect, authority and power from Govt; and superiors. Early life will be unhappy. He/She will have gain and happiness, kingly ranks and comforts wealthy and famous, gain of wealth, comforts and happiness, long journeys, secret missions during middle life. If afflicted reverse are the results.

Moon: He/She will have happy relations with brothers and relatives, well respect, wealthy and honest, rank and authority from Govt.; last part of life will be troublesome, gain through long journeys and friendship with high ups.

Mars: He/She will be fond of journeys, death of parents in old age, loss through father, gain due to one's hard efforts, wealthy, respected and man of rank, death in a distant land, long journey, gain through honour or renown through writings and journeys with mother.

Mercury: He/She will be intelligent, learned and respected, loss through father, no gain from his education, average finances and comforts, long journeys, religious or psychic, inventions and prophetic dreams, gain through in-laws, gain through learning, writing, publishing etc.

Jupiter: He/She will be blessed with long life, respected, birth in a good family. From middle life there will be gain, comforts, authority and kingly life, marriage to a stranger and gain, foreign travel, gain through in laws, science, religion and long journeys.

Venus: He/She will be intelligent, wealthy and fortunate, blessed with property, marriage in early age, respect, comforts and authority like a king, honourable voyage, gain through learning, writings, publishing, research, and a good mother.

Saturn: He/She will be lucky to some extent, early part of life difficult, loss of wealth and property, last part of life happy, sickness while travelling, dutiful children and pleasure in foreign land.

Rahu or Ketu: He/She will be proud, intelligent and respected, devoid from comforts of parents in early age, residence and gain in foreign land. There will be troubles and worries as the age advances.

Tenth House:

Sun: He/She will be respected, wealthy and famous, brave and scholar or Astrologer, will lead a pious life, power and authority from Govt.; more respect after middle age if well aspect, otherwise reverse results.

Moon: He/She will be respected and generous, many servants, gain from agriculture land, a good profession, conveyance and comforts from parents, friendship with those in good position and gain there from.

Mars: He/She can not work in a subordinate position. He/She will have changeable profession, strained relations with parents, self made man, death of mother, rise due to inheritance, renowned through writings when well aspect, otherwise reverse result.

Mercury: He/She will have changeable profession and worries in early age, wealthy, gain through profession, office of Govt.; and employees, favour from mother, respected and enjoy comforts of life after middle age.

Jupiter: He/She will be intelligent, respected and wealthy, a professor or lecturer and gain through this profession, trade or Govt.; work, gainful and obedient to parents.

Venus: He/She will be religious, preceptor and wealthy teacher, kingly status, wealthy, power and authority. In middle life he will be at helm of affairs, gain by occupation, trade, Govt.; and in-laws, success in foreign affairs.

Saturn: He/She will be non religious, anti to parents, licentious and adulterous, low type profession and dishonest, cruel to subordinates and a cheat, honourable children renown in speculation, honour through national activities.

Rahu or Ketu: He/She will be skilled in many arts and will earn respect and fame, intelligent and generous, helpful to others, journeys to foreign land.

Eleventh House:

Sun: Lord of 12th house is in a friend's, house and in 11th house. This will bestow fame, respect, wealth and prosperity on the native, wealthy but extravagant, commanding speech, comforts from conveyance and birth in family if well aspect otherwise reverse result.

Moon: He/She will be respected and famous, a good profession, wealthy and comforts from conveyance, power and authority from Govt.; generous, charitable and helpful to others, gain through friends, friends among unique, ingenious, original or radical people, well defined hopes and wishes.

Mars: He/She will have loss of wealth and property, under debt, downfall in life, non-religious, liar and a cheat, friends through journey and gain from them, successful hopes, gain and legacies through friends.

Mercury: He/She will have happy and comfortable life, wealthy, respect and regards from rulers, knowledge of many languages, more prosperous in old age, friends in high ups and gain from them, honourable future, ambitious ideal, successful hopes and wishes.

Jupiter: He/She will have kingly status and comforts, lucky and respected, long life, respect and gain through brothers, generous, wealthy, comforts from conveyance and charitable, a native of power, rank and authority, if, afflicted reverse results.

Venus: He/She will have a comfortable and joyous life, blessed with good palatial buildings and houses, comforts from conveyance, not so close to relations with opposite sex, a good and successful, argumentative, disgrace and worries in old age, fortunate friendship on voyage and with foreigner, accidental fortune.

Saturn: He/She will have quarrelsome wife, unhappy life, disturbed married life, disrespected, under and defamation during middle part of life, hopes attained through acquaintances, interest in politics and social welfare, love of children.

Rahu or Ketu: He/She will be cautious for his respect, forceful talk, wealthy, respected and famous, a good rank, charitable, generous and happy old age, but death of elder brother and if alive grief and troubles to him on this account.

Twelfth House:

Sun: He/She will be respected, famous a man of good rank and authority, cautious for the respect of family, devoid from comforts of parents in early age, more respect and fame after middle age, victory over enemies, occult science, power of inner understanding.

Moon: He/She will be fond of travelling, voyages, extravagant, short life, relations with opposite sex, comforts and gain in old age, deceitful friends, gain from occult, pleasure in peaceful, quite, harmonious or secluded places.

Mars: He/She will have loss through enemies, no permanent source of income, loss of land and many reversals in life, difficulty over inheritance, great sorrows, secret sufferings, danger of enmity and imprisonment while travelling.

Mercury: He/She will be intelligent and learned but no gain from education, worried and sad grief of parents. In middle age one will earn well and blessed with all comforts of life, fear of imprisonment, secret unhappiness, occult gain and if afflicted reverse results.

Jupiter: He/She will have residence in foreign land, a man of power, authority from early age, may be a judge or magistrate, wealthy, generous and honest, many servants. In old age he will also enjoy more respect, comforts and wealth, afflicted health, occult learning, unhappy marriage, secret sorrows, fear of imprisonment.

Venus: He/She will have food of conveyance and blessed with the same, wealthy and owner of property and palatial residences, conveyances and houses, extravagant, occult, gain by affairs of secret nature.

Saturn: He/She will have middle part of the respectful, victory over enemies, unhappy life on the whole, poor will be troublesome, under debt and troubled, may remain out of his home, ups and downs in life, a worried life, happy old age.

Rahu or Ketu: He/She will have psychic and spiritual experiences, unwanted expenditure, worries, loss of wealth in disputes and litigations, disfavour from Govt., and superiors.

11.7 Libra (Tula) Lagna

Jupiter, Sun and Mars are malefic. Auspicious are Saturn and Mercury. Moon and Mercury will cause Raja Yoga. Mars is a killer. Jupiter and other malefic will also acquire a disposition to inflict death. Venus is neutral.

Sun: Surya is L-11 as measured from Thula lagna, Thula Chandra, or partner-effects from Thula navamsha. Ravi (L-11) is enemy of lagnesha Shukra and also enemy of Shani the natural 11th lord and so is malefic planet and Double-trouble maker for Thula lagna unless Surya is well placed like in own Sign or is in exaltation in the Kundali. Surya's selfishness or autocratic behaviour causes trouble in relationship to marketplace earnings, friendly networks, the elder sibling or father's brothers. Remedial Ratna for Thula lagna Surya (Manika - pure Ruby): He can wear a Ruby when there is loss of wealth or obstructions in acquisition of wealth on account of ill disposition or affliction of the Sun in any way. If the Sun is in his own sign, or in exaltation, the wearing of Ruby will make the native very fortunate and he will become wealthy, famous, honoured and will own conveyances and property. The wearing of Ruby will also improve his health and longevity. The beneficial results will be more apparent during the major period and sub-periods of Sun.

Moon: Chandra is L-10 measured from Thula lagna, Thula Chandra, or Thula navamsha. Chandra L-10 is enemy of lagnesha Shukra and enemy of Shani the natural 11th lord and is double-trouble maker. Presuming Chandra is well placed and strong, it will provides Leadership capabilities and Style, a natural capability to lead and govern, to make decisions which provide social order and lawfulness with special emphasis upon matters of home and homeland, folkways customs and ways one's people live upon the land, property ownership, houses and schools, farms and fisheries, parenting and protection of the weak, physical-emotional security and grounded ness in a fixed place and provides him with servants and employees who bring good fortune and also good relationships with social workers, followers, employees, and members of the marginalized classes including refugees, criminals, the dispossessed, chronically ill and addicted persons, and will form the basis of a creative strategy for his leadership. He helps the underclass and the underserved classes who will contribute to his positive public reputation and empowerment to make decisions affecting the downstream like Mahatma Gandhi's career similarly was based in his creative solutions to the chronic exploitation (6) and disenfranchisement of the Indian masses and Adolph Hitler who famously created from the massive discontent of the German working classes who felt victimized by the Treaty of Versailles, after WWI. Mahadasha of Chandra may signify an exceptionally strong engagement with public organizations such as church-temple, children's school, or charitable group for a person with an overall public-service oriented nativity. His mother will have a strong career in karakatva of the bhava of Chandra and she is role model in his developing professional identity. Remedial Ratna for Thula lagna Chandra (Mukta/moti - pure natural Pearl): Moon is lord of 10th house is not a friend of Venus or Venus friend, Shani, but wearing of a Pearl brings the name, fame, honours, advancement in sign career and wealth, if the Moon is in his own sign in the 10th, when it gives rise to a powerful Raja Yoga or is exalted. The results will be more pronounced if the Pearl is worn in the major period of Moon. If Moon is not auspiciously disposed or is afflicted by aspect or conjunction, causing obstruction in the advancement of profession or career, Pearl is the gem stone to remove such evil effects. It will also give proper equilibrium to the mind.

Mars: Kuja is L-2 + L-7 measured from Thula lagna or Thula Chandra; partner-effects from Thula navamsha. Mars is the lord of two death inflicting houses (maraka houses), the 2nd and the 7th and also is an enemy of Venus, the lord this Ascendant and so is malefic planet, unless it occupies its own sign or sign of exaltation. Presuming Mars is well placed or in bhava-2 and strong, it can energize the Treasury (2) of money and knowledge. Mars in 7th can very good marriage alliances, and make him advisers such as physicians, attorneys, brokers. The negotiation skills are fuelled and sharpened like a fierce warrior's sword. Mangala periods indicate fleshly death for the mother, father, and other family members in his early age. Mangala can signify the end of a romantic relationship or a cutting of family ties during its period. L-2 Mangala facilitates a split with a family and brings him to the dentist's chair, or may signal a new food diet. As lord of both 4th-from-11th and 9th-from-11th, Mars is a highly auspicious Yogakaraka graha in regard to his ability to set and attain his goals. When Mangala occupies Kumbha, it engages him and his life-partner toward earnings (L-7 in 11th from swakshetra). If Mangala in a Rashi of Shani in 5, it suggests miscarriage or therapeutic abortion of at least one conception; however if Guru is strong there may be other live births when Mangala is not the time lord. There may be a still-birth, which will cause much sorrow. He may be cruel, ruthlessly striving to attain the objects of desire and causing terror to his adversaries. Remedial Ratna for Thula Mars (Munga - flawless red coral): Mars is the lord of two death inflicting houses (maraka houses), the 2nd and the 7th. Moreover, Mars is an enemy of Venus, the lord this Ascendant. Wearing a Red Coral should be generally avoided. The second house is the house of wealth, if Mars is in his own sign in the 2nd, Red Coral can be worn with advantage during the Major period of Mars. As far as the 7th house is concerned, if Mars is there in his sign, this disposition will give rise to a powerful Panchamahapurusha yoga known as Ruchaka yoga, and he will have a strong physique, and will be famous, well versed in ancient lore, king or equal to king, conforming to traditions and customs. He will have a ruddy complexion, attractive body, charitable disposition, wealthy, long-lived, and leader of an army. Thus, if there is Ruchaka yoga present in the nativity, the Red Coral can be worn with great advantage, particularly in the Major and sub-period of Mars.

Mercury: Budha is L-9 + L-12 measured from Thula lagna or Thula Chandra; partner-effects from Thula navamsha. Mercury is lord of the 9th, the house of Bhagya, (a trine), and is a friend of the ascendant, Venus and will prove very beneficial. Budha as lord of 12th from 10th, drains energy away from public leadership roles so that the native can study philosophy and deliver wisdom teachings to world. Budha gives a negative influence on career, and a strong Budha is often the signal of a weak career for Thula-1. If Budha periods occur during his young age or at adult years, it gives unfavourable results career-wise. Good results begin to accrue in regard to the native's ability to communicate in positions of leadership and dominion, but only with the passage of time. Budha as lord of both 3rd from 7th and 6th from 7th is a major trouble-maker in human partnerships, and is a general miscreant in marriage. Budha may excite internal argumentation that can damage partnership negotiations and harm an otherwise good marriage. Remedial Ratna for the Thula lagna Budha (Flawless Emerald - panna): Mercury is lord of the 9th, a trine, the house of Bhagya, and the 12th. It will do well to the native if he wears an Emerald in the major or sub-periods of Mercury. As lord of the ascendant Venus is a friend of Mercury; it will prove very beneficial if an Emerald is worn always along with a Diamond. **Jupter:** Guru is L-3 (Malefic Bhava) + L-6 (Malefic Bhava) measured from Thula lagna or Thula Chandra; partner-effects from Thula navamsha and is unfriendly to Venus and is, therefore, not considered an auspicious planet for this Ascendant and is a problematizer. This combination can however help a self-owned business in religious trainings such as conferences and seminars, and this combination is also beneficial for academic publishing. Jupiter in this position activates his mind so much that his mental balance may be disturbed and he suffers from a mental disability due to L-3+L-6 Guru which makes awkward and uncomfortable to him. He sometimes face danger, such as, few children as daughters only or partner is barren or ailing, dispute or separation with partner or 2nd marriage, serious illness to the elder brother, poverty, indebtedness, imprisonment, loss in business, disrepute, leaving birth place are some of the unfavourable results seen under this Jupiter. Remedial Ratna for Thula lagna Guru (Pukhraj - yellow sapphire): Jupiter as lord of the 3rd and 6th houses is, therefore, not considered an auspicious planet for this Ascendant. It will, therefore, be advisable for him not to wear Yellow Sapphire. It may, however, be worn if necessary, in the major period of Jupiter if Jupiter is in his own sign in the 3rd or the 12th.

Venus: Shukra is L-1 + L-8 measured from Thula lagna or Thula Chandra; partner-effects from Thula navamsha. Shukra as lagnesha is excellent good or benefic planet for for Thula lagna. As Lord of Lagna, Venus gives powerful results for Vaanija (the trader) and gains of income (11) in career. He earns an adequate income (unless there are very difficult graha in bhava-8) like others' monies, such as inheritance, insurance settlement, and in particular the assets of the spouse. Education and Properties: 5th-from-4th (Vrishabha) and 10th-from-4th (Thula), Shukra provides support in matters of education and property, acquisition of properties; and ascending to a respected social position, if Shukra is a well-placed. Shukra provide the education of his children, parenting or other culture-sustaining service that expresses individual intelligence. He may experience a burst of creativity and ingenious speculative behaviours in response to the need to sustain personal security. Shukra in 11 provides marketplace and profits, a very distinctive wealth and luxury from marketplace transactions within a network of buyers and profits. Swakshetra Shukra in Thula lagna provides the social personality embedded in the iconic material appearance and a cult of personality. Shukra in 9 provides affiliations with temples and universities; He is fundamentally oriented toward attractiveness, wealth, making agreements, living in harmonious balance with other humans, sensual pleasures from art, music, and the experience of beauty. Periods of the L-1 Shukra are self-defining. 8th-from- lagna provides cycles of birth, death, and rebirth for Thula-1 during periods of Shukra. Shukra periods bring an apparent catastrophic collapse or destruction of the physical and social identity, followed by a rebirth. Shukra periods may also bring inheritance money, financial gifts from unexpected or mysterious sources, or access to the suddenly revealed (or suddenly offered) monies of the spouse or business partner. Occasionally a severe and mysterious illness (8) is required to affect the conditions of death, release, reconstitution, and rebirth at a cellular level. 2nd-from-7th (Vrishabha) and 7th-from-7th (Thula) Shukra is a wealth-and-pleasure agent in the world of human relationships like marriage, partnership, peer and advising relationships. As 2nd-from-7th, Periods of Shukra bring increased accumulation and hoarding empowerment to increase the holdings of the joint assets of marriage (8). Relations with in-laws move to prominence but if Shukra is well-disposed the family of the spouse (8) are pleasurable companions and wealth-agents. As 7th-from-7th, Shukra periods define his role in partnership, which is generally as an agent of negotiation, balanced design, diplomatic alliance making, and artistic satisfaction. The Mahadasha and Antardasha is the favourable period for marriage and business agreements. One is inclined to support the mutual interests of both self and partner. Family history, Lineage Values, Acquisition of Money and Knowledge: As lord of 7th-from-2nd (Vrishabha) and 12th-from-2nd (Thula) Shukra controls the mutually supportive agreements made between the native and one's family. 7th-from-2nd = Shukra periods express new attempts to balance one's relationships with the family as a unit. On the basis of one's own knowledge, acquired and accumulated learning of history and languages, one's own values (originally from the family) and one's speech and facial appearance, one may generate new relationships of mutual benefit. Shukra = the native looks attractive for one's face, voice, wealth, and learning. 12th-from-2nd= loss and dissolution of attachment to stored wealth, accumulated collections of art or knowledge, and to the lineage values of the family history. One strikes off on one's own. Attachment to the Self is strengthened. Whether the family history interest is fundamentally positive or negative, one dissolves the lineage bond a bit, and moves forward into a deeper focus on individual self. Siblings, Teamwork, Communication, Publications: As lord of 6th-from-3rd (Vrishabha) and 11th-from-3rd (Thula) Shukra is overall an inauspicious influence on sibling relationships and scripted daily communications in general. 6th-from-3rd - periods of Shukra may signal illness or poverty for the younger sibling, conflict with team-mates, and imbalance in the tasks of making announcements, publications, writing, and producing messages in all communications media. the native may experience a temporary deficiency of courage and general downswing in mental health due to daily conversations becoming somewhat adversarial. However Shukra is such a strong natural benefic that personal sweetness will compensate for most of the crabby conflicts. 11th-from-3rd = on the basis of team work, mental process, writing and generating messages, conversations and narrations, the native is able to attain goals related to the development of the social personality. **Children and Divine Creativity:** As lord of 4th-from-5th (Vrishabha) and 9th-from-5th (Thula) Shukra gives a highly auspicious a Yogakaraka effect upon children And creative intelligence. 4th-from-5th = Shukra periods support the education of one's children. However the base house is bhava-8, which means that Thula-1 tends to be somewhat mystified by the education which one's children are receiving. Nevertheless, good results in the children's education are to be expected. If no children are presently being socialized, the outcome will be related to security obtained, licensing or social approval given to one's creative endeavours, for example on the basis of speculation (5) one may purchase a home or vehicle at a favourable price (Shukra). The native with Shukra in 8 can do this every day. If Thula lies elsewhere, wait for periods of Shukra to trigger good luck in acquisition (Shukra) of properties and vehicles. 9th-from-5th = good fortune and higher education for one's children. Children may enjoy global travel, religious enlightenment, and participation in philosophical discussions. Through the auspice of the philosophical development level of the parent, the children may begin to develop their own wisdom-awareness during a bhukti of Shukra. Conflict, Imbalance, Disease, and Debt: As lord of 3rd -from-6th (Vrishabha) and 8th-from-6th (Thula) Shukra provides a particularly pernicious influence upon matters of animosity, debt, and disease. Which is a good thing? 3rd-from-6th = Shukra controls the mental narrative regarding conflict in human relationships, poverty, disease, exploitation and oppression, and every other manifestation of inequality (6 = 12th-from-7th, loss of equality). Bhava Effects: If Shukra is well disposed, the native may be a highly skilled and sensitive diplomat able to produce and develop an attractive, balanced narrative of conflict. Shukra in bhava-1 environment = pleasure in one's own body, the social personality identity, sweet winning of competitions and fights. Shukra in bhava-2 environment = sweet money, sweet death, sweet knowledge of history, sensuality of the second marriage. Shukra in bhava-3 environment = delights in the interior mental narrative, loves to write, pleasure in conversations, charmed by publications, love of business administration, media production. Shukra in bhava-4 environment = pleasure in home and security, sweet patriotism, sweet love of cars and houses, delight in cultural roots and established customs, enjoys schooling and basic education. Shukra in bhava-5 environment = sweet winning of speculative challenges and games, charmed by children, delights in creative genius, the romantic lover. Shukra in bhava-6 environment = pleasure in the interior mental argument, love of conflict, leading to breakdown of contractual agreements, imbalance in the social relationships, imbalance in the physical body due to excessive sweet. Shukra in 6 exacerbates the extreme situation of drug-alcohol addictions, a classic effect of, no doubt even further. Shukra in bhava-7 environment = karako bhavo nashto, if Shukra is strong. The sensuality of marriage and the marriage partner have agency of constant upheaval and change. Shukra in bhava-8 environment = secret information, pleasure in emergencies, magic and magicians, sweet spouse's money, the in-laws, taxes. Shukra in bhava-9 environment = religious awakening, pleasure in world travels, philosophical study, universities, priests, temples. Shukra in bhava-10 environment = pleasure in corporate leadership and responsibility to impose an orderly environment, particularly amongst women or in regard to feminine cultural artefacts; sweet power. Shukra in bhava-11 environment = pleasure in social networking and progressive political movements with goals to expand the opportunity to network. Jimmy Carter's Shukra in Simha is particularly political in nature; he has undergone countless radical changes of lifestyle ranging from a very surprising election to USA president through many dangerous democratic election-reform campaigns worldwide. Shukra in bhava-12 environment = pleasure in long residence in foreign land, monastic withdrawal from the social ego, love of hospitalization or sanctuary space, private sweet "bedroom-based" relationships. Remedial Ratna for Thula lagna Shukra (Flawless Diamond - heera): Venus is lord of the Ascendant and the 8th house. He is considered as an auspicious planet for this Ascendant and wearing of a Diamond promises good health, long life and success in life. It will be immensely profitable to wear it in the major and sub-periods of Venus. Actually Diamond is a life time protective charm for the Libra natives. If Venus is in Libra in the Ascendant he causes Malavya yoga and wearing of diamond with such disposition will enhance the good effects of the yoga.

Saturn: Shani is L-4 + L-5 measured from Thula lagna or Thula Chandra; partner-effects from Thula navamsha and so Shani are Yogakaraka for Thula Lgna Thula natives will successfully complete schooling and earn a diploma. Good children are born. There will be legacy educational diploma, license to practice a profession, scholarly performances, well-structured speculative ventures, political empowerments earned by accepting responsible duties, and lawful production of well-raised children. Shani is Yogakaraka for Thula lagna, but this fact does not make Shani an "easy results" graha. Thula natives have such a favorable prognosis for both cultural education (4) and individual intelligence (5) because Shani provides the lifetime discipline to shape, polish, and perfect the process. For Thula lagna, Shani, the dark force of ignorance and unnamed fears, does His work through the agency of cultural education (4) and intelligence (5), and gives discipline and structure, not oppression and slavery. Whatever domain Shani occupies, Shanaicarya will bring the security of bhava-4 and the creativity of bhava-5 along with Him. The lord of bhava-5 is especially beneficial. If the worst threat to human happiness is indeed doubt, the lord of bhava-5 which brings confidence is the antidote to life's worst distress. This single trait is responsible for the extraordinary number of Thula lagna nativities in positions of executive power and authority in the world. Examples: Rudolf Steiner has yogakaraka Shani in 11 and parivartan yoga with Surya, which provides great achievements in progressive social change movements (11) via education (4) and creative intelligence expressed through literature (5) with many threats and setbacks both personal and social. Beatle George Harrison has yogakaraka Shani in 8, which provides great achievements in mystical rebirth and discovery of secrets (8) via education (4) and creative intelligence expressed through stage performance and song writing (5) with many threats and setbacks both personal and social. Partner of USA Pres-35 Jacqueline Kennedy has yogakaraka Shani in 3, which provides great achievements in media communications and small-group structure (3) via education (4) and creative intelligence expressed through politics and new media (5) with many threats and setbacks both personal and social. Mahatma Mohandas Gandhi has yogakaraka Shani in 2 which provides great achievements in true knowledge and true wealth, (2) via education (4) and creative intelligence expressed through politics and literature (5) with many threats and setbacks both personal and social. French General and President France Gen. Charles de Gaulle has yogakaraka Shani in 11 which provides great achievements in social progress movements (11) via education (4) and creative intelligence expressed through politics (5) with many threats and setbacks both personal and social. USA Sec of State Hillary

Clinton has yogakaraka Shani in 10 which provides great achievements in leadership and governance (10) via education (4) and creative intelligence expressed through politics and literature (5) with many threats and setbacks both personal and social. (Shani + neechcha-bhanga Mangala)

Thula makes plenty of mistakes in life, perhaps even more than average for an intelligent human being. However Yogakaraka Shani compensates for deficiencies in personal intelligence "content" by imposing the conceptual structure from educational "form", and as a result the native actually learns from one's mistakes. Remedial Ratna for Thula lagna Shani (Blue Sapphire - Nilam): Saturn is lord of the 4th and the 5th houses and is, therefore, a Yogakaraka planet. Venus, the lord of this Ascendant is also a friend of Saturn. Consequently, the native of this Ascendant will enjoy very beneficial results by wearing a Blue Sapphire. The wearing of a Blue Sapphire during the major period of Saturn will prove very advantageous. The good effects of the Blue Sapphire will be fortified if it is worn along with Diamond, the gem stone of Venus, the lord of this Ascendant and Emerald the gem stone of Mercury, the lord of the 9th house.

Rahu: Rahu is co-lord of bhagya bhava-5 measured from Thula lagna or Thula Chandra; partner-effects from Thula navamsha and so Rahu a benefic planet and friendly to Venus and Saturn. Rahu in dhana-bhava or yuvati-bhava is associated with the end of life or death. Example: Thula lagna Jacqueline Kennedy died of lung cancer from lifetime cigarette smoking age 64, in Rahu-Chandra period. Chandra = L-10 but both Rahu + Chandra occupy the Makara yuvati bhava and Rahu owns the dominant degree even if the target is enhancement of creativity and acquisition of children, the Rahu-ratna should be considered only if the lord of bhava-5 and its occupants are highly auspicious. As co-lord (with Shani) of Kumbha/5, Rahu gives passionate and unusual children. Due to austerity-karaka Shani's rulership of the house of children, Thula natives typically have few offspring. However those few produced are complex, adventurous, ambitious, scientific, and passionate taboo-challengers by nature. [cf. Hillary Clinton's daughter Chelsea Clinton]. Rahu and Ketu are said to give positive results in the 3rd, 6th, and 11th bhava from the lagna or from Chandra. In addition, some authorities posit that since R-K are exalted in Vrishabha-Vrischika (per BPHS), R-K will give good results when Mangala and Shukra are well-disposed. Remedial Ratna for Thula lagna Rahu (Gomedha): Gomedha may become a beneficial amplifier-gem for self-elevating promotion of creative intelligence, procreativity, politics and entertainments, speculative investment and gambling.

Ketu: Ketu is co-lord of dhana bhava, bhava-2 (Vrischika) measured from Thula lagna or Thula Chandra; partner-effects from Thula navamsha and so is an auspicious or benefic planet for Thula lagna. Ketu amplifies the effect of any graha sharing Ketu house; and also Ketu scatters the effects of the lord of His occupied Rashi. Ketu contributes detachment and misfit qualities to the family identity. Ketu is beneficial to provide wealth and language, acquisition, knowledge of history, and relations with the family history, if the lord of bhava-2 and its occupants are auspicious. Ketu effects depend significantly on Ketu bhava, Rashi, and drishti. Special case: If Ketu in Vrishabha-8, it arouses supernatural powers, and his intuitive understanding of relationships will be surprising and will acquire much fame. Remedial Ratna for Thula lagna Ketu (Cat's Eye - Vaidurya): Vaidurya could be a beneficial gem in the case that Shukra and Mangala are both well disposed. Ketu, even if auspicious, shall certainly scatter the marital and other alliances. Ketu-ratna may prove beneficial in those rare activities wherein disconnection, un-gluing, and surrender is the aim. Ketu-ratna may be a beneficial gem for matters of wealth and language acquisition, knowledge of history, and relations with the family history, if the lord of bhava-2 and its occupants are auspicious.

Planet Effects in Twelve Houses for Libra (Tula) Lagna

First house:

Sun: Sun is in debilitated sign in Lagna. Sun gives love of justice, peace and harmony. The native is courteous, pleasant and agreeable. They are even tempered, affectionate and generous. There will be many ups and downs in life, disrespect in old age. He/She will have worried and of doubtful nature, loss through Govt., victory over enemies, successful hopes and fortunate actions.

Moon: He/She will have preferment and dignities, honours through merits, success through industrious efforts, high ambitions, gain through mother and Government; relations with opposite sex, comforts from conveyance, respectful to elder, medium age, may be afflicted with rheumatic pains.

Mars: He/She will be handsome, wealthy; love of women and benefit through them, connections with the law, public enemies, also in unions and partnership, money comes readily. When afflicted, there will be obstacles through money which may be hard to accumulate, loss through enemies, bad stomach and afflicted eyes.

Mercury: Lords of 9th and 12th houses when posited in Lagna or Ascendant, indicates long journeys, learning, fortunate with strangers, foreigners, wife's relatives and voyages. He/She will have animosity, heavy troubles, sorrow, and loss through enemies, danger of imprisonment or disablement, requiring hospitalisation, loss through woman, wealthy and licentious. Eyes will be afflicted.

Jupiter: He/She will be comforts and gains from children, blessed with long life, religious, wealthy and respected, pilgrimage, comfortable life, concern with affairs of brethren and neighbours, writings, learning and accomplishments, loss through servants and sickness through irregularity.

Venus: He/She will be fond of music, dance, poetry and perfumes, relations with opposite sex, gain through women, bestows power over enemies, wealthy, respected and comforts from children, self made man, good health, harmony in married life if not afflicted, legacies and money, death by irregularity.

Saturn: Saturn is a Yoga Karka for Libra ascendant and it is exalted in Lagna. It will bestow fortunate inheritance, gain through land, property or parentage, makes the native, wealthy, respected, of good habits and nature. In middle age adulterous, unhappy. He/She will have loss through enemies, average life span, pleasure seeking, healthy and cordial relations with people.

Rahu or Ketu: He/She will have long life, many enemies, changeable residence and of bad thoughts, journeys in life, two marriages or women, danger from reptiles, good luck in old age.

Second House:

Sun: He/She will have money through personal efforts, gain through friends and acquaintances and hard work, devoid of comforts from brothers.

Moon: He/She will have gain through trade, industry and Govt.; or parents, gain of ancestral property, wealthy and respected, knowledge of many languages, less progeny.

Mars: He/She will have gain by marriage and own hard work, heavy losses in early life, losses through unions, partnerships, contracts, law-suits, public enemies, loss through women and death of partner.

Mercury: He/She will have gain of money through foreign merchants or the sea, learning, science publications, inventions and banking, gain through secrets and the occult and large animals, worries and loss through enemies and generally fortunate conditions.

Jupiter: He/She will have gain through educational affairs, short journeys, writings, music, losses through sickness; servants and animals, if, well aspect, gain through service, employees and small animals, pleasure or young people. Command over foreign language, a few brothers, respected and comfortable old age.

Venus: He/She will have gain through women and hard work, benefit through industrious activity, gain by partnership, collecting debts, and gain through the deceased. If afflicted losses and wants, loss through partnerships. One will be licentious and sweet spoken.

Saturn: He/She will be disrespected, dishonest, devoid of progeny and if blessed gain through them, gain through investments, pleasure and children, gain through deals in land, property and estates, interest in domestic science. If afflicted, there will be poor education, low type profession, diseased mother, and long life, strained relations with elder brothers, weak eyesight and bad teeth, subject to throat disease.

Rahu or Ketu: He/She will have gain of money through unauthorised sources, a cruel and quarrelsome native, swelling in eyes. In latter part of life, there will be comforts and wealth. The native will feel happy but will not be respected.

Third House:

Sun: He/She will have friendship among kindred and neighbours, friends through writings and journeys, wealthy, respected and fond of music, dance, drama, number of brothers, gain from maternal relations.

Moon: He/She will be respected, gain in honours and advancement through partner's relatives, brave and sanguine temperament, many enemies, honours through short journeys, writings and accomplishment, an average financial condition, no gain from brothers, gain through business trips and Governmental commission.

Mars: Lords of second and seventh houses is posited in third house. It denotes gain through education, writings brother, neighbours and short journeys, but enmity with some of the brothers or neighbours, brave and or rash temperament, annoyed over triples, many enemies, and difficulty through writings or contracts. He/She will have religious or legal disputes, troubles or short journeys, an average financial condition.

Mercury: He/She will be learned, intelligent, soft and sweet spoken, accomplishments and progress through research, travel, investigations, explorations, travels or writings, successful in hopes and plans. He/She will have faith in religion, many brothers. When Mercury is weak or combust, there will be reverse results, Disappointments, sorrows, and troubles through some of the kindred, friends or neighbours, troublesome short journeys and writings, interest in knowledge of occult science.

Jupiter: He/She will have benefits through learning, writings, and short journeys, learned and intelligent. Many brothers and gain through them, knowledge of many languages, pilgrimage, interest in studies and methods of healing and industrial economy. There may be some strained relations with brothers or kindred and sickness also.

Venus: He/She will be beautiful or handsome native, respected, sweet and soft spoken, comforts from wife, mental development, short journeys, less gain from brothers and relations. When Venus is afflicted, there will be break in educations, restricted developments, troublesome relations. Since 8th house lord is in third house it may indicate danger of death on short journeys, death of brothers, psychic and mysterious experiences.

Saturn: He/She will be poor, gloomy, strained relations and loss through brothers, last part of life happy, general life full of ups and downs and turmoil, powerless enemies and under debt when Saturn is ill placed and badly aspect. When strong being a Yoga Karka for Libra ascendant, it will confer pleasure through kindred, liking for travel, drama, sports and adventure, journeys with or through children or young people. Gain through brothers.

Rahu or Ketu: He/She will be quarrelsome and obstinate, many brothers and gain through them, litigations and disputes in life, fickle minded and of bad temperament.

Fourth House:

Sun: He/She will have inheritance and gain, fortunate in property, respected and famous, wealthy, love for father, success and accomplishment of hopes, worried and Owner of house. The parents may be separated in early age.

Moon: He/She will have wealth, respect and comforts from conveyance, land and property, gain and honour through parentage land and property, success at close of life, troubles and worries from internal relations, interest in reclamation, colonization, cooperative movements, horticulture, mining, architecture and archaeology.

Mars: He/She will have estate or benefit through parents, land or mines, property by marriage, a chaste wife and happy married life. When Mars is afflicted, it indicates litigations over property, robbery in house, loss by investment in property and mines, afflicted health of mother and may be disrespected or ill famed.

Mercury: He/She will have ecclesiastical or scientific, inheritance, possessions through wife's relations, travel on account of family affairs or wife's mother. May there be loss or great difficulty through property, father, wife's secret sufferings, restrictions or limitation at the end of life, a worried or troubled life.

Jupiter: Jupiter is in debilitated sign. He/She will be generous and well respected, intelligent and learned impressionable and sweet speech, lucky and wealthy, builder of temples, inns, and charitable places. Since it is 6th lord it will give sickness of father through worry and changes, sickness of native through anxiety and troublesome home or domestic affairs. He/She will have troubles through servants in the home, travelling and writing in connection with home affairs and property. If afflicted reverse are the results and may deny birth of male child.

Venus: He/She will be respected and wealthy, fond of music, knowledge of law and other languages, comforts from conveyance, limited progeny, gain through lands, mines, inheritance and possessions, home connection and affairs with the father, success late if life, occult investigations, gain in property through the deceased, probably by the death of parents (being 8th Lord) When Venus is afflicted, it will give death of parents, danger through falls and falling buildings, floods and storms, troubles over inheritance, Lord and property, few children and difficulty through them.

Saturn: Saturn is Yoga Karka planet and is in fourth house, in his own sign, denotes gain from immovable property, acquisition of property and conveyance, gain from superiors and Govt.; loss of money, good health, ill health to mother. He/She will have kidney trouble, assistance of father, benefits and profit in the place of birth. His/Her children get profit through gifts from their grandfather, love of home.

Rahu or Ketu: He/She will have a worried and troublesome life, residence or travel to foreign land, devoid from the comforts of parents, under debt and poor, enmity with relations, unwanted expenditures in life.

Fifth House:

Sun: He/She will have a propensity to pleasure through gambling and children, many love affairs, respected and famous, grief of children may result in untimely death of a progeny, more troubles and turmoil in middle part of life.

Moon: Children may suffer from sickness but rise to honours. He/She will have gain and honour through speculations, pleasure, young people, sports or the stage, wealthy, respected and owner of land and property, gain from children in life. He/She will be lucky and happy in last part of life.

Mars: He/She will have gain by speculations, investment, pleasure, entertainment, young people and children, abortions. The native marries and enjoys many pleasures thereby, loss of earned and accumulated money and has a native of rash temperament.

Mercury: He/She will be liberal or of unconventional ideas, free living, learned, fond of opposite sex, takes pleasure in science, voyages, philosophy, air flights, foreign investments and speculations, secret sorrows and difficulty through love. There can be loss through speculations also.

Jupiter: He/She will be respected, famous and wealthy, pleasure journeys, mental pleasure through children but they will remain sick. He/She will have pleasure through brothers, reading, study accomplishments and travel to pleasure places, illness due to over indulgence in pleasure and sports resulting in loss of accumulated wealth, death of some nearest relative.

Venus: He/She will be wealthy and learned. He/She will have unfortunate children or death of some one, danger through excessive pleasures, speculations, sports and recreations.

Saturn: His/her Middle part of life is troublesome, not well educated, death of child, worried life, defamed due to some accusation and eye trouble, but when aspect beneficially reverse results.

Rahu or Ketu: Rahu frees the native from many troubles calamities and dangers. He/She will have favourable for children with land, life and are fortunate. The native gains some public employment or office and fond of pleasures. Ketu in 5th house is not good. It may deny the native from any issue or may portend their destruction suddenly or violently and may become disobedient, excessive or irregular pleasures produce much harm.

Sixth House:

Sun: He/She will be brave and victory over enemies. He/She will have faithful servants and beneficial circumstances, friends among working and army, navy or service people, grief from parents, loss of ancestral property and through law suits.

Moon: He/She will have modest worldly position, gain and honour through service, employment, army or navy affairs, many enemies, may be disrepute, may face worries and troubles up to middle of life and face debts.

Mars: He/She will have gain through inferiors, hygiene and services, marriage below status, has sickly wife and evilly disposed employees, strained relations with relatives, loss of ancestral property, weak stomach.

Mercury: He/She will have many enemies, troubles through employees, difficulty through work in foreign lands or in connection with exporting limitations on account of sickness.

Jupiter: He/She will have some severe illness, command over foreign languages, difficulty through brothers, interest in study of social economy, financial condition not sound, grief and worries from wife and children, many enemies and losses through relatives and friends, if well aspect, reverse results are indicated.

Venus: He/She will have dangerous sickness, loss of money earned by employment and through those who may keep it, licentious and adulterous, many enemies and loss through women, disease of wind are indicated.

Saturn: He/She will have stomach trouble, diseased, disputes with relations, under debt, loss of agricultural land, disrespected, and loss of professions through sickness, servants and animals. When well aspect reverse results.

Rahu or Ketu: He/She will be worried or troubled through enemies but victory over enemies, loss of property and respect through the Govt.; change of residence, a worried and troubled.

Seventh House:

Sun: He/She will have a loving partner with desirable friendship and social connections, success in law suits, respected, famous and wealthy, blessed with land and property, gain through brothers.

Moon: He/She will have gain through law suits and dealing with the public generally, honours and reputation assisted by an honourable marriage and partnerships in responsible concerns, connections with many women. Respect and gain through mother's relations, intelligent and learned, travel in foreign country, gain from Govt.; and business.

Mars: He/She will have marriage in a good family, gain by marriage, contract business and dealing with others, especially with women, brave and courageous, victory over enemies, success in law suits, a progressing partner, but likely one who grows cold or proves untrue or hostile. Brothers will be prosperous and gain of wealth and respect through them is also indicated.

Mercury: He/She will be learned and intelligent, knowledge of many languages, public enemies through religious, scientific or sea faring people, marriage to a stranger be opposed to the native, grief from wife in middle age, troubles through deceit and treachery through unions, contracts, partnerships and law suits, sickness, discord and opposition of partner, trouble through women generally.

Jupiter: He/She will have marriage as a result of journey, writing or with one of kin, long life, generous and truthful, average life up to middle age, under debt, worried, quarrels through servants difficulties with disrepute women and sickness, troubles with employees, last part of life will be comfortable. If afflicted reverse results.

Venus: He/She will have a rich partner or one to whom money comes unexpectedly, association with other women and fond of opposite sex, generous, gain and comforts from conveyance, favoured, respected, a jolly and submissive nature, a man of rank and powers. If afflicted there will be loss and trouble through law, union and open enemies, marital happiness, may there be death of partner, danger of death by violence, suicide, accident or war.

Saturn: Saturn is in enemy's camp. He/She will have ugly women, difficulties in childhood, rash temperament greedy, slow workers, many enemies, and influence of opposite sex, mainly remain in foreign country, diseased wife, and disturbed family life. He/She will have no gain from journey and business, comforts in old age.

Rahu or Ketu: Rahu in 7th house lessens the number of enemies, increased profit dealing with others and portends delight and gain through opposite sex whose influence may benefit the native. It is testimony for a wise and wealthy partner.

Ketu denotes many enemies, calumnies raised by enemies or competitors, contentions and difficulties with partner but is also portends the death or destruction of enemies.

Eighth House:

Sun: He/She will have death of friends, gifts or legacies from friends, an easy demise. Loss of ancestral property, law suits, danger to respect and may face punishment, many enemies and opponents, devoid from the comforts of parents.

Moon: He/She will have gain and honours in handling the estate or money of others, also through law suits, legacies, inheritance, insurance etc. of deceased person, many enemies, difficulties in early part of life, short life (subject to afflicted or weak Moon), troubles through relations, last part of life will be very lucky.

Mars: He/She will have death of wife of partner, some inheritance but difficulty over it, loss of money through marriage and partnerships if afflicted, gain by legacy and goods of the dead or through money of the partner. If afflicted loss of legacy, money through the financial loss of others, as bank failures etc.

Mercury: He/She will have an unsatisfactory end and medium life, many misfortunes, secret enemies die, trouble over inheritance, persecution regarding religious, scientific or educational convictions also through publications, gain by long journeys concerning legacies or good of deceased persons, death of partner's kindred, psychic experiences, strained relation with learned persons.

Jupiter: He/She will have long life, grief of brother and friends, honest and respected man, troubled early life, dangerous illness, death of servants, animals, poultry, financial rewards for faithful work and service, false accusations, or trouble on account of death or bequests, death at the hand of a cruel enemy.

Venus: He/She will have birth of sons, respected, concern over affairs of dead and with money of partner and finances of others, death through irregularity, occult experiences, mediumistic, a comfortable fixed partner, spiritualistic experiences.

Saturn: He/She will have short life, more sons than daughters, and diseases in early life, respected, and reputed, comfortable middle life, foreign journey, diseased and comforts from conveyance, suffers through children who may die before the native, loss through speculation and gambling, gain by legacy or estate.

Rahu: Rahu in 8th house promotes health and conduces to longevity, testimony for gifts and legacies and gain by those deceased.

Ketu: Ketu denotes loss of goods through deception, sudden or violent death, and danger from poison, worried and troubled.

Ninth House:

Sun: He/She will have gain and success through long journeys, respected and famous, gain from parent's property, and friendship through travel or learning. Friends among educators, ministers, writers, explores and inventors.

Moon: He/She will be Honourable voyages, professional journeys, and honours through learning, writings, publishing, research or philosophy, a religious or intellectual, mother's care, training and efforts, wealth and honest.

Mars: He/She will have marriage through stranger, from a far, gain by partner's relatives, partner's journey to foreign lands, and gain by books, trading at sea, long journeys, philosophy, religion, science and wife's kindred. If afflicted reverse the results.

Mercury: He/She will have pilgrimage, travelling for education, scientific or religious purposes, prophetic dreams, many fine qualities, splendid possibility through culture and development, long journeys, secret missions, investigations, exploration, secluded, research, learned, intelligent and generous.

Jupiter: He/She will have long life, journey to foreign land, gain through philosophy, publishing and science, danger through over study, generous, wealthy, intelligent, gain through relations, sickness abroad at sea while travelling, work in connection with foreign affairs or universities, expenditure on religious and charitable institution.

Venus: He/She will have ancestral property, respect, favour and gain from Govt.; wealthy, long journeys, religious or psychic experience, a beautiful and well versed wife, respect and gain through her, prophetic dreams, interest in science, invention, law, philosophy and

matters of higher mind, old age will be comfortable and lucky.
Saturn: He/She will have acquisition of money, happy, respected, high status, comfortable during young life, good and harmonious relations and gain from father and mother's brothers, dutiful children and gain from them, pleasure in foreign land.
Rahu: Rahu indicates the improvement in the mental qualities and gives success in educational, legal or studies, favourable for voyages and foreign affairs, true dreams and prophetic intuition.
Ketu: Ketu afflicts the faculty of faith and portends miserable or unfortunate voyages, curious dreams, unreliable promotions, trouble or danger of imprisonment in foreign lands.

Tenth House:
Sun: He/She will have friendship among those of good position. Gain and honours through those of high standing in social, business or Govt.; circles, respected, wealthy and famous, a native of power and authority, profit from ancestral property, happiness and gain from brothers, respect from officers and rulers.
Moon: He/She will have gain and honour through profession, honorary office or Govt.; employee, success aided by mother's care, training and efforts, comforts after troubles and worries. Reverse results when afflicted or weak.
Mars: He/She will have gain by occupation, profession, and merchandising with Govt.; wife's parents, an honourable partner beneficial to the professional carrier.
Mercury: He/She will have honour, credit, influence of opposite sex, adulterous and esteem through science, literature or travel, success in foreign affairs, sorrows through mother or in-laws, liability to disgrace, discredit and preservation from superiors and authorities.
Jupiter: He/She will be wealthy, intelligent and famous, professional and honourable journeys and gain through them, renown also through writings or other accomplishments, intelligent and gain through brothers, respect and authority from Govt.; comforts from conveyance.
Venus: He/She will have gain and comforts from conveyance, successful in hope, merit, honour, preferential success, and rise to high social or professional position, other person will lose money by him, according to the planet's aspect. If afflicted reverse the results.
Saturn: He/She will have honourable children, renown in speculations, gain in possession by trade, profession, public or Govt., work or by fruits of earth, gain from in laws. When afflicted it makes the native anti-religious, cheat, liar, irritable temperament, disrespect and litigations.

Rahu: Rahu confers on the native honours, credit and high position by merit of industry and ability.

Ketu: Ketu denotes loss of position through deception, treachery and adverse public conditions such as sudden depressions, changes or failures.

Eleventh House:

Sun: He/She will have respect, regards and wealth well defined hopes and successful wishes, wealthy and a native of rank and power. Gain of parental property, brave and courageous, many friends and beneficial acquaintances, friend among unique, ingenious, original or radical people, gain through brothers.

Moon: He/She will have an honourable fortune, a comfortable and happy life after middle age, eminent friends among legislators and those in high or professional position, a law suits but recovery from worries. The native is helpful to his associates and others, ambitious and ideal.

Mars: He/She will have gain or loss by friends and accidental fortune, fond of company of religious persons. A native will lead an independent life, grief from the death of brother in the middle age, friends become public opponents or enemies, journeys to lands in the company of religious persons, marriage to a widow or widower with children but liable to trouble through them.

Mercury: He/She will have fortune friendships on voyages and friends among foreigners and foreign lands, acquaintances among travellers, scientists and legislators, adulterous and fond of opposite sex, waste and loss of money through women, under debt, sufferings on account of religious or other convictions.

Jupiter: He/She will be fortunate through brothers, intelligent, learned and generous, comforts from conveyance, a wealthy and respected native, hopes and wishes accomplished through progressive studies, correspondences with friends, through journeys and gain through them, respect, good rank and favour from Govt.; sickness amongst friends and family, hopes depend too much on acquaintances, interest in legislative activities, political and social welfare. In middle of life, one becomes under debt.

Venus: He/She will have no dearth of enemies, rheumatic pain, danger of defamation, loss of wealth, reversal in business and profession, strained relations with brother, litigation when afflicted. But when well aspect then there will be great attachment with children and friends, happy conclusions to the hopes and wishes, friends through speculation or pleasure among legislators, ambassadors and sportsman, many hopes and wishes attained in old age.

Saturn: He/She will be learned, dear to people, be an author of treatises, be very skilful, and be endowed with many sons and wealth, fear of secret diseases and he will be liberal, virtuous, charitable, and helpful to others.

Rahu: Rahu in eleventh house indicates meritorious friendships, acquaintances which assist in the realisation of hopes and wishes.

Ketu: Ketu denotes undesirable associations, loss of opportunities, and of hopes, wrong advice and false friends. When well aspect reverse result are denoted.

Twelfth House:

Sun: He/She will be intelligent, respected and will hold a power and authority, gain from Govt. generous and of sanguine thinking, deceitful friends who may cause much sufferings, sorrowful friends. One may become worried due to unwanted expenditure but will soon control them, kind to everybody, good friendship amongst occult people, and pleasure in peaceful, quiet, harmonious or secluded places.

Moon: Position of 10th lord in 12th house is derogatory. It indicates loss of office or honour, dignity etc. through business associates who become secret enemies. He/She will have unfortunate environments and conditions, professional secrets, difficulty in employment, limitations relieved by the studies of recondite science or metaphysics. If it is well aspect and strong, then there will be favourable results.

Mars: He/She will have independent view and living, many enemies, unhappy marriage, secret sorrows, jealous, vexation and sickness, law suits after middle age, partner or opponents may cause imprisonment or fear of it, danger of death at the hands of enemies if 8th house lord is afflicted, gain by affairs of secret nature through hospitals, occult investigations and by large animals.

Mercury: He/She will have difficulty and loss in business, sorrows through religion, science, and journeys. If a writer or inventor has a hard time to complete his work and put it to the public. In middle or latter part of his life seeks seclusion for development, occult learning etc., and makes long journeys for the same, powerless enemies, secret investigation, fondness of animals, limitations, afflictions and adversities prove to be blessing in disguise by developing inner growth of understanding, early life full of difficulties.

Jupiter: He/She will be a teacher or professor. He/She will have great sorrows through brothers, secret sufferings, occult learning, seclusion or estrangements from brothers or kindred, many enemies who will be source of worries, danger of enmity and imprisonment while travelling, respected, unwanted and uncalled for expenses, sickness or work in large institutions, work of secret or mysterious nature, interest in archaeology and submarine life, grief from children.

Venus: He/She will have loss through opposite sex, difficulty over inheritance, death of secret or private enemies, great sorrows, fear or anxiety due to death or imprisonment, less respect, a happy man in spite of many adversities, gain and benefit through understanding of the occult and secret missions, success in middle life.

Saturn: He/She will be gentle, difficulties in early life, loss through enemies, loss of money and expenditure more than income, unhappy life, afflicted health, end of life in seclusion or devoted to the study of occult subjects, children may cause secret sorrows, ruin through speculations, pleasure through the investigation of things of mysterious and research nature.

Rahu: Rahu denotes gain by secret methods or in seclusion, and success in occultism.

Ketu: Ketu Indicates frequent harassment by the mechanisations of the secret enemies, liability to imprisonment or restraint and restrictions, unfavourable to health, inclines to self undoing.

11.8 Scorpio (Vrischika) Lagna:

Venus, Mercury and Saturn are malefic. Jupiter and Moon are auspicious. Sun, as well as Moon is Yoga Karakas. Mars is neutral. Venus and other malefic acquire the quality of causing death.

Sun: Surya is L-10 measured from Vrischik lagna or Vrischik Chandra; Vrischik navamsha and Sun is a friend of Mars the lord of Scorpio Ascendant. Therefore, Surya is benefic planet for Vrischik Lagna. L-10 Surya is a double career karaka. Presuming Ravi is well-disposed and strong; he is capable of channelling a remarkable amount of divine intelligence and accepts public roles and professional leadership responsibilities of a large corporations and government bureaucracies. When [Surya] the lord of the 10th house occupies the Navamsha of Jupiter, he finds his living with the help of Brahmins, deities, or through state favour, recitation of Puranas, studying Sastras, preaching or giving religious instructions, or lending money. Remedial Ratna for Vrischik lagna Surya (Manika - pure Ruby): Sun is lord of the 10th house. Sun is a friend of Mars the lord of Scorpio Ascendant. Therefore, wearing of a Ruby will bestow him with honours, promotions and great political power. He will achieve great success in his ventures if he is a businessman or industrialist. If

Sun as lord of the 10th is in 10th this will give rise to a powerful Raja Yoga and by wearing a Ruby, the native can even become a high dignitary, Minister, Prime Minister or President. The wearing of Ruby will prove very helpful if there are obstructions in the achievement of success on account of inauspicious disposition or affliction, in any way, of Sun in the birth chart. The results will be more pronounced in the major period and sub-periods of Sun."

Moon: Chandra is L-9 measured from Vrischik lagna or Vrischik Chandra; Vrischik navamsha and Chandra is a friend of Mars the lord of Scorpio Ascendant. Therefore, Moon will prove very beneficial or benefic planet for Vrischik lagna. Presuming Moon is well-disposed and strong, L-9 Chandra provides religious ritual and priestly duties, auspicious travel, spiritual advancement through religious fellowship and a life long emotional affinity for universities, seminaries, and teaching in temples. Moon will promote religious and charitable inclinations and purity of mind, brings good fortune and will prove good for the longevity of father. Remedial Ratna for Vrischik lagna Chandra (Mukta (moti - pure natural Pearl): Moon is the lord of the ninth house; the house Bhagya (fortune). Therefore, wearing of a Pearl to the native s of this ascendant would prove very beneficial. Wearing of a Pearl always will prove very beneficial. If Moon is not well disposed or afflicted in any way, the wearing of Pearl with a Red Coral will do immense good to them.

Mars: Kuja is L-1+L-6 measured from Vrischik lagna or Vrischik Chandra; Vrischik navamsha. Mars is lord of the Ascendant and also of the 6th and therefore, lordship of the Ascendant will prevail over the lordship of the sixth house and so Mars is beneficial or benefic planet Vrischik Ascendant. Kuja character in Rashi and bhava will determine the field of action for the perpetual competition and combat which propel this highly vital native into life and work. He may be a ground-breaking researcher, a brilliant healer, a genius financier, or a fierce warrior and may hunt with physical weapons, or with the mind. Conflict in his life can become intensely psychological and even overtly violent due to the maha-dushtastana being ruled by an aggressive malefic graha. Deadly significations are intensified. It may give the death of a child (2nd-from-5th) or gains from confidential insider information (11th-from-8th) Kuja. If Kuja is exalted, in own sign, in parivartamsha with his lord, or in Kendra, he will enjoy worldly success. The stronger Kuja's traits provide the strong the native's performance in his specialty. Example: if Kuja occupies Makara in Sahaja bhava, he marks excellence in business and a propensity to create self-made wealth. He will be exalted in the house of sales, marketing, administration, meetings, teamwork, colleagues/siblings, and mental tasking. When Kuja = uchcha and in bhava-3, the native flourishes in all varieties of commerce and business administration, with a persistently competitive thrust toward domination and control. If there is also a parivartan yoga with Kuja lord Shani rising in Scorpio (every 29 years a batch of these people are born) expect self-made wealth (3) through determined hard work, dominating aggressiveness, and sharp business judgment. By contrast, if Kuja occupies Kumbha, the sign of oppressive Shani, in bandhu bhava, the individual may find that the life force tends to implode into domestic disputes, conflict with educational system, and private scientific interests. Remedial Ratna Vrischika lagna Mangala (Munga – red Coral): Mars is lord of the Ascendant and also of the 6th. But his lordship of the Ascendant will prevail over the lordship of the sixth house. The Red Coral will be beneficial to the native in the same manner as for the Aries Ascendant.

Mercury: Budha is L-8+L-11 measured from Vrischik lagna or Vrischik Chandra; Vrischik navamsha. Budha and Mangala are enemies and therefore, Mercury is not considered an auspicious planet for this Ascendant. Remedial Ratna for Vrischika Lagna Budha (Flawless Emerald - Panna): Mercury is lord of the 8th in the 11th. Therefore, Mercury is not considered an auspicious planet for this Ascendant. If Mercury as lord of the 11th is in the Ascendant, 2nd, 4th, 5th, 9th, 10th or 11th, wearing of Emerald in the major period of Mercury will help the native to acquire wealth and have financial gains in his profession.

Jupiter: Guru is L-2+L-5 measured from Vrischik lagna or Vrischik Chandra; Vrischik navamsha. Guru is friend of Mars, lord of this Ascendant. Being lord of 5th house, a Trine, Jupiter is considered an auspicious or benefic planet for this Ascendant. L-2+L-5 Guru benefits him with family lineage and children and bring childbirth, publications of literary works, recognition of the value of one's intelligence, success in speculation, and wealth. Remedial Ratna for Vrischik lagna Guru (Flawless Pukhraj -yellow sapphire): Jupiter will be lord of the 2nd and 5th houses. Being lord of 5th house, a Trine, Jupiter is considered an auspicious planet for this Ascendant. Apart from this, Mars, lord of this Ascendant, is a friend of Jupiter. Consequently Yellow Sapphire is a very suitable stone for him and more so in the major period of Jupiter and his sub-periods. It will prove more profitable if worn along with Red Coral -- the gem stone of the lord of this Ascendant."

Venus: Shukra is L-7+L-12 measured from Vrischik lagna or Vrischik Chandra; Vrischik navamsha. Venus is lord of the 7th (Kendradhipatidosha) and the 12th. Moreover, Mars, lord of this ascendant, is not a friend of Venus. Venus is malefic planet for this Ascendant. Shukra provides diplomatic agreements and meditation, prayer, a good deal of sleep, a beautiful architectural sort of imagination, and long separations due to residence in foreign lands. L-7+L-12 Shukra cause foreign travel and contractual partnerships of all kinds. Periods of Shukra are unfavourable for the affairs of one's children, as Shukra is L-3+L-8 vis-à-vis putra bhava. Shukra is weakened by conjunction with Ketu or Rahu. Shukra + Rahu are a risk-taker who self-elevates via partnerships. Shukra+ Ketu is apathetic toward partnership and prone to unsatisfied abandonment of contractual unions. According to BPHS ch 43, the period of Chandra-Shukra can be fatal. Remedial Ratna for Vrischika Lagna Shukra (Flawless Diamond -hieea): Venus is lord of the 7th and the 12th. Moreover, Mars, lord of this ascendant, is not a friend of Venus. It will, therefore, not be advisable for the native s of this Ascendant to wear Diamond even when Venus is in his own sign in the 7th and the 12th.

Saturn: Shani is L-3+L-4 measured from Vrischik lagna or Vrischik Chandra; Vrischik navamsha. Moreover, Mars, lord of this ascendant, is not a friend of Saturn. Saturn is considered as a neutral planet for this Ascendant. L-3 + L-4 Shani are good for practical education and social connections leading to increased social security. Periods of Shani give educational confirmation such as diplomas and certificates. Homes are gained and lost during Shani period. Favours all kinds of training which forms foundation for later accomplishments. Remedial Ratna for Vrischika Lagna Shani (Blue Sapphire - Neelam): Saturn is lord of the 3rd and the 4th houses. Saturn is considered as a neutral planet for this Ascendant. If as lord of the 4th, Saturn is in his own sign in the 4th (causing Sasa Yoga) or is in the 5th, 9th, 10th or 11th Blue Sapphire can be worn, if necessary, in the major period of Saturn. But Red Coral, Pearl and Yellow Sapphire are much more appropriate gem stones for this Ascendant.

Rahu: Rahu is co-lord of bhava-4, Kumbha, therefore Rahu-ratna (gomedha) may become a beneficial gem for matters of home-ownership, schooling in the ethnic culture, diploma completion, licensing and passing examinations, obtainment of vehicles and shelters, parents and care-taking including stewardship of the land and patriotic defence. However Rahu in either Rashi of maraka Shukra is dangerous for Vrischika lagna. Even for the security-and-stability target, the Rahu-ratna should be considered only if the lord of bhava-4 and its occupants are highly auspicious. Rahu and Ketu are said to give positive results in the 3rd, 6th, and 11th bhava from the lagna or from Chandra. In addition, some authorities posit that since R-K are exalted in Vrishabha-Vrischika (per BPHS), R-K will give good results when Mangala and Shukra are well-disposed.

Ketu: Ketu is co-lord of Vrischika; therefore Ketu may be a beneficial or benefic palnet. Ketu's effects for Vrischika lagna depend significantly on Ketu bhava, rashi, and drishti. Ketu amplifies the effect of any graha who are sharing Ketu's house; and also Ketu magnifies the effects of the lord of His occupied Rashi. Ketu is co-lord of Vrischika; therefore Ketu-ratna may be a beneficial gem for matters of personality development and the vitality of the physical body, if the lord of bhava-2 and its occupants are auspicious.

Planet Effects in Twelve Houses for Scorpio (Vrischika) Lagna

First House:

Sun: Sun is the ruler of 10th house when posited in Ascendant indicates honour through merits, success and high ambitions. One will be troublesome for the parents; he will be cruel, of rash temperament and quarrelsome. He/She will have danger from fire and weapon is indicated. In youth, one will be blessed with power and authority in Govt.; or in military, well respected, preferment and dignities and gain through mother.

Moon: The Moon, lord of ninth house in Ascendant is in good position. One is learned, wise, prudent, and fortunate with strangers, foreigners, wife's relatives and voyage. He/She will have long journeys in life are indicated. Such persons have interest in science, invention, law, philosophy or political economy.

Mars: Lord of Lagna and 6th house in Lagna indicates powers, dignity and victory over enemies, long fortunate life, good health, harmony, and triumph over difficulties. He/She will be an intelligent, strong willed person, rash temperament and energetic. Comforts from children and wife, can be a military officer or renown, danger from poison, fire or weapon.

Mercury: Lords of 8th and 11th in Lagna denotes gain from legacies and money through deceased. He/She will have accumulation of money and will overcome his enemies and obstacles through friends and acquaintances, successful ambitions and hopes, will meet real friends and supporters, fortunate actions.

Jupiter: He/She will be blessed with long life wealthy, respected, and learned and a man of power and authority, many love affairs. He/She will have generous, sweet speech and propensity to pleasure through gaming and children; many face some troubles in last part of the life.

Venus: He/She will have secret sorrows, troubles, limitation, fond of occults and new methods of study, disablement requiring hospitalization and danger of imprisonment. Many love affairs, partnership and gain from opposite sex, public enemies, unions, connection with the process of law. In middle of his life, he may face troubles due to a woman, licentious, disrespected and extravagant.

Saturn: It is posited in enemy's house. He/She will have whitish complexion, rash temperament, cruel, not good health, strained relations with brothers, loss of money, danger from poison and enemies. Journeys and removals, fortunate inheritance, gain through land, property and interest in domestic science.

Rahu: He/She will be blessed with long life, houses, wealth and favours through religion, education and scientific affairs. Adds power to personality and gives opportunity for self expression.

Ketu: He/She will have a short life, troubles and tribulations, loss and scandals, endangers the face and eyes.

Second House:

Sun: He/She will have gain in money by industry, trade, profession or Govt; office. One will be wealthy, respected, honourable, gain by ancestral property and gain of other's property, full comforts from children is not indicated, loss through vile habits, of rash temperament and true to his words.

Moon: He/She will have gain of money through foreigners, sea, science, learning, publications, travel, invention or banking, little gain from ancestral property, gain through efforts, a glutton, talkative, diseases of watery nature, eyes, cough or gout are indicated.

Mars: He/She will have gain through ancestral property but will remain worried, brave and clever, losses and limitations and financially strained through sickness, servants and animals, gain through hard work, benefits through industrious activity.

Mercury: He/She will have gain of money by means of friends, acquaintances partners and others, fortunate in collecting debts, gain through deceased, legislators, interests or through own creations, knowledge of many languages, learned and intelligent, gain through

Jupiter: Lord of Second and 5th in 2nd house and in own sign indicates wealthy and renowned person, gain form ancestral property. He/She will have many subordinates, victory over enemies, prosperous children and comforts from them, gain from profession, good power and authority. If afflicted reverse results are indicated than above.

Venus: He/She will have gain by marriage and in-laws, also secret and occult ways, popular and generous, many friends, ambitions to lead a pleasurable life, relations with opposite sex and will do unwanted expenditure, possibility to come under debt. When afflicted, there will be loss through women and death of partner.

Saturn: He/She will have loss of wealth, cheat and liar, gloomy, poor, remains under debt, not well respected, will liquidate his assets, gain through deals in land and property and by the estate, short journeys, also gain through educational affairs, writings, music and art.

Rahu: He/She will have gain through ancestral property, legacies and gifts, inflow of money through science and learning. Increase in possession.

Ketu: There will be misfortunes in finance, under debt, loss and damage to property, sorrows, fear and worries.

Third House:

Sun: He/She will be much worried for his brothers and may remain upset and worried on this account. He/She will have many journeys, danger from ruler and Govt., respected amongst kindred and neighbours, gain in honour and advancement through wife's relatives, honour through short journeys, writings, business trips, Govt.; commission and fulfilment or hopes.

Moon: He/She will be blessed with more sisters, many professions, accomplishments and progress through research, travel, investigations, explorations, travel or writings, journeys and removals on account of beliefs and convictions, a good pointer, musician and other arts, early and young are comfortable ages but in old age may face some financial troubles and losses.

Mars: He/She will have sickness of some relatives and brothers and journeys on this account, interest in studies and methods of healing and in industrial economy, devoid from comforts of brothers, gain through legacies, comforts in middle age, mental development, when afflicted break or afflicted education, troublesome relations, journeys and writings.

Mercury: He/She will have friendship amongst relations and neighbours, friends through writings and journeys, strained relations with brother, unfortunate brothers and death of same, psychic and mysterious experiences, many brothers, devoid from comforts of children, well versed with arts and trained in hard work.

Jupiter: There will be many brothers and gain through them also from learning, writings and short journeys. Gain through dear ones. He/She will have loss of ancestral property. The native will have average wealth.

Venus: He/She will have many daughters. Respected and learned, wealthy, inflow of money without efforts, marriage with some one of kin or neighbour, legal or religious disputes, troubles through short journeys, disappointments, sorrows through relations or neighbours, occult learning.

Saturn: He/She will be believer in many religions, strained relations with brothers, residence in foreign land, unhappy, promotion and good income during middle age, successful efforts and hopes, gain from learning, writings and short journeys. If afflicted the father is put to sorrows or difficulty by one of the brothers.

Rahu: He/She will add quality to the mentality and conduces to the interest in spiritual or educational matters, gain from brother, journeys and neighbours, writings or publishing.

Ketu: He/She will have mental anxiety, troubles with brothers and neighbours, unprofitable journeys, downfall in business or service, worried through enemies.

Fourth House:

Sun: He/She will have gain and honour through parents, land and property, success at the close of life, may be devoid from parents in early age, many enemies, not to the interest of relations, will have interest in reclamations, colonisations, co-operative movements, mining, architecture and archaeology.

Moon: He/She will be intelligent, have scientific mind, gain from relatives of wife, may enter into business, devoid from comforts of the parents in the early age, blessed with property, a good trader and businessman, respect and gain through learning.

Mars: He/She will have gain through lands, mines, inheritance and possessions, success late in life, occult investigations, sickness of father through changes and worries, death of mother in early age, troublesome home or domestic affairs, trouble through servants.

Mercury: He/She will be learned, intelligent and respected, fond of yoga and spiritual life, voyages, journeys in mountains, gain of property through deceased, probable by the death of parents, inheritance through friends, blessed with property.

Jupiter: He/She will have gain of money and property through father, gain from children and love for home, wealthy and learned, comforts from conveyance, gain from mother's relatives.

Venus: He/She will have property by marriage, a chaste partner and happy ending to wedded life, litigations and loss of property may come under debt, more journeys, and may be voyage, grief from untimely death of wife, trouble through wife's mother, secret sufferings, restrictions or limitations at the end of life.

Saturn: He/She will have gain from agricultural land, earnings through unfair means, not submissive to parents, not well educated and intelligent, blessed with immovable property and conveyance, gain through old people and antiquities, assistance of father, loss through storms and floods.

Rahu: He/She will have gain through property and in an unexpected manner, fortunate in discovery or findings, long lived and trustworthy.

Ketu: He/She will have loss or confusion with land and buildings, waste of patrimony, not so much respected, turmoil in the lives of ancestors and family discord.

Fifth House:

Sun: He/She will be respected and a man of authority but even then there will be troubles and worries, grief through death of children, gain through speculations, pleasures, young people, sports of stage, drama, devoid from the comforts of parents in early age, loss of ancestral property, litigations, and eye disease.

Moon: He/She will be intelligent, learned, creative and of liberal and unconventional ideas and believe in free living. Journeys, take interest in science, philosophy, voyages, air flights, sports, foreign travels, speculations or investments, many daughters, success in service.

Mars: He/She will be licentious, destroyer of ancestral property and fond of occult science, gain through children, speculations, sickly children and illness through pursuits of sex or over indulgence in pleasure or sports, accumulations of wealth through unfair means, pleasure and gain in young age.

Mercury: He/She will be intelligent and learned, will have children but relations with children will not be cordial, over sexed and secret sorrows and difficulty through love, children, speculation and gambling, respect and gain in old age.

Jupiter: He/She will have knowledge of law, learned and intelligent, generous and God fearing. One will be respected, wealthy and will lead a pious life, gain by speculations, investment, and entertainment and through young people and children, love affairs and prosperous children and pleasure through them. In author's view Jupiter in own house posited in 5th house denies children.

Venus: The native will be intelligent and well respected. His wife will be beautiful and gain from her is indicated. He/She will be blessed with long life, voyages and residence on river or sea side, relations with other women, death or wife in young age is indicated and secret sorrows through love, children and gambling.

Saturn: He/She will be devoid of children's comforts, non-educated, anti to parents, irreligious, dishonest, earns by unfair means, happy in licentious deeds. In middle age may get disrepute, travels in life, gain and pleasure through father's good fortune and mental pleasures through children, brothers, reading and study.

Rahu: He/She will be saved from troubles, natural calamities and dangers, long lived and fortunate children, fond of recreations, pleasure and sports.

Ketu: This position either denies children or there are sudden or violent abortions. Children will be disobedient and source of adversity, excessive or irregular pleasures.

Sixth House:

Sun: He/She will have gain through law suits and legacies, death of parents in early age, law suits for ancestral property, modest worldly position, gain and honours through service, employment and practice of healing, employed in Navy or Army.

Moon: He/She will have many enemies, sickness through travelling and quarrels due to women, study of hygiene, medicine, difficulty through work in foreign lands or in export business, wife will be sickly and may there be death of mother in early age.

Mars: The native is not lucky for maternal relations, some severe illness, and gain through humanitarianism, food and employees. He/She will have surgical operation, fond of animals, pets and work, may be a source of trouble for relatives, will repute in middle age, victory over enemies. If Mars is well aspect then the native will have usually a good health, success in service and with employees, poultry, medicines, healing or social work.

Mercury: He/She will have dangerous sickness, little gain from education and art, popular in public, wound through fall from high place, learning, and scar on the body, loss of money and also through those in whose custody it may be. Friends among army, navy or air force persons, faithful servants and interest in social welfare.

Jupiter: He/She will have sickness among children, short life, many enemies and may remain under debt, chest disease is indicated, money through careful speculations or by children's earnings, gain through inferiors, poultry and small animals.

Venus: He/She will have marriage below status, sickly wife and evilly disposed employees, difficulty, trouble and loss through small animals and employees, limitations on account of sickness, sex relations with other women, syphilis disease is denoted, fond of travel, changeable residence and worried from opponents.

Saturn: He/She will have rheumatic pains, devoid of comforts from children, good wealth and owns property, and victory over enemies.

Rahu: He/She will have muscular body and good health, faithful and honest employees, gain through service and fortunate through father, residence in foreign land.

Ketu: He/She will have sickness due to various diseases, deceived and loss through servants, danger of illness through bites by insects, reptiles or animals, a worried life.

Seventh House:

Sun: He/She will have gain through lawsuits and dealing with public generally, honours and reputation assisted by honourable marriage and partnership in responsible concerns, authority and power as age advances, gain from respected and persons in position, troubles through enemies, wife of wicked nature, early life will be in law suits, law suits for ancestral property.

Moon: He/She will have marriage to a stranger of education or refinement taste whose relations will oppose the native, power and authority as age advances, possibility of two wives, first wife may die due to illness, gain from business, old age will be comfortable, beautiful but with a scar on the face, public enemies through religious, scientific or sea fearing persons, respected and wealthy.

Mars: He/She will have marital unhappiness, two marriages if afflicted, partnership and close association with others, quarrels of lawsuits with servants, difficulties with disrepute women and sickness, fond of opposite sex, loss of wealth due to enemies.

Mercury: He/She will be a loving partner with desirable friendships and social connections, success in law suits, a rich partner or one to whom money comes unexpectedly, coward and relations with other women. If afflicted, then death of partner and public enemies, danger of death by violence, suicide or accident, indigestion and kidney troubles, may get short life.

Jupiter: He/She will have long life, obedient to parents, wife will be faithful and of adjustable temperament, help from brothers, old age will be comfortable, gain by marriage, contracts, business and dealing with others especially with women, pleasure, close association or understanding with partner but discord with children, loss by theft, early age will be in poverty, middle age in worries, no gain from business but will become under debt, worried through enemies.

Venus: He/She will have success in lawsuits, a beautiful but licentious wife, intelligent and learned, fond of poetry and debates, journey in foreign lands. Old age will be troublesome, troubles through deceit and treachery concerning unions, partnerships, contracts and law suits, sickness, discord and opposition with partner, troubles generally through women.

Saturn: He/She will have disturbed married life, worried and unrest in nature, more than one wife, no comforts from first wife, abortions, a good orator, average life, rash temperament and loss of money, marriage as a result of journeys, writings or to one of kin, property by marriage, delayed marriage, gain by land and through women.

Rahu: It lessens the number of enemies, increases profit through dealings with others, gain through women. It is a testimony for a wise and wealthy partner, travel to foreign countries.

Ketu: He/She will have many enemies or competitors, contention and difficulty with partner, sickly wife, worries and troubles, loss through business.

Eighth House:

Sun: He/She will have gain and honours in handling the estate and money of others, also through law suits, legacies, inheritance, insurance of deceased persons, death of parents in early age, unhappy and diseased, loss of ancestral property, law suit loss worried through enemies. In middle age one will be troubled due to opponents.

Moon: He/She will have death of parents in early age, short life if afflicted, loss of money, unhappy and worried in middle age, loss and troubles through wife, prosecution regarding religion, scientific or educational conviction also through publications, and gain by long journeys concerning legacies or goods of deceased persons, psychic experiences and death of partners, relative.

Mars: He/She will have financial rewards for faithful service. Death through irregularity, occult experiences and dangerous illness and of muscular built, strained relations with relatives, loss through enemies, a difficult death.

Mercury: He/She will be an intelligent and wise doctor dealing with herbs. In early age danger of fall from high place and may lead to loss of one body part, death of friends, gifts and legacies from parents, gain by the dead, a natural and easy demise, a comfortably fixed partner, interest in matters relating to future life, spiritual experiences.

Jupiter: He/She will have long life span, anti to religion and change of that, worried and loss through children. They may die before the native. In middle age the native will have some good rank, gain of ancestral property, loss through speculation, gain by legacies or goods of the deceased.

Venus: He/She will have money or property by marriage, death of partner, some inheritance but difficulty over it, an unsatisfactory end, many misfortune, death of secret enemies, faithful and chaste wife but trouble and grief due to her death, danger of fall or drowning, loss of money. Old age will be happy due to gain of ancestral property.

Saturn: He/She will have long life, many enemies and danger from them, early life full of difficulties, unhappy, under debt, sickly and danger from reptiles, a happy end. Old age will be happy and full of enjoyment, sickness and death of brother, journeys due to troubles and false accusations, danger to mother and loss of inherited property.

Rahu: He/She will have long life and good health, gifts. Gain of the dead, journeys and travels, danger from thieves.

Ketu: He/She will have sudden or violent death, loss of goods through deception and enemies, disrespected and ill famed, troubles due to women.

Ninth House:

Sun: He/She will be learned, intelligent and well respected. He/She will have knowledge of many languages, a man of authority and power, honourable voyages, professional journeys and religions, honour through learning, writing, publishing, research or philosophy, a religious person or intellectual and builder of many temples and sacred places. No permanent residence and many journeys, devoid from comforts of parents in early age. A part of ancestral property in wasted in early age.

Moon: He/She will be wealthy and prosperous, gain from other's property, religious and learned, successful hopes and more happy in old age. Honour, credit and esteem through science, literature or travel, success in foreign affairs, voyages and pilgrimage, a good company in life.

Mars: He/She will have long journeys, religious, or psychic experiences, liking for surgery, science, invention, law, philosophy and all matters connected with higher mind, prophetic dreams and visions, gain through partner's relations danger through over study, sickness abroad at sea or while travelling, work with foreign affairs or universities, devoid from comforts of brothers and parents, a native of doubtful nature.

Mercury: He/She will be learned and intelligent, gain and success through long journeys, friendship through travel and learning, friends among educators, ministers, writers, explorers and inventors, a religious preceptor, interest in poetry, death in a distant land, danger of death by drowning or while on voyage.

Jupiter: He/She will have pleasure in foreign land, dutiful children who travel and give pleasure and learning to the native. Children may become scientists or explorers, a religious preceptor, founder of educational institutions, many professions, gain by books, trading at sea, long journey, philosophy, religion and science, a native of good conduct, fewer comforts from wealth, well versed in many trades and jobs.

Venus: He/She will have marriage with a stranger from a far and gain through that, also gain through partner's relatives, learned and intelligent, partner's journey to foreign lands, wealthy and blessed with all comforts of life. Early life will be troubled and worried, gain of respect and rank in foreign land, secret mission, investigation, exploration, secluded research, and sacrifice for science or religions.

Saturn: He/She will be wealthy, gain of ancestral property, gain of other's wealth and property and long journeys, clear, intelligent, religious but shrewd, gain by occupation, profession, science, religion, and merchandising, Govt.; and wife's relation.

Rahu: He/She will have improves mental tendencies and gives success in education, legal or religious studies, favourable for voyages and foreign affairs, true dreams and prophetic intuition.

Ketu: He/She will have curious dreams, afflict the faculty of faith and portends miserable or unfortunate voyages, unreliable premonitions, trouble and danger of imprisonment in foreign lands.

Tenth House:

Sun: He/She will have gain and honours through profession, office or Govt; employees, success aided by mother's care training and efforts. The native will be wealthy, respected and famous, comforts of conveyance and property, brave and courageous, fond of religious ceremonies, journeys to mountains and association with persons in authority.

Moon: He/She will have honour, credit and esteems through science, literature and travel, success in foreign affairs, good conduct, gain through other's property, changeable professions.

Mars: He/She will have merit; honour, preferment and success, rise to high social and professional position, honour through service, municipal or national activities, strained relations with relative, worried and waste of property.

Mercury: He/She will be broad minded, learned and intelligent, an author, poet and associations with learned persons, gain through inheritance, danger of violent death, gain through persons in authority and high position.

Jupiter: He/She will be learned, well respected, religious and of good deeds, comforts to parents through native, gain by occupation, profession and merchandising, government or wife's parent, gain from speculation, business or theatrical work. The children will be honourable.

Venus: He/She will have an honourable partner beneficial to the professional career, blessed with property, a man of authority and power, jolly nature and will lead a pleasurable life, liability to disgrace and discredit through superiors, Govt. or persons in authority, retirement or seclusion, sorrows through parents.

Saturn: He/She will have journey by sea, disrepute, harsh temperament, low profession, inclined to adulterous and licentious deeds, mental anguish through brothers, professionals and honourable journeys and gain through them, honours and renown, through writings, gain through trade, public or Govt.; work or by fruits of the earth.

Rahu: The native achieves honour, credit and high position by merit of industrious work and ability.

Ketu: He/She will have loss of position through deception, treachery and adverse public conditions such as sudden depressions, changes of failures.

Eleventh House:

Sun: He/She will have birth is a respected family, respected and friends of legislature and persons in authority, comforts from conveyance and property, intelligent and blessed with power and authority. Elder brothers will enjoy a good status, an honourable fortune, and ambitious ideals. The native is helpful to his associates.

Moon: He/She will be a good mathematician, sufferings on account of religious or other convictions, fortunate friendship on voyages and foreigners and gain through them and in foreign lands, acquaintances among travellers, scientists and legislators, elder sisters.

Mars: He/She will have comforts from brothers, over sexed, fond of opposite sex, many enemies, disrespected due to some accusation, sickness among friends and family, fulfilment of hopes through friends, interest in municipal or national activities, large circle of friends, much pleasure in life, gain by success of employer.

Mercury: He/She will have gain in legacies through friends, well defined hopes and wishes, respected and popular among friends, interest in mathematics, astrology and medicines, wealthy and will like truth, last part of life will be troublesome.

Jupiter: He/She will have accidental fortune, fond of children and pain through them, gain through friends, speculations, or pleasure and among legislators, ambassadors and sportsman, a wealthy and lucky native, generous and religious, founder of religious places, comforts from property and conveyance, wife will be from rich family, a respected and pleasurable life and owner of palatial buildings.

Venus: He/She will have unfortunate undertakings, peculiar hopes, great disappointment and obstacles, deceitful friends and losses through the advice of friends; friends will become public opponents or enemies, troubled married life. No comforts from relations, litigations, many enemies, death of brothers and sisters.

Saturn: He/She will be suspicious, stomach indigestion, death of child, wealthy but latter part of life unhappy, insincere friends and loss through them, loss of money, gain through brothers, hopes and wishes accomplished through progressive studies, many hopes and wishes attained in middle part of life.

Rahu: He/She will be meritorious, friendships acquaintances, assist in realisations of hopes and wishes.

Ketu: Undesirable associations, loss of opportunities and frustration of hopes, wrong advice and false friends.

Twelfth House:

Sun: He/She will have trouble in eyes, litigations, journeys, a worried and troubled life, loss of office or honour, dignity through profession associates who become secret enemies, unfortunate environments and conditions, difficulty in employment, limitations relieved through the study of recondite, science or metaphysics.

Moon: He/She will have voyages, eye troubles, loss from ancestral property, death of parents in early age, loss of wealth due to being extravagant, difficulty and sorrows through religion, science and journeys. The writer or inventor will not be able to get the same published without a hard time. In middle or latter part of his life, the native may seek seclusion for development, occult learning.

Mars: He/She will have Service in Army, loss through enemies, litigations, worried and unhappy, cautious for his respect, fear of imprisonment, secret unhappiness, enmities, gain through occult, secret missions, success in middle life, diseases during middle life, possibility of two wives, work of secret or mysterious nature like espionage, C.I.D., C.I.B or detector etc.

Mercury: He/She will have effective speech, many opponents and enemies, more unwanted expenditure, difficulty over inheritance, death of secret or private enemies, great sorrows, imprisonment or death through fear or anxiety.

Jupiter: He/She will have intelligent and balanced talks, wealthy, many professions, gain by secret affairs occult, secret sorrows through children, speculations cause ruin, pleasure through investigations and research of mysterious nature, danger of imprisonment through gaming if afflicted.

Venus: He/She will be generous, expenditure for good causes, wealthy and relations will be strained, secret sorrows, jealously, sickness, loss through opponents, occult abilities, powerless enemies, secret investigations, limitations, afflictions and adversities prove to be blessings in disguise due to inner growth of occult power.

Saturn: He/She will be miser, low tactics through profession for livelihood, wealthy, devoid from comforts of elders, disrespected and worried and unhappy, loss through treachery, end of life in seclusion or devoted to occult science, great sorrows through brother, danger of enmities and imprisonment while travelling.

Rahu: He/She will have gain by secret methods or in seclusion, success in occultism, earnings through unfair means, long journeys.

Ketu: He/She will have loss and troubles through secret enemies, imprisonment or restrain and restrictions and unfavourable for health, worried and unhappy.

11.9 Sagittarius (Dhanu) Lagna

Only Venus is inauspicious. Mars and Sun are auspicious. Sun and Mercury are capable of conferring Yoga. Saturn is a killer, Jupiter is neutral. Venus acquires killing powers.

Sun: Surya is L-9 measured from Dhanu lagna or Dhanu Chandra; Dhanu navamsha. Sun is the lord of the 9th house, the house of Bhagya (fortune). Sun is also a friend of Jupiter, the lord of the Ascendant and so Surya is, therefore, a very suitable or benefic planet for natives of this Ascendant. His father is very philosophical, although strong-willed; and confident. He receives public recognition, and gets high professional leadership responsibilities within universities, temples (9) and world travelling religious or philosophical groups (9). Presuming a strong Surya, it produce a celebrity in the fields of religion, philosophy, moral teachings like ritual priesthood, university professor, specialties of sacred literature and wisdom teaching, temple-based education, religious missionary, preacher, moral talks or international travel pursuits. But Ravi is L-12 from 10th, so is a culprit for loss of professional dignity or "job loss" or other public indignities.

Remedial Ratna for Dhanu Lagna Surya (Manika - pure Ruby): Sun is the lord of the 9th house, the house of Bhagya (fortune). Sun is also a friend of Jupiter, the lord of the Ascendant. Ruby is therefore a very suitable, agreeable and profitable gem stone for natives of this Ascendant particularly when the Sun is in his own sign in the ninth house. Ruby will give the same beneficial results to the Sagittarius natives as it gives in the case of Scorpio Ascendant. In addition it prolongs the life of the father of the native.

Moon: Chandra is L-8 measured from Dhanu lagna or Dhanu Chandra; Dhanu navamsha. Moon is the lord of the 8th and so is a malefic planet. However, Moon is friendly with Lagnesha, Jupiter. Presuming psycho-emotional Chandra (L-8) is strong and posited well, he is the happiest (not always) in mysterious roles that express healing behaviours like psychotherapist, surgeon, psychic intuitive, money launderer, silent business partner, secret sexual partner, or as the holder of highly confidential information in secure positions such as executive secretary or military officer. Chandra's Radix and Navamsha Rashi and any drishti to Chandra will show the native's specific style of pursuing security and emotional fulfilment. Example: Moon ruled by Shukra gives confidential roles in beauty and justice; ruled by Shani gives political and military roles handling government and corporate secrets; ruled by Surya a priestly role in communicating religious secrets, etc. Both he and his/her mother may suffer mysterious, medically not diagnosable illnesses which a psychic would know to have emotional causes. Chandra being L-8 provides sudden, apparently unprovoked eruptions of violence.; secrets; disasters; upheavals

Remedial Ratna for Dhanu lagna Chandra (Mukta/moti- pure natural Pearl): Moon is the lord of the 8th, he should avoid wearing a Pearl unless Moon is in his own sign in 8th and that too only in the major period of Moon. It will not give any adverse effects if it is worn along with yellow sapphire, the gem stone of Jupiter, the lord of this ascendant, who is a friend of Moon.

Mars: Kuja is L-5 / L-12 measured from Dhanu lagna or Dhanu Chandra; Dhanu navamsha. Mars is the lord of the 5th, a trine house, and so Mars is auspicious or malefic planet for Dhanu lagna as lordship of the 5th a trine will prevail over the lordship of the 12th. If Mars is in his own sign in the 5th, it will be more benefic. Mars will blessed with children, good fortune, name and fame, performance arts, politics, brilliant acts of genius, as well as long-term foreign residence, contemplative arts, and dissolution of the physical identity.

Remedial Ratna for Dhanu lagna Mars (Munga – Red Coral): Mars is the lord of the 5th and 12th houses. The auspicious lordship of the 5th a trine will prevail over the lordship of the 12th and by wearing Red Coral, the native of this ascendant will be blessed with children, good fortune, name and fame. It will give best result and more pronouncedly if Mars is in his own sign in the 5th.

Mercury: Budha is L-7 / L-10 measured from Dhanu lagna or Dhanu Chandra; Dhanu navamsha. L-7 / L-10 Budha suffers from Kendradhipati dosha on account of his lordship of two Kendra and is so the badhaka graha or malefic planet for Dhanu lagna or "harming" planet.

Remedial Ratna for Dhanu lagna mercury (Emerald – Panna): Mercury is the lord the 7th and 10th. It is said Mercury suffers from Kendradhipati dosha on account of his lordship of two Kendra. Still if Mercury is in the Ascendant, 2nd, 4th, 5th, 7th (in his own sign) 9th, 10th, (in his own sign) and 11th; wearing of Emerald in the major period of Mercury will prove advantageous. If Mercury (as lord of the 10th) is disposed in the 6th, 8th or the 12th, he suffers set backs in his profession. Wearing of Emerald at that time will help him to overcome his difficulties. If as lord of 7th Mercury is not well placed, we recommend that Yellow Sapphire should be used instead of Emerald. That will help to overcome the set backs in conjugal life.

Jupiter: Guru is L-1 / L-4 measured from Dhanu lagna or Dhanu Chandra; Dhanu navamsha. Jupiter is lord of the Ascendant and the 4th house and so is a benefic planet of this Ascendant. Wise and expansive Guru defines the physical body level of social personality identity, as well as the native's approach toward matters of education and property ownership and is a natural humanistic educator.

Remedial Ratna for Dhanu lagna Jupiter (Yellow Sapphire – Pukhraj): Jupiter is lord of the Ascendant and the 4th house; he can always wear Yellow Sapphire as a protective charm. The beneficial results will be felt pronouncedly in the major and sub-periods of Jupiter. If Yellow Sapphire is worn along with Ruby, the gem stone of Sun, the lord of the 9th (Bhagya) house, the results will be more beneficial.

Venus: Shukra is L-6 + L-11 measured from Dhanu lagna or Dhanu Chandra; Dhanu navamsha. Venus is lord of the 6th and 11th houses. Both lordships are inauspicious according to the principles of astrology and so is malefic or inauspicious planet for Dhanu lagna. Apart from that, Venus is an enemy of Jupiter, the lord of this Ascendant. L-6 Shukra has generally a problematic influence for the Dhanushya lagna. Shukra provides both conflict (6) and gainfulness (11). Shukra brings both the hostility of imbalanced relationships which produce enemies and the structured network of association through the marketplace of goods and ideas which creates a great circle of friends.

Due to Shukra's inauspicious lordship he has quite a bit of trouble (especially during Shukra periods) with the range of Venus significations like partnerships and alliances of all kinds. He should in general avoid business partnerships, or at least be very careful to have the terms of agreement made perfectly. He is prone to divorce. Shukra is an especially difficult graha for Dhanushya nativities in which Shukra is L-6 from both radical lagna (material animosity) and Chandra lagna (emotional imbalance).

Remedial Ratna for Dhanushya Lagna Venus (Diamond - Heera): Venus is lord of the 6th and 11th houses. Both lordships are inauspicious according to the principles of astrology. Apart from that, Venus is an enemy of Jupiter, and the lord of this Ascendant. Still if Venus in his own sign in the 6th or 11th Or as lord of the 11th is in the Ascendant, 2nd, 4th, 5th, 9th or 10th wearing of Diamond in the major period of Venus will bring financial gains to the native. If Venus is in the 10th, it will give rise to Amala yoga and Diamond in the major period of Venus will enhance the good effects of this yoga."

Saturn: Shani is L-2 / L-3 measured from Dhanu lagna or Dhanu Chandra; Dhanu navamsha. Saturn is an enemy of Jupiter, the lord of this Ascendant. Saturn is a malefic planet for Dhanu lagna. Self-made wealth from bhava-3 depends on the disposition of Shani. if Saturn occupies the Ascendant at a person's birth, he will be poor, sickly, love stricken, very unclean, suffering from diseases during his childhood and indistinct in his speech. But an exception can be made if Saturn is posited in the Ascendant. Good results of such disposition of Saturn are described by Brihat Jatak. However, if any of the signs, such as, Sagittarius, Pisces, Aquarius, Capricorn and Libra is the Ascendant and Saturn occupy it at birth, the person concerned will be equal to a king, the headman of a village or the mayor of the city, a great scholar and will be handsome.

Remedial Ratna for Dhanu lagna Shani (Blue Sapphire - Nilam): Saturn is lord of the 2nd, the death inflicting house (maraka house) and the 3rd. Moreover, Saturn is an enemy of Jupiter, the lord of this Ascendant. It will, therefore, generally not be advisable for the native of this Ascendant to wear a Blue Sapphire. Therefore, in a Sagittarius nativity, if Saturn is in the Ascendant, Blue Sapphire will enhance the good effects of the yoga caused by Saturn if it is worn in the major period of Saturn.

Rahu: Rahu is co-lord of Kumbha (bhava-3) measured from Dhanu lagna or Dhanu Chandra; Dhanu navamsha. Rahu is malefic planet for Dhanu lagna. Rahu and Ketu are said to give positive results in the upachaya 3rd, 6th, and 11th bhava from the lagna or from Chandra. In addition, some authorities posit that since R-K are exalted in

Vrishabha-Vrischika (per BPHS), R-K will give good results when Mangala and Shukra are well-disposed.

Remedial Ratna for Dhanu lagna Rahu (Gomedha): Rahu is co-lord of Kumbha (bhava-3), therefore Rahu-ratna (gomedha) may become a beneficial gem for matters of business administration, commerce, the neighbourhood, the work team, siblings, publications, writing, communications media, and brief travels, may be at the requisite risks of life and limb. However Rahu in either Rashi of maraka Shani is dangerous for Dhanushya lagna. The Rahu-ratna should be considered only if the lord of bhava-3 and its occupants are highly auspicious.

Ketu: Ketu is co-lord of Vrischika (bhava-12) measured from Dhanu lagna or Dhanu Chandra; Dhanu navamsha. Therefore is a malefic planet for Dhanu lagna. Ketu effects for Dhanushya lagna depend significantly on Ketu bhava, rashi, and drishti. Ketu amplifies the effect of any graha who are sharing Ketu house; and also Ketu magnifies the effects of the lord of His occupied Rashi.

Remedial Ratna for Dhanu lagna Ketu (Cat's Eye - Vaidurya): Ketu is co-lord of Vrischika (bhava-12). Cat's Eye should be used with caution to prevent excessive psychic permeability and destructive (Vrischika) scattering of the mental attention. To merit prescription of the vaidurya, lord of bhava-12 and occupants of 12 should be quite auspiciously disposed.

Planet Effects in Twelve Houses for Sagittarius (Dhanu) Lagna

First house:

Sun: Sun is lord of 9th Trikona house and is posited in friend's house. It indicates long journeys, learning, wisdom, and prudence, fortunate with strangers & foreigners, wife's relations and voyages, interest in science, invention, law, philosophy or political economy. Such persons are wealthy, respected, and blessed with power and authority, long life, strained relations with friends, gain from ancestral property but if Sun is afflicted may destroy the same.

Moon: Eighth lord in Lagna if not afflicted, one will be blessed with long life. He/She will have affliction of stomach is indicated, fond of art, adulterous, sweet spoken but spend thrift, gain through law suits and matters connected with the deceased. He/She will have accumulation of wealth and gain through business of others, likelihood of some wounds on the body, devoid of comforts from parents till young age. If Moon is afflicted, then it indicates short life, death by irregularity, trouble and loss through others.

Mars: The lord of 5th and 12th houses is in friend's camp. His/Her complexion will be like copper, many enemies and love affairs, wealthy and man of power and authority. He/She will have an afflicted health, a propensity to pleasure, a trouble life on the whole, secret sorrows, and limitations. All these difficulties can be minimised through occult or spiritual practice, danger of imprisonment or disablement.

Mercury: Mercury is lord of 7th house posited here in Lagna in enemy's camp aspecting its own 7th house. Lord of 10th house in Lagna is good and will confer dignities, honours through merit, intelligent, good natured and of good behaviour, wealthy and blessed with comforts of life, high ambitions, gain through Govt.; unions, partnership, love of women and gain through them, connection with the process of law. Such persons are found well versed in Mathematics and are connected with medical Science. Since Mercury is lord of two Kendra and so malefic but when afflicted will give the reverse results as indicated above.

Jupiter: Jupiter is also lord of two Kendra houses, 1st and 4th but being Lagna lord it is benefic, when posited in own house in the Ascendant will confer wealth, respect, rank and authority to the native. One will be blessed with long life, fortunate inheritance, and gain through land, property, and power over enemies, owner of property and conveyance, good health, harmony and triumph over difficulties, religious, generous and charitable, judge the reverse if afflicted or combust.

Venus: Lord of 6th and 11th houses (both malefic) in the Ascendant is in enemy's house. This position being in Lagna will give average length of life. Will hold a rank under the Govt.; kidney trouble in old age is indicated. Difficulties and reversals in life and may face debt. But the native meets with real friends and supporters can overcome enemies and obstacles through the support of acquaintances. Fortunate actions and hopes are attained.

Saturn: Saturn in Ascendant is in enemy's camp and lord of 2nd and 3rd houses. "Being lord of 2nd house gives good results but lord of 3rd malefic. Saturn weak in this case is very benefic and makes the persons millionaire by writing book.

So it will make one wealthy, of good position in life, comforts from wife and children, respected but devoid from comforts of brothers, worried and gloomy face. Average intelligence helpful to others, any part of body will become defective and worries on this account. Gain through writings and learning.

Rahu: Rahu in Lagna of this ascendant confers honours, wealth and favour through religious, educational or scientific affairs. It adds power to the personality and gives opportunity to self expression, foreign travel, a good progeny, respect and honour in foreign countries in old age.

Ketu: Ketu in the Ascendant is not good and gives short life unless aspect by Jupiter. One is devoid from comforts of ancestral property, tribulations and difficulties in life may endanger face and eyes.

Second House:

Sun: Sun in 2nd house indicates gain of money by foreign merchants or the sea, science, learning, publications, travels, invention or banking. This will give strained relations with relatives and disrespected by them, a diseased and troubled early life, of rash temperament and defective eyes, last part of life will be happy and comfortable.

Moon: 8th lord in 2nd house indicates gain in finances by the money of partner and others, fortunate in realising the debts. Gain through deceased. If Moon is weak and afflicted, there will be loss and troubles in life, defective eye sight, soft spoken but strained relations with relatives, changeable professions.

Mars: Lord of 5th and 12th house posited in 2nd house indicates gain through secrecy and occult, an engineer by profession or building contractor, less comforts from wife, of loud voice but greedy, pleasure through children, should not speculate at all.

Mercury: He/She will have sanguine temperament and speech, skilled in arts, gain by marriage, average financial position, may be afflicted in lower part of body through some disease, gain of money through trade, industry, profession or Govt.; office. If afflicted or combust, loss of money through partnership contracts, unions, law suits or through women.

Jupiter: He/She will have gain through property and land, hard work, considerable time strenuous and persistent given to the efforts to obtain money, benefits through industrious activity, an intelligent person, good mediator and has a weight and force in his speech, an average span of life will be pleasurable but last part of life will be troublesome.

Venus: He/She will have business and money through friends and acquaintances, through legislative interest or development of his own creation, but loss through enemies, defective eye, more daughters than sons, marriage in good family, debts in middle age, finances will be average.

Saturn: He/She will have gain of money through personal efforts, educational affairs, short journeys, writings, and music, devoid of children's comforts, blue eyes, and proud liar, wealthy through ancestral property.

Rahu: Rahu is good in second house, promoted, fortunate heredity or gain by legacy and gifts, gain by science and learning and increases the professions, long journeys undertaken to remove want and to bestow affluence.

Ketu: Misfortune in finances, loss and damage to estate, sorrows, fears and worry concerning money matters, loss of parents in early age, in average life, there will be troubles and debts.

Third House:

Sun: He/She will be bold and courageous, successful in hopes, respected and famous, learning, accomplishments and progress through research, travel, investigations, or writings. Brothers will be on good posts and wealthy but no gain through them.

Moon: He/She will have psychic and mysterious experiences, less comfort from brothers and death of the same, unfortunate brothers, loss of power of virility and ear troubles, strained relations with relatives and loss through wife.

Mars: He/She will have disappointments, sorrows and troubles through relations and friends, troublesome short journeys and writings, occult learning, liking for travel, sports, drama and adventure, short life, trouble in ears, diseased and troubled, many opponents and loss through thieves.

Mercury: He/She will have good relations with relatives and friends, loss of virile power, respect among relatives, neighbours, gain in honour and advancement through partner's relations, honour through short journeys, writing, accomplishments, business trips and Govt.; favour, company of adulterous and disrepute persons, habit of theft, last part of life will be happy and comfortable.

Jupiter: He/She will be a learned, intelligent, religious and well respected person, comforts from property and conveyance, many brothers, power and authority in middle age, gain from short journeys, a comfortable life full of pleasure and power. But if afflicted, reverse results and restricted or broken education.

Venus: He/She will have friends through writings and journeys on account of sickness, mental disturbances is indicated when sick, unhappy through brothers, fond of opposite sex and loss due to the same, trouble in ears, urinary disease, last part of life will be unhappy.

Saturn: He/She will have many brothers but no gain through them, adulterous, pleasure seeker, and deafness of ears in middle age. Gain through education, writings and short journeys.

Rahu: He/She will be bold and advance in spiritual and educational matters, gain through brothers and neighbours and through journeys, writings or publishing.

Ketu: He/She will have mental anxiety, unprofitable journeys, troubles through brothers and neighbours, adulterous, fond of opposite sex, company of ill reputed and dacoits, loss through standing a surety, a few brothers. Last part of life will be a bit comfortable.

Fourth House:

Sun: He/She will be intelligent and learned but no gain from that, blessed with property, gain of wealth by industry, business and trade. Gain of wealth in old age, loss of ancestral property, many enemies and opponents, profession through partner's relatives, journeys on account of family affairs or wife's mother, journey's home to die.

Moon: He/She will be learned, sweet speech and intelligent, blessed with property in middle life, a religious man, lower part of body or legs will be afflicted with pain, gain of property through deceased, probably by the death of parents. The death of native will be at home, but if Moon is in conjunction of Sun, then abroad. If Moon is weak or afflicted, then death of parents, danger through falls and falling buildings, floods and storms, trouble over inheritance, land and property.

Mars: He/She will be a good skilled, craftsman, voyages, loss of ancestral property, bold but quarrelsome, gain through children, and gifts from grandfather, love of home, secret sufferings restrictions and limitations at the end of life, trouble through father or wife's mother.

Mercury: He/She will have property by marriage, a chaste wife, gain and honour through parentage, lands and property, success in last part of life, interest in reclamations, colonization, cooperative movements, horticulture, mining, architecture or Archaeology, changeable profession, gain through women and influence, thereof.

Jupiter: He/She will have gain through lands, mines, inheritance and possessions, occult investigations, learned, intelligent and knowledge of other languages, assistance of father and gain through old people and antiquities. In middle of life may be defamed due to connections with women, generous and comforts from mother, gain from other property, loss through storms and floods.

Venus: He/She will be learned, intelligent, a good astronomer, faithful to parents, fond of opposite sex and troubles through them, sickness of father through changes and worry, troubled married life, inheritance through friends, fortunate in property, love of father and skilled in many arts.

Saturn: He/She will have changeable life, property and residence, benefits through parents, household goods, land or mines, quarrelsome, suspicious, loss of property, joking in nature, devoid of mother's affection, dishonest and accusation there from, gain from other's property and money, travelling and writings in connection with home affairs and property.

Rahu: He/She will have gain of property in an unexpected manner, fortunate in discovery or in findings, ancestry, noble, long lived and trustworthy.

Ketu: He/She will have loss of confusion with land and buildings, waste of patrimony, turmoil in the life of ancestors, family discord and law suits, changeable professions.

Fifth House:

Sun: Sun is exalted in this sign. Lord of 9th house (Trikona house) in another Trikona house is very beneficial and benefic. It will confer on the native good intelligence, one will learned, well respected, bold and with fortune, long journey, learning, wisdom, prudence, fortunate with strangers, foreigners, wife's relations and voyages, interest in science, inventions, law, philosophy or political economy, wealthy and comforts of life, property and conveyance, gain through secrets. In early age one may be deprived from comforts of parents.

Moon: He/She will have average financial position, unfortunate children or death of same, danger through excessive pleasure, speculations, children and their affairs, medium span of life.

Mars: He/She will be lucky and man of power and authority, may be a magistrate or some army officer, danger from fire, of rash temperament.

Mercury: He/She will be an intelligent and learned person, pleasure through wife, relations with children will be cordial and comforts from them. Native children will suffer from sickness but rise to honours, gain through speculation, pleasures, young people, sports and stage, a good writer and a God fearing man.

Jupiter: He/She will have gain through father, a learned, intelligent and generous native, God fearing and blessed with children and gain through them, helpful to people, knowledge of a few languages, comforts from children in old age, delight in pleasure, amusement, sports, speculations and children with tendency to success through these, interested in art, literature and poetry. If afflicted reverse results.

Venus: He/She will be very benefic, happiness and pleasure in life through children, friends and beneficial circumstances, average financial status, may face some financial difficulty and debt in life, illness through pursuits of or over indulgence in pleasure of sports, troubles through enemies and danger of poison, troubles through women.

Saturn: Saturn is posited in debilitated sign in 5th house. It devoid any profit from children, cheat, disrespected, average intelligence, back bitter, adulterous, devoid from parent's affliction, gain by speculation, investment and pleasure, pleasurable journeys, pleasure through children, reading, study and travels.

Rahu: He/She will be frees from many troubles, calamities and danger, long lived and fortunate children, fond of recreation, pleasure and sports, gain through employment.

Ketu: It may deny issue subject to other checks, disobedient and crossed children, excessive or irregular pleasures produce much harm.

Sixth House:

Sun: He/She will have victory over enemies, sickness, travelling, and study of hygiene, medicine, healing, and difficulty through work in foreign lands, or in connection with exporting, change of profession in middle of life, diseased and troubled life, danger of thefts during journey, and loss to maternal relations.

Moon: He/She will have gain through maternal relations, dangerous sickness, death by pets, small animals and servants, loss of money earned and through one in whose custody it may be, short life (subject to other checks), many enemies and opponents, residence in mountainous region, unsuccessful hopes, worried and troubles on these counts.

Mars: He/She will be troubles through enemies and loss of property through them and one may be forced to leave the country, sickness among children, pleasure through hygienic methods, difficulty, troubles and loss through employees, limitations on account of sickness, troubled married life, travel in foreign land and fond of wealth but generally unaccomplished hopes.

Mercury: He/She will be lean and weak constitution, educated but happy, loss, gain from education; wife may be in service and source of financial help, happy and gain through children in old age, modest worldly position, gain and honour through service and employment. The native may marry below his status, mentally or socially has sickly wife and evilly disposed employees.

Jupiter: He/She will be a man of strong determination, a good healer, gain through humanitarianism, food, clothing and employees, loss of possessions through sickness or servants, may build his own house, loss of respect in middle age, many enemies and loss through them, loss of money on account of being a surety of someone. Friends may acts as enemies, danger of some punishment in middle age.

Venus: He/She will have many enemies and relations with women and will therefore suffer at their hands, severe illness, success in service and with employees, poultry, medicine or social services, friends among working people and those in Army, Navy or Air Force, faithful servants and interest in social welfare, disrespected due to connections with other women.

Saturn: He/She will be untrustworthy, adulterous, non-religious, skin disease, poor, dishonest, earn money through illegal ways loss through animals, fall from height and fracture in leg and may become lame, gain through inferiors, sickness or injury during journey, difficulties through brothers and interest in study and methods of healing or social economy.

Rahu: He/She will have favours, the healthy and strengthens the body, faithful and honest employees, gain through service and fortunate through father.

Ketu: He/She will have law suits and litigation, danger of illness through bite of reptiles, insects or animals, suffers due to physical afflictions, changeable residence, and loss through deception from servants.

Seventh House:

Sun: He/She will be learned, respected and popular, legal ways of livelihood, marriage to a stranger opposed by relations, public enemies through religious, scientific or sea-fearing people. Inventor of new items, gain from business or trade, knowledge and well versed in law of the land, fond of opposite sex and bickering in married life.

Moon: He/She will have voyages, loss through business, fond of opposite sex, disrespect and loss through them. Wife will be rich to whom money comes unexpectedly. If Moon in weak and afflicted, may deny the marriage or death of the partner, public enemies and loss through them, danger of death by violence, suicide, accident or war, loss and spending of money on vices.

Mars: He/She will have sickness among children, gain of money through speculation and children's earnings, pleasure of hygienic methods, troubles through deceit and treacherous ways, concerning unions, partnerships, contracts and law suits, sickness, discord and opposition of partner, troubles generally through women, many enemies and opponents, loss in business and extravagant spending.

Mercury: He/She will have gain through law suits and dealing with public generally. Honour and reputation assisted by marriage and partnership in a responsible concern, comforts from wife, comfortable old age, comforts from conveyance, success in law suits and a pre-possessing partner, but one is likely to grow cold or proves untrue or hostile, gain through business and trade.

Jupiter: He/She will be learned, respected and knowledge of law, careful of his respect, marriage in good family, gain by marriage, land and as a general, fond of opposite sex, partnerships and close association with others, power and authority in foreign country and gain of money. If afflicted reverse results.

Venus: He/She will have late marriage, adulterous, disrespected, of loose morals and vile acts in early age, quarrels or law suits, difficulties with disrepute women and sickness, trouble with employees or physicians, success in law suits also, a loving partner desirable friendship and social connections.

Saturn: He/She will have gain by marriage, contracts, business and dealing with women and others, marriage as a result of journey, writings or to one of kin, but adulterous, disturbed married life, journeys, hatred of religion, a good orator, lecturer and will earn through this profession but average profession.

Rahu: He/She will have reduces the number of enemies, increases profit through dealings with others and portends delight and gain through woman whose influence may benefit the native. It is a testimony for a wise and wealthy partner.

Ketu: This is not a happy position. He/She will have many oppressors, calumnies raised by enemies or competitors, contenders and difficulty with partners but also it indicates death and destruction of enemies. He/She will have changeable professions, sudden death, and a troubled and worried life.

Eighth House:

Sun: He/She will have gain through long journeys concerning legacies, or goods or deceased persons, psychic experiences, death of wife's nearest relations, scientific or educational pursuits also through publications, painful death, diseased and troubled.

Moon: He/She will have eighth lord is posited in own house, gain by the dead, a comfortably fixed partner, interest in matters relating to future life, spiritual experiences. If Moon is weak and afflicted short life, devoid from ancestral property, poor, danger of drowning and loss through women.

Mars: He/She will have suffers through children who may die before him, loss through speculations, an unsatisfactory end, many misfortunes, secret enemies may die, troubles over inheritance, short life, a troubled native, many enemies and loss through law suits.

Mercury: He/She will be not so well educated unless aspect by Jupiter, gain and honours in handling the estate and money of others, also through law suits, legacies, inheritance, insurance of deceased persons, money or property by marriage, death of partner, some inheritance but difficulty over it, unsuccessful hopes and worries.

Jupiter: He/She will be wealthy and learned many enemies and blessed with long life. Jupiter is exalted in this sign and is a Lagna lord. He/She will be proud, obstinate and a man of strong determination, last part of life will be comfortable, death through irregularity, occult experiences and mediumistic, concern over affairs of dead with the money of others and partners, gain by legacies or estate, when afflicted, death of mother, loss of inherited property, disappointments over legacies and troubles in financial matters.

Venus: He/She will have financial reward for faithful service, intelligent, gift or legacies from friends and an easy demise, death of friends, dangerous, illness and death of servants, animals, poultry, unhappy due to wife, quarrels over ancestral property, danger in 6th years of age, less comfort from parents.

Saturn: He/She will have short life, cheat, liar, bad company and habits, licentious and adulterous, devoid from comforts of progeny, sickness and death of brothers, journeys on account of trouble, false accusations or trouble on account of death or bequests, gain by legacy and goods of the dead or through money of the partners.

Rahu: He/She will have promotes health and conduces to longevity, testimony for gifts and legacies and gain by those deceased.

Ketu: He/She will have loss of goods through deception, sudden or violent death.

Ninth House:

Sun: He/She will be respected, generous, learned, intelligent and good advisor, travelling for education, scientific or religious purposes. Prophetic dreams, many fine qualities, splendid possibilities through culture and development, careful for his respect, company of high ups and wealthy persons and gain through them, unhappy due to brothers.

Moon: The partner has trouble with relatives over money matters. He/She will have death in a distant land, danger of death by drowning or while on voyages, a drunkard and of loose morals.

Mars: He/She will be bold, stone hearted and a man of strong WILL power. Beautiful children, who travel and give pleasure to the native, children can become preachers, scientists or explorers, pleasures in foreign lands, blessed with landed property, worries and grief in young age due to loss of some property and death of a child. If afflicted shipwreck or isolation, difficulty with relatives, men of science and religion, long journeys, secret mission, investigations, explorations, secluded research. Sacrifice for science or religion.

Mercury: He/She will have marriage to a stranger from far off, gain by partner's relatives, partner's journeys to foreign, honourable, voyages, professional journeys, honour through learning, writing, publishing, research or philosophy, of sanguine temperament, a religious or intellectual father. If afflicted contentions with religious or scientific people.

Jupiter: He/She will be well respected and generous, builders of charitable institutions, long journeys, religious, or psychic experiences, liking for science, law, philosophy and all matters connected with higher mind, prophetic dreams or visions, gain through partner's relation, science religion and long journeys. If afflicted loss of foreign property, trouble with foreigners and religion, legal or educational affairs, fruitless and dangerous voyages.

Venus: He/She will be wealthy and reliable, gain and success through long journeys, friendship through travel and learning, friends among educators, ministers, writers, explorers and inventors, sickness abroad at sea or while travelling, illness of wife's relations, danger through over study, work in connection with foreign affairs or universities.

Saturn: He/She will be obstinate, anti to parents, unhappy, miser, inclined toward bad deeds, diseased and ill reputed. Gain by books, trading at sea, long journey, philosophy, religion, science and wife's kindred, brothers are likely to proceed to journey to foreign countries and marry.

Rahu: He/She will have improves the mental qualities and gives success in educational, legal or religious studies, favourable for voyages and foreign affairs, true dreams and prophetic intuition.

Ketu: He/She will be unfortunate or miserable voyages, afflicts the faculty of faith. Curious dreams, unreliable promotions and trouble and danger of imprisonment in foreign lands, devoid from the comforts of parents in early age.

Tenth House:

Sun: He/She will have honour, credit and esteem through science, literature or travel, success in foreign affairs, respected and blessed with power and authority, knowledge of law and owner or property, religious and generous. If afflicted reverse results.

Moon; He/She will have rise to high position through an inheritance, other persons will gain or loose money by him, according to the planets aspects. If weak and afflicted, there will be then danger of violent death probably due to discharge of Govt. order or through war, business through matters connected with dead.

Mars: He/She will have pleasure of honour through wife's father or his own mother, honourable children, renown in speculation, business enterprises and in the theatrical world, but liable to disgrace, discredit and persecution from superiors, Govt.; or persons in authority resulting in a long journeys or retirement and seclusion, sorrows through mother or troubles with in laws.

Mercury: He/She will be skilful in all jobs, valorous, truthful, and devoted to elders, will beget a disobedient wife, will be religious and endowed with wealth, sons.

Jupiter: He/She will be wealthy, respected, famous and learned, author of books, psychic and prophetic nature. Rise to high, social and professional position, merit, honour, preferment and success, religious and generous, gain in profession by trade, Govt.; work or by the fruits of the earth.

Venus: He/She will be well respected, honour through service and healing or through municipal or national activities, friendship among those of good positions and gain through them, benefits from persons of social standing, business or Govt. officers.

Saturn: He/She will have ancestral property, many professions, troublesome early life, happy in old age, owners of other's property, life above average, gain by occupation, profession, merchandising, Govt. and wife's parents, professional and honourable journeys and gain through them, honour or renown through writings and other accomplishments.

Rahu: He/She will have achieves honours, credit and high position by merit of industry and ability, knowledge of Vedic literature.

Ketu: He/She will have loss of position through deception, treachery and adverse public conditions such as sudden depressions, changes of failure.

Eleventh House:

Sun: He/She will be lucky and blessed with property, conveyance and orchards, fortunate friendship on voyage and friends among foreigners and in foreign Banks, acquaintances among travellers, scientists and legislators. If afflicted sufferings on account of religious or other convictions, relations with opposite sex, and may be unhappy in old age.

Moon: He/She will have gain and legacies through friends, wealthy and skilled in arts. If weak and afflicted, may give two marriages and unhappy life.

Mars: He/She will have great attachment with children, friends through them, successful hopes and wishes, friends through speculations, pleasure and among legislators, ambassadors and sports men. If afflicted, there will be unfortunate undertakings, peculiar hopes, great disappointments and obstacles, deceitful friends and losses through advice of acquaintances.

Mercury: He/She will be blessed with wealth and property, trouble through wife and children, eminent friends among legislators and those in high Govt., or professional positions, an honourable fortune. The native is helpful to his associates and others, ambitious ideal.

Jupiter: He/She will be well respected, wealthy and famous, blessed with comforts of life and conveyance, brothers will hold good position, large circle of friends, assistance to and from them, much pleasure in life, hopes and wishes often attained, gain by success of employer. If afflicted then friends are detriment, hopes are often defeated and reverse results as indicated above.

Venus: He/She will be fond of scents and perfumes, interest in legislature activities, political and social welfare, many friends and beneficial acquaintances, well defined hopes and wishes, friends among unique, ingenious original or radical people, blessed with comforts from opposite sex.

Saturn: He/She will be wealthy, proprietor of many immovable properties but disrepute, no respect from masses and public, venereal disease and devoid from children's comforts, accidental fortune, friends through journey or vice versa, fortunate conditions through brothers, hopes and wishes accomplished through progressive studies.

Rahu: He/She will have meritorious friendships; acquaintances assist in the realization of hopes and wishes, gain from some older man in age than the native.

Ketu: He/She will have undesirable associations, loss of opportunities and frustration of hopes, wrong advice and false friends, disrespected and grief from children.

Twelfth House:

Sun: He/She will have journeys to distant lands, difficulty and sorrow through religion, science, and journeys. If a writer or inventor, has a hard time to complete his work and get it to the public. In middle or latter part of life seeks seclusion for development, occult learning and takes long journeys for the same, respected and hard worker, company of wealthy persons and high dignitaries, may face sudden death.

Moon: He/She will have voyages, may have change of religion, difficulty over inheritance and death of secret or private enemies, great fear or anxiety connecting death or imprisonment, death while in an institution, worried and restless, downfall in business.

Mars: He/She will have occult abilities, powerless enemies, service in police or army; foreign travel, secret investigations, limitations, afflictions and adversities in life prove to be blessings in disguise by developing inner growth and understanding. Children may cause sorrows, speculation cause ruin, pleasure through investigation of things of mysterious or research nature.

Mercury: He/She will be generous and knowledge of law, unhappy marriage, secret sorrows, jealousy, vexation and sickness, partners or opponents cause imprisonment of fear of it, partnership in business is not gainful, loss of office of honour, dignity through business associates who become secret enemies, unfortunate environments and conditions, professional secrets, difficulty in employment, gain from occult science or metaphysics.

Jupiter: He/She will have fear of imprisonment, secret unhappiness, gain through occult, and enmities, afflicted health, sorrows through partners and loss through treachery, end of life in seclusion or devoted to the study of occult subject.

Venus: He/She will have loss through enemies and opponents, imprisonment, work of secret or mysterious nature such as in C I D or C I B department, sorrowful and deceitful friends, but good friendship amongst occult persons, pleasure in peaceful quiet, harmonious or secluded places.

Saturn: He/She will have gain through secret affairs, occult investigations, sorrow through brothers, secret sufferings, seclusion or estrangement from brothers, danger of enemies and imprisonment while travelling, disrespectful profession, litigation, licentious, adulterous,
worried, loss of money, and property.

Rahu: He/She will have gain by secret methods or in seclusion, and success in occultism.

Ketu: He/She will have worries through enemies, liable to imprisonment or restraint and restrictions, unfavourable to health and of rash temperament.

11.10 Capricorn (Makara) Lagna

Mars, Jupiter and Moon are malefic, Venus and Mercury are auspicious. Saturn will not be a killer on his own. Mars and other malefic will inflict death. Sun is neutral. Only Venus is capable of causing a superior Yoga.

Sun: Surya is L-8 measured from Makara lagna, Makara Chandra, Makara navamsha. Sun is lord of eighth (a very inauspicious Bhava) and a bitter enemy of Saturn, the lord of Capricorn Ascendant and so Sun is most malefic planet for this Ascendant. As a fierce enemy of lagnesha Shani, ego-promoting Surya causes social awkwardness and psychological frustration for the careful, cautious, law-abiding Makara native. Sun gives a blazing hot sudden jolt and an unexpected upsurge of an inner clarion call to him. L-8 Sun manifest as personality disturbance, even anti-social behaviour. If Surya is weak by placement, it may generate secret or covert self-assertion behaviours like prostitution. If L-6 Budha (prostitution) + L-9 Budha (public religion) is exacerbated by sexuality-karaka L-7 (Chandra) in hidden Simha (Bhava-8), she will get indulge more likely in prostitution.

Remedial Ratna for Makara lagna Surya (Manika -pure Ruby): Sun is lord of eighth Bhava. He should avoid wearing Ruby as far as possible because not only is Sun, lord of the 8th a very inauspicious Bhava, but also the Sun is a bitter enemy of Saturn, the lord of Capricorn Ascendant. If however the Sun is in his own sign in 8th, Ruby may be worn in the major period of Sun. That will give long life to the native and he might benefit from an inheritance. Ruby should not be worn if the major period is not likely to be operative during the life time of the native.

Moon: Chandra is L-7 measured from Makara lagna, Makara Chandra, and Makara navamsha. Moon is the lord of the 7th (Kendradhipatidosha) and Moon is also an enemy of Saturn, and so Moon is a malefic planet for Makara lagna. The highly fluctuating and impressionable Chandra regulates contracts, agreements, alliances, and peer-to-peer relationships (bhava-7). Chandra's condition will determine the quality of counselling and advising relationships in the native's personal and professional life. After marriage, a man will find his wife to be his most influential advisor in all matters. For a woman, her own mother remains her closest counsel. Excellent for women in legal and diplomatic practice (presuming Chandra is well disposed).

Remedial Ratna for Makara lagna Chandra (Mukta/moti - pure natural Pearl): Moon is the lord of the 7th, the Bhava of death (maraka Bhava) and is also an enemy of Saturn, the lord of the Ascendant. He should avoid wearing a Pearl. However, pearl can be worn if necessary, in the major period of Moon, [only] if Moon is in Cancer, her own sign.

Mars: Kuja is L-4+ L-11 measured from Makara lagna, Makara Chandra, and Makara navamsha. Mars is lord of the 4th and 11th. Mars is also an enemy of Saturn, and so Mars is a malefic planet for Makara lagna. Badhaka (L-11) provides trouble from the elder sibling, friendly association, the marketplace, the nature or method of setting goals, achievements and in profits.

If Shani occupies Vrischika or Mangala-Ketu are damaged in kundali, he has considerable troubles with the network of friends and associates and may suffer isolation or the identity is oppressed by sudden upheavals in the network or selfish actions of a mentoring friend.

Remedial Ratna for Makara lagna Mangala (Red Coral – Munga): Mars is lord of the 4th and 11th Bhava. He should avoid wearing Munga. However, by wearing of Red Coral in the major period of Mars, the native will acquire land, property and conveyances will get domestic harmony, happiness from mother and gains of wealth.

Mercury: Budha is L-6+ L-9 measured from Makara lagna, Makara Chandra, and Makara navamsha. Lagnesha Shani is a good friend of Budha. Mercury is lord of the 6th and the 9th, a trine house, which is the Moola trikona sign of Mercury and therefore is very auspicious. Venus and Mercury together are Yoga Karaka for Makara lagna. Budha is an especially strong influence in His natural domain bhava-6. As lord of bhava-6, Budha brings social conflict, exploitation, personal imbalance, mental resistance to marriage fidelity and a quick-witted avoidance of the terms of agreement. In particular, Budha can signify medical and mental issues with substance addiction and the psycho-physical consequences of sex (Budha) addictions. (Addiction indicates a behaviour which becomes so compulsory that it interferes with other necessary functions in work and family life.

As lord of bhava-9, Budha brings a mental engagement with religious principles and practices, affinity for sacred space, and personal priesthood. If Budha is well-disposed, the native has knowledge of the sacred scriptures of one's own religious tradition like Swami Vivekananda whose Budha in 1 had parivartana Yoga with L-1 Shani and like Jeddu Krishnamurti whose Budha in 5 had parivartana Yoga with L-5 Shukra. If Budha occupies a dualistic Rashi such as Mithuna or Meena, one may grasp the meanings offered by the sacred writings of traditions outside one's own, as well. If Budha is prominently placed, such as in lagna or with Chandra, the native's personality and behaviour may appear contradictory or at least enigmatic.

Remedial Ratna for Makara lagna Budha (Flawless Emerald – Panna): Mercury is lord of the 6th and the 9th, a trine house and is the Moola trikona sign of Mercury and therefore Budha is very auspicious. Consequently, wearing of an Emerald will be beneficial to him, particularly in the major and sub-periods of Mercury. As Mercury is a friend of Saturn, the lord of this Ascendant, the latter's gem stone Blue Sapphire can be worn always with advantage by him.

Jupiter: Guru is L-3+ L-12 measured from Makara lagna, Makara Chandra, and Makara navamsha. Lagnesha Shani is mutually neutral toward Guru. Jupiter will be lord of the 3rd and 12th Bhava, both inauspicious Bhava, and therefore Jupiter is not an auspicious planet and is malefic planet for this Ascendant. Guru L-12 is 6th from 7th, Badhaka or enemies of the marriage alliance.

Remedial Ratna for Makara lagna Guru (Pushkaraja - yellow sapphire): Jupiter is lord of the 3rd and 12th Bhava, and will, therefore, not be an auspicious planet for this Ascendant, and it will generally not be advisable to wear this gem stone by him. It can, however, be worn, if necessary, in the major period of Jupiter if he is in his own sign in the 3rd or the 12th.

Venus: Shukra is L-5 + L-10 measured from Makara lagna, Makara Chandra, and Makara navamsha. Lagnesha Shani is a good friend of Shukra. L-5+L-10 Shukra is a powerful Yogakaraka for Makara lagna. (Shukra is also L-4+L-9 yogakaraka for Kumbha lagna). Shukra yogakaraka L-5 (politics) + L-10 (governance) has natural capability to lead and govern, either by royal entitlement (5) or by charismatic charm that facilitates successful democratic election. He is empowered to make decisions which provide social order and lawfulness like Barack Obama. Shukra gives excellent results in regard to matters of children, politics, speculative ventures, romance, professional respect, leadership roles, and public duties. He will enjoy the company of women. Depending on Shukra full character (Rashi, drishti, etc.) periods of Shukra are usually very beneficial for the Makara native. A good position of Shukra in the Makara lagna chart can compensate for many other difficulties. Shukra is best in any Kendra, but especially in lagna where it makes the person physically attractive and brings much happiness from profession and children.

Remedial Ratna for Makara lagna Sukra (Flawless Diamond - Heera): Venus will respectively be lord of the 5th and 10th and the 4th and 9th Bhava and, therefore, Venus is considered an excellent and Yogakaraka planet for these Ascendants. He will do well in every sphere of their life by wearing a Diamond. In the major period of Venus, wearing of a Diamond is a 'must' for those who have faith in the divine power of gem stones. The results will become pronounced if Diamond is worn along with Blue Sapphire.

Saturn: Shani L-1+ L-2 measured from Makara lagna, Makara Chandra, and Makara navamsha. Shani is L-1, a trine house, and is Lagnesha and so a benefic planet. If Shani is well placed, he will lead a disciplined life and accrue stores of material wealth over time. The Maraka woman looking forward to her second marriage may consider wearing a beautiful blue sapphire on the longest finger of the left hand in order to encourage qualities of dignity, self-discipline, respect for order, and conventional behaviours in her forthcoming alliance with a second spouse.

Remedial Ratna for Makara lagna Shani (Blue Sapphire – Nilam): Saturn is the lord of the Ascendant and Blue Sapphire should be worn by him as a protective charm all their life. It will bestow good health, long life, prosperity, wealth, happiness and success in their ventures. The effects of Blue Sapphire will be strengthened if a Diamond, the gem stone of Venus, who is a Yogakaraka for these two Ascendants, is worn along with it. As L-2, Shani a natural maraka also becomes a temporal maraka; therefore one who wishes to postpone death should avoid the Shani-ratna and practice the Shani-scarcity austerities such as conservative food diet.

Rahu: Rahu is L-2 (Kumbha) measured from Makara lagna, Makara Chandra, and Makara navamsha. Rahu is co-lord of bhava-2 (Kumbha), therefore Rahu is a beneficial or benefic planet for Makara lagna and for matters of wealth acquisition, stock-piling and hoarding, learning in history and languages, second marriage, and expressions of voice and face such as song and appearance in story-telling imagery such as films. However Rahu in the Rashi of maraka Chandra is dangerous for Kumbha lagna. Rahu and Ketu are said to give positive results in the 3rd, 6th, and 11th bhava from the lagna or from Chandra. In addition, some authorities posit that since R-K are exalted in Vrishabha-Vrischika (per BPHS), R-K will give good results when Mangala and Shukra are well-disposed.

Remedial Ratna for Makara lagna Rahu (Gomedha): Rahu-ratna (gomedha) is a beneficial gem for matters of wealth acquisition, stock-piling, second marriage, and expressions of voice and face such as song and appearance in story-telling imagery such as films. Even for the wealth-and-knowledge target, the Rahu-ratna should be considered only if the lord of bhava-2 and its occupants are highly auspicious.

Ketu: Ketu is L- 10 (Vrischika) measured from Makara lagna, Makara Chandra, and Makara navamsha. Ketu is co-lord of bhava-10 Vrischika, and so Ketu is a benefic planet for Makara lagna. Ketu's effects for Makara lagna depend significantly on Ketu bhava, rashi, and drishti. Ketu amplifies the effect of any graha who are sharing Ketu Bhava; and also Ketu magnifies the effects of the lord of His occupied Rashi.

Remedial Ratna for Makara lagna Ketu (Cat's Eye – Vaidurya): Ketu-ratna may be a beneficial gem for matters of professional recognition and leadership responsibility but only if the lord of bhava-10 and its occupants are auspicious.

Planet Effects in Twelve Houses for Capricorn (Makara) Lagna

First House:

Sun: Lord of 8th house in Lagna, in enemy's camp indicates assistance in the accumulation of money and business of others. He/She will have legacies and gain through affairs and matters connected with the deceased, of rash temperament and whimsical nature, becomes angry when anything is said against him, disease of headache and eye troubles. The native will be greedy, disrespected and devoid from comforts of father in early age. If afflicted loss of money and reverse results of above good points.

Moon: Lord of 7th house in Lagna, in enemy's house aspecting own house denotes that one will gain through unions, partnership and love of women. He/She will be fond of music, blessed with pleasurable pursuits, average financial conditions and urinary troubles and diseases. The Moon is weak and afflicted, reverse results will be experienced, may enjoy short life.

Mars: Mars is exalted in Lagna, and is lord of 4th and 11th houses, will confer upon a native wealth, power and authority, of strong constitution, many servants, fortunate, inheritance, gain through land and property of parents. He/She will have real friends and supporters, victory over enemies and obstacles through acquaintances, fortunate actions and successful hopes, danger of fall from height during middle age, wound or scar mark on body, quarrelsome and inclined to quarrels, war or wrestling and may receive wound through them.

Mercury: He/She will have long journeys, learning, wisdom, prudence, fortunate with strangers and foreigners, wife's relations and voyages, grief from enemies, interest in science, invention law or political economy including philosophy, sickness and poor of health, unsuccessful business at times, may be devoid from the comforts of parents in early age subject to strength of Mercury and aspects on it.

Jupiter: He/She will have reversals in life from good position, under debt, ear trouble, worried and troublesome periods in life, journeys and removals, worries on account of brothers, neighbours, writings, learning and accomplishment, heavy troubles, secret sorrows and limitations. But all these trouble can be mitigated through occult science and spiritualism, danger of imprisonment or disablement requiring hospitalisation, medium span of life.

Venus: Venus is a Yoga Karka for Capricorn Ascendant being lord of

5th and 10th houses. He/She will be blessed with long life, dignity, power, honour and authority through merits and success, through industrious efforts, high ambitious, gain through mother, many love affairs, wealthy and licentious, victory over enemies, blessed with property and other's belongings, gain through children and games.

Saturn: The native is of sober temperament, high rank and position, endowed with property, engineer by profession, journeys, good company with wealthy persons, fond of scents, long life, power over enemies, good health, harmony, triumph over difficulties, money comes readily.

Rahu: He/She will have bestows wealth, honour, and favour through religion, educational or scientific affairs. It adds to the personality and gives opportunities for self expression.

Ketu: He/She will have short life, troubles and worries, loss of wealth in youth, scar or wound on the body, a worried life, scandals and it endangers the face and eyes.

Second House:

Sun: Lord of 8th house is posited in enemy's camp aspecting its own house indicates affliction of eyes, gain through partner and money of others. He/She will be fortunate in collecting debts, gain through deceased, no gain from relations, and disrepute in middle age.

Moon: He/She will have gain by marriage, contracts, law suits, little education. If Moon is weak and afflicted then loss through women, death of partner, public enemies, trouble in the eyes, enmity with relations and friends.

Mars: He/She will be bold and of rash temperament, gain and owner of agriculture land, scar or wounds or boils on the body, gain through deals in land and property, business and money by means of friends and acquaintances, through legislative interest or through one's new ideas.

Mercury: He/She will have money and gain through foreign merchants, science, learning, publication, travel, invention or banking, knowledge of foreign languages, well versed in mathematics and medicine. If it is combust or ill aspect, there will be reverse results.

Jupiter: He/She will be learned, intelligent and of strong will power, gain through educational affairs, short journeys, writings, music, also gain through secrecy and the occult, knowledge of law, respected and reputed and blessed with wealth.

Venus: He/She will be wealthy, blessed with mimicker, licentious and company of ill reputed people, gain of money by industry, the profession or Govt.; office, investment, pleasure and children. If afflicted, loss through speculations, pleasure and reverse results than above.

Saturn: He/She will have money through personal ingenuity and industry. One has to do hard work, and benefit through industrious efforts, loss of wealth through enemies, may become lunatic in middle of life. One becomes mischievous and liar, wife from poor family and loss during middle life also.

Rahu: He/She will have fortunate heredity, gain by legacy and gifts, science and learning, increase in possessions and bestows affluence.

Ketu: He/She will have misfortune in finances, in debt, loss and damage to estate, sorrows, fears and worry concerning money matters.

Third House:

Sun: He/She will have dangerous for parents and younger brothers, victory over enemies, disease of syphilis, danger of death on short journeys, unfortunate brothers and death of some, psychic and mysterious experiences.

Moon: He/She will have many sisters, low of brothers and sisters, fond of dance and music, ear trouble, enmities with some of brothers and neighbours, marriage to a relative or neighbour, difficulty through writings or contract, legal or religious disputes, troubles on short journeys.

Mars: He/She will be bold and brave, trustworthy and of rash temperament, fond of narcotics, gain through brothers and friends, cordial relation with relatives and neighbours, friends through writings and journeys.

Mercury: He/She will have many brothers, learning, accomplishments and progress through research, travel, investigations, explorations, travel or writings, sickness of some brothers and relatives, interest in studies and method of healing and industrial economy, likelihood of mental disturbance when sick.

Jupiter: He/She will have benefits through education, writings, short journeys and brothers, many brothers, many journeys in early age and sweet speech, disappointments, sorrow and troubles through relations, friends or neighbours, occult learning, may become the victim of impotency disease, brothers will be wealthy and respected, troublesome short journeys.

Venus: He/She will be careful for his respect, respect among relations and neighbours, gain and honour and advancements through partner's relations, honour through short journeys, writings, business trips and Govt;, or commissions, fond of music, dance and drama, gain through opposite sex, many sisters who will be prosperous and wealthy.

Saturn: He/She will be agriculturist, average relations with brothers, average profession, and victory over enemies, mental development, opportunities delayed, and gain through education, writings, brothers, neighbours and short journeys. If afflicted losses through above.

Rahu: He/She will have many brothers, careful for his respect, respected and skilled in many arts, early life will be comfortable, in middle age, and one will meet with some accident or grief, old age be will troublesome, conduces to spiritual or educational matters, gain through brothers, neighbours and journeys.

Ketu: He/She will have mental anxiety, troubles with brothers and neighbours, unprofitable journeys.

Fourth House:

Sun: He/She may be a building contractor, gain in property through the deceased, probably by the death of parents, less comforts from conveyance, troublesome for parents, and of suspicious nature. If afflicted death of parents, danger through fall and falling buildings, storms and floods, troubles over inheritance, land and property.

Moon: He/She will have Property by marriage, a chaste partner and happy married life, gain through partnership. If afflicted and weak then short life, early death of mother, litigation over property and robbery of house.

Mars: He/She will have gain in property and through old people or antiquities benefit and profit at birth place, gain from mother, inheritance through friends and fortunate in property, love for brothers.

Mercury: He/She will have gain through partner's relations, journey's home to die, sickness of father through journeys, sickness of native due to troubles and domestic unhappiness and trouble through servant, incomplete education, skilled in many arts and changeable professions, dependent on others.

Jupiter: He/She will be blessed with long life, travelling and writing in connection with home affairs and property, troubles through mother, secret sufferings, restriction or limitations at the end of life.

Venus: He/She will be cheerful, owner of land and orchards, victory over enemies, property and blessed with long life, gain and honour through parents, land and property, old age will be happy, interest in reclamation, colonization, horticulture, mining or architecture.

Saturn: He/She will be restless, not well educated, gloomy, and unhappy, without property and conveyance, no comforts from parents in middle life, ancestral property may be destroyed, success late in life. Occult investigation, gain through lands or mines.

Rahu: He/She will have travel of foreign countries; gain through property and in an unexpected manner, fortune in discovery, long lived and trustworthy.

Ketu: He/She will have loss of confusion with land and buildings, waste of patrimony, jeopardizes the esteem and credit, family discord and troubles in life of elders.

Fifth House:

Sun: This is an unfortunate position of Sun and indicates unfortunate children or death of some. He/She will have danger through excessive pleasures, speculation, children and their affairs, gain through ancestral property, loss through Govt.

Moon: He/She will have many daughters, scholar of occult science, wealthy and will lead a comfortable life, learned and intelligent, religious and of sanguine temperament.

Mars: He/She will have gain through parents, much happiness and gain through children, friends and beneficial circumstances, skilled in use of fire arms, military service and fond of wrestling. If afflicted reverse results.

Mercury: He/She will be liberal and unconventional ideas, free living, child by strange consort, pleasurable journeys, learned and intelligent, voyages, air flights, sports, foreign investments and speculations, many enemies, sickly children and illness through pursuits of or over indulgence in pleasure and sports, skin diseases.

Jupiter: He/She will be learned, intelligent, respected and famous, gain from Govt; and persons in authority, religious and pleasure journeys, mental pleasure through children, brothers, reading, study and travel to pleasure resorts.

Venus: He/She will have much pleasure, love affairs and adventures, speculation and investment, prosperous children and pleasure through them; children will rise to rank and honour, gain and honour through speculations, pleasure, sports or stage.

Saturn: He/She will be delight in pleasure, amusements, sports, speculation and children with tendency to success through these gain by speculations, investment, pleasure entertainment, young people and children. If afflicted, no comforts from children loss and grief through women, early life good, middle average, third part of life unhappy and full of troubles, disturbed life due to wife and children.

Rahu: He/She will be fortunate for children and blessed with long life, wards off many troubles, calamities and dangers. The native gains some public appointment or office and is fond of recreations and sports, wealthy and respected.

Ketu: He/She will have many enemies and opponents, may deny children or gives abortion, the children will be disobedient and will face adversity, excessive or irregular pleasures produce much harm, limited education, gain of wealth through unfair means, the native may be quarrelsome.

Sixth House:

Sun: He/She will have dangerous sickness, death, of pets and servants, loss of money earned by employment and through those in whose custody it may be, troubled and worried life, worried and under debt, a diseased native. The parents of the native will face sickness, loss in profession.

Moon: He/She will have marriage below status, sickly and evilly disposed wife, many enemies and opponents, loss through thieves, partnership and contracts, loss and disrespected due to opposite sex in middle of life.

Mars: He/She will have troubles through enemies, danger of wound, loss of profession through sickness, and servants, may build his own house, friends among working people, army, Air or Navy, faithful servants, interest in social welfare, loss of agriculture land, may be there is some effect in arms or feet, many troubles and turmoil, danger of being bitten through reptiles, a surgical operation.

Mercury: He/She will have a weak and diseased constitution, troubles through law suits, danger to respect, sickness through travelling, study of science, medicine, healing, difficulty in work in foreign land or in export business, some severe illness. When aspect by Jupiter, there will be gain through service, employment, poultry, medicine and social service. The native will be popular and will be respected by others.

Jupiter: He/She will have sickness or injury through journeys, difficulty through brothers, interest in study or social economy, difficulty and troubles in profession and through employees, limitations on account of sickness, troubles and worries due to law suits, danger to respect may be entangled in some criminal suits and under debt.

Venus: He/She will have modest worldly position, gain and honour through service, employment, the practice of healing, army or navy affairs, sickness among children, money through careful speculation or by income of children, many enemies and opponents, pain in chest and a scar mark on the body, danger from water, vice habits. Old age will be pleasurable.

Saturn: He/She will have victory over numerous enemies, thievish and a liar, adulterous, mischievous, owns a factory and may be a shareholder in many factories, a good healer, gain through humanitarianism, food, clothing and employees, gain through inferiors.

Rahu: He/She will be good health and strong body, faithful and honest employees, gain through service, fortunate through service, and father's relations and by small animals.

Ketu: He/She will have various afflictions to health and sickly, crossed and deceived by servants, gain through unfair means, loss through small animals, danger through bite of insects, reptiles or animals, danger to respect or punishment in middle age.

Seventh House:

Sun: He/She will have a rich partner or one to whom money comes unexpectedly. If afflicted death of parents and wife, public enemies, danger of death by violence, suicide, accident or war, less comforts from relatives and friends, licentious, obstinate and of rash temperament, danger to respect in middle age, loss in profession.

Moon: He/She will have success in law suits, marriage in good family, a good partner but likely to become cold or proves untrue or hostile. If Moon is weak or afflicted then reverse results. Also one makes many journeys, windy or cold diseases.

Mars: He/She will have gained through marriage, land and women generally, a loving partner with desirable friendships and social connections, success in law suits. If afflicted then wife will be adulterous, unfaithful and make divorce or separation will take place, loss in partnerships, law suits and loss through marriage, diseases of blood are indicated.

Mercury: He/She will have quarrelsome, law suits with servants, difficulties with disrepute women, sickness and troubles with employees, public enemies through religious or scientific persons, and marriage to a stranger whose relatives may oppose. If afflicted or combust reverse results.

Jupiter: Jupiter is exalted in this house. He/She will be marriage as a result of journeys, writings or to one of relatives, troubles through deceit, contracts and law suits. But the native will be educated, learned and respected. The wife will be faithful, learned and helpful. He/She will have comforts from brothers, rank and authority through Govt.; if afflicted reverse results are indicated.

Venus: He/She will be pleasure, close association or understanding with partner but discord with children, loss through theft. He/She will have gain through law suits and dealing with public, honour and reputation through marriage and partnership in a responsible concern, knowledge of law and gain through that.

Saturn: He/She will have partnerships and close association with others, fond of opposite sex, gain by marriage, contract, business and dealing with others especially with women. If afflicted, loss in business and profession will be sickly and of opposite temperament, adulterous, suspicious, loss of money through law suits, unions and open enemies, marital unhappiness.

Rahu: He/She will have decreases the number of enemies, increases profit through women. It is a testimony for a wise and wealthy partner.

Ketu: He/She will have many enemies, opponents and loss through them, contentions and difficulty with partner, changeable profession, happy after middle age.

Eighth House:

Sun: He/She will have gain by the dead, blessed with long life, intelligent, lucky but of average rank and power. Early part of life will be happy whereas old age will be comparatively unhappy, may face troubles and turmoil, spiritualistic experiences, a comfortably fixed partner, a natural death, interest in matters of future life.

Moon: He/She will have money or property by marriage, may there be death of partner, some inheritance but difficulty over it. If weak and afflicted then short life, unhappy from wife, loss of money through marriage and partnership.

Mars: He/She will have gain through inheritance, death of friends, gift and gain through legacies from friends, an easy death. If afflicted then indicates short life, death due to sudden accident, fall or through some wound, danger to respect, loss through above.

Mercury: He/She will have broken and incomplete education, gain through long journeys, psychic experiences, prosecution regarding religious, scientific or educational connections, dangerous illness like bronchitis, death of servants, financial gain and reward for faithful service, a lean and weak constitution. If afflicted reverse results.

Jupiter: He/She will have journeys due to troubles, false accusations, sickness or death of brothers, an unsatisfactory end, many misfortunes, secret, enemies die, troubles over inheritance, long life, under debt and troubles in life.

Venus: He/She will have loss through enemies, speculations or gambling, suffers through children who may die before the native, gain and honours in dealing the estate and money of others, also through law suits, legacies, inheritance, of dead persons, unhappy from wife, diseased in 6th year and unsuccessful hopes.

Saturn: He/She will have occult experiences, death through irregularity gain by legacy and goods of the dead or through money of partner, many difficulties, assumption on many counts, displeasure from the Govt., and officers, grieved from children, disrepute, short life if afflicted and death at unknown place.

Rahu: He/She will have promotes health and conduces to longevity, testimony for gifts and legacies, and gain by the dead, intelligent and clever.

Ketu: He/She will have loss of money through deception, sudden or violent death, danger through bite of insects or reptiles, disease and trouble in middle age and also unhappy, danger of disrepute.

Ninth House:

Sun: He/She will have generous, charitable, respected and blessed with power through Govt., or officers, no gain from ancestral property, no gain from parents. The partner has troubles with relatives over money matters, foreign travel, death in a distant land, danger of death by drowning or while on voyage.

Moon: He/She will have gain through marriage and through partner's relatives, partner's journeys to foreign land, fond of business and agriculture, gain through inheritance, contracts, and partnership, fond of travels, careful of self respect and of changeable moods. If Moon is weak or afflicted then reverse results.

Mars: He/She will have gain through science, religion, long journeys and wife's relations, success and profit through long journeys, friendship through travels and learning, friends among educators, ministers, writers, explorer and skilled, disrespected and dishonest, may commit fraud in charitable funds.

Mercury: He/She will have prophetic dreams, many fine qualities, good in mathematics and an artist, a good painter and an architect, gain through ancestral property, travelling for education, scientific or religious purposes, a respected and learned native, danger through over study, fond of opposite sex, comforts in last part of life, sickness abroad, at sea or while travelling, illness of wife's relatives, work in connection with foreign affairs or universities. If it is weak, afflicted or combust then indicates reverse results.

Jupiter: He/She will have professional or honourable journeys, respected and learned, honour and gain through journeys, writings and other accomplishments, long journeys, secret missions, investigation, research, secluded work, sacrifice for science or religion, rank and authority, may be a founder of charitable institutions and inns.

Venus: He/She will have honourable voyage, professional journeys, honours through learning, writings, publishing, research or philosophy, dutiful children who travel and give pleasure and learning to the native, children may become preachers, scientists or explorers, pleasure in foreign land. If afflicted reverse results.

Saturn: He/She will have long journeys religious or psychic experiences, liking for science, law invention and all matters connected with higher mind, prophetic dreams or visions, gain through books, trading and export work. If it is afflicted then, there will be waste of ancestral property, adulterous and disrespected, death of parents in childhood and irreligious.

Rahu: He/She will have success in educational, legal or religious studies, good mental qualities, favourable for voyage and foreign affairs, prophetic institution and true dreams.

Ketu: He/She will have unfortunate voyages, curious dreams, unreliable premonitions, trouble and danger of imprisonment in foreign lands, not happy relations with parents and relatives, not religious, troubled and diseased.

Tenth House:

Sun: He/She will have rise to high position in religion, others will gain through him, gain through inheritance, be careful that Sun in this house as it is in his debilitated sign and will not be so beneficial. If afflicted danger of violent death through Govt., order or war, business through matters of dead and reverse results.

Moon: He/She will be fond of travel, respect and careful for that, an honourable partner gainful to professional career. If it is weak and afflicted then, there will be troubles by opponents through Govt., or officers, honour or enemy, no substantial gain through inheritance, may remain under debts.

Mars: He/She will have gain through trade, profession, public or Govt. work, or by fruits of the earth, friendship among those in good position, gain and honour through them, also through social standing, business of Govt. circles, and many enemies. If afflicted reverse results.

Mercury: He/She will have victory over opponents and enemies; skilled and intelligent, honour, credit and respect through science, literature and travel, success in foreign affairs, honour through service, municipal or national activities, gain through ancestral property, a good status in life, respected and famous. If afflicted, weak or combust, then results like sickness through honour, sorrows or afflictions due to mother or father-in-law and reverse results or above.

Jupiter: He/She will have gain by occupation, power, rank and authority, obedient to parents, also through profession, merchandising or Govt. or wife's parents, but of suspicious nature, respected and famous, a good comfortable life. If it is weak or afflicted, there will be loss through employer, superiors and Govt. resulting in disgrace and disrepute, all this will result in long journeys or retirement and seclusion.

Venus: He/She will have gain and honour through profession, honorary office or Govt. employment, success through parents and opposite sex, honourable children, renown in speculations, business or theatrical world, gain through in-laws, a native of good authority and power, learned and intelligent, will spend on charitable and noble causes, wealthy and blessed with property, favourite of people.

Saturn: He/She will have gain of high position, successful administrator and wealthy, rise and high social status, by occupation, profession Govt. or in-laws. When weak or afflicted then financial loss through employer displeasure of superiors, or suffers dignities, limitation because of rights withheld, liable to disrespect loss of parents or trouble through them.

Rahu: The native will achieve honours, credit and high position by merit of industry and ability, fond of travel.

Ketu: He/She will have loss of position through deception, treachery and adverse public conditions such as sudden depressions, changes or failures.

Eleventh House:

Sun: He/She will have gain and legacies through friends, death among friends. One will be greedy, adulterous and devoid of ancestral property, careful for his respect, average profession.

Moon: He/She will have friends become public opponents or enemies, marriage to a widow or widower with children but liable to trouble through them, unsuccessful hopes, no gain through parents, an unlucky native.

Mars: He/She will have large circle of friends and gain through them, much pleasure in life, successful hopes and wishes in old age, good relations with brothers, and friends among unique, ingenious, or radical people. If afflicted and weak, there will be reverse results like separation and strained relations with brothers, enemies among friends, loss through trade and ancestral property.

Mercury: He/She will have fortunate friendship on voyages and friends among foreigners and foreign lands, acquaintances and gain through travellers, scientists and legislators and interest in political social and legislative activities. If afflicted, combust or weak then there will be reserve results like sickness among family and friends, grief through brothers and unsuccessful hopes.

Jupiter: He/She will be respected and famous, religious and God fearing and will earn money through righteous means, friends through journeys, gain through brothers, progressive studies, hopes and wishes will be attained. If afflicted, there will be loss through advice of friends, unfortunate undertakings, great disappointment and obstacles, deceitful friends etc.

Venus: Native of sweet spoken and intelligent, comforts of conveyance and property. There will be gain through opposite sex, eminent friends and legislators and persons in authority. He/She will have an honourable fortune, ambitious ideals, successful hopes, love of children. If afflicted reverse results.

Saturn: He/She will have gain by friends and accidental fortune, wealthy, large circle of friends, much pleasure in hopes and gain by the success of employment. If afflicted, friends will be detrimental and defeated hopes, short life, bad temperament, doubtful nature, licentious, disrepute, no gain from education and loss in business.

Rahu: He/She will be meritorious, friendships, acquaintances, successful hopes and wishes, voyages, intelligent and learned, skilled in a few arts, rise as life advances and good promotions and comforts in life.

Ketu: He/She will be undesirable associations, loss of opportunities, defeated hopes, wrong advice and false friends.

Twelfth House:

Sun: In early age, one will lead a comfortable life, respected and long journeys, in middle age, grief from children, downfall, financial troubles, loss of property and disrepute. He/She will have difficulty over inheritance, death of secret or private enemies, great sorrows, and fear of death or imprisonment.

Moon: He/She will be weak and diseased, unhappy marriage, secret sorrows, jealous, vexation, sickness through opponents or fear of imprisonment. If Sun is afflicted, there will be death at the hands of enemies.

Mars: He/She will have many journeys, loss of ancestral and self earned property, afflicted health, end of life in seclusion or devoted to the study of occult science, deceitful friends, good friendship through occults, pleasure in peaceful, quite, harmonious or secluded places.

Mercury: He/She will have many journeys, careful for his respect, loss of money; disease of bronchitis in old age is indicated, fear of imprisonment, sickness, work of secret nature, difficulty and sorrow through religion, science and journeys. In middle or latter part of life seek seclusion for self development, occult science and may take long journey for the same. If a writer or inventor, he will hardly complete his work due to many obstacles, and get it to public.

Jupiter: He/She will have occult abilities, learned, sweet speech, pilgrimage, generous and well reputed, may be poet, charitable, powerless enemies, disappointments and obstacles, adversities will enable him to develop inner mind growth and understanding, grief through brothers, secret sufferings, seclusion and danger of enmities or imprisonment.

Venus: He/She will have unexpected gain of rank and position, sorrow through children, speculations, mysterious research, friends may be secret enemies and cause loss of position, office or honour, unfortunate environments, benefic use through occult science.

Saturn: He/She will have gain through occult science and secret missions, success in middle life, no fixed profession, wastage of money in licentious deeds, service or income through law people and illegal means, irreligious, fear of imprisonment, secret unhappiness, enemies but benefit and gain through occult.

Rahu: He/She will have gain by secret methods or in seclusion and through occultism, fond of journeys, when afflicted reverse results.

Ketu: He/She will have loss and troubles through secret enemies, liability to imprisonment or restraint and restrictions, unfavourable to health, relations with low people, grief through children and a troubled life.

11.11 Aquarius (Kumbha) Lagna

Jupiter, Moon and Mars are malefic, while Venus and Saturn are auspicious. Venus is the only Planet that causes Raja Yoga. Jupiter, Sun and Mars are killers. Mercury gives meddling effects.

Sun: Surya L-7 measured from Kumbha lagna, Kumbha Chandra, and Kumbha Navamsha. Sun is L-7, a Maraka house (house of death), and is also an enemy of Saturn, the lord of the Ascendant, so Sun is malefic for Kumbha lagna. Ravi as L-7 controls the marriage house, therefore for Kumbha lagna the condition of Ravi in the nativity becomes a strong indicator of the character of the spouse.

Career: The dominant indicator of career is lord of 10th navamsha, and lord of 10th radix; however Surya = a lesser karaka for career and public recognition.

If L-7 Surya is strong, he undertakes career roles in consulting, advising, counselling, arbitration, mediation, and coordinated work such as partner dancing, partner skating, professional tennis, and various types of business partnerships.

Remedial Ratna for Kumbha lagna Surya (Manika - pure Ruby): Sun is lord of the 7th, a Maraka house (house of death) and Sun is also an enemy of the lord of the Ascendant. It will be, therefore, advisable for him not to wear Ruby. However, if the Sun is lord of Seventh in his own sign, Ruby may be worn in the major period of Sun. According to 'Bhawarth-Bhavam' theory of Hindu Astrology, the seventh house is 10th to 10th house and wearing of Ruby in addition to enhancing the good effects of the Seventh house, will also help.

Moon: Chandra is L-6 measured from Kumbha lagna, Kumbha Chandra, and Kumbha Navamsha and so Moon is the highly inauspicious agent and is a Problematizer planet. Chandra is enemy of lagnesha Shani, and unfortunate lord of the maha-dusthamsha Ripu bhava, so Chandra is the most malefic planet for Kumbha Lagna. The divorce is a common occurrence for Kumbha, due to the L-6 Chandra. Moon as L-6, 12th-from-7th (alliances, agreements and trusts) indicates dissolution of contract, divorce, disagreement and distrust. Most acute results will be during Chandra Mahadasha or Vimshottari periods of graha in shad-ashtaka 6-8 angle to Chandra. He may feel often beset with jealousy and mood swings, digestive problems and body pain and financial problems due to indebtedness against taking of loans. The child may be accused of a parental strategy to control over a child being used as a servant. If Rahu in 6 or Rahu involved with Chandra, higher likelihood of criminal extremes such as human's trafficking.

Marriage and Alliances: Kumbha lagna and Kumbha Chandra unbalances relationships between individual partners, between individuals and various groups, and between the individual and society.

Social isolation or Fringe identity: The Kumbha native feels that he does not belong to the society. Chandra occupies swakshetra-6, leading to an exceptional career in maternal-child medicine or the care of the service class. If Chandra is in the uchcha Vrishabha (bhava-4), he may become a happy and prosperous owner of a great many real-estate properties. However no matter what material contentment the Kumbha native may find, it will fluctuate according to the nature of Chandra. One tends to get in and out of relationships very quickly, due to the chronic emotional need to find one's place in life but followed shortly by the emotional reaction to the inevitable loss of agreement that is the hallmark of the lord of Ripu Bhava. Chandra is an especially difficult graha for Kumbha Chandra = L-6 from both radix lagna (material animosity) and Chandra lagna (emotional imbalance). Kumbha natives are emotionally oriented toward providing service in exchange for security and protection.

Remedial Ratna for Kumbha lagna Chandra (Mukta moti - pure natural Pearl): It not recommended wearing natural Pearls for remedial purposes, and avoiding wearing of the same. Chandra is enemy of lagnesha Shani, plus Chandra is unfortunate lord of the maha-dusthamsha Ripu bhava. Moon is the highly inauspicious agent of "emotional discord" as L-6 and a karaka for disagreements with or exploitation by the Mother. He should scrupulously avoid wearing Pearl.

Mars: Kuja is L-3 / L-10 measured from Kumbha lagna, Kumbha Chandra, and Kumbha Navamsha. Kuja as L-10 (Kendradhipatidosha) is reasonably auspicious at birth and is a benefic planet for **Kumbha lagna.** Vital and competitive Kuja controls daily mental process and leadership roles and he has strong administrative business inclinations.

Remedial Ratna for Kumbha lagna Kuja (Red Coral – Munga): Mars is lord of the 3rd and 10th houses. Kuja as L-10 (Kendradhipatidosha) is reasonably auspicious graha at birth. It may be OK to apply coral and other gems of Kuja. Red Coral can be worn in the major period of Mars for promotion and success in the professional field. If, however, Mars is in his own sign in the 10th, this disposition will give rise to Ruchaka Yoga and he can wear a Red Coral with advantage whenever necessary.

Mercury: Budha is L-5 / L-8 measured from Kumbha lagna, Kumbha Chandra, and Kumbha Navamsha. Saturn is a friend of Mercury and on account of his lordship of a trine; Mercury is accepted mostly as an auspicious and is a benefic planet for **Kumbha lagna**. Well disposed and strong Budha provides children, intelligence, occult, politics, charisma, games and gambling - winning prizes, awards and trophies, fortune and luck, creative arts, speculation, prestige, fame, self-expression and Theatre centre-stage roles. If Budha is not well disposed, he may develop blustering and boasting nature about knowing occult secrets, detecting hidden resources and making transformative changes.

Remedial Ratna for Kumbha lagna Mercury (Flawless Emerald - Panna): Mercury is lord of the 5th and 8th houses. On account of his lordship of trine Mercury is accepted mostly as an auspicious planet for this ascendant. The lord of this Ascendant Saturn is a friend of Mercury. Therefore, if Emerald is worn with blue sapphire it will prove very beneficial to the native. Similarly beneficial will be the combination of Emerald and Diamond as Venus being a friend of Mercury and a yoga karaka for this Ascendant on account of his lordship of the 4th and the 9th houses. The same applies in the case of Capricorn Ascendant where also Venus is a yoga karaka planet on account of his lordship of the 5th and 10th houses.

Jupiter: Guru is L-2 + L-11 measured from Kumbha lagna, Kumbha Chandra, and Kumbha Navamsha. Jupiter is lord of the 2nd (Maraka) and the 11th (inauspicious house) and Saturn, the lord of this

Ascendant, is neutral toward each other, however, Jupiter is accepted as an inauspicious and is a malefic planet for **Kumbha lagna. However** Jupiter is lord of the 2nd (Wealth house) and is posited in own house, it forms a wealth yoga. There are four main wealth houses: 2 = banked savings; 5 = gambling or speculative wins; 9 = assets provided to be used for acts of faith; and 11 = earned income.

Wealth: Presuming Guru is fairly well disposed or in own sign, periods of L-2+L-11 Guru generate considerable wealth via family-of-origin hoarded assets and knowledge-values, as well as via earned income from the marketplace of goods and services.

Remedial Ratna for Kumbha lagna Guru (yellow sapphire – Pukharaja): Apply the pukharaja ratna with caution, appreciating that L-2 is a powerful maraka. Although a maraka graha, Guru will normally not kill the native early in life, unless Brihaspati is also lord of 2nd or 7th from Chandra. Therefore Kumbha natives should avoid the "Pukh-raj" when Chandra occupies Vrischika, Kumbha, Kanya, or Mithuna. Otherwise, According to BPHS ch 43, the period of Chandra-Guru can be fatal for the Aquarius Ascendant. Saturn, the lord of this Ascendant, is an enemy of Jupiter. Still yellow sapphire can be worn in the major period of Jupiter particularly when it is in his own sign in the 2nd or the 11th, to gain wealth, children and other comforts. However, those who are apprehending death because of old age or some illness or on account of indication of short life in their horoscopes, should not wear Yellow Sapphire as Jupiter being lord of the 2nd house, is a death inflicting planet for this Ascendant.

Venus: Shukra is L-4+ L-9 measured from Kumbha lagna, Kumbha Chandra, and Kumbha Navamsha. L-4+L-9 Shukra are a powerful Yogakaraka for Kumbha lagna. As a natural benefic L-4+L-9, Shukra becomes a potent and positive Yogakaraka in control of schooling (4), attainments in arts and music (Shukra), property and vehicle ownership (4); relationships with priests and professors (9), and privilege of access to sacred teachings (9). A good position of Shukra in the Kumbha lagna chart can compensate for many other difficulties. Shukra is best in any Kendra, but especially in lagna where it makes the person physically attractive and brings much happiness from property ownership, interior decorating, architectural practice and also from global scholarly travels in the context of alliance-building and acquisition of higher knowledge.

If strong Shukra occupies bhava-5, he obtains the most attention from an admiring public who are entertained and amused by his drama (acting).

Remedial Ratna for Kumbha lagna Shukra (Flawless Diamond - Heera): Venus is lord of the 4th and 9th houses; therefore, Venus is considered an excellent benefic and Yogakaraka planet for Kumbha lagna. He will do well in every sphere of their life by wearing a Diamond. In the major period of Venus, wearing of a Diamond is a 'must' for those who have faith in the divine power of gem stones. The results will become pronounced if Diamond is worn along with Blue Sapphire. White Diamond is very auspicious for this native, particularly good for schooling, diplomas, certificates, licenses, deed of title to property.

Saturn: Shani is L-1+ L-12 measured from Kumbha lagna, Kumbha Chandra, and Kumbha Navamsha. Shani as lagnesha controls both identity and loss of identity and is the agent of acquisition of a great inner peace. He is prone toward meditation and pilgrimage. Even in middle age, undertaking long sojourn abroad during Shani period is easy for the Kumbha native due to the identity-dissolving effects of 12 combined with the new-clothes or new-attributes effect of 1.

Remedial Ratna for Kumbha lagna Shani (Blue Sapphire - Neelam): Saturn is the lord of the Ascendant. Blue Sapphire should be worn by him as a protective charm all his life. It will bestow good health, long life, prosperity, wealth, happiness and success in their ventures. The effects of Blue Sapphire will be strengthened if a Diamond, the gem stone of Venus, who is a Yogakaraka for this Ascendant, is worn along with it.

Rahu: Rahu is co-lord of Kumbha; L-1 measured from Kumbha lagna, Kumbha Chandra, and Kumbha Navamsha and is friendly with Saturn, lord of Kumbha. Rahu is a benefic planet for Kumbha Lagna. Rahu effects depend significantly on Rahu bhava, rashi, and incoming drishti. Rahu amplifies the effect of any graha who are sharing Rahu house; and also Rahu magnifies the effects of the lord of His occupied Rashi. Rahu and Ketu are said to give positive results in the 3rd, 6th, and 11th bhava from the lagna or from Chandra. In addition, some authorities posit that since R-K are exalted in Vrishabha-Vrischika (per BPHS), R-K will give good results when Mangala and Shukra are well-disposed.

Remedial Ratna for Kumbha lagna Rahu (Gomedha): Rahu is co-lord of bhava-1; Kumbha, therefore Rahu-ratna (gomedha) may be a beneficial gem for matters of personality development and the vitality of the physical body, if and only if the lord of bhava-1 (Saturn) and its occupants are auspiciously placed.

Ketu: Ketu is L-8 + L-10, i. e. co-lord of 8th and 10th measured from Kumbha lagna, Kumbha Chandra, and Kumbha Navamsha. Ketu as

L-10 (Kendradhipatidosha) is reasonably auspicious at birth and is a benefic planet for **Kumbha lagna.** Ketu effects depend significantly on Ketu bhava, rashi, and drishti. Ketu un-connects the effect of any graha sharing Ketu house. Ketu scatters the effects of His planetary lord. Ketu is the natural co-regulator of bhava-8.

Remedial Ratna for Kumbha lagna Ketu (Cat's Eye - Vaidurya): Under certain carefully assessed circumstances, if the lord of bhava-10 and occupants of karma bhava are auspiciously placed, and a "scattering" non-discriminating effect could be perhaps helpful, the Cat's Eye might aid special cases of iconic public recognition (10). These are rare circumstances but they do occur.

Planet Effects in Twelve Houses for Aquarius (Kumbha) Lagna

Sun: He/She will have sun is lord of 7th house when posited in Lagna and aspects own sign of the 7th house indicates late marriage, long life, wealthy reputed and respected, connections with the process of law, public enemies, unions, partnership, love of women and benefit through them, enmity with people, no permanent residence and will remain in journeys and diseased. If afflicted, reverse results than good.

Moon: He/She will have very careful for his respect, financial position will be average, many enemies, may get short life, diseased and little education. If weak and afflicted, then sickness through irregularity, servants prove troublesome or unprofitable.

Mars: He/She will have preferment and dignities, honours through merit, success through efforts, high ambitions, gain through mother and Govt. journeys and removal, concern with affairs of brothers and neighbours, writings, learning and accomplishments. If Mars is afflicted, there will be reverse results.

Mercury: He/She will have contented, learned and intelligent, skilled and fond of arts, weak health and of short age, pleasure through children and speculation, many love affairs, loss of wealth in youth, limited ambitions, accumulation of money, and gain through the matters of dead. If weak afflicted and combust, reverse results are expected.

Jupiter: He/She will have gain through real friends and supporters and can overcome obstacles and enemies, fortunate and successful in hopes. Money comes readily. He/She will be respected, reputed and blessed with long life. Last part of life will be happy, trouble through disease. If afflicted and weak, then reverse results.

Venus: Venus is a Yoga Karka planet for this ascendant and when posited in Lagna, He/She will be fortunate inheritance, gain through land, property and parents, interest in domestic science, long journeys, learning, wisdom and prudence, fortunate with strangers, foreigners, voyages and wife's relations, interest in science, law or political economy. If Venus is afflicted, reverse result will be realised.

Saturn: He/She will have blessing with dignity and victory over enemies, a long fortunate life, good health, harmony, victory over difficulties, secret sorrows, vile deals, gain through occult power and studies, danger of imprisonment or disablement requiring hospitalisation. He/She will be gloomy, sad, licentious and adulterous, early life will be happy, loss through enemies, devoid from comforts of brothers in middle life, average length of life.

Rahu: He/She will have bestows honour and wealth, favour through religious, educational and scientific affairs. It adds power to the personality. But he/she is inclined to malicious sex acts, amorous and fond of company of women.

Ketu: He/She will have short life, tribulations, loss and scandals, eyes and face and be afflicted, of not commendable character and of vile habits, avaricious, industrious, voluptuous and worried, no domestic happiness.

Second House

Sun: He/She will have gain by marriage, wealthy but with afflicted eyes, loss and trouble through enemies and opponents, last part of life will be happy. If afflicted, then loss of money through unions, partnership, contracts, law suits and women, death of partner and public enemies.

Moon: He/She will have money through service, employees and small animals, clever and intelligent, author of books, acquisition of property. If Moon is weak and afflicted, then there will be financial losses through sickness and servants.

Mars: He/She will have money through educational affairs, short journeys, writings, and music, gain through industry, and trade profession of Govt. office. If Mars is afflicted, reverse results add troubles with relations over money matters.

Mercury: He/She will have Gain through investment, pleasure, children and other's money, fortunate in realising debts and gain through the dead. If afflicted loss and trouble through other matters.

Jupiter: He/She will have money and business through friends, legislative interest or through own hard work and planning. When Jupiter is afflicted, there will be heavy losses and much work but little gain.

Venus: He/She will have fond of music and dance, sweet and soft spoken, licentious, gain through women in middle part of life, gain through dealings of land and property, foreign merchants, science, learning, publication, travel or banking, if afflicted, reverse results.

Saturn: He/She will have gain of money through hard work and industrious efforts. Also gain through secrecy and occult, blue eyes and sexy by nature, average education and wealth, destroyer of ancestral property.

Rahu: He/She will have bestows fortunate heredity, gain by legacy and gifts, science and learning and increase in possessions, respect and comfort in old age. He/She becomes untruthful, insincere and talkative, opposes others, may expect assault from enemies, early life will be unhappy.

Ketu: He/She will have unfortunate in finances, indebtedness, loss and damage to estates, sorrow, fears and worry concerning money matters, may cause some defects in eyes, mouth diseases.

Third House

Sun: He/She will be bold and energetic and has difficulty through writings or contracts, legal or religious disputes, trouble on short journeys, strained relations with brothers, licentious and under the influence of opposite sex, limited brothers and he/she may become cause of worry for them especially for the elder brothers.

Moon: The health of brothers will be afflicted and native has to make many journeys on this account. He/She will have Interest in studies and in industrial economy, more sisters than brothers unless Moon is not aspect by Jupiter. He/She will remain happy unless Moon is not weak and afflicted.

Mars: He/She will have gain through learning, short journeys, writings and brothers, gain in honour and advancement through partner's relatives, respected among relations and neighbours, honours through short journeys, writings and accomplishments, business trips, Govt. commission, many brothers and the native will be under the influence of opposite sex.

Mercury: He/She will have pleasure through children and relations, liking for travels, sports, drama and adventure, psychic and mysterious experiences. Early age will be comfortable, pain in the ears, unhappy in old age, investigation about death and the continuity of life, strained relations with brothers but gain form a few, and will leave the house at the age 10 or 12 years, may become victim of bronchitis or impotency. If afflicted or combust, reverse than good.

Jupiter: He/She will have Friends and gain through writings and journeys and friendship through relations and neighbours. Also gain through education, brothers, writings, neighbours and short journeys.

The native will be learned and intelligent. Wealthy and comfortable life, respected and famous and will have successful hopes. If afflicted, losses and reverse results.

Venus: He/She will have comforts from relations, good health and wealth, many brothers and sisters and fond of music and dance, learning, accomplishment, progress through travel, investigations, explorations, travel or writings. If Venus afflicted, reverse results will be experienced by the native.

Saturn: He/She will have mental development, short journeys, happy relations with brothers and relatives, opportunities delayed obstinate, ups and downs in life, religious instinct. As Saturn is lord of 12th house posited in 3rd house it will give some disappointments and sorrows, occult learning.

Rahu: He/She will have spiritual up-lift and advancement in quality to the mentality and educational matters, gain through brothers neighbours, journeys, writings and publishing. He/She has fixity of purpose, proud, courageous, intelligent, optimistic and valorous, profitable business attains good fortune and long-lived.

Ketu: He/She will have mental anxiety, obstinate, less comforts from brothers and unprofitable journeys. He/She will be intelligent, courageous and destroyer of enemies, good for finances, loss of brothers is indicated, long life but injury and wound to arm is indicated.

Fourth House

Sun: He/She will have gain by marriage, a good partner, happy end of married life, gain through maternal relations, learned, famous and man of power and authority, grief of children in middle age, many opponents and enemies, gain through other's property and conveyance.

Moon: He/She will be devoid from benefits through parent, sickness of parents through changes and worries, troublesome home or domestic affairs and troubles through servants.

Mars: He/She will have travelling and writing in connection with home affairs and property, gain and honour through parents, land and property, success and comforts in old age, interest in reclamation, colonization, co-operative movements, horticulture, mining, architecture and archaeology. If Mars is afflicted, there will be reverse results.

Mercury: He/She will be Intelligent, learned and famous, blessed with wealth and property, gain through parents in latter part of life and from children, love of home, gain through the property of the dead, death at home. In case Mercury is conjunct with Venus then the death will take place abroad. If combust, weak or afflicted then death of parents, danger through fall of buildings, flood and storms, trouble over inheritance, land and property.

Jupiter: He/She will be famous, respected, and intelligent, a man of power, authority and rank, knowledge of law, comforts and gain from fortunate children, all comforts of life, gain through land, mines or household goods also through parents, fortunate in property, love for father and inheritance through friends, a kingly status in life.

Venus: A yoga Karaka planet for Aquarius Lagna posited in own house. He/She will have gain in property, land and conveyance, intelligent, learned and blessed with land and orchards, fond of opposite sex, gain and pleasure through women, gain in property and through old people and antiquities, assistance of father and benefit at birth place. Scientific inheritance, gain and profit through partner's relations, journeys home to die. If afflicted, then will be adverse results.

Saturn: He/She will have troubles through father or mother-in-law, secret sufferings and sorrows, restriction and limitations at end of life, gain through mines, land, inheritance, occult investigations, may cause, if afflicted, early death of parents, worried, no gain of property, loss through opponents and enemies, disrepute on account of women, worried due to debts of father.

Rahu: He/She will have unexpected gain and also through property, fortunate in discovery or in findings, blessed with life, trustworthy. No happiness from brothers and sisters, hard hearted, deceptive and worried stomach disease, gain from Government.

Ketu: He/She will have loss through land and buildings, family troubles, may be devoid from parental property, many secret enemies, loss, difficulties and troubles through them, many unproductive journeys. If Ketu is exalted and well-aspect, all 4th house matters will be benefited.

Fifth House

Sun: He/She will have learned, famous and a native of authority, pleasure through wife. The relations with children will be strained and grief through them during middle age, loss by speculation or gambling.

Moon: He/She will have diseased children, sickness and ill health due to over indulgence of pleasure and sports, financial condition will not be good, devoid from worldly comforts, death or grief of children, less comforts from parents.

Mars: He/She will have pleasure through children, journey, brothers, reading study and travel to pleasure resorts. The children will remain sick but rise to honour, gain through speculations, young people, sports or stage. If afflicted, death of young son and reverse results.

Mercury: He/She will be intelligent, learned and respected native, blessed with wealth and property and gain of others' property, much pleasure, love affairs and adventures, speculations and investments, prosperous children and gain through them, danger through excessive pleasures and there may be death and grief if children. If Mercury is afflicted and combust, reverse results will be experienced.

Jupiter: He/She will have respected and intelligent, an important person in family. The mother will be from good family, a man of power, comforts and pleasure from children, blessed with comforts of life, gain by speculation, investments, pleasure, entertainments, young people and children. If Jupiter is afflicted, reverse will be results.

Venus: He/She will have gain and pleasure through parents, children and acquaintances, respected and learned, fond of poetry or may be a poet, fond of opposite sex and occult science, birth of more daughters than sons. Pleasure and gain through women, child by strange consort, journeys on account of children, free living, native of unconventional ideas in regard to unions, takes pleasure in science, voyages, philosophy, air flights, sports, foreign investments and speculation. If Venus is afflicted, reverse will be results.

Saturn: He/She will have delight in pleasures, amusements, speculations and children and success through them, secret sorrows and difficulty through love, no comforts from son remain worried on this account, strained with father, brothers and sisters, unsatisfied with wife's uneducated and abortions.

Rahu: He/She will have relief form many troubles and calamities, pleasure, sports and other recreations, gain through public employment or office, timid but hard hearted, stomach, abdomen or belly diseases, to female troubles of uterus, sons late in life, and disturbed peace of mind.

Ketu: He/She will have denial or shortage of children, little happiness through them as they will be disobedient, excessive or irregular pleasure will bring much harm, evil nature, stomach disease, influence of evil spirits, good education, intelligent but worried.

Sixth House

Sun: He/She will have marriage below status, sickly and ill-disposed wife, troubles through employees, many enemies and opponents, of rash temperament, father will be diseased. In middle of life, one will face troubles and worries from enemies.

Moon: He/She will have many opponents and enemies, illness, troublesome for maternal relations, diseased and troubled mother. If Moon is strong and well-aspect, the/she will have usually good health, success in service and employment, medicine or social service.

Mars: He/She will have wound or scar mark on body, disrespect in middle age, sickness or injury through journeys, difficulty through brothers, modest worldly position, gain and honour through service and employment, the practice of healing, army or navy affairs.

Mercury: He/She will have money through careful speculations, weak health and constitution, a man of magisterial authority, pleasure in hygienic method and hobbies, loss of money, and through those in whose keeping it may be.

Jupiter: He/She will have respect and authority but disliked by his superior friends among working people and those in army, navy or air service, interest in social welfare, concerned with sickness among friends, faithful servants and gain through them. If afflicted, reverse will be results.

Venus: He/She will have danger from water but escape from drowning, loss of possession through sickness, may build his own house, difficulty through work in foreign lands, or in export business, diseased body, diabetes is indicated, grief through wife, may become under debt.

Saturn: He/She will have gain through food, clothing and employees, limitation on account of sickness, worried and under debt, rheumatic pains, loss through enemies, may be imprisoned, ill health and danger from bite of reptiles.

Rahu: He/She will be valorous and long-lived but is oppressed by enemies and evil planetary influences. He however overcomes his enemies and evil planetary influences, suffers from teeth, ulcer, and wound and also from annuls diseases, early period of life unhappy, faithful and honest, gain through service, fortunate by means of father's kindred.

Ketu: He/She will suffer because of various physical afflictions, loss through servants, many enemies, but one will overpower them, danger through reptiles or animals, eye and teeth troubles, famous, strong, fixity of purpose, over all good health, early life will be unhappy, liked by relatives but bitter relations with his maternal uncles.

Seventh House

Sun: He/She will have success in law suits, marriage in good family, a prepossessing partner, but likely one who grows cold or proves untrue or hostile. He/She will be respected and learned, careful for respect, respect in middle age, and gain in trade. If the sun is debilitated or aspect by malefic, there ill be reverse results.

Moon: He/She will have gain from land, average life and quarrels or law suits with servants, trouble with employees, difficulties with disreputable women and sickness, unprofitable journeys, loss through trade and business.

Mars: He/She will have rash temperament, diseased and worried, many women, loss in trade and business, marriage as a result of journeys, writings or to one of relations, gain through law suits, and dealing with the public generally, honours and reputation assisted by honourable marriage and partnership in responsible concern. If afflicted reverse results.

Mercury: He/She will have love with wife, fond of travels, knowledge of laws, gain through business, discord with children, loss by theft, dangerous sickness, subject to other check more than one marriage is indicated. If afflicted, reverse will be results and troubles and loss through love affairs.

Jupiter: He/She will have marriage in a respected family, intelligent and wife of sanguine temperament, wealthy and respect, comforts from conveyance, gain by marriage, contracts, business and dealings with others especially with women. Success in law suits, desirable friendships and social connections. If afflicted, reverse will be results.

Venus: He/She will have gain through marriage, land and women generally, marriage to a stranger, whose relation may oppose, a thoughtful, changeable professions, average financial conditions, relations with women, danger to respect, grief through wife, public enemies through religious, scientific people. If afflicted, reverse will be results.

Saturn: He/She will have partnership and close association with others, fond of opposite sex, troubles through deceit and treachery concerning unions, partnerships, contracts and law suits, sickness, discord and opposition of partners, generally trouble through women, abortion, short life, will be of rash temperament and of opposite views to others, under debt and loss in business particularly in export.

Rahu: He/She will be self-willed, proud and independent, reduces the number of enemies, increases profit through dealings with others and pretends delight and gain through women, whose influence may benefit the native. It is a testimony for wise and wealthy partner. He/She is clever, fickle minded and inclined to have liaison but in male nativity, may lose money through contracts, or liaison with women, conduct not commendable, no peace of mind. If afflicted, reverse will be results.

Ketu: He/She will have troubles through enemies and opponents, troubled married life, destruction of enemies, aimless travelling, sickly wife, may cause separation from wife, illicit connections with women avaricious, adulterous, worried and heavy expenditure, loss of vitality and fear from thieves.

Eighth House

Sun: He/She will have money or property by marriage, death of partner, difficulty over inheritance, cruel, rash temperament but obliging the people, will face many troubles, may be devoid from comforts of parents in old age and disobedient, diseases of bloods are indicated. If Sun in afflicted, loss of money through marriage and partnership.

Moon: He/She will have mental trouble, short life, financial reward for faithful service, dangerous illness, death of servants, troubled, worried life, may be under debt due to extravagance, average financial position.

Mars: He/She will have rash temperament, unsuccessful hopes, proud, unhappy and ill health, losses and troubles through enemies, death of brothers, journeys on account of troubles, false accusation, gain and honour in handling the estate or money of others, also through law suits, inheritance, legacies, insurance etc. of deceased persons.

Mercury: He/She will have many enemies, diseased and troubled, grief from children, incomplete education, a natural death, and gain by the dead, a comfortably fixed partner, spiritual experiences, and loss through speculations.

Jupiter: Early part of life will be comfortable, under debt, troubled and worried. He/She will have gain by legacy and goods of dead or through money of partner, death of friends, an easy demise, gifts or legacies from friends, grief due to death of wife, birth of son after middle age from second wife.

Venus: He/She will have psychic experiences, gain by long journeys concerning legacies and goods of dead, persecution regarding religious, scientific or educational convictions, also through publications. If afflicted, loss of inherited property, death of father, troubles through wife, danger to mother during confinement.

Saturn: He/She will be blessed with long life, experiences, mediumistic, death through irregularity, concern own affairs of dead and with finances of others and money of partner, unsatisfactory end, many misfortunes, secret enemies die, troubles over inheritance, early death of parents, worried, turmoil, litigations and poor.. Last part of life will be financially happy.

Rahu: He/She will be healthy and conducive to longevity, testimony for gifts and legacies and gain from the deceased, an evil position for mental aptitude and physical health. The native is dilatory in work, resorts to unhappy acts and inclined to much sexual activity. Diseases of wind, stomach, enlargement of glands are indicated. He has a few sons. He will earn name and respect but his reputation gets blemished. Financial conditions are mixed. He will earn name a respect but his reputation gets blemished. Financial conditions are mixed. He is oppressed by enemies.

Ketu: He/She will have loss of goods through fraud, sudden or violent death affecting the longevity, wounds and separation from dear ones, stomach troubles and opposition of enemies. He/She has desires for other's property and wives. If aspect by benefic, the native gets wealth and is long-lived. Gain of money after obstruction.

Ninth House

Sun: He/She will have marriage to a stranger from afar. Gain by partner's relatives, generous and respected, gain through trade and business, partner's journeys to foreign lands, pilgrimages and long journeys, grief from parents in early age, blessed with landed property. If Sun is weak or afflicted, benefic results will be reversed.

Moon: He/She will work in connections with foreigner's affairs, builder of inns and property, generous and may give his wealth in charity, death of relations of father, loss of wealth if Moon is weak or afflicted.

Mars: He/She is cruel and rash temperament, long journeys, gain through philosophy, publishing and science; brothers may travel to foreign countries and marry there, honourable voyages, professional journeys, and honour through learning, writing, publishing or research, religious or intellectual parents. If afflicted, there will be enmity with relations, waste of ancestral property, grief from parents, extravagant, connection with criminals and decocts.

Mercury: He/She will be Learned, dutiful children who travel and give pleasure and learn offspring, become preachers scientists or explorers, pleasure in foreign lands. The partner will have troubles on money matters with wife's relatives, death in distant land by drowning or on voyages, will impart education to other, builder of charitable institutions and helpful to others. When Mercury is afflicted or weak, reverse results will be realised.

Jupiter: He/She will have intelligent and knowledge of law, comforts from children in old age specially, gain by books, export business, long journeys, religion, science and wife's relations, friendship through travel and learning, educations, ministers, writers and inventors and gain through them. If afflicted waste of ancestral property and reverse results.

Venus: He/She is religious and charitable, gainful to people, happy married life, learned and intelligent, fond of religion, travelling for education scientific or religious purposes, prophetic dreams, many fine qualities, gain through culture and development, comforts through conveyance and blessed with property. If weak and afflicted, results will be reversed.

Saturn: He/She will have long journeys, religious or psychic experiences, liking for law, religion, science, invention and all matters connected with higher mind, prophetic dreams or visions, gain through partner's relative, secret missions, secluded research, sacrifice for science or religion. If afflicted or weak then turns a cheat, miser, greedy, gain of money through illegal means, shipwreck, difficulties with relations, religion and men of science, troubles with foreigners and religious, legal or educational affairs, fruitless and dangerous voyages and experiences in foreign land.

Rahu: He/She improves mental qualities and gives success in education, legal or religious studies, favourable for voyages and foreign affairs, true dreams and prophetic intuition, intelligent, opposed to father and less happiness from him. He has fixity of purpose, love for brothers, but is not fortunate in respect of sons. For worldly prospects like name, fame, wealth and splendour this is an excellent position.

Ketu: He/She will have unfortunate voyages, curious dreams, trouble and danger of imprisonment in foreign land, worried and religious, hypocrite, and intelligent, proud, courageous and wealthy, pilgrimage, unhappiness from brothers. He will suffer in respect of father. If well-aspect, the evil effects will be mitigated.

Tenth House

Sun: He/She will have an honourable partner beneficial to the professional career, respected and famous, blessed with property, troubles through enemies, devoid from the service of parents at death-bed. If afflicted public disgrace or scandal through a union or enemy.

Moon: He/She will have honours through service and leaning or through municipal or national activities, respected and wealthy, hard earned ancestral property will be destroyed but he will regain after much hard work. If afflicted and weak, then sickness through dishonour, sorrows, or worried on account of mother or father-in-law.

Mars: He/She will have gain and honours through profession, honorary office of Govt. employment, success aided by mother's care, training and efforts, professional and honourable journeys and gain through them, honours or renown through writings or other accomplishments, bold and determined, fond of travels, gain through commodities of iron if afflicted reverse results.

Mercury: He/She is learned, intelligent, mathematician, religious and scholar of Astrology, wealthy and comfortable, religious preceptor and preacher, honourable children, renown in speculations, business enterprises, pleasure and honour through father-in law or mother-in-law, rise through inheritance. If weak, afflicted or combust then there will be death of mother and employer, danger of violent death, brothers will cause mental anguish.

Jupiter: He/She will have gain by occupation, profession, business, Govt. or in-laws, friendship and gain from those in good position, intelligent learned but averagely respected, comforts and respect in latter part of life. If afflicted, the results will be reverse.

Venus: He/She will be fond of music, dance and theatrical companies, blessed with property, wealth, position, rank, and authority as the age advances, success in foreign affairs, benefits and honour through science, literature or travels, gain by travels. Gain by trade, public or Govt., work or by fruits or earth. If afflicted, reverse will be results.

Saturn: He/She will rise to high, social and professional position. Merit honour and success, liability to disgrace, discredit from superiors and persons in authority, long journeys, retirement or seclusion, troubles with in-laws, miser, and no gain to any body from him, life full of turmoil and difficulties.

Rahu: He/She will achieve honour, credit and high position through hard work, industry and ability, proud and fearless, fond of struggle and strafes, does not have reliable friends, powerful enemies, economical, happiness from parents is impaired, a few sons, and worried mind. In male birth charts, there will be liaison with widows and fond of opposite sex.

Ketu: He/She will have loss of position due to deception, treachery and adverse public conditions, sudden depressions, changes or failures, courageous, pushing, popular and brilliant, philosophical bent of mind. No happiness from father or vice-versa, questionable character, many journeys and travels, obstacles but will overcome them, accident from conveyance and animals, economical and accumulates money. The enemies will be crushed.

Eleventh House

Sun: He/She will be careful for his respect, gain of lost ancestral property, fall from a high place, many troubles and turmoil. Friends will become enemies or opponent, marriage to a widow or widower with children but trouble through them.

Moon: He/She will have interest in legislative activities, political and social welfare, property, strained relations with elder brothers. If Moon is weak or afflicted, reverse will be results.

Mars: He/She will be fortunate through brothers, friends through journeys, wealthy and comfortable, successful hopes and wishes through progressive studies, eminent friends among legislators and from persons in authority and gain through them, ambitious ideals, helpful to associates and others. If afflicted, reverse will be results.

Mercury: He/She will have knowledge of many foreign languages, blessed with property, conveyance, respect and fame and will be cause of advancement of his family, a comfortable life, great attachments between the native and his children, friends, successful hopes and wishes, gain through speculations, legislators, ambassadors and sportsmen, also through legacies and friends. If afflicted or combust, reverse results.

Jupiter: He/She will be lucky and learned, wealthy, native of power and authority, a good Astrologer and debater, blessed with long life and God fearing, gain through friends, well-defined hopes and wishes, accidental fortune. If afflicted, reverse will be results.

Venus: He/She will be long lived, intelligent and will have good handwriting, fond of pleasures, blessed with property and orchards, authority to the rank of judge, a very comfortable life, and gain through friends, many hopes and wishes attained in old age. Gainful voyages, friends among foreigners and in foreign lands, acquaintances among scientists, legislators and travellers. If afflicted, reverse will be results .and. sufferings on account of religious or other convictions etc.

Saturn: He/She will have gain through large circle of friends, many pleasures in life, hopes and wishes often attained, gain by success of employer. If afflicted, then loss through deceitful friends and losses through their advice, un-gainful journeys, limited number of brothers, loss of respect and wealth etc.

Rahu: He/She will be wealthy and long-lived, meritorious, friendships, acquaintances, assistance from them in realization of hopes and wishes, intelligent, few sons, lot of travels, famous, and industrious, courageous, avaricious and will appropriate other's wealth, gain through there's advice, heroic in strife and overcomes his enemies, a good health on the whole. In advanced age ear trouble or deafness is indicated.

Ketu: He/She will have undesirable associations, loss of opportunities and failure of hopes, false friends, wealthy, popular, good authority gain and success, performs good deeds and commands respect, few sons, worries from children. He will become licentious; diseases of stomach and anus are indicated. If afflicted, reverse will be results.

Twelfth House

Sun: He/She will have average financial position, profession in foreign concerns or lands or export business, loss and troubles through enemies, secret sorrows, unhappy marriage provided Venus is also afflicted, partners or opponent may cause fear of or imprisonment, danger of death at the hands of enemies if Mercury afflicts.

Moon: 6th lord in 12th house indicates worried mind, good service but subordinates may irk you, afflicted health, may remain under debts. If Moon is well aspect, effects of Vipareeta Raja yoga will be realised.

Mars: He/She will have occult learning, seclusion or estrangements from brothers. Danger of enmities and imprisonment while travelling, great sorrows through brothers, secret sufferings, difficulty in employment, will serve in army etc. or deal with arms, loss of office, honour and prestige through business associates who become secret enemies, unfortunate environments and conditions, difficulty relieved through occult science.

Mercury: He/She will have pleasure through investigations and research work, difficulty over inheritance death of secret enemies, great sorrows, but gain through the business, generous and charitable nature, and many travels. If combust or afflicted with Sun, death or imprisonment through enemies etc.

Jupiter: He/She will have gain through secret affairs, good friends through occultisms, pleasure in peaceful, quite, harmonious or secluded places. If afflicted, sufferings through deceitful friends, under debt, loss of money and ancestral property etc.

Venus: He/She will have many long journeys, sweet speech, blessed with land and orchards, devoid from ancestral property, eye troubles in middle of life, difficulty and sorrow through religion, journeys and science. In middle or latter part of life will seek seclusion for the development and advancement of occult science and will undertake long journeys for the same, afflicted health, suffers at the hands of parents.

Saturn: He/She will have gain through occult affairs and secret mission, success in middle life, powerless enemies, secret investigations, obstacles and adversities will be blessings in disguise, politician, not very wealthy and average status in life. If afflicted, fear of imprisonment, secret unhappiness, enmities, sufferings and misfortune etc.

Rahu: He/She will have gain by secret methods or in seclusion, success in occultism, afflicted health, weak eye-sight, windy complaints or heart trouble, frustrated initially but success finally, courageous, unfavourable for position of pleasures of beds.

Ketu: He/She will have loss and trouble through secret enemies, bad health secretive, sinful deeds, and more expenditure, victorious in disputes, worried, diseases of feet, naval, anus or eye-diseases. He is weak and voluptuous, liability of improvement and restrictions.

11.12 Pisces (Meena) Lagna

Saturn, Venus, Sun and Mercury are malefic. Mars and Moon are auspicious. Mars and Jupiter will cause Yoga. Though Mars is a killer, he will not kill the native (independently). Saturn and Mercury are killers.

Sun: Surya is L-6 measured from Meena lagna, Meena Chandra, and Meena Navamsha and is "The Problematizer". Sun is lord of the 6th, an inauspicious house and does not rules any other house and so Sun is malefic planet for Meena lagna. Ravi controls the most evil dushthamsha, bhava-6. Naturally, most of the significations of Ravi like father and father's figures become problematic. His father has been born into the service class (6). The father will be restless and discontent with his social position. He must be inordinately cautious of hisown independent, egoistic instincts, since self-assertion tends quickly toward conflict.

Remedial Ratna for Meena lagna Surya (Manika - pure Ruby): No-no-no. He should avoid Ruby. However, if Sun is in 6th house in his own sign, a Ruby can be worn in the major period of Sun. This will prove useful as Sun is a friend of Jupiter the lord of the Pisces Ascendant. Wearing of this gem stone, as advised above will ensure freedom from enemy troubles, debts and diseases.

Moon: Chandra is L-5, a Trine house, measured from Meena lagna, Meena Chandra, and Meena Navamsha, a very auspicious house and so Moon is a benefic planet for Meena lagna. He is the best emotionally satisfied in creative social roles that express individual intelligence and self-determination, and allow him to take speculative risks. He is the best in parenting, literary author, speculative investor, independent educator, and almost any political or entertainment role including "genius" innovative teaching styles. He loves children, and expects to profit by raising them. He is prone toward lifelong self-improvement and often involved in the consciousness industry and will achieve some degree of celebrity or notoriety.

Remedial Ratna for Meena lagna Chandra (Mukta/moti - pure natural Pearl): Yes-yes-yes. Moon is lord of the 5th house, a very auspicious house. By wearing a Pearl he will be blessed with children, will get name and fame, his good fortune will be accelerated. According to Bhavath Bhavam principle, 5th being 9th to 9th is also treated as a house of Bhagya (fortune). If Moon is in his own sign in the 5th, Pearl can be worn always with beneficial results. The good effects will be more pronounced in the Major period of the Moon. If Moon is afflicted or ill-disposed, wearing of Pearl is a 'must.

Mars: Kuja is L-2 + L-9, a Trine house, measured from Meena lagna, Meena Chandra, and Meena Navamsha. Kuja is a money-maker, as He controls two important wealth/values houses out of four wealth houses like, 2, 5, 9, and 11.

For Matsya Lagna, presuming that Mangala is fairly well disposed, Mangala generates considerable wealth via family-of-origin hoarded assets and knowledge-values, as well as via the higher philosophical education and world travel.

Remedial Ratna for Meena lagna Mars (Red Coral – Munga): Mars is lord of the 2nd, the house of wealth and 9th the house of Bhagya (fortune). Mars is, therefore, a very auspicious planet for this Ascendant and wearing of Coral will bring great success in the life of the native. Red Coral is a 'must' for him if Mars is ill disposed or afflicted in any way in the birth chart. Wearing of Red Coral with Yellow Sapphire will prove exceedingly beneficial.

Mercury: Budha is Kendra lord L-4 + L-7, measured from Meena lagna, Meena Chandra, and Meena Navamsha. Here also on account of ownership of two Kendra Mercury suffers from Kendradhipati dosha and so Mercury is inauspicious or malefic planet for Meena lagna. Budha controls two of the most basic relationships, like parents (4) and spouse (7). Budha in 11th (Makara) and 12th (Aquarius) ruled by Shani (L-11+ L-12) will have his quiet and conservative early childhood home and marriage. Budha in 3rd (Vrishabha) and 8th (Libra) ruled by Shukra (L-3+L-8) will have a luxury-loving spouse and parents; however the pursuit of that luxury may become problematic. Budha in Simha ruled by the L-6 Surya will naturally have some struggles regarding personal sovereignty, both in the early childhood home and the marriage. If Surya is well-disposed, these conflicts can be managed.

Remedial Ratna for Meena lagna Mercury (Emerald – Panna): He should avoid wearing Emerald stone. Still, if Mercury is in his own sign in the 4th or 7th or be in the 2nd, 5th, 9th, 10th or 11th houses, emerald can be worn with profit in the major and sub-periods of Mercury. In this connection it should be kept in mind that natives of this ascendant, who are apprehending death on account of old age or some illness, should never wear Emerald as Mercury's ownership of the 7th house makes him a very strong death inflicting planet.

Jupiter: Guru is L-1+L-10 measured from Meena lagna, Meena Chandra, and Meena Navamsha and is a superb career graha for Meena lagna. Guru is L-1, a Trine house, a very auspicious house and so is an auspicious or a benefic planet for Meena lagna. He benefits from a leadership responsibilities. Presuming that Guru is a healthy graha, he enjoys fairly high levels of social recognition and a favourable public reputation. Guru is not helpful for money, because of L-10 (12th from 11th and L-1 is (12th from 2nd). Guru typically raises the prestige and lowers the bank accounts.

Remedial Ratna for Meena lagna Jupiter (yellow sapphire – Pukharaj): Guru is, therefore, an auspicious planet for this Ascendant. He can fulfil all his desires by wearing a yellow sapphire. If Jupiter is in his own sign in the Ascendant or the 10th house, it will give rise to Hamsa Yoga and the Yellow Sapphire will help to enhance the beneficial effects of the yoga. The results will be felt more pronouncedly in the major period of Jupiter. Wearing of a Yellow Sapphire with Red Coral, the gem stone of Mars, the lord of the 9th (house of Bhagya), will be an ideal combination for achieving success.

Venus: Shukra is L-3 + L-8 measured from Meena lagna, Meena Chandra, and Meena Navamsha. Venus is an enemy of Jupiter, the lord of this Ascendant. Shukra is lord of 3rd and 8th, an upachaya house and a dushthamsha (two inauspicious houses), and so Sukra is an inauspicious or malefic graha for the Meena lagna. Shukra is unfavourable for men, because Shukra describes his female partners. He may attract a female partner who is handicapped by her suspicious mind, particularly in regard to her husband's hidden monies or inheritance, and issues of marital fidelity. Harmonious relationships are hard to come by. In fact, Shukra is nearly as problematic for Meena lagna.

Remedial Ratna for Meena lagna Jupiter (yellow sapphire – Pukharaj): Venus will be lord of the 3rd and 8th, two inauspicious houses. Moreover, Venus is an enemy of Jupiter the lord of this Ascendant. It will not be advisable for him to wear a Diamond even when Venus is in the 3rd or the 8th in his own sign in a birth chart.

Saturn: Shani is L-11 + L-12 measured from Meena lagna, Meena Chandra, and Meena Navamsha. Shani is an enemy of Jupiter, the lord of this Ascendant. Shani is lord of 11th and 12th, an upachaya house and a dushthamsha (two inauspicious houses), and so Shani is an inauspicious or malefic graha for the Meena lagna. Shani gives profit and takes it away. Shani creates loneliness, isolation, and a profound loss of identity. It is the great irony of the turning of the cycle of birth and death, that at the moment of accomplishment of a goal, the karma is completed, and one tends to lose interest in the activity. If Shani is favourably disposed, the L-12 Shani may create foreign travel, ashram retreat, healing hospitalization, and relaxing spas, rather than marital conflict or wandering.

Remedial Ratna for Meena lagna Shani (Blue Sapphire - Nilam): Saturn is lord of the 11th and the 12th houses. He is not considered auspicious for this Ascendant. Moreover, Saturn is an enemy of Jupiter, the lord of this Ascendant. Still, if Saturn is in his own sign in the 11th or in the Ascendant, Blue Sapphire can be used with advantage in the major period of Saturn.

Rahu: Rahu is co-lord of Kumbha (dushthamsha bhava-12) measured from Meena lagna, Meena Chandra, and Meena Navamsha. Rahu is an enemy of Jupiter, the lord of this Ascendant. Rahu is lord of 12th, a dushthamsha (inauspicious house), and so Rahu is an inauspicious or malefic graha for the Meena lagna. Rahu in either Rashi of maraka Mangala is dangerous for Meena lagna. Rahu and Ketu are said to give positive results in the 3rd, 6th, and 11th bhava from the lagna or from Chandra. In addition, some authorities posit that since R-K are exalted in Vrishabha-Vrischika (per BPHS), R-K will give good results when Mangala and Shukra are well-disposed.

Remedial Ratna for Meena lagna Rahu (Gomedha): Rahu is co-lord of bhava-12 (Kumbha), therefore, Rahu-ratna (Gomedha) is not a beneficial gem *except* for matters of meditation and withdrawal into the hermit's cave of reflective awareness, and even then only if the lord of bhava-12 and its occupants are auspicious. In general Rahu-ratna (gomedha) is to be avoided as an amplifier of the negative attributes of personality which can cause withdrawal and isolation from the vitality of life. Rahu in either rashi of maraka Mangala is dangerous for Meena lagna.

Ketu: Ketu is L-9; co-lord of Vrischika (bhava-9) measured from Meena lagna, Meena Chandra, and Meena Navamsha. Ketu is lord of 9th, a Trine house (an auspicious house), and so Ketu is an auspicious or benefic graha for the Meena lagna. Ketu amplifies the effect of any graha who are sharing Ketu house; and also Ketu magnifies the effects of the lord of His occupied Rashi.

Remedial Ratna for Meena lagna Rahu (Gomedha): Ketu is co-lord of bhava-9 Vrischika, therefore Ketu-ratna may be a beneficial gem for matters of acquisition of wisdom-teaching and priestly rituals, if the lord of bhava-9 and its occupants are auspicious.

Planet Effects in Twelve Houses for Pisces (Meena) Lagna

First House

Sun: He/She will have gain through trade and profession and through relations and friends, not a favourite of his superiors, loss of ancestral property, average span of life, weak and afflicted health, and sickness through irregularity, fond of pets. When it is afflicted and weak gives troubles and loss through servants and small animals etc.

Moon: Trikona lord in Lagna will indicate, he/she will have middle stature and fair complexion, average span of life, good and lucky children, bestows wealth, a propensity to pleasure through children, speculations, gaming etc. Many love affairs. When weak or afflicted, loss of wealth, unlucky and loss of children, less comforts from parental property etc.

Mars: He/She will have money comes readily; skilled, learned, intelligent and wealthy, respected and careful for his respect, rash temperament, fond of travels, long journey, learning, wisdom prudence; fortunate with stranger and foreigners, wife's relations and voyages, interest in science, invention, law, political economy and philosophy, avoid relations with opposite sex, law suits for property, hard hearted and worried, inclined towards vice acts. But when Mars is weak and afflicted, there will have obstacles and difficulties through money which will be hard to accumulate, loss through Govt., unlucky through family and loss of money through father.

Mercury: He/She will be fortunate in inheritance, respected, intelligent skilled but of weak constitution, medium age, and many women in relations, religious, and gain through land, conveyance and property, Interest in domestic science, benefit and gain through unions, partnership, and love of women, and connection with the processes of law. When weak or afflicted or combust indicates malefic for family and conveyance, unsuccessful hopes, worried and unreligious etc.

Jupiter: He/She will be lucky, intelligent, famous, respected, a native of power, rank and authority, blessed with property, generous victory over enemies, a long fortunate life, good health, harmony and triumph over difficulties, honour through merits and industry, gain through Govt. and superiors, fair complexion and handsome, a comfortable if lucky life. When Jupiter is weak or afflicted then it will give displeasure of officers, loss from Govt., and health and of wealth and reverse results.

Venus: He/She will have accumulation of wealth and gain through wealth of others, handsome, wealthy lucky and a man of good authority, fond of opposite sex victory over enemies' pleasurable life extravagant and concern with affairs of brothers and neighbours. Writing, learning and accomplishment and gain through them, legacies and money through affairs of dead. When afflicted, death by irregularity and losses through above.

Saturn: He/She will have happy married life and through children, generous and charitable. A business man and gain through the same, success in worldly affairs, real friends and supporters and gain through them, victory over enemies and obstacles. Fortunate actions, successful hopes, occult experiences and interest in that. When afflicted or weak indicates ugly face and dark complexion, ancestral property will be destroyed, gloomy and adulterous, worried up to middle age, danger of imprisonment, heavy troubles and secret sorrows etc.

Rahu: He/She will have honours, wealth, favour through religious, educational or scientific affairs, adds power to personality and gives opportunity for self expressions. But he/she is inclined to malicious sex acts, amorous and fond of company of women.

Ketu: He/She will have tribulations, loss and scandals, endangers face and eyes. Not conducive to longevity, devoid of happiness, avaricious, industrious, voluptuous and worried, impaired domestic happiness.

Second House

Sun: He/She will have gain and money through service, employees and small animals, eye diseases, troubled and diseased father, losses and limitations financially through sickness, servants and animals, loss of ancestral money, troubles through rash temperament and rash actions. If afflicted, troubles, worries and losses etc.

Moon: He/She will have gain through investment, pleasure and children, lucky children and gain through speculations, average finances, and knowledge of different languages, worried and of weak constitution, watery eyes. If Moon is weak or afflicted, there will be losses through above and adverse results.

Mars: He/She will be bold and courageous, rash temperament, lucky and wealthy in middle age, money through personal ingenuity and industry, through merchants, foreign trade, sea, science, learning, publications, travel, invention or banking. When weak and afflicted losses, through troubles and above matters.

Mercury: He/She will have gain in business, sweet speech, learned and of sanguine nature, gain by marriage, deals in land, property and by the estate or condition of parents. If combust or afflicted indicates defective eye sight, loss of money through unions, partnership, contracts, law suits, public enemies, loss through women. Death of partner and public enemies and may stammer in his speech.

Jupiter: He/She will have gain through hard work and industrious activity, trade, profession or Govt. office, good orator and truthful, will hate vice acts, early life will be unhappy but the latter part of the life will be happy and blessed with wealth. If afflicted reverse results.

Venus: He/She will have knowledge of many languages, money through educational affairs, short journeys, writing, music etc. Gain through other's money, fortunate in realising debts and gain from the dead. If afflicted and weak then, there will be trouble and losses through above, troubles over money matters with relatives.

Saturn: He/She will have gain through occult and secrecy, large animals, gain and profit in business and money through friends and acquaintances, through legislative interest or development of his own creation. If afflicted then eye trouble, loss of wealth, poor, change of religion and loss through enemies and generally unfortunate conditions.

Rahu: He/She will have position for fortunate heredity or gain by legacy and gifts, science, learning, increases the possessions, removes want and bestow affluence. He/She becomes untruthful, insincere and talkative, opposes others. He may apprehend danger or assault from enemies. Period of early life will be unhappy.

Ketu: He/She will have misfortune in finances, indebt ness, loss and damage to property, sorrows, fear and worries, loss through Govt., and relations, mouth disease.

Third House

Sun: Sun, the lord of 6th house when posited in 3rd house indicates respect, famous, bold and interest in studies, he/she will have grief and sickness of brothers, likelihood of mental trouble when sick, may remain under some control and latter on will act independently.

Moon: He/She will have pleasure through children and relatives, liking for travel, sports, drama and adventure, comfortable and respected, changeable professions, skilled, fond of opposite sex, more sisters than brothers, comfortable and happy life in middle age. If weak or afflicted, reverse will be results.

Mars: He/She will have gain through education, writings, brothers and short journeys, learning, accomplishments and progress through research, travel, investigation, exploration, travel or writings, late marriage and that too with difficulty, many travels, removals on account of belief and convictions.

Mercury: He/She will be intelligent, weak body, many brothers, may be an artist, comforts of life, legal or religious disputes, troubles on short journeys difficulty through writings and contracts, gain through brothers. If afflicted or combust then enmity with brothers and other reverse results.

Jupiter: He/She will have religious lucky wife, gain in honour and advancement through wife, respected among friends, relations and neighbours, honours through short journeys, writings, accomplishment, business trips and Government, commissions, mental development, good relation with brothers. If afflicted restricted development or education, trouble through some relatives, journeys and writings etc.

Venus: He/She will have benefits through learning, short journeys and brothers. Licentious, blessed with property, a pleasurable and comfortable life, gain through business due to one's intelligence and dignity, psychic and mysterious experiences. When afflicted, death of brothers, danger of death on short journeys.

Saturn: He/She will have friendship with relatives and neighbours, friends through writings and journeys, occult learning. When afflicted disappointments, he/she will have sorrows and troubles through relatives, friends and neighbours, trouble in some short journeys, coward, flatterer, late marriage, few brothers, in middle age becomes poor and worried.

Rahu: Rahu adds quality to the mentally and conduces to interest to spiritual or educational matters. He/She will have gain through brothers, neighbours and journeys, writings or publishing, fixity of purpose, proud and courageous, intelligent, optimistic and valorous, profitable business. Long lived and attains good fortune.

Ketu: He/She will have mental anxiety, troubles with brothers and neighbours, unprofitable journeys. He/She will be intelligent, courageous, destroyer or enemies and good for finances, loss of brothers is indicated, long life but injury and wound to arm is indicated.

Fourth House

Sun: He/She will have journeys, respected and famous, careful for his respect; will remain separated from his family due to his profession, sickness of mother, troublesome home or domestic affairs and through servants, less comforts from home, troubles through enemies.

Moon: He/She will have in latter part of life gain through parents, property and land, love of home. His/Her children will be benefited through his grandfather. When weak and afflicted, there will be reverse results and weak constitution, devoid from the comforts of parents.

Mars: He/She will be blessed with property and house, wealthy, respected mines, gain through parents, household, goods, land or mines, gain through partner's relations. Journeys home to die, travel on account of family affairs or wife's mother. When afflicted, no gain and worries, many enemies, disrepute in middle age etc.

Mercury: He/She will have gain of property through old people and father, benefits and profits in birth place, property by marriage, long lived, and mother will be blessed with long life, a good and happy life, intelligent, skilled religious and owner of good houses. If combust or afflicted, litigations over property, robbery of household goods, loss through storms etc.

Jupiter: He/She will have comforts from conveyance, intelligent, learned and with successful hopes and wishes, lucky sisters, bless with property, lucky and comfortable life, gain through lands, mines, inheritance, success late in life, occult investigations, honour through parents, interest in reclamation, colonization, co-operative movements, horticulture and mining etc. If afflicted reverse results.

Venus: He/She will have comforts through comfortable conveyance, travelling and writing in connection with home affairs and property. Gain through dead and death of parents, licentious, and amorous, relations with women, owner of buildings and property. If afflicted, grief through children, troubles from enemies and loss of ancestral property etc.

Saturn: He/She will have inheritance through friends, fortunate in property. Love for father, loss of property, secret sufferings, restrictions and limitations at end of life, diseased and worried mother, death in relations, and renunciation in middle life and may set out on long journeys.

Rahu: He/She will have gain through property in unexpected manner, fortunate in discovery and findings, ancestry noble, long-lived and trustworthy, no happiness from brothers and sisters, hard hearted, deceptive and worried and stomach disease, also denotes gain from Government.

Ketu: He/She will have waste of patrimony, turmoil and family discord, adverse for land, conveyance, mother, and loss of parental property and malicious. If Ketu is well aspect, all matters relating to 4th house are benefited.

Fifth House

Sun: He/She will have respected and judicious, man of power and authority, many enemies and opponents, unsuccessful hopes, worried through children, loss of ancestral property, careful for his respect, sickly children and illness through over indulgence in pleasure and sports.

Moon: He/She will have much pleasure, love affairs, gain through investments and speculations, prosperous children and pleasure through them. Intelligent, learned and blessed with powers and authority, wealthy, many daughters. If afflicted and weak, reverse will be results.

Mars: He/She will have gain by speculations, pleasure and entertainment, conventional ideas, free living, journeys due to children, pleasurable journeys, takes interest in science, speculations, sports, voyages, air fights and foreign investments. If afflicted then, there will be reversal in professions, loss through above, worried due to brothers, loss of property and grief through children.

Mercury: He/She will be learned and wise, gain through ancestral property, religious preceptor, gain through speculations, and pleasure through wife but troubles with children. If afflicted reverse results.

Jupiter: He/She will be learned and famous, gain of wealth and comforts through children, careful for his respect, knowledge of Astrology and good orator, knowledge of many languages, gain through speculations, pleasure, sports and stage. If afflicted reverse results.

Venus: He/She will have pleasurable journeys, more daughters, relations with women and fond of music and dance, avaricious, respected and wealthy, mental pleasure through children, brother, reading, study and travel to pleasure resorts. If afflicted unfortunate children or death of some, dangers through excessive pleasure, speculations, children and their affairs.

Saturn: He/She will have much happiness and gain through children, rank and power, beneficial circumstances, also secret sorrows and trouble through love, and speculation, proud, rash temperament, non-religious, disrespectful in old age, worried due to death of children.

Rahu: He/She will be freed from many troubles, calamities and danger, favourable for children who are long lived and fortunate, gains through public employment and office and is fond of social recreation, pleasure and sports, timid but hard hearted, stomach, abdomen or belly diseases, for females' troubles of uterus, sons late in life and disturbed peace of mind.

Ketu: He/She will have denial or shortage of children, little happiness as they will be disobedient, excessive or irregular pleasure and will bring much harm, evil nature, and stomach disease, influence of evil spirits, good education, intelligent but worried.

Sixth House

Sun: He/She will have Good health, success in service and with employees, poultry, medicine or social service, Victory over enemies, respected, will travel extensively. Under debt, litigations and law suits, worried. If afflicted service illness, loss and difficulties in employment.

Moon: He/She will have money through careful speculation or by children's earnings, pleasure in hygienic methods through pets and hobbies. But when weak or afflicted, there will be urinary diseases, loss of property, loss through thieves, diseased mother, worries from enemies and opponents.

Mars: He/She will have many enemies and loss through them, accusation, gain through inferiors, poultry, fond of animals, sickness through travelling, difficulty through work in foreign lands or export work, diseases of blood, worried, unhappy in middle age, danger for disrepute and loss of property.

Mercury: He/She will have Marriage below status, has sickly and evilly disposed wife, skilled, troubles and worries in life, gain through service, and may build his own house. If combust or afflicted then loss through sickness, or servants and under debt, a worried and poor in life.

Jupiter: He/She will be a good healer, gain through clothing, employees and small animals, modest, worldly position, gain and honour through service, employment army or navy affairs. If afflicted by luminaries, much sickness and short life, if by Mars, surgical operation, gift through children, diseased and worried.

Venus: He/She will have grief due to wife, sudden self death, many enemies, many travels, difficulty through brothers, injury or sickness through journeys.

Saturn: He/She will have friends among working persons, and in army, navy or air service, faithful servants, interest in social work, limitations on account of sickness. If afflicted loss through above, loss through employees and enemies, ill respected, may be imprisoned in middle part of life, litigations, diseased, unsuccessful hopes and underground activities.

Rahu: He/She will be valorous and long lived but oppressed by enemies, whereas he will overcome enemies, suffers from teeth, ulcer, and wounds and annul diseases, early period of life unhappy, faithful and honest. Gain through service, fortunate through father's kindred. Increases physical as well as mental stamina and the native will be victorious after struggle and strife.

Ketu: He/She will suffer on account of various afflictions, loss through servants, many enemies but victorious over them, danger through reptiles or animals, eye and teeth troubles, famous, strong, fixity of purpose, good health on the whole, much liked by his relatives but bad relations with maternal uncles.

Seventh House

Sun: He/She will have quarrels or law suits with servants, difficulties with disrepute women and sickness, troubles with employees. Wife will be of rash temperament and sickly, poor and under debt, unhappy in middle age, latter part of life will be happy.

Moon: He/She will have pleasure, close associations or understanding with partner but discord with children, loss by theft. If afflicted or weak, troubles and loss through love affairs. Delayed marriage, loss in partnerships, contracts etc. In latter part of life, difficulties will be reduced, respect will increase and one will feel happy.

Mars: He/She will have rash temperament, many enemies, and family discord, unhappy in early age, gain by marriage, contracts, law suits and dealing with women specially, marriage to a stranger, of education opposed by relatives. If afflicted losses through above, public enemies through religious or scientific people, troubles and open enemies, competition and theft.

Mercury: He/She will have property and gain through marriage, land and women generally, success in law suits, marriage in good family, a good wife but likely to grow cold or proves untrue or hostile, learned, intelligent, knowledge of law, many journeys, cordial and good relations with wife, wealthy, good profession, less results will be indicated.

Jupiter: He/She will be learned, intelligent and well respected, religious and will have many disciples, wealthy and blessed with all comforts, long lived, a good and chaste partner. In last part of life will become wealthier, gain through partnerships with others, fond of opposite sex, gain through law suits and dealing with public generally, honours and reputation through honourable marriage and partnership in a responsible concern. If afflicted or combust reverse results will be indicated.

Venus: He/She will have a rich partner to whom wealth comes unexpectedly, marriage as a result of journeys or writings etc. If afflicted many worries, loss through robbery, union will be unfortunate. After middle age he/she will be blessed with wealth and comforts, faces reversals in life, profession in medicines or explosive material.

Saturn: He/She will have a loving partner with desirable friendship and social connections, success in law suits, will hold power and authority, wife will be intelligent, learned and from honourable family, unhappy after middle age, may be grieved through death of first wife, weak health. If afflicted, difficulty and loss through deceit and treachery concerning, unions, partnerships, contracts and law suits, sickness, discord and opposition of partner and troubles through women generally.

Rahu: He/She will be self willed, proud and independent, reduces enemies, increases profits through dealings with others and portends delight and gain through women. It is sure sign of wise and wealthy partner. He is clever, fickle minded but in a male nativity adverse for virility, connections with women, impairs conjugal happiness, may loss money through contrast or liaison with women, conduct not commendable. No peace of mind.

Ketu: He/She will have troubles through enemies and opponents, troubled married life, destruction of enemies, aimless travelling, sickly wife and separation from wife, illicit relations with other women, avaricious and adulterous, worried and more expenditure, loss of vitality and fear from thieves and more expenditure, loss of vitality.

Eight House

Sun: Sun is in debilitated sign of Libra in eighth house and indicates dangerous illness, death of servants, animals, poultry etc. He/She will have financial reward for faithful service, unlucky and worried life, loss through enemies, disrespect in middle of life, may face sudden death.

Moon: He/She will be unhappy and unlucky, sudden death, suffers through children, who may die before the native, loss through speculation or gaming, may leave his house and reside somewhere else.

Mars: He/She will have death abroad, short life, many enemies, disrespected, gain by legacy and goods of the dead or through money of partner, preservation regarding religion, scientific or educational connections also through publications, gain by love, journeys concerning, legacies or goods or dead, psychic experiences. If afflicted, loss of legacy, money etc. and reverse results.

Mercury: He/She will have incomplete education, gain by legacy, money or property by marriage, death of partner, some inheritance but difficulty over it. If afflicted or combust, loss of money through partnership and marriage, short life may be the cause for ruin of maternal uncles, etc.

Jupiter: He/She will have gain and honours in handling the estate and money of others, also through law suits, legacies, inheritance insurance etc. of dead persons, occult experiences, mediumistic. If afflicted disappointment over legacy, troubles in financial matters.

Aspect of Sun will shorten the life, other wise average span of life, no gain from education.

Venus: He/She will have gain through wife or ancestral property, danger of drowning or wound on the body, sickness and death of brothers, troublesome journeys, false accusation, gain by the dead, a comfortable fixed partner, spiritualistic experiences. Middle part of or life will be happy.

Saturn: He/She will be long-lived, many misfortunes, secret enemies who may die, troubles over inheritance, death of friends, gift or legacies from friends, worried, unhappy, residence in foreign lands, loss of property, gain of wealth and comfortable life.

Rahu: He/She will have good health and conducive to long life, gifts and legacies and gain by the dead, an evil position for mental aptitude and physical health. He/She is dilatory in work, unholy acts inclined to much sexual activity, wind, stomach, enlargement of glands, diseases are indicated. He has a few sons. He will earn fame, name and respect but his reputation is blemished, mixed financial conditions, oppressed by enemies.

Ketu: He/She will have loss of goods by fraud, sudden or violent death affecting the longevity, wounds, separation from dear ones, stomach troubles, and opposition by enemies. He/She has desire for other's property and wives. If aspect by benefic, there will be wealth and long life and gain of money after obstruction.

Ninth House

Sun: He/She will be bold and industrious, gain of wealth through hard work and industry, respected and famous, diseased health, unhappy and worries in middle of life, sickness abroad or while travelling, danger through over study, work in connection with foreign affairs or universities.

Moon: He/She will have gain through ancestral property, dutiful and prosperous children, wealthy and comfortable, pleasure in foreign lands. If weak or afflicted death of children and losses through above.

Mars: He/She will be wealthy, bold and courageous, gain by books, publications and writings, long journeys, export business, religion, science and wife's relations, prophetic dreams, many fine qualities, splendid possibilities through culture development. If afflicted losses and reverse results.

Mercury: He/She will be a good advisor, disrepute, and gain through science, religion, long journeys and wife's relations, marriage to a stranger, gain from in-laws, and partner's journey to foreign lands. If weak or afflicted loss in foreign countries and reverse results of the above.

Jupiter: He/She will be learned, rank in middle age, wealthy and will lead religious life and good deeds, honourable, voyages, professional journeys, gain and honour through learning, writings, publishing or research work, psychic experience, prophetic dreams, gain through parents relations, science, long journeys, inventions, law and matters of higher mind. If afflicted troubles with foreigners and in religious, legal or educational affairs, fruitless and dangerous voyages, loss and troubles in foreign land.

Venus: He/She will have long journeys, gain through science and publishing, brothers are likely to go to foreign lands and marry, adulterous, death in a distant land, danger of death while on voyage and by drowning. The partner has trouble with relatives over many matters.

Saturn: He/She will have gain and success through long journeys, respected by Govt., and officers' friendship through travel and learning, friends among educated, ministers, writers or inventors, secluded research, sacrifice for science or religion. If afflicted worried and unlucky, adulterous, very unhappy and unlucky, doubtful nature and enchorial relations with people.

Rahu or Ketu: He/She will be unfortunate voyages, curious dreams, trouble and danger of imprisonment in foreign lands, worried, religious, hypocrite intelligent, proud; courageous, wealthy, pilgrimage, unhappy for brothers. He suffers in respect of father.

Tenth House

Sun: He/She will have honour through service, municipal or national activities, respected, famous and native of authority and power, sorrow and afflictions. If afflicted then reversals, cheat and liar, not good thinking. In last part of life may face grief from children.

Moon: He/She will have honourable children, renown in speculations and business also in theatrical work, pleasure or honourable through in laws and own mother, skilled, blessed with long life, gain through children. If afflicted reverse results.

Mars: He/She will be wealthy and respected, magisterial power or army office, bold and valorous, gain by occupation, profession, merchandising, Govt., and from wife's relations and parents, honour, credit and esteem through science, literature and travel, success in foreign affairs, reverse affects if badly aspect and malefic.

Mercury: He/She will be learned, intelligent, good education, wealthy and gain through fruits of the earth, an honourable partner, beneficial to the professional career. If afflicted, there will be troubles through rivals, public disgrace, and scandals through union or enemy.

Jupiter: He/She will be blessed with long life, knowledge or law, intelligent, good deeds, wealthy and respected, also owns property, builder of charitable places, generous and pilgrimage, rise to high social and professional position. If afflicted; displeasures of officers dishonour and reverse of above.

Venus: He/She will be respected, charitable, fond of relation with women, professional or honourable journeys and gain through writings or other accomplishment, danger of violent death, business through matters of dead, rise to high position through in inheritance. If afflicted reverse results.

Saturn: He/She will have friendship with those in high-position, in high social standing persons in business and Govt. circles. If afflicted liability to disgrace, discredit from superiors, Govt., long journeys or retirement and seclusion, sorrows through father and troubles through partner's parents. Middle part of life will be unhappy, non-religious and unhappy.

Rahu or Ketu: He/She achieves honour, credit and high position through merits, industry and ability, proud and fearless, fond of struggle and strife, does not have reliable friends, economical, powerful enemies, happiness from parents is impaired. He may have a few sons, worried mind, in male charts, his liaisons with widows.

Eleventh House

Sun: He/She will have short life, harsh temperament, grief from wife and children, sickness among friends and family, hopes to be realised through acquaintances, interest in legislature, political or social welfare, and reversal in profession. Troubles and worries through enemies, loss of property.

Moon: He/She will have great attachment with wife and children, friends through them, realised hopes and wishes, friends through speculation, or pleasure among legislators, ambassadors and sportsman, happy and good temperament, fond of travels, comforts and gain from children, loss of money by grief from wife.

Mars: He/She will have Gain by friends and accidental fortune. Wealthy respected and famous, blessed with rank, power and authority. Voyages, gain from foreigners. If afflicted grief from wife, loss of wealth and sickness and troubles through above.

Mercury: He/She is a good mathematician, successful hopes, gain through friends, many hopes and wishes attained in old age, marriage to a widow or widower with children, but liable to troubles them, friends become opponents or enemies, gain of money through industry, average wealth.

Jupiter: He/She will have honourable fortune gain from friends among legislator, officers in power or professional positions, ambitious ideals, helpful to associates, much pleasure in life, hopes and wishes often attained. If afflicted detriment friends, hopes often defeated and reverse results of above.

Venus: He/She will have friends through journeys and gain from them, honours and renown through writings, death among friends. Gain through legacies and friends, connection with many women. If afflicted then troubles, grief from brothers and children, unsuccessful hopes, less comforts from friends, etc.

Saturn: He/She will have many friends and beneficial friends, well defined hopes and wishes. If afflicted unfortunate undertakings, unsettled profession, unhappy, non-religious, middle part of life unhappy, great disappointment and obstacles, deceitful friends and losses through them.

Rahu: He/She will be wealthy and long lived, meritorious, friendship, assistance in realization of hopes and wishes, intelligent, few sons, travels, famous, industrious and courageous, avaricious, gain through other's advice, good health, and deafness in old age.

Ketu: He/She will have undesirable associations, loss of opportunities and failure of hopes, false friends, wealthy, popular good authority, gain and success, performs good deeds, commands respect, few sons and worries from them. He will become licentious. Disease of stomach and anus are indicated.

Twelfth House

Sun: He/She will have imprisonment or private enemies, work of mysterious nature, interest in archaeology or submarine life, may suffer from bronchitis, unhappy in old age, may be an engineer or connected with land measurements, many travels.

Moon: He/She will be diseased and worried, children will cause secret sorrows, and speculations may cause ruin, pleasure through mysterious work of research nature, worried and troubled, loss of ancestral property, unlucky for the parents.

Mars: He/She will have early life troublesome, unhappy and worried, all earned money will be destroyed through loss. In middle part of life seeks seclusion for study of occult learning long journeys, difficulties and sorrows through religion, science or journeys, gain by affairs of secret nature.

Mercury: He/She will have afflicted health, long journeys, loss through treachery, end of life in seclusion, devoted to occult science, unhappy marriage, secret sorrows, jealous vexation and sickness, partner or opponents cause imprisonment or fear of it, danger of death at the hands of enemies, if Venus afflicts.

Jupiter: He/She will have secret unhappiness, generous, charitable, respect in profession in foreign lands, gain through occult affairs and secret missions, success in middle life. If afflicted unfortunate environments and conditions, loss of honour, office of dignity through secret enemies, difficulty in employment, gain from occult science.

Venus: He/She will have great sorrows through brothers, secret sufferings, occult learning, seclusion, strained relations with brothers, danger of enmities and imprisonment while travelling, difficulty over inheritance, great sorrows.

Saturn: He/She will have deceitful friends who may cause sufferings, sorrowful friends, good friendship among occults, pleasure in peaceful, quite harmonious and secluded place, occult abilities, powerless enemies, secret investigations, limitations, affliction and adversities prove to be blessings in disguise by developing growth and understanding, worried, greed, spendthrift, non-religious, residence at other than permanent place, quarrels with brothers and friends, unsuccessful hopes, attachment with women, many enemies and adulterous.

Rahu: He/She will have gain by secret method or in seclusion, success in occultism; afflicted health, weak eye sight, windy complaints, heart trouble and such like diseases. He becomes frustrated at the initial stage but successful at last, courageous, unfavourable for pleasures of bed as Rahu is of separative nature.

Ketu: He/She will have loss and troubles through secret enemies, liability of imprisonment and restrictions, bad health, secretive sinful deeds, more expenditure, victorious in disputes, feet, naval, anus, eye diseases. He will be weak and voluptuous.

12

Divisional Charts

12.1 Predictions by Divisional Charts:

The events signified by a house depend on the divisional chart in which we are finding that particular houses. Each divisional chart throws light on event of a specific nature. Again, that colours the meaning of a house. The 4th house from lagna in D-16 may mean something and the 4th house from lagna in D-24 may mean something else. We have already listed the list of life events to be predicted from various divisional charts. We choose the divisional chart to analyze the matter we are interested to predict. This is the key to correct chart analysis, such as, **Career:** we analyze the Dasamsa chart (D-10). **Luxuries and pleasures:** we analyze the Shodasamsa (D-16). The Planets, Sign and Houses show a particular matter of interest and find links between them in the divisional chart.

1) D-1 (Lagna Chart- Deha, physique, body): This is the Lagna / Natal Chart. This is main Varga showing circumstances and actualization. It provides all about our general attitude, fortune and its overall picture of all situations. Rasi is like Head and it describes perspective of individual and resources for living.

2) D-2 (Hora Chart- Sampati, Wealth, money): This division is about dividing the Rasi (sign) into two equal parts. Hora Chart is used to judge the Financial Status & Wealth of the individual. The first half of an odd Rashi or Hora is ruled by Sun. The second half of an odd Rashi or Hora is ruled by Moon. There is reverse in the case of an even Rashi, i.e. the first half of an even Rashi or Hora is ruled by Moon and the second half of an even Rashi or Hora is ruled by Sun. Kashinath Hora is special type of Hora chart which is use to see matters of finances.

3) D-3 (Drekkana Chart- Sukha, happiness from co-born, siblings): One third of a Rashi is called Drekana. It is used to judge the happiness through sibling or co-born happiness, such as their ties with you. In short, this chart predicts all the family skeletons and the family conflicts. The Drekana or Decanate chart is also used to predict the nature of death. Third Bhava lord for odd Lagna or eleventh lord for even Lagna of D-3 will show fortune and work of sibling. Rahu and Ketu will show amount of brothers/sisters.

4) D-4 (Chaturthamsa- Bhagya, fortune, houses/Land properties): Each Chaturthamsa is one fourth of a Rashi. 4th Bhava in D4 will show our house and Lagnesh will show our attitude in matters of house (s). The Lords of the 4 Kendra from a Rashi are the rulers of respective Chaturthamsa of a Rashi, commencing from Mesh. The deities, respectively, are Sanak, Sanand, Kumar and Sanatan. Chaturthamsa is used for assessing the immoveable properties and prosperity. Eighth lord in D4 shows us blockage to our fortune.

5) D-5 (Panchamsa- Fame, authority and power): It is One-Fifth Sign. It is used to predict the fame, authority and power as well as the spiritual inclinations. It also reveals the spiritual evolution of the native and whether one has a leaning for religion, philosophy or even atheism or moral fibre.

6) D-6 (Shashtiamsa- Health & Diseases): It is One-Sixth Sign. It is used to predict the diseases and hereditary diseases, if any, and general health of the native. The indications about the diseases help to take the remedial measures as suggested by the chart.

7) D-7 (Saptamsa- Putra, sons and grandsons): It is One-Seventh Sign. The Saptamsa counting commences from the same Rashi in the case of an odd Rashi. It is from the seventh Rashi for an even Rashi. The names of the seven divisions in odd Rashi are Kshaar Ksheer, Dadhi, Ghrith, Ikshu, Ras, Madhya and Suddh Jal. These designations are reversed for an even Rashi. Saptamsa is an interesting chart. Saptamsa is used for study of children, grand children or childbirth, prosperity of issues and childlessness and happiness through children. Fifth Bhava (ninth for even Rasi of Lagna, or odd Rasi with Ketu) will show character, sex and fortune of first child. Relation between Lagnesh of D7 and this lord will show the relations with our kid.

8) D-8 (Ashtamsa- Sudden and unexpected troubles, litigation): It is One-Eighth of Sign. It is used to predict the accidents and the longevity of individual with a fair accuracy.

9) D-9 (Navamsa- Kalatra, spouse, Marriage): It is One-Ninth of Sign. It is used for study mainly about the spouse, the couple compatibility by matching the two Navamsa Charts of the couple instead of Natal Chart, the quality of spouse and the time of marriage, longevity of the marriage life. It reveals the temperament, mental and moral character of spouse, the number of the marriages and whether one's married life will be happy or otherwise.

10) D-10 (Dasamsa- Mahat Phalam, Power, Position, Profession, Career, and achievements): It is One-Tenth of Sign. Dasamsa is used for study the profession, carrier & allied matters and success, educational, vocation and power and position of the native. Dasamsa gives the destiny and the field in which native's good fortune resides. This also gives indication of wealth. Second, sixth and tenth Bhava of D-10 will show type of work (profession). Venus will show private sector, while Luminaries will give work in politics.

11) D-11 (Ekadasamsa- Rudramsa- Death and destruction): It is One-Eleventh Sign. It is about gain, cure, inheritance, the legacies, and sudden inflow of money in chance games, speculations and gambling.

12) D-12 (Dwadasamsa- parents, uncles, aunts, grand-parents): It is One-Twelfth Sign. It is about Parents, details about parents, their length of life and higher consciousness and links the past and future lives. Dvadasamsa can also predict skill in astrology and philosophy and in depth knowledge of the native correctly.

13) D-14 (Chaturadasamsa- Vahana): It is One-fourteenth Sign. Chaturadasamsa helps in study of Conveyance and general happiness.

14) D-16 (Shodasamsa- Sukha/Asukham- happiness/distress from Vahana-Vehicles): It is One-Sixteenth Sign. It indicates the General happiness with conveyance (Vehicle) and timings of potential accidents in the native's life. The House, whose Lord is in a benefic Shodashamsa will flourish.

15) D-20 (Vimsamsa- Upasana, worship, Sadhna, spirituality): It is One-Twentieth Sign. Lagnesh in that Varga will show our approach to spirituality. Surya here is the real source/ and core of spirituality while Jupiter shows our Jnana. Mangal in Lagna can show focus on Ahimsa. Twelfth and Forth lords show meditation/dhyana while Kona's are connected to Jnana Marg (fifth is bhakti).

16) D-24 (Chaturvimshamsa- Vidya- Learning, knowledge, education): It is One-Twenty fourth Sign. Moon will show how we study. If it is connected to night-planets then we like to study in night. Tenth lord shows how we teach.

17) D-27 (Saptavimshamsa/Nakshatramsa- Strengths and weaknesses, inherent nature): It is One-Twenty seventh Sign. It is about strength of native. The Saptavimshamsa Lords are, respectively, the presiding deities of the 27 Nakshatra. These are for an odd Rashi. Count these deities in a reverse order for an even Rashi.

18) D-30 (Trimsamsa- Vistha Phala- Evils effects, punishment, diseases): One thirty of a Rashi is called Trimsamsa. Trimsamsa shows knowledge about vices and diseases. It is also used to time death and punishment. It is about Malefic Effects and Bad luck. These are totally 360. Trimsamsa indicates the Evil effects.

19) D-40 (Khavedamsa- Subha/Asubha- auspicious/inauspicious effects): One forty of a Rashi is called Khavedamsa. It is about Auspicious and inauspicious effects.

19) D-45 (Akshavedamsa- All indications): It is One-forty-fifth Sign. It is a general indication about good and bad effects and Conditions of the native in all areas.

20) D-60 (Shashtiamsa- Karma of past life and its effects on current life): It is One-Sixtieth Sign or half a degree each. It is a general indication about good and bad effects in all areas of life. Shashtiamsa indicates the All Conditions. The House, whose Lord is in a malefic Shashtiamsa will diminish. Graha in Benefic Shashtiamsa produces auspicious, while the opposite produce inauspicious effects.

Navamsa Chart Predictions

Navamsa Chart Making: The very simplified method of making Navamsa Chart is that we divide the total measure of the angle or degree including their minutes of arc of position of the Ascendant or Planets or Points by 3^0 20', i.e. 3.333. We get some quotient and some fraction in the form of decimals. Consider a digit 1 for the fractions. We add 1 for every fraction with the quotient to get the total number. If the total number is less than twelve (12), then count the number from Aries, i.e. the first Sign of the Lagna - Chart. Then place that Ascendant or Planets or Points in the house where it falls. If the total number is more than twelve (12), we divide the total number by 12 and we get some remainder digit from 1 to 11 or some time the remainder will be zero. If the remainder is zero, we consider it as 12. Then, we count the remainder digit number from Aries, i.e. the Sign number one (1) of the Lagna Chart and place that Ascendant or Planets or Points in that house where it falls.

Aries Lagna (Rising) Navamsa: Following are the effects of births in the nine Navamsa, such as, Effects of Aries Ascendant, First Navamsa **(Up to 3^0 20'):** He/She will have a face, resembling that of a he-goat, with nose and shoulders not being very prominent. He/She will have a fierce voice, ugly appearance and narrow eyes. His/Her body will be thin, but free from defects. Effects of Aries Ascendant,

Second Navamsa **(3^0 20' to 6^0 40')**: He/She will be dark in complexion; will have broad shoulders and long arms, small forehead, strong collar bones, sharp sight and prominent face and nose. He/She will be an affable speaker and will possess weak legs. Effects of Aries Ascendant, Third Navamsa **(6^0 40' to 10^0 00')**: He/She will suffer loss of hair, be fair in complexion, will have irregular (defective) arms, charming eyes and nose, will be a scholar in direct poetic ability and will have weak thighs. Effects of Aries Ascendant, Fourth Navamsa **(10^0 00' to 13^0 20')**: He/She will have an erratic sight, be irascible, short-nosed, and wandering-natured, will have rough legs and coarse hair, and be bereft of co-born and emaciated. Effects of Aries Ascendant, Fifth Navamsa **(13^0 20' to 16^0 40')**: He/She will be fierce and will have eyes, resembling that of a supreme elephant, a fat nose, and thick eye brows, wide fore face, fat body and coarse hair. Effects of Aries Ascendant, Sixth Navamsa **(16^0 40' to 20^0 00')**: He/She will be dark in complexion, soft in disposition, will have eyes, akin to that of a deer, be tall in stature, will have irregular (defective) stomach and hands, be a eunuch, and be timid and garrulous. Effects of Aries Ascendant, Seventh Navamsa **(20^0 00' to 23^0 20')**: He/She will have complexion akin to green sprout, be fickle-minded, will possess white eyes, will marry an unchaste lady, be malicious and will have a broad physique. Effects of Aries Ascendant, Eighth Navamsa **(23^0 20' to 26^0 40')**: He/She will have a face, akin to that of a monkey, be a good speaker, will have an afflicted and tawny body, will suffer from secret diseases, be torturous, be a liar, be fond of friends and be fierce. Effects of Aries Ascendant, Ninth Navamsa **(26^0 40' to 30^0 00')**: He/She will be tall, emaciated, wandering, will have defective fore face and ears, will have a face, akin to that of a horse, will possess many names and be crooked.

Taurus Lagna (Rising) Navamsa: Following are the effects due to births in the Navamsa belonging to Taurus Ascendant: Effects of Taurus Ascendant, First Navamsa **(Up to 3^0 20')**: He/She will have an even and dark coloured physique, be hard-hearted, will perform obsequies in the beginning and ending parts of his life, be base, will indulge in unnatural acts and will have crooked sight. Effects of Taurus Ascendant, Second Navamsa **(3^0 20' to 6^0 40')**: He/She will be endowed with majestic looks, be indolent, will have a bent body, be not very intelligent, will indulge in hostile acts and be a great liar. Effects of Taurus Ascendant, Third Navamsa **(6^0 40' to 10^0 00')**: He/She will be erratic in sight, be fierce, will have a short nose, be wandering natured, will have coarse legs and hair will have soft limbs, be beautiful, will have broad eyes and big limbs, be interested in

Sacrifices and will have stiff legs and hands. Effects of Taurus Ascendant, Fourth Navamsa **(10° 00' to 13° 20')**: He/She will be short in stature, will be wandering natured, be easily irritable, will have eyes, akin to that of a he-goat, is tawny in complexion, be poor and will steal others' wealth. Effects of Taurus Ascendant, Fifth Navamsa **(13° 20' to 16° 40')**: He/She will be vicious, will have a well-elevated nose, and will appear, like a giant ox, will have crooked hair, be sportive, will have large shoulders and hips. Effects of Taurus Ascendant, Sixth Navamsa **(16° 40' to 20° 00')**: He/She will have beautiful eyes and hair, be firm, be endowed with a fair complexioned physique, will speak sweetly, be pre-eminent, be fond of amusements, and be emaciated and skilful. Effects of Taurus Ascendant, Seventh Navamsa **(20° 00' to 23° 20')**: He/She will be interested in females, who lost their sons, will have somewhat elevated nose and prominent eyes, will possess a strong physique, will hate his own men and will have stout feet and exquisite hair. Effects of Taurus Ascendant, Eighth Navamsa **(23° 20' to 26° 40')**: He/She will have eyes, akin to that of a tiger and charming teeth, be unconquerable, will possess a full-blown nose, will work sparingly, will have curly and bluish hair and sharp nails and be garrulous. Effects of Taurus Ascendant, Ninth Navamsa **(26° 40' to 30° 00')**: He/She will be honourable, will not be very strong, and be timid, given to anger, will possess an even and charming body, be a rogue (or cheat), will gather money, be famous, will have a thin lower body.

Gemini Lagna (Rising) Navamsa: Following are the effects of births in the all Navamsa of Gemini Ascendant. Effects of Gemini Ascendant, First Navamsa **(Up to 3° 20')**: He/She will have hair on the shoulders, will possess charming and dark eyes and an elevated nose, be akin to green (Durva) grass in complexion and will possess thin legs and thin hands. Effects of Gemini Ascendant, Second Navamsa **((3° 20' to 6° 40')**: He/She will have a pot-like head, will do dirty acts, be fond of torturous deeds, will have depressed nose, will speak much, will work much and will lead in strife and quarrels. Effects of Gemini Ascendant, Third Navamsa **(6° 40' to 10° 00')**: He/She will be fair in complexion, will possess blood-red eyes, charming nose and even physique be very intelligent, will have a long face and dark eye-brows and is a skilful speaker. Effects of Gemini Ascendant, Fourth Navamsa **(10° 00' to 13° 20')**: He/She will possess charming eye brows and forehead, be lustful, will possess a physique with the splendour of a blue lotus, will be broad-chest and white teethed, be soft in speech and will have attractive hair. Effects of Gemini Ascendant, Fifth Navamsa **(13° 20' to 16° 40')**: He/She will have a broad face, strong chest and big head, be wicked, cunning and will possess charming and friendly looks. Effects of Gemini Ascendant, Sixth Navamsa **(16° 40' to 20° 00')**: He/She will possess eyes with the hue of honey, be garrulous, will possess a broad fore face, even body and charming lips, be a rogue, be fickle-minded and be strong. Effects of Gemini Ascendant, Seventh Navamsa **(20° 00' to 23° 20')**: He/She will possess a copper coloured physique, copper red and prominent eyes and a broad chest, be skilful in teaching and arts and be jocular in disposition. Effects of Gemini Ascendant, Eighth Navamsa **(23° 20' to 26° 40')**: He/She will be dark in complexion, be great, intelligent, soft in disposition, sweet in speech, will have a broad and tall physique and large and black eyes and will be an expert in arts. Effects of Gemini Ascendant, Ninth Navamsa **(26° 40' to 30° 00')**: He/She will have round and dark coloured eyes and charming body, be successful, very intelligent and be fond of sexual cohabitation, poetry and worldly knowledge.

Cancer Lagna (Rising) Navamsa: Following are the effects for births in various Navamsa related to Cancer Ascendant: Effects of Cancer Ascendant, First Navamsa **(Up to 3° 20')**: He/She will have a clean, charming and fair coloured physique, beautiful hair, broad belly, impressive face, prominent eyes, thin body and thin shoulders. Effects of Cancer Ascendant, Second Navamsa **(3° 20' to 6° 40')**: He/She will be blood red in complexion, be fierce in quarrels, will like fine arts, will possess face and eyes, akin to that of a cat, will be well disposed to scarify for others and will have weak knees and shanks. Effects of Cancer Ascendant, Third Navamsa **(6° 40' to 10° 00')**: He/She will be fair in complexion, will possess beautiful eyes, be an eloquent speaker, will have a soft body, akin to that of a female, be intelligent be a sparing and light worker and be indolent. Effects of Cancer Ascendant, Fourth Navamsa **(10° 00' to 13° 20')**: He/She will be black in complexion, will have pressed eyebrows, be graceful in appearance, will have charming eyes and nose, be courageous, liberal, will perform acts prescribed for superior caste-men and be crafty. Effects of Cancer Ascendant, Fifth Navamsa **(13° 20' to 16° 40')**: He/She will possess voice akin to the sound of bell, crooked, or stooping face, allied eyebrows and very long arms, be interested in worship, be bereft of dutifulness, will injure others and be not very intelligent. Effects of Cancer Ascendant, Sixth Navamsa **(16° 40' to 20° 00')**: He/She will have a long and broad physique, charming eyes and great courage, be fair in complexion, be a good speaker and will possess beautiful nose and big teeth. Effects of Cancer Ascendant, Seventh Navamsa **(20° 00' to 23° 20')**: He/She will have scattered hair, big body and sinewy knees, will be disposed to protect others' families and will be akin to a crow in appearance. Effects of Cancer Ascendant, Eighth Navamsa **(23° 20' to 26° 40')**: He/She will have a head with bell shape, charming face, shoulders and limbs, will be a degraded artisan (i.e. an infamous worker), will have the gait of a tortoise and crooked nose and be dark in complexion. Effects of Cancer Ascendant, Ninth Navamsa **(26° 40' to 30° 00')**: He/She will be fair complexioned, will possess eyes resembling fish, be great, be soft-bellied, broad-chest, and will have prominent chins and lips, large, but weak knees and similar ankles.

Leo Lagna (Rising) Navamsa: Following effects will mature in the various Navamsa prevailing in Leo Ascendant at birth: Effects of Leo Ascendant, First Navamsa **(Up to 3° 20')**: He/She will have an even belly (like a lion), be fierce, will have sharp and blood-red nose and a big head, be valorous and will possess a prominent and fleshy chest. Effects of Leo Ascendant, Second Navamsa **(3° 20' to 6° 40')**: He/She will have a prominent and broad fore face, a square body, broad eyes, broad chest, long arms and big nose. Effects of Leo Ascendant, Third Navamsa **(6° 40' to 10° 00')**: He/She will have hairy and broad arms, eyes, akin to that of (Greek) partridge (said to live on moon-beams), be fickle-minded, charitable, will have an elevated nose, pure white physique and round neck. Effects of Leo Ascendant, Fourth Navamsa **(10° 00' to 13° 20')**: He/She e will have (ash-coloured) body akin to ghee, large and black eyes, soft hair, peculiar voice, big hands and legs and stomach, resembling that of a frog. Effects of Leo Ascendant, Fifth Navamsa **(13° 20' to 16° 40')**: He/She will have a bell-shaped head with limited hair, charming nose and eyes, hairy body and long belly, be fierce and will have unsightly teeth and strong and broad cheek. Effects of Leo Ascendant, Sixth Navamsa **(16° 40' to 20° 00')**: He/She will have limited, but soft hair on his physique, white and large eyes, be tall in stature, dark in complexion, will have proven skill (only) among females, be swaggering and be learned. Effects of Leo Ascendant, Seventh Navamsa **(20° 00' to 23° 20')**: He/She will have a long face, be sinewy, will have a prominent physique, be unfortunate in the matter of wife (or females in general), dark-complexioned, fierce, hairy-bodied and be cunning and harsh in speech. Effects of Leo Ascendant, Eighth Navamsa **(23° 20' to 26° 40')**: He/She will be endowed with excellent speech, firm limbs, charming and majestic looks, be undutiful, poor and crafty. Effects of Leo Ascendant, Ninth Navamsa **(26° 40' to 30° 00')**: He/She will have a face akin to a donkey's, will possess dark eyes, long arms, and charming legs and be troubled by breathing disorders.

Virgo Lagna (Rising) Navamsa: The various Navamsa emanating from Virgo Ascendant at birth will give the following effects: Effects of Virgo Ascendant, First Navamsa **(Up to 3^0 20')**: He/She will possess eyes, akin to that of an antelope, be a good speaker, be charitable, will enjoy sexual pleasures, be very rich, dark in complexion and large hearted. Effects of Virgo Ascendant, Second Navamsa **(3^0 20' to 6^0 40')**: He/She will have a charming face, charming eyes and fair complexion, be soft, argumentative, fickle minded and long-bellied. Effects of Virgo Ascendant, Third Navamsa **(6^0 40' to 10^0 00')**: He/She will have blown nose, prominent feet, long arms, pure speech, and fair complexion and be friendly. Effects of Virgo Ascendant, Fourth Navamsa **(10^0 00' to 13^0 20')**: He/She will be learned, will be sportive with the fair sex, is beautiful, sweet, blood red in complexion, sharp, intelligent, emaciated and will have charming eyes and face. Effects of Virgo Ascendant, Fifth Navamsa **(13^0 20' to 16^0 40')**: He/She will have large lips and hands, big body, broad chest, strong ankles and will depend on others. Effects of Virgo Ascendant, Sixth Navamsa **(16^0 40' to 20^0 00')**: He/She will have charming appearance, impressive speech, and splendour body and is an exponent of Shashtra, be very intelligent, skilful in writing and fine arts, be good hearted and will take pleasure in walking, or roaming. Effects of Virgo Ascendant, Seventh Navamsa **(20^0 00' to 23^0 20')**: He/She will have a small face, elevated nose, compact arms, very fair complexion, prominent belly, hands and legs and will have fear for water. Effects of Virgo Ascendant, Eighth Navamsa **(23^0 20' to 26^0 40')**: He/She will be very beautiful, fair in complexion, tall in stature, will have charming eyes, be fierce, honourable and will have long and stout arms and brown hair. Effects of Virgo Ascendant, Ninth Navamsa **(26^0 40' to 30^0 00')**: He/She will be famous, will have charming physique, broad eyes, incomparable vigour, be skilful, be with stooping shoulders and be learned a writer.

Libra Lagna (Rising) Navamsa: The various Navamsa out of Libra Ascendant at birth will produce following effects: Effects of Libra Ascendant, First Navamsa **(Up to 3^0 20')**: He/She will be fair complexioned, broad eyed, praiseworthy, long-faced, skilful in business, happy and famous. Effects of Libra Ascendant, Second Navamsa **(3^0 20' to 6^0 40')**: He/She will have squint and round eyes, elevated (ill formed) teeth, depressed waist, charming neck, large (physical) heart, ugly body and compact brows. Effects of Libra Ascendant, Third Navamsa **(6^0 40' to 10^0 00')**: He/She will be fair in complexion, will have a face, akin to that of a horse, be thin-bodied, famous, long-haired and long-nosed and will have beautiful legs. Effects of Libra Ascendant, Fourth Navamsa **(10^0 00' to 13^0 20')**: He/She will have weak hands, be timid, will have ill-formed teeth (some placed over others), weak body, rolling eyes, small nails, dark complexion and be devoid of virtues and be miserable. Effects of Libra Ascendant, Fifth Navamsa **(13^0 20' to 16^0 40')**: He/She will have majestic looks, be firm disposition, be not proud, rough haired, even-eyed and will possess a beautiful nose. Effects of Libra Ascendant, Sixth Navamsa **(16^0 40' to 20^0 00')**: He/She will have fleshy limbs, be fair in complexion, will have broad eyes, beautiful nose and white nails, be diplomatic and be learned in Shashtra. Effects of Libra Ascendant, Seventh Navamsa **(20^0 00' to 23^0 20')**: He/She will be blood-red in complexion, be intelligent, will have long physique and long arms and a big head, and be miserly, fierce and intelligent. Effects of Libra Ascendant, Eighth Navamsa **(23^0 20' to 26^0 40')**: He/She will have elevated shoulders and prominent neck, will enjoy pleasures, will have a coarse physique, long and dark brows, is a polite speaker and will have a beautiful chest and bruised head. Effects of Libra Ascendant, Ninth Navamsa **(26^0 40' to 30^0 00')**: He/She will have charming eyes, pleased mind, be fair complexioned, even and beautiful bodied, be skilful, fond of arts, and be charitable and jocular.

Scorpio Lagna (Rising) Navamsa: The Ascendant Scorpio will produce various following effects according to the nine Navamsa thereof. Effects of Scorpio Ascendant, First Navamsa **(Up to 3^0 20')**: He/She will be short in stature, will have prominent lips and nose, charming forehead, strong and fair complexioned body with belly, akin to that of a frog and will act, as a marriage broker (ascertaining genealogies and negotiating marital alliances). Effects of Scorpio Ascendant, Second Navamsa **(3^0 20' to 6^0 40')**: He/She will be fair in complexion, will possess a strong and broad chest and shoulders and reddish eyes, and will conquer his enemies be valorous and will have abundant hair. Effects of Scorpio Ascendant, Third Navamsa **(6^0 40' to 10^0 00')**: He/She will be learned, will have strong shoulders and arms and beautiful hair, be endowed with clear speech, fair complexion and charming lips. He is born of a virgin. Effects of Scorpio Ascendant, Fourth Navamsa **(10^0 00' to 13^0 20')**: He/She will be intent upon joining others' wives, will induce others to be active, be valorous, tall, dark in complexion, dark-haired and dark eyed. Effects of Scorpio Ascendant, Fifth Navamsa **(13^0 20' to 16^0 40')**: He/She will be majestic, will possess copper-red eyes, depressed nose, be courageous, proud, will perform fearful acts, be famous and will have a strong physique. Effects of Scorpio Ascendant, Sixth Navamsa **(16^0 40' to 20^0 00')**: He/She will be impudent, intelligent of a high order, will have elevated nose and great strength, and will be endowed with knowledge of justice (or be diplomatic), be skilful, will possess less hair and compact brows. Effects of Scorpio Ascendant, Seventh Navamsa **(20^0 00' to 23^0 20')**: He/She will have a split face, strong body, and teeth in various sizes, depressed belly, and squint sight and be very splendour. Effects of Scorpio Ascendant, Eighth Navamsa **(23^0 20' to 26^0 40')**: He/She will have blown nose, be dark in complexion, devoid of virtues, dirty in appearance, will possess stiff hair and be foolhardy. Effects of Scorpio Ascendant, Ninth Navamsa **(26^0 40' to 30^0 00')**: He/She will have a fair coloured physique, be beautiful, like a deer, will possess calm and tawny eyes and similar hair and strong body and be amiable to elders.

Sagittarius Lagna (Rising) Navamsa: The native born in Sagittarius Ascendant, but in its various Navamsa will obtain following effects: Effects of Sagittarius Ascendant, First Navamsa **(Up to 3^0 20')**: He/She will have charming big nose, will possess sight, akin to that of a goat, be a gifted speaker, will have charming teeth and hair, be fair in complexion, will have inset testicles and be fierce. Effects of Sagittarius Ascendant, Second Navamsa **(3^0 20' to 6^0 40')**: He/She will have a prominent head, be firm in disposition, will have large eyes, strong waist and knees, ugly nose, tall stature and firm cheeks. Effects of Sagittarius Ascendant, Third Navamsa **(6^0 40' to 10^0 00')**: He/She will have skill in educating and in fine arts, be majestic, just, fond of females, intelligent and jocular. Effects of Sagittarius Ascendant, Fourth Navamsa **(10^0 00' to 13^0 20')**: He/She will be skilful, be tawny in complexion, will possess round eyes, fair-coloured physique and a belly akin to tortoise, be intelligent, wandering-natured, will have charming hair and charming appearance. Effects of Sagittarius Ascendant, Fifth Navamsa **(13^0 20' to 16^0 40')**: He/She will have large ears, eyes and face, will possess a (majestic) physique, like a lion, widely spread eye brows, strong shoulders and arms, hairless physique and firm disposition. Effects of Sagittarius Ascendant, Sixth Navamsa **(16^0 40' to 20^0 00')**: He/She will have beautiful and large eyes, broad forehead and broad face, be a poet, and be mean and interested in scholarly discussions. Effects of Sagittarius Ascendant, Seventh Navamsa **(20^0 00' to 23^0 20')**: He/She will be dark in complexion, soft in disposition, will keep up his promise, will have a prominent head, will be interested in accumulating savings, be tall in stature, will possess broad eyes and be liberal. Effects of Sagittarius Ascendant, Eighth Navamsa **(23^0 20' to 26^0 40')**: He/She will be flat-nosed, broad headed, inimical, and erratic sighted, garrulous and be dear to elders. Effects of Sagittarius Ascendant, Ninth Navamsa **(26^0 40' to 30^0 00')**: He/She will be fair in complexion, will have a face, akin to that of a horse and broad and dark eyes, will speak sparingly, be truthful, miserable and will possess crooked walking limbs.

Capricorn Lagna (Rising) Navamsa: The various Navamsa resulting in the Ascendant Capricorn will emanate the following effects: Effects of Capricorn Ascendant, First Navamsa **(Up to 3^0 20')**: He/She will have interstice and outwardly visible teeth, be dark in complexion, will speak with broken words, or be stammering, will have coarse hair, be famous, be interested in music and amusement, be emaciated and will have fluctuating wealth. Effects of Capricorn Ascendant, Second Navamsa **(3^0 20' to 6^0 40')**: He/She will be indolent, crafty, crooked nosed, fond of music, broad bodied, be interested in many females, will prattle much and be skilful. Effects of Capricorn Ascendant, Third Navamsa **(6^0 40' to 10^0 00')**: He/She will have lust for music, be famous, fair complexioned, be endowed with superior looks and charming nose, will be fond of many friends and relatives and achieves fulfilment of desires. Effects of Capricorn Ascendant, Fourth Navamsa **(10^0 00' to 13^0 20')**: He/She will have round eyes with a mix of blood-red and black hue, large forehead, emaciated body and thin arms, scattered hair, interstice teeth and broken speech. Effects of Capricorn Ascendant, Fifth Navamsa **(13^0 20' to 16^0 40')**: He/She will have prominent neck, nose and belly, will enjoy pleasures, be attached to women, be dark in complexion, will have round knees and arms and will attain successful beginnings in his undertakings. Effects of Capricorn Ascendant, Sixth Navamsa **(16^0 40' to 20^0 00')**: He/She will possess a splendorous body, will attire charmingly, be libidinous, will possess small and even teeth, is a good speaker and will have big cheeks and large forehead. Effects of Capricorn Ascendant, Seventh Navamsa **(20^0 00' to 23^0 20')**: He/She will be dark in complexion, be indolent, be an eloquent speaker, be short-haired, big-bodied, be harsh in disposition, will possess soft hands and legs, and be intelligent and very virtuous. Effects of Capricorn Ascendant, Eighth Navamsa **(23^0 20' to 26^0 40')**: He/She will be endowed with majestic sight and charming nose, reddish face, uneven nails and hair, grotesque body and forehead protruding, like a pot. Effects of Capricorn Ascendant, Ninth Navamsa **(26^0 40' to 30^0 00')**: He/She will have broad chest and large eyes, high intelligence, fully developed face, interest in musical studies, be endowed with sweetness and strength, be gentle and diplomatic.

Aquarius Lagna (Rising) Navamsa: The various Navamsa ascending, while Aquarius is on the East will produce following effects: Effects of Aquarius Ascendant, First Navamsa **(Up to 3^0 20')**: He/She will be dark in complexion, be soft in disposition, will have an emaciated body and prominent cheeks, be learned in poetry and Shashtra, be libidinous, interested in carnal pleasures and be splendour. Effects of Aquarius Ascendant, Second Navamsa **(3^0 20' to 6^0 40')**: He/She will possess coarse skin, nails, sight and hair, be kind to the helpless, be gentle, be tall in stature, foolish and will have a 'distinct' head. Effects of Aquarius Ascendant, Third Navamsa **(6^0 40' to 10^0 00')**: He/She will have a compact body (limbs sticking close to each other), be fond of females, will possess the splendour of lapis lazuli, be learned in the meanings of Shashtra and will act accordingly. Effects of Aquarius Ascendant, Fourth Navamsa **(10^0 00' to 13^0 20')**: He/She will be fond of women, be fair in complexion, will have a split face, will destroy enemies, be majestic, courageous and be fond of pleasures and sexual enjoyments. Effects of Aquarius Ascendant, Fifth Navamsa **(13^0 20' to 16^0 40')**: He/She will be learned in Shashtra and in fine arts and dark in complexion and will have coarse hair on legs, concealed (not prominent) neck and ears. Effects of Aquarius Ascendant, Sixth Navamsa **(16^0 40' to 20^0 00')**: He/She will have a face, resembling that of a tiger, be bold, short-haired, will have unchanging aims, will kill living beings viz. tiger, deer, snake etc. and be dear to king. Effects of Aquarius Ascendant, Seventh Navamsa **(20^0 00' to 23^0 20')**: He/She will have eyes and face, resembling that of a goat, be fierce in disposition, be delighted in village life, insulted by females, will suffer diseases of bilious imbalances and be endowed with strength and courage. Effects of Aquarius Ascendant, Eighth Navamsa **(23^0 20' to 26^0 40')**: He/She will possess a no diminishing strength, firm be disposition and affection, be a warrior with the king, or be a king himself, be beautiful and will have strong teeth and broad eyes. Effects of Aquarius Ascendant, Ninth Navamsa **(26^0 40' to 30^0 00')**: He/She will be dark in complexion, will possess unclean and elevated teeth, be disunited from his wife, children and wealth, be an affable speaker and be famous and skilful.

Pisces Lagna (Rising) Navamsa: Should Pisces ascend at birth, the various Navamsa thereof will yield following specific effects: Effects of Pisces Ascendant, First Navamsa **(Up to 3^0 20')**: Though he/she may be white in complexion, his body will reveal the splendour of blood-red hue; he will be soft in disposition, be akin to a female in mental makeup, (i.e. will act, like a female), be fickle-minded and will have a short neck and emaciated waist. Effects of Pisces Ascendant, Second Navamsa **(3^0 20' to 6^0 40')**: He/She will have a big nose, be skilful in his assignments, will eat meat and the like, be endowed with a charming physique, will wander in forests and hills and will have a big head. Effects of Pisces Ascendant, Third Navamsa **((6^0 40' to 10^0 00')**: He/She will be white in complexion, be crafty, will possess beautiful eyes, and be beautiful, righteous, learned, courteous, modest and charming in appearance. Effects of Pisces Ascendant, Fourth Navamsa **(10^0 00' to 13^0 20')**: He/She will have praiseworthy attributes, will fall into adversity, will serve aged people, be skilful in his assignments, learned, very strong versed in justice and will have elevated nose. Effects of Pisces Ascendant, Fifth Navamsa **(13^0 20' to 16^0 40')**: He/She will be tall in stature, dark in complexion, be valorous, be not peaceful, will have a small nose and charming eyes, be fond of torturing others, be impatient, and will possess beautiful teeth and prattle. Effects of Pisces Ascendant, Sixth Navamsa **(16^0 40' to 20^0 00')**: He/She will be self-respected, righteous, excellent, strong, miserable, crafty, and unsteady and be a minister. Effects of Pisces Ascendant, Seventh Navamsa **(20^0 00' to 23^0 20')**: He/She will be self-respected, will show interest in other religions, be excellent, and be a minister, strong, miserable, cruel and unsteady. Effects of Pisces Ascendant, Eighth Navamsa **(23^0 20' to 26^0 40')**: He/She will be tall in stature, will have a big head, be emaciated, indolent, will have uneven (or dirtied) eyes and hair, will have foolish children (or a few children), be interested in earning money and be skilful in war (or quarrels). Effects of Pisces Ascendant, Ninth Navamsa **(26^0 40' to 30^0 00')**: He/She will be short, soft in disposition, courageous, broad chest, broad eyed, big nosed, be bright, will have a broad physique, be intelligent, virtuous and famous.

13

Dasha Period

The Dasha is period's timingg gives the effects of different events of an individual. Malefic actually give excellent results if they are well placed, well aspect, or ruling good houses, i.e. the first, fifth and ninth houses. Benefice planets give terribly bad results when badly associated or occupying or ruling bad houses (6th, 8th and 12th) or poorly placed in sign of Debilitation.

Every human being is scheduled to experience nine major periods within a lifetime, assuming one's life is 120 years. Each period is called a Maha Dasa and corresponds to one of the original seven heavenly bodies and Rahu (North Node) and Ketu (South Node). Uranus, Neptune, and Pluto are excluded as the Dasa system because it was found after the discoveries of Dasa System. The system of 'Dasha' or period predicts the past, present and future. One particular planet influences the native more than other planets during Dasha Period, and then the other planet takes over and so on in a cyclic order. Birth chart does not predict when the indicated events will take place.

13.1 Vimshottary Dasa Systems

The Vimshottary (""vim-sho-tree") Dasha Systems are the most widely used as it is far more popular and believed to be more accurate. Each period is called a Maha Dasha and comprises of 120 Years period, which is based on the fact that normal span of life for all persons was considered to be 120 years. Beginning from Krittika, the Lords of Dasha are Surya, Chandra, Mangal, Rahu, Guru, Sani, Buddha, Ketu and Sukra in that order. Thus, if the Nakshatra from Krittika to the Janma Nakshatra are divided by nine, the remainder will signify the Lord of the commencing Dasha. The remaining Dasha will be of the Graha in the order, given above. The periods of Dasha of Surya, Chandra, Mangal, Rahu, Guru, Sani, Buddha, Ketu and Sukra are 6, 10, 7, 18, 16, 19, 17, 7 and 20 in that order. The Dasha that a person is born into is based upon the degree and minute of the natal Moon position. In Vimshottary Dasha system, the complete cycle of Dasha is of 120 years. The most important charts for reconciliation will be Birth Chart (Rasi Chart) and Divisional Charts.

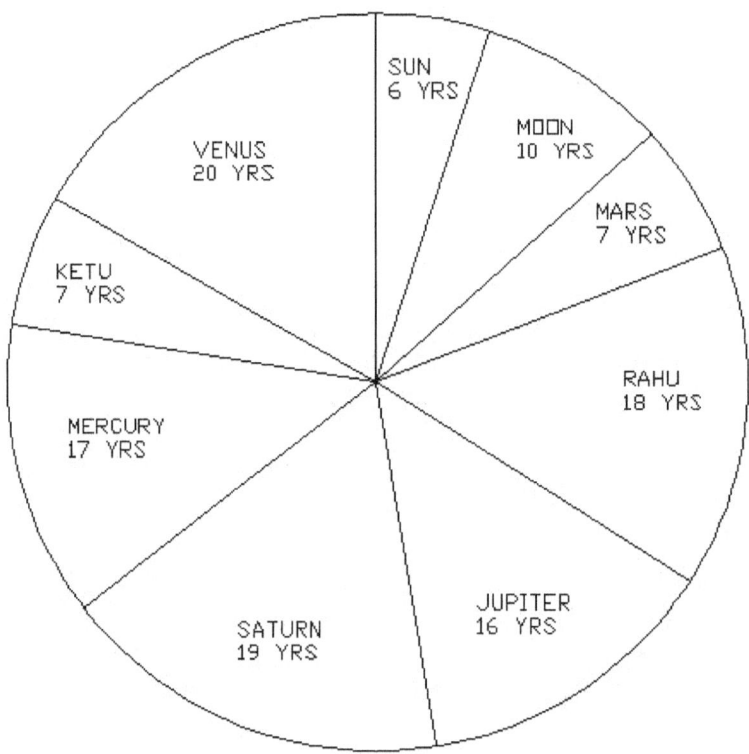

Balance Dasha: At the time of birth, the constellation occupied by the Moon (Nakshatra) determines which planet will rule at that time. The fraction of the constellation balance at that time determines the "balance of Dasa". "Balance of Dasha" is the time period, which the planet (Dasha Lord) will rule the native after his birth. After the end of the ruling period of that planet, other planets rule one by one in the cyclic order of the Dasha as listed in table 1. It is a fact that the planets are pure effective and influencing during their major (Main) Dasha and Sub-Period (Antradasa) Dasha and Pratyantar Dasha. The potential possessed by the planet is effective in the native life, either for good or bad effects as per the planet's influence depending whether the planet is exalted or debilitated, Vargottama or Yoga karaka, benefic or malefic, and aspect by other friendly or inimical planets. Period or Sub periods become malefic in the association of malefic planets

Calculation of Maha Dasa Balance: Maha Dasa balance is the period equal to the total Maha Dasa Period of the running planet at the birth time minus the elapsed Maha Dasa Period of that planet at the time of birth. We give following methods for calculation of Vimshotari balance Maha Dasa:

1. Degree Method: The Maha Dasha at the time of birth of a person depends upon the position of the Moon in the Nakshatra at the time of his birth. The boundary of each Nakshatra is marked by a particular fixed star located near the zodiacal belt. It begins at the same point in the Rashi of Mesh or Aries, which marks the beginning of the Zodiac, and each extends over an arc of 13 degrees 20 minutes. The Ruling Planet (the Lord of the Particular Nakshatra) in which Moon is positioned at the time of birth is the Planet ruling Maha Dasha of the individual. The exact number of years, months and the days of the Maha Dasha passed is calculated by the proportion of the total angles of $13°\ 20'$ and the degrees, already, passed or covered by the Moon in that particular Nakshatra at the time of birth of the individual.

Formula: Suppose, Maha Dasha Period already passed or elapsed = Y years
Degrees already covered or lapsed by Moon in that Nakshatra = $a°$
Where,

$a°$ = (Total Degree position of Moon as per Nirayana Planetary Longitude – Starting Degree Position of Nakshatra). The Maha Dasha Period of that particular Planet = B years,
Then,

$$Y = \frac{a°}{13°\ 20'} \times B \text{ years}$$

The balance period of the Maha Dasha is equal to the Planet Maha Dasha Period minus already elapsed or passed period of the planet ruling that Nakshatra.
So, the balance period of the Maha Dasha = $Z = (X - Y)$ years

2. Time Method: The Dasa balance is calculated according to the time the moon has spent in the birth Nakshatra at the time of birth. We must know the time of entry of Moon into birth Nakshatra and the time of entry to the next Nakshatra. The time of birth has to be taken as proportional value in relation to the total time the moon spends in the birth Nakshatra. Formula: So, let's say birth time is T_1, entry time is T_2, and entry time to next Nakshatra is T_3. The balance b can be calculated with the formula

$b = (T_1 - T_2) / (T_3 - T_2)$

Antradasa/Sub period:

Antradasa is the sub period or shorter intervals within each Maha Dasa of each planet during the Maha Dasa of main planet. Antradasa Period is very useful for day-to-day living, organising oneself and goal setting for future tasks. The Antradasa is calculated with the help of the Main Dasa or Maha Dasa. The Antradasa period is proportionate to the Graha of the Antradasa. Total period of the nine Antradasa of each planet is equal to the period of one Main Dasa in that particular Main Dasa period. The Antradasa Graha and the Main Dasa Graha are the same at the start of each Main Dasa. The total period of nine Antradasa, in the same order of Vimshottary System, is over along with the end of the Main Graha period. Accordingly, the applicable Antradasa is calculated at the time of the native birth. During the planet's Dasa, Antar Dasa starts from the Dasa-planet itself, followed by the same sequence as given in the Maha Dasa. The period is also subdivided into the same fractions as assigned in table No. 1. To compute the Antar Dasa, the basic formula is:

Antar Dasa of a planet = (period of Maha Dasa planet x period of Antar Dasa planet/120 years).

Since 1 year contains 12 months, the above formula reduces to

Antar Dasa of a planet = (period of Maha Dasa planet x period of Antar Dasa planet x 12/120) in months.

The whole number so obtained gives Antar Dasa of a planet in the months. We multiplied the fraction in decimal by 30 days (1 month) to get the whole number as days of the Antar Dasa of a planet.

Pratyantar Dasha/Sub-Sub Periods:

For further refining the results, it is further division of the Antar Dasa into sub-sub periods (Pratyantar Dasha) as per following formula: Sub-Sub Periods (Pratyantar Dasha) = (Maha Dasa Period of Maha Dada planet x Maha Dada Period of Antar Dasa planet x Maha Dada Period of Pratyantar Dasha planet) / (120 x 120) years = (Period of Maha Dasa planet x Maha Dada Period of Antar Dasa planet x Maha Dada Period of Pratyantar Dasha planet) x 12 / 120 x 120 months. Example 1: Sub-Sub period (Pratyantar Dasha period of Sun) in Sun Maha Dasha and Sun Antar Dasha = 6 x 6 x 6/120 x 120 = 216/14400 years = (216 x 12 x 30)/14400 days = 5.4 days. This is the smallest sub-sub period (Pratyantar Dasha) = 6 x 6 x 6/120 x 120 years = 5.4 days. Example 2: Sub-Sub period (Pratyantar Dasha period of Sun) in Sukra Maha Dasha and Rahu Antar Dasha = 6 x20 x 18/120 x 120 = 2160/14400 years = (2160 x 12)/14400 months = 1 month 24 days.

Sookshma Dasha/sub-sub-sub periods:

By adopting the same procedure as for sub-sub period (Pratyantar Dasha), we can get sub-sub-sub periods (Sookshma Dasha). Prana dasha/Sub-sub-sub-sub period: If we go one division further, we get Sub-sub-sub-sub period (Prana Dasha) and so, we get the smallest sub-sub-sub-Sub period = 6 x 6 x 6 x 6 x 6/120 x 120 x 120 x 120 years = 19.44 Minutes.

13.2 Predictions by Maha Dasha

There are two kinds of effects of Maha Dasha, such as (1) General Effects and (2) Distinctive Effects. The natural characteristics of the Graha cause the general effects and the placement of the planet in the house cause the distinctive effects. The effects of the Maha Dasha of the Graha are in accordance with their strength. The effects of a Graha in the first Drekkana are realized at the commencement of the Dasha. The Graha in the second Drekkana makes its effects felt in the middle of the Dasha. The effects of the Graha in the third Drekkana are experienced at the end of the Dasha. If the Graha is retrograde, these effects would be in the reverse order. The Maha Dasha effects of Rahu and Ketu, who are always retrograde, will always be realized in the reverse order.

Favourable/Auspicious Dasa effects: Planets in Exaltation, Own House, Trikona, Friendly Sign, in a Shant Rashi, or the Lord of a Kendra in a Trikona or the Lord of a Trikona in a Kendra, or the Graha yuti with either of them, Graha hemmed between benefic, Graha possessing full rays (between 6^0-25^0 longitudes) and those with full strength (Vargottama, Raja Yoga Karaka or Benefic) are auspicious in their Dasa and yield good results. The Graha, receiving a Drishti from Lord of a Kendra, or Trine produces favourable Dasha effects. Those exalted in the Ascendant confer ruler ship of a province during Dasa, while those in exaltation and in Angle or Trikona will bestow wealth, conveyances and ruler ship of a country. He/She will have abundant income, happiness in the major period of a benefic planet and in the sub period of another benefic planet. The planet, that is in the 10th house will confer kingdom to him/her during his Maha Dasa. If there is none in the 10th, the strongest of the planets will give the said effect. The planet in the 3rd or 11th will give good result during his Dasa. He/She will be opulent, if Mercury, Jupiter and Venus occupy the Ascendant, Saturn the 7th and the Sun the 10th. If Angles are occupied by benefic, while malefic planets are not there, he/she will be wealthy becoming leader of forest tribes. The Ari, Randhra, or Vyaya Lord, if get associated with the Lord of a Trikona, produces favourable Dasha effects. If Dharma Lord is in Lagna and Lagna Lord is in Dharma, the Dasha of both of them will produce extremely beneficial results. If Karma Lord is in Lagna and Lagna Lord is in Karma, there will be acquisition of a kingdom in the Dasha of Lagna Lord and Karma Lord. Rahu and Ketu, if are in Sahaj, Ari and Labh Bhava, they give favourable results.

Unfavourable/Inauspicious Dasa effects: The planet in its Sign of fall, Sign ruled by an enemy, Navamsa of an enemy, combust in the Sun, devoid of splendour (between $0°$-$5°$ or $26°$-$30°$ Longitude), or in an Ugra Rashi or is in aspect to malefic will yield inauspicious effects in its Dasa. If a planet has departed from its exaltation, its Dasa is known, as Avarohini (stepping down) Dasa. If, however, it is in own, or friendly divisions, the effects will be medium, i.e. neither good nor bad. The Dasa of a planet approaching its exaltation, departing from debilitation, is known, as Arohini (climbing) Dasa. If the said planet is in inimical, or debilitation divisions, it will inflict difficulties. The Sun is the drier (or baker) of the effects of a Dasa, while the Moon is the sustaining agent. That is why the Moon's transit position at the time of the commencement of a Dasa should be duly considered. If a planet is in the 6th, he/she will be subjected to evils, misfortunes. The planet placed in the 8th will inflict death to him/her. There will only be confinement in the Dasa of a planet in combustion. The Dasa of a retrograde malefic planet will cause wanderings and gives an evil and enmity effects. Though he/she may be a king, he/she will become a servant of his own servants in the Dasa of a planet in fall, a planet in deep fall, a planet in inimical Sign and a planet in dreaded enemies. In the Dasa of a planet placed in enemy's Sign, there will be circumstances to leave his/her own country (in distress), diseases, and loss of position, frequent quarrels and confinement. Even a king will surely wander in difficulty in the Dasa of a planet occupying an inimical Sign. In the Dasa of the planets placed in the 6th, or 12th Houses, there will be troubles to limbs and secondary limbs like the nose, while the Dasa of the planet in the 8th House will cause beheading or afflictions to head, fear from enemies, wanderings in foreign places, troubles of confinement, diseases. The Dasa of a planet in fall will provide humiliation from royal sources. A weak planet's Dasa will be futile. Even though a planet may be in friendly Sign, if it occupies simultaneously a Navamsa belonging to its enemy, the effects in the Dasa will be mixed. If a planet is combust, during such planet's period, he/she will have a dirty body and will incur destruction of wealth, happiness, valour, beauty and enthusiasm. In the Dasa of planets inimical to Ascendant Lord, or to the Moon Sign Lord, he/she will be very foolish, will be deprived of kingdom, will be insulted by enemies and will depend on others. The Dasha of Sahaj, Ari and Labh Lord, or the Graha placed in Sahaj, Ari and Labh or the Graha yuti with the above Graha, will produce unfavourable Dasa effects. The Dasha of the Graha, associated with the Lords of Maraka Bhava, namely Dhan and Yuvati, if in Dhan, or Yuvati and or placed in Randhra, produce unfavourable Dasha effects.

Sun Maha Dasha: If Sun is in his own Rashi, in his exaltation Rashi, in a Kendra, in Labh, be associated with the Lord of Dharma or the Lord of Karma or strong in his Varga, there is acquisition of wealth, great felicity and honours from the Government. He/She will be blessed with a son (children), if Surya is with the Lord of Putra. He/She will acquire elephants and other kinds of wealth, if Surya is associated with the Lord of Dhan. He/She will enjoy comforts of conveyances, if Surya is associated with the Lord of Bandhu. He/She attains a high position, like that of Army Chief, by the beneficence of the king and enjoys all kinds of happiness. Thus, during the Dasha of a strong and favourable Surya there are acquisitions of clothes, agricultural products, wealth, honours, and conveyances. If at the time of birth or during the Dasha, Sun is strong, un-afflicted and well placed, he/she will feel strong, enjoy power and status, authority, position, favours from elders and from Government. There will be success in all undertakings, prosperity to father and good health. If the Sun is favourably placed, there will be gains from forests, medicines, travels, poison, evil friendship, forests, ivory, skin, fire and royal association. Courage, diligence, intelligence, fame and valour will be on the increase and excellence and kingship will be achieved by him/her. If the Sun is exalted, there will be gain of kingdom with royal insignia in such Dasa. Apart, there will be acquisition of elephants, steeds. The Dasa of the Sun placed in his Moolatrikona will remove all kinds of miseries and gives leadership of large provinces.

If the Sun is adversely placed, there will be grief on account of servants, wealth, theft, eyes, weapons, fire, water and king and the native will be tormented by children, wife and relatives He will be interested in sinful acts and be troubled by hunger, thirst or morbid fever, heart ailments, bilious diseases and bodily damages. The Sun in Ascendant always gives kingdom during his Dasa; while posited in other Kendra ruled by inimical planets, there will be disorders of hip, throat and eyes. Debilitated Sun causes destruction of eyes, fever and head disorders, confinement, leprosy and other disorders. In the Dasa of the Sun placed in inimical Sign, he/she will incur affliction of eyes, hump-back, disorders caused by fire, facial diseases and humiliation due to misconduct. The Sun placed in 8th House will cause, in his Dasa, affliction to the entire body and wanderings in various countries. If Sun is weak and afflicted or not well placed, there will be miseries, impediments, obstacles in the way of progress of life; period of uncertainties, bad reputation; forced to live away from home; wander like beggars for living; deterioration in financial status and losses in permanent assets of life. If Sun is weak and afflicted or not well placed, it will cause ill health, diseases like tension, heart ailment, eye troubles, burning of body, loss of reputation, criminal proceedings, torture by Government, threat of imprisonment, danger to father and leading to quarrel in family. If Sun is in Movable Sign, he/she may be aimlessly wandering. If Surya is in his Rashi of debilitation, is weak, is in Ari, Randhra, or Vyaya, or is associated with malefic Graha, or with the Lord of Ari, Randhra, or Vyaya during the Dasha of Surya or at the time of birth, there will be anxieties, loss of wealth, punishment from Government, defamation, opposition by kinsmen, distress to father, inauspicious happenings at home, distress to paternal and maternal uncles, anxiety and inimical relations with other people for no reason whatsoever. There will be some favourable effects at times, if in the above situations, Surya receives a Drishti from benefic Graha. The effects will always be unfavourable, when malefic Graha give a Drishti to Surya.

Moon Maha Dasha: If Chandra be in her exaltation, in her own Rashi, in Kendra, in Labh, Dharma, or Putra, be associated with, or receives a Drishti from benefic, be fully powerful and is associated with the Lord of Karma, Dharma, or Bandhu, there will be opulence and glory, good fortune, gain of wealth, auspicious functions at home, dawn of fortune, attainment of a high position in Government, acquisition of conveyances, clothes, birth of children and acquisition of cattle. There will be extraordinary gains of wealth and luxuries, if such a Chandra is in Dhan Bhava. If the Moon is favourably placed, there will be gains through female association, soft disposition, journey, water, ice, milk, juice, sugar, sports, initiation into Mantras through Brahmins, flowers, robes and sweet food. He/She will achieve success in his undertakings out of his intelligence, will receive honours from elders and the king and will be endowed with increased courage and wisdom. If the Moon occupies a friendly Sign, exaltation Sign, or an Upachaya House 3rd, 6th, 10th, or 11th House or 5th, 9th, or 7th counted from Moon, good effects will come to pass in the Dasa.

The exalted Moon gives kingdom, debilitated gives death, inimical Sign Moon gives confinement, friendly Sign Moon gives association with own men; decreasing Moon gives fever and disorders of abdomen, head and eyes. If the Moon is full and is endowed with strength, he/she will be surrounded by thousands of superior females. If Moon is relegated to the 8th House, he/she will be humiliated by his own men; will live through machinery, grass, firewood, cow dung, bamboo, medicinal plants, fruits and water. He/She will acquire only deteriorated food and rags and even a king will have to resort to servitude. If the Moon receives the aspect of Mars or Saturn there will be loss of relatives and of wealth followed by diseases. Mercury's aspect confers learning, while the aspects of other planets will give effects according to their nature.

If the Moon is evil in the chart, there will be fear in family, misunderstanding with his/her family members, difficulties, loss of wealth, enmity with royal assembly, sleep, indolence, evils through women and grief. If Moon is weak and afflicted or not well placed, it gives ill health, low spirit; and he/she may suffer due to mental agony; troubles at the working place; loss from quarrel or opposition to women; displeasure of the functionary of Government to misunderstanding; lethargic and sleepy, lazy and mother may suffer from ailment. Strength of Moon is to be checked with utmost care. If Moon is Strong and Lord of good Houses, it will enhance the prosperity of that house. If Moon-Rahu is conjunction or if Moon/Rahu is aspect by malefic planets then, then during Rahu Antar Dasha under Moon period or Moon Antar Dasha under Rahu Period, it normally does not give good results and worries will multiply and strong Rahu may give rise to many unexpected situations in life. He/She will be tending towards cunning, sinful, black market, tax evasion and will be running after many unpleasant things of life and change of residence. If Moon is waning, or in her debilitation Rashi, there will be loss of wealth in her Dasha. If Moon is in Sahaj, there will be happiness off and on. If Moon is associated with malefic, there will be idiocy, mental tension, trouble from employees and mother and loss of wealth. If waning Moon is in Ari, Randhra, or Vyaya, or is associated with malefic, there will be inimical relations with Government, loss of wealth, distress to mother and similar evil effects. If a strong Moon is placed in Ari, Randhra, or Vyaya there will be troubles and good times off and on. If the Moon, at the commencement of a Maha Dasa or at the time of birth, is in Virgo, he/she, throughout the period, will be related to courtesans and be associated with them. If Moon is in Cancer at the beginning of Dasa, he/she will gain wealth through virtuous ladies. If Moon is in Aries, or in Scorpio, he/she will injure the honour of virgins and also bring evils to his/her own female/male. If the Moon is in the Sign of Mercury, knowledge of Shashtra and acquisition of friends will come to pass. In a Sign of Venus, it brings plenty of food and drinks, happiness and destruction of enemies. In a House of Jupiter, it brings happiness, wealth, honour and command ability. Saturn's Sign endows with association with old ladies. With Leo denotes movements in forts and forests, agriculture and misunderstanding with wife and children.

Mars Maha Dasha: If Mangal is in his exaltation, in his Moolatrikona, in his own Rashi, in Kendra, in Labh, or Dhan Bhava with strength, in a benefic Navamsa and is associated with a benefic, there will be during his Dasha acquisition of kingdom, attainment of a high administrative, or political position in Government, gain of wealth and land, recognition by Government, gain of wealth from foreign countries and acquisition of conveyances and ornaments. There will also be happiness and good relations with co-born. If Mangal with strength is placed in a Kendra, or in Sahaj, there will be gain of wealth through valour, victory over enemies, happiness from wife and children. There will, however, be a possibility of some unfavourable effects at the end of the Dasha. During the Dasa of favourable Mars, he/she will gain through kings, fire, thieves, enemies, snakes, poison, weapons, confinements, fictitious articles, lands, goats, buffaloes, copper, gold, prostitutes, gambling, intoxicants, pungent articles/juices, wealth and grains. If Mars is strong, un-afflicted and well placed, there will be gain from brothers and help from them; favours from persons in authority; defeat of enemies; joining jobs of adventure; acquisition of land and property; gain from travel in Southern direction; optimistic, vigorous, dashing, courageous; reward through act of courage.

If Mars is weak and afflicted or not well placed, native may fall from his prestigious job; Loss to body through accidents; Blood related diseases may appear; H/She becomes victim of his own ill tempered behaviour; displeasure with persons in authority; fractures, wounds, threat to life. If Mangal is in his debilitation Rashi, weak, in an inauspicious Bhava, or is associated with, or receives a Drishti from malefic, there will be in his Dasha loss of wealth, distress and similar unfavourable effects. During the Dasa of evil Mars, there will be enmity with friends, wife, children and brothers, sufferance from thirst, swooning, bloody defects, and loss of limbs, wounds, and affliction to limbs and enmity with children and servants. There will be attachment to others' wives and irreligious acts, followed by aversion to elders and to truth. Bilious excesses will trouble the subject. Mars, if Lord of Ascendant or 8th House or its aspect on them or in Moon signs, it will inflict accidental nature of problem. Mars associated with 2nd, 4th and 6th house, then, there will be injury due to his/her mistake.

Mercury Maha Dasha: Mercury is called a Kumar (in his teens) amongst all the Graha. If Mercury is 9th or 10th Lord or Exalted or related to Lagna, there will be Good benefits and he/she will get involved in educational activity. If Buddha is in his exaltation, in his own, in a friendly Rashi, or in Labh, Putra, or Dharma, there will be during his Dasha acquisition of wealth, gain of reputation, improvement in knowledge, benevolence of Government, auspicious functions, happiness from wife and children, good health, availability of sweetish preparations and profits in business. If Buddha receives a Drishti from a benefic, is in Dharma, or is the Lord of Karma, the aforesaid beneficial results will be experienced in full and there will be great felicity all-round. In the Dasa of favourable Mercury, there will be gains via friends, children, wealth, king, businessmen, priests, gambling. He/She will acquire horses, lands, gold, fame, ambassadorship, limitless happiness, fortunes, high intelligence, fame, success in religious activities, fond of amusement, destruction of foes and enthusiasm in mathematics and drawings. If Mercury is evil, he/she will be tormented by imbalances of the three temperaments, viz. bile, phlegm and wind; hard disposition, imprisonment, mental grief and excitement. If Mercury is strong, un-afflicted and well placed, he/she will gain in business transactions; engaged in intellectual pursuits; increase in educational attainments; publishing of writings; gain through friends and near relatives; success in examination and attainment of higher education.

If Mercury is weak and afflicted or not well placed, he/she suffers from nervous disorders; bad liver; troubles through friends; victim of fraud and cheating; involved in act of forgery; tampering of documents and throat troubles. If Mercury is in conjunction with malefic, the good or bad results are immediately felt. If Buddha is associated with a malefic, there will be during his Dasha punishment by Government, inimical relations with kinsmen, journey to a foreign country, dependence on others and the possibility of urinary troubles. If Buddha is in Ari, Randhra, or Vyaya, there will be loss of wealth, due to indulgence in lascivious activities, possibility of suffering from rheumatism and jaundice, danger of thefts and malevolence of Government, loss of land and cattle etc. At the commencement of the Dasha of Buddha, there will be gains of wealth, betterment in the educational sphere, birth of children and happiness. In the middle of the Dasha, there will be recognition from Government. The last part of the Dasha will be distressful.

Jupiter Maha Dasha: If Jupiter is in his exaltation, his own Rashi, his Moolatrikona, in Karma, Putra, or Dharma Bhava, in his own Navamsa, or in his exalted Navamsa, there will be during his Dasha: acquisition of kingdom, great felicity, recognition by Government, acquisition of conveyances and clothes, devotion to deities and Brahmins, happiness in respect of his wife and children and success in the performance of religious sacrifices (oblations). If Jupiter is favourable, he/she will gain through ministers or advisers, kings, dances and states man ship. He/She will gain honours, virtues, courage, wide friend ship, life partner, gold, quadrupeds, conveyances, auspicious acts, destruction of enemies, royal acquaintance and will obtain wealth through public, king, businessmen and elders. Guru is the great benefic and preceptor of the Gods. If Guru is in his debilitation Rashi, combust, associated with malefic, or in Ari, or Randhra, there will be during his Dasha loss of residential premises, anxiety, distress to children, loss of cattle and pilgrimage. If Jupiter be inauspicious he/she will have physical defects, excessive grief, lameness, diseases of the stomach and ears, loss of semen and marrow and fear from kings. The Dasha will give some unfavourable effects at its commencement only. During the later part of the Dasha there will be good effects, like gain of wealth, awards from and recognition by Government.

There will be signification of the house/Rashi, where it is placed or associated or aspecting and normally good results are expected during its major.

Venus Maha Dasha: If Sukra is in his exaltation, in his own Rashi, or in a Kendra, or a Trikona, there will be during his Dasha acquisition of fancy clothes, ornaments, conveyances, cattle and land etc., availability of sweet preparations every day, recognition from the sovereign, luxurious functions of songs and dances by the benevolence of Goddess Lakshmi. If Sukra is in his Moolatrikona, during his Dasha there will definitely be acquisition of a kingdom, acquisition of a house, birth of children and grandchildren, celebration of marriage in the family, attainment of a high position, like the Commander of an Army, visits of friends, recovery of lost wealth, property, or kingdom. If Venus is auspicious he/she will be successful and will gain lands, residences, bed comforts, wife, garlands, robes, good food, fame, happiness and treasure. He/She will have liking for music, will develop acquaintance with females, will gain through kings and lands, will derive wisdom, desired happiness and friends and gains through articles used for sexual enjoyments. Venus Dasa is also a trend setter. Venus can become a Yoga Karaka planet. Venus in 7th, associated with Rahu, Saturn, Mercury, gives a difficult life. If Venus is Strong, un-afflicted and well placed, native will acquire things of art; promotion, success in undertakings; gain from relations with opposite sex; and self marriage or in family may take place. If Venus is weak and afflicted or not well placed, then domestic disharmony; lack of cooperation from others; low vitality and onset of certain diseases. Sukra is the incarnate of intoxication, ecstasy, delight and pride amongst all the Graha. If Sukra is in Ari, Randhra, or Vyaya, there will be during his Dasha inimical relations with kinsmen, distress to wife, losses in business, destruction of cattle and separation from relations. If Sukra is in Bandhu, as Lord of Dharma, or Karma, there will be during his Dasha attainment of ruler ship of a country, or village, performance of pious deeds, like building of reservoirs and temples and giving grains etc. in charity, availability of sweet preparations every day, vigour in work, name and fame and happiness from wife and children. If Venus is inauspicious the native will have litigations with virtuous people, will have troubles from conveyances, furniture, females and kings and will develop intimacy with unsocial elements. If Sukra is Lord of Dhan, or Yuvati, there will be during his Dasha physical pains and troubles. To get alleviation from those troubles the native should perform Shatarudriya or Mrityunjaya Japa in the prescribed manner and give in charity a cow, or female buffalo.

Saturn Maha Dasha: The Sani is considered the vilest and most inferior amongst all the Graha. Among all the planets, Saturn is very powerful. Unfortunately people and the pundits ignore the positive side of Saturn and project it only as an evil planet. If Sani is in his exaltation, in his own Rashi, or in Moolatrikona, or friendly Rashi, Yoga Karaka, Vargottama or in Sahaj, or Labh Bhava, there will be during his Dasha recognition by Government, opulence and glory, name and fame, success in the educational sphere, acquisition of conveyances and ornaments, gain of wealth, favours from Government, attainment of a high position, like Commander of an Army, acquisition of a kingdom, benevolence of goddess Lakshmi, gain of property and birth of children. He/She, having a strong and favourable Saturn in the horoscope, can accomplish tough mental tasks and he can be a successful writer, musician, astrologer, politician, Tantra expert, actor, policeman and administrator. He can be a very successful industrialist and might manufacture chemicals or explosives. If Saturn is favourable he/she will gain camels, asses, association with unchaste or old ladies, gain through pulses, sesame and grains eaten by poor people, leadership over villages and iron etc. He/She will acquire a permanent position. If Saturn is adverse, there will be loss of conveyances, excitement, enmity, and separation from wife and other family members, loss in battles, liking for intoxicants and gambling, windy diseases, loss of meritorious acts, confinement and lassitude. If Saturn is Strong, un-afflicted and well placed, it gives fruits of our own Karmas and will get benefits from agriculture of related activities.

Indira Gandhi got re-elected as PM, she was going through the Maha Dasa of Saturn. When Amitab Bacchan was at his peak, he was also going through Saturn Maha Dasa. If Sani is in Ari, Randhra, or Vyaya, in his debilitation Rashi, or combust, there will be during his Dasha ill effects from poison, injury from weapons, separation from father, distress to wife and children, disaster, as a result of displeasure of Government, imprisonment etc. If Sani receives a Drishti from, or is associated with a benefic, is placed in a Kendra, or in a Trikona, in Dhanu, or in Meena, there will be acquisition of a kingdom, conveyances and clothes.

Rahu Maha Dasha: Malefic Rahu leads to seclusion in life, suffer losses, troubles from Government, scandal, misbehaviour; and disruption in education and career. If Rahu is in his exaltation in his Moolatrikona, in his own Rashi, in Kendra, there will be during the Dasha of Rahu great happiness from acquisition of wealth, agricultural products, acquisition of conveyances with the help of friends and Government, construction of a new house, birth of sons (children), religious inclinations, recognition from Government of foreign countries and gain of wealth, clothes etc. If Rahu be associated with, or receives a Drishti from benefic, be in a benefic Rashi and be in Thanu, Bandhu, Yuvati, Karma, Labh, or Sahaj, there will be during his Dasha all kinds of comforts by the beneficence of the Government, acquisition of wealth through a foreign Government, or sovereign and felicity at home. If Rahu is in Randhra, or Vyaya Bhava, there will be during his Dasha all kinds of troubles and distress. If Rahu is associated with a malefic, or a Maraka Graha, or is in his debilitation Rashi, there will be loss of position, destruction of his residential house, mental agony, trouble to wife and children and misfortune of getting bad food. There will be loss of wealth at the commencement of the Dasha, some relief and gain of wealth in his own country and distress and anxieties during the last portion of the Dasha.

Ketu Maha Dasha: Ketu being Moksha Karaka planet, there will be development of religious instinct, domestic comforts and luxurious life, freedom from diseases and she/she gets relieved from many diseases and ailments. If Ketu is in a Kendra, a Trikona, or in Labh, in a benefic Rashi, in his exaltation, or in his own Rashi, there will be during his Dasha cordial relations with the king, desired headship of a country, or village, comforts of conveyances, happiness from children, gain from foreign countries, happiness from wife and acquisition of cattle. If Ketu is in Sahaj, Ari, or Labh, there will be in his Dasha acquisition of a kingdom, good relations with friends and opportunities for the acquisition of elephants. At the commencement of the Ketu Dasha there will be Raja Yoga. During the middle portion of the Dasha there will be possibilities of fearfulness and in the last part there will be sufferings from ailments and journeys to distant places. If Ketu is in Dhan, Randhra, or Vyaya, or receives a Drishti from a malefic, there will be imprisonment, destruction of kinsmen and residential premises and anxieties, company of menials and diseases. If Ketu is weak and afflicted or not well placed, Ketu, during its major period gives troubles from silent enemies, wounds from weapons, stigma in family, Ill health, body pain and anguish of mind and mental dissatisfaction.

13.3 Predictions by Antar Dasha:
13.3.1 Predictions by the Antar Dasha in the Dasha of Sun:

Sun Dasa-Sun Bhukti: This is not so much a favourable period. In general, there will be royal displeasure, quarrel in family, ill health, travel, mental uneasiness during period. If Sun is exalted, in Labh, in a Kendra, or in a Trikona, good effects like acquisition of wealth and grains will be experienced. Adverse results will be experienced, if Sun is debilitated, or in an inauspicious Bhava, or Rashi. Medium effects will be realized, if Surya is in other houses. If Surya is the Lord of Dhan, or Yuvati, there will be danger of premature death, or death-like sufferings. The remedial measures to be adopted are Mrityunjaya Japa, or the worship of Surya (by recitation of appropriate Mantras, charity).

Sun Dasa-Moon Bhukti: This is a better period. In the major period of the Sun, the Moon's sub period will subdue the valour of his/her enemies. He will gain health, wealth and happiness. There will be promotion in job, expansion of business, new ventures, respect among relatives, and if moon is badly placed ill health and danger from water. Functions like marriage, gain of wealth and property, acquisition of a house, of land, cattle and conveyances will be the effects of the Antar Dasha of Chandra in the Dasha of Surya, if Chandra is in a Kendra, or in a Trikona. There will be his/her marriage, birth of children, beneficence of and favours from kings and fulfilment of all ambitions, if Chandra is in his exaltation Rashi, or in his own Rashi. Distress to wife and children, failures in ventures, disputes with others, loss of servants, antagonism with the king and destructions of wealth and grains will be the effects, if Chandra is waning, or is associated with malefic. The effects like danger from water, mental agony, imprisonment, danger from diseases, loss of position, journeys to difficult places, disputes with coparceners, bad food, trouble from thieves, displeasure of the king, urinary troubles, pains in the body will be experienced, if Chandra is in Ari, Randhra, or Vyaya. Luxuries, comforts, pleasures, dawn of fortune (Bhagyodaya), increase in the enjoyment from wife and children, acquisition of kingdom, performance of marriage and religious functions, gain of garments, land and conveyance and birth of children and grandchildren will be the auspicious effects, if there are benefic in the 1st, the 9th, or a Kendra from the Lord of the Dasha. Unpalatable food, or course food, exile to outside places will be the effects in the Antar Dasha, if Chandra is in the 6th, the 8th, or the 12th from the Lord of the Dasha, or, if Chandra is weak. There will be premature death, if Chandra is the Lord of a Maraka Bhava. Giving in charity of a white cow and a female buffalo will help to acquire peace and comfort.

Sun Dasa-Mars Bhukti: This is not so much a favourable period. In the sub period of Mars and in the major period of the Sun, he/she gains corals, gold, and success in battle, splendour and happiness. However, there will be mental tension, ill health, litigation, miss-understanding with relations, loss, and worry. When Mars is in favourable position with beneficial aspect, there will be purchase of property, promotions or appointment. Auspicious effects, like acquisition of land, gain of wealth and grains, acquisition of a house will be derived in the Antar Dasha of Mangal in the Dasha of Surya, if Mangal is in his exaltation Rashi, in his own Rashi, in a Kendra, or in a Trikona. All-round gains, attainment of the position of a Commander of the Army, destruction of enemies, peace of mind, family comforts and increase in the number of co-born will be the effects, if Mangal is yuti with the Lagna Lord. Brutality, mental ailment, imprisonment, loss of kinsmen, disputes with brothers and failure in ventures will result, if

Mangal is in the 8th, or in the 12th from the Lord of the Dasha, if Mangal is associated with malefic, or, if Mangal is without dignity and strength. Destruction of wealth by the displeasure of the king will be the effect, if Mangal is in his debilitation Rashi, or be weak. Diseases of the mind and body will result, if Mangal is the Lord of Dhan, or Yuvati Bhava. Recitation of Vedas, Japa and Vrashotsarg, if performed in the prescribed manner, will recover from ill health, increase in longevity and success in adventures.

Sun Dasa-Rahu Bhukti: If Rahu is in a Kendra, or in a Trikona from Lagna, there will be in the first two months loss of wealth, danger from thieves, snakes, infliction of wounds and distress to wife and children. After 2 months inauspicious effects will disappear and enjoyment and comforts, sound health, satisfaction, favours from the king and government will be the favourable effects, if Rahu is yuti with benefic, or, if Rahu is in the Navamsa of a benefic. Recognition from the king, good fortune, name and fame, some distress to wife and children, birth of a son, happiness in the family will be derived, if Rahu is in an Upachaya from Lagna, if Rahu is associated with a Yoga Karaka, or is placed auspiciously from the Lord of the Dasha. Imprisonment, loss of position, danger from thieves and snakes, inflection of wounds, happiness to wife and children, destruction of cattle, house and agricultural fields, diseases, consumption, enlargement of the skin, dysentery will be the results, if Rahu is weak, or is in the 8th, or in the 12th from the Lord of the Dasha. Adverse effects, like premature death and danger from snakes will be derived, if Rahu is in Dhan, or Yuvati, or, if Rahu is associated with the Lords of either of these Bhava. When Rahu is in 6, 8, 12 place or combined with evil planet there will be mental worry, loss, failure in undertakings, separation in family, children affected, and food poison. When it is in 10 or 11 houses, there will be royal favour, extra income, free from disease.

Worship of Goddess Durga, Japa, giving in charity of a black cow, or female buffalo are the remedial measures for alleviation of the above evil effects, or total escape from them.

Sun Dasa-Jupiter Bhukti: Marriage of the his own, favours by the king, gain of wealth and grains, birth of a son, fulfilment of the ambitions by the beneficence of the sovereign and gain of clothes will be the auspicious effects, derived in the Antar Dasha of Guru in the Dasha of Surya, if Guru is in a Kendra, or in a Trikona to Lagna, in his exaltation Rashi, in his own Rashi, or in his own Varga. When Jupiter is placed in his own house or exalted, there will be marriage and benefits from friends and relatives, increase of wealth, royal favour, promotion in job, contact with saintly persons, pilgrimage to holy places, victory in court cases. Acquisition of a kingdom, comforts of conveyance, like palanquin (motor car in the present times), gain of position will result, if Guru is the Lord of Dharma, or Karma. Better fortune, charities, religious inclinations, worship of deities, devotion to preceptor, and fulfilment of ambitions will be the auspicious effects, if Guru is well placed with reference to the Lord of the Dasha. There will be destruction of diseases, enemies, sins and poverty and gain of virtues and happiness. Distress to wife and children, pains in the body, displeasure of the king, non-achievement of desired goals, loss of wealth, due to sinful deeds, mental worries will result in his Antar Dasha, if Guru is in the 6th, or in the 8th from the Lord of the Dasha, or is associated with malefic. Giving in charity gold, a tawny-coloured cow, worship of Ishta Lord (Ishta Dev) are the remedial measures to obtain alleviation of the evil effects and to achieve good health and happiness.

Sun Dasa-Saturn Bhukti: There will be royal wrath, poverty and defeat from enemies. There will be sickness to wife and children, loss of money, royal displeasure, transfers in job as punishment, misery, debt. Destruction of foes, full enjoyment, some gain of grains, auspicious functions, like marriage at home will be the good effects, derived in the Antar Dasha of Sani in the Dasha of Surya, if Sani is in a Kendra, or in a Trikona from Lagna. Well-being, acquisition of more property, recognition by the king, achievement of renown in the country, gain of wealth from many sources will be the effects, if Sani is in his exaltation, in his own, in a friendly Rashi and, if Sani is yuti with a friendly Graha. Rheumatism, pains, fever, dysentery-like disease, imprisonment, loss in ventures, loss of wealth, quarrels, disputes with co-partners, claimants will be the effects in the Antar Dasha, if Sani is in the 8th, or the 12th from the Lord of the Dasha, or is associated with malefic. There will be loss of friends at the commencement, good effects during the middle part and distress at the end of the Dasha. In addition to other evil effects there will be separation from parents and wandering, if Sani be in his Rashi of debilitation. If Sani is the Lord of Dhan, or Yuvati, there will be danger of premature death. Giving in charity black cow, buffalo, goat and Mrityunjaya Japa, are the remedial measures for obtaining relief from the evil effects of the Antar Dasha. These measures help to achieve happiness and gain of wealth and property.

Sun Dasa-Mercury Bhukti: He/She will suffer from itch, leprosy etc. and will face increase of enemies. There will be purchase of ornaments and clothes favourable trend in education, expansion of business, marriage, pilgrimage. When Mercury is in bad position 3, 6,

8, 12 or lord of 4 and 7, mental depression, court troubles, scandal, ill health, quarrel, unnecessary travels. Acquisition of a kingdom, enthusiasm and vivacity, happiness from wife and children, acquisition of conveyance through the beneficence of the sovereign, gain of clothes, ornaments, pilgrimage to holy places, acquisition of a cow will be the good effects in the Antar Dasha of Buddha in the Dasha of Surya, if Buddha is in a Kendra, or in a Trikona from Lagna. Buddha becomes very beneficial, if he gets associated with the Lord of Dharma. Reverence from and popularity amongst people, performance of pious deeds and religious rites, devotion to the preceptor and deities, increase in wealth and grains and birth of a son, will be the auspicious effects, if Buddha is in Dharma, Putra, or Karma. Marriage, offering of oblations, charity, and performance of religious rites, name and fame, becoming famous by assuming another name, good food, becoming happy, like Indra, by acquiring wealth, robes and ornaments will be the effects, if Buddha is in an auspicious Bhava, like Trikona from the Lord of the Dasha. Body distress, disturbance of peace of mind, distress to wife and children, will be the evil effects in the Antar Dasha of Buddha, if he is in the 6th, the 8th, or the 12th from the Lord of the Dasha (Note: Buddha cannot be in the 6th, or the 8th from Surya). There will be evil effects at the commencement of the Antar Dasha, some good effects in the middle part of the Antar Dasha and the possibility of displeasure of the king and exile to a foreign country at the end of the Dasha. If Buddha is the Lord of Dhan, or Yuvati, there will be pains in the body and attacks of fever. For relief from the evil effects and to regain good health and happiness the remedial measures are the recitation of Vishnu Sahasranam and giving in charity grains and an idol, made of silver.

Sun Dasa-Ketu Bhukti: There will be mental worry, change of residence, family troubles, disease, poison due to insects biting, headache, misunderstanding among friends. Ketu in 11th house with the aspect of Jupiter or Venus will give advancement in education name and fame. Body pains, mental agony, loss of wealth, danger from the king, quarrels with the kinsmen will be the effects of the Antar Dasha of Ketu in the Dasha of Surya. If Ketu is associated with the Lord of Lagna, there will be some happiness at the commencement, distress in the middle part and receipt of the news of death at the end of the Antar Dasha. Diseases of teethe, or cheeks, urinary troubles, loss of position, loss of friends and wealth, death of father, foreign journey and troubles from enemies will be the results, if Ketu is in the 8th, or the 12th from the Lord of the Dasha. Beneficial effects, like happiness from wife and children, satisfaction, increase of friends, gain of clothes and renown will be derived, if Ketu is in Sahaj, Ari, Karma, or Labh. If Ketu is Lord of Dhan or Yuvati or is in any of those Bhava, there will be danger of premature death. The remedial measures for obtaining relief from the evil effects are recitation of Mantras of Goddess Durga (Shat Chandi Path) and giving a goat in charity.

Sun Dasa-Venus Bhukti: He/She will suffer from head throat diseases, fever and gout, followed by defeat from enemies and leaving the native country. When Venus is in Trikona or 2nd house, there will be benefits from women or marriage, Royal favour, promotions, contact with saintly person, mind towards intense prayers, and travel to holy places. When Venus is not in favourable places or combined with evil planets like Rahu, Ketu, Mars or Saturn, there will be contact with immoral ladies, displeasure in office, or loss in business, court trouble, and ill health. Marriage and happiness, as desired from wife, gain of property, travels to other places, meeting with Brahmins and the king, acquisition of kingdom, riches, magnanimity and majesty, auspicious functions at the home, availability of sweet preparations, acquisition of pearls and other jewels, clothes, cattle, wealth, grains and conveyances, enthusiasm, good reputation are the auspicious effects of the Antar Dasha of Sukra in the Dasha of Surya, if Sukra is placed in a Kendra, or in a Trikona, or, if Sukra is in his exaltation Rashi, in his own Rashi, in his own Varga, or in a friendly Rashi. Displeasure of the king, mental agony and distress to wife and children will be the effects in the Antar Dasha of Sukra, if he is in the 6th, the 8th, or the 12th from the Lord of the Dasha. The effects of the Antar Dasha would be moderate at its commencement, good during the middle portion and evil effects, like disrepute, loss of position, inimical relations with kinsmen and loss of comforts, will be derived at the end. If Sukra is the Lord of Yuvati (and Dhan), there will be pains in the body and the possibility of suffering from diseases. There will be premature death, if Sukra is associated with Ari or Randhra Lord. The remedial measures for obtaining relief from the evil effects are Mrityunjaya Japa, Rudra Japa and giving in charity a tawny cow, or female buffalo.

13.3.2 Predictions by the Antar Dasha in the Dasha of Moon:

Moon Dasa-Moon Bhukti: This is a favourable period. There will be marriage or birth of child, seeing saints or holy persons, pilgrimages, good food, success in undertakings, gain of lost property, change in the living place, good health. Acquisition of horses, elephants and clothes, devotion to deities and preceptor, recitation of religious songs in praise of God, acquisition of a kingdom, extreme happiness and enjoyment and name and fame will be the beneficial results in the Antar Dasha of Chandra in her own Dasha, if she is placed in her exaltation Rashi, her own Rashi, in a Kendra, or in a Trikona, or is associated with the Lord of Dharma, or Karma. Loss of wealth, loss of position, lethargy, agony, antagonism towards the king and ministers, distress to mother, imprisonment and loss of kinsmen will be the evil effects in her Antar Dasha, if Chandra is in her debilitation Rashi, if Chandra is associated with malefic, or, if Chandra is in Ari, Randhra, or Vyaya. If Chandra is the Lord of Dhan, or Yuvati, or is associated with Randhra or with Vyaya Lord, there will be pains in the body and danger of premature death. The remedial measures are giving in charity of a tawny-coloured cow, or female buffalo.

Moon Dasa-Mars Bhukti: Normally this is not a favourable period. There will be bilious complaints, diseases of the blood, fear from fire, evils, ill health and fear from thieves. There will be misunderstanding in the family, loss of wealth, extra expense, and ill health due to fire or instruments, blood troubles. Otherwise, there will be Royal favours, expansion of activities, new vehicles, benefits from brothers, new ornaments purchased and general mental easiness. Advancement of fortune, recognition by the government, gain of clothes and ornaments, success in all efforts, increase in agricultural production and prosperity at home and profits in business will be the favourable effects of the Antar Dasha of Mangal in the Dasha of Chandra, if Mangal is in a Kendra, or in a Trikona. Great happiness and enjoyment of comforts will be derived, if Mangal is in his exaltation Rashi, or in his own Rashi. Distress to the body, losses at home and in agricultural production, losses in business dealings, antagonism, or adverse relations with servants (employees) and the king, separation from kinsmen and hot temperament will be the evil effects in the Antar Dasha of Mangal, if he is placed in Ari, Randhra, or Vyaya from Lagna, be associated with, or receives a Drishti from malefic in the 6th, the 8th, or the 12th from the Lord of the Dasha.

Moon Dasa-Rahu Bhukti: It is a very taxing period. There will be loss of wealth, displeasure with boss or higher authorities, ill health to father, or family members, mental agony. When Rahu is aspect by benefic, the results are positive in purchasing new houses, new vehicle, new venture and political victory. There will be some auspicious results at the commencement of the Antar Dasha of Rahu in the Dasha of Chandra, but later there will be danger from the king, thieves and snakes, distress to cattle, loss of kinsmen and friends, loss of reputation and mental agony, if Rahu is placed in a Kendra, or in a Trikona. Success in all ventures, gain of conveyances, garments from the king in the South-West direction will be derived, if Rahu in his Antar Dasha receives a Drishti from benefic, if Rahu is in Sahaj, Ari, Karma, or Labh, or, if Rahu is yuti with a Yoga Karaka Graha. Loss of position, mental agony, distress to wife and children, danger of diseases, danger from the king, scorpions and snakes will happen, if Rahu is weak and is placed in the 8th, or the 12th from the Lord of the Dasha. Pilgrimage to holy places, visits to sacred shrines, beneficence, and inclination towards charitable deeds will be the results, if Rahu is in a Kendra, in a Trikona, or in the 3rd, or the 11th from the Lord of the Dasha. There will be body troubles (physical afflictions), if Rahu is in Dhan, or in Yuvati. Rahu Japa and giving a goat in charity are the remedial measures for obtaining relief from the evil effects in the Antar Dasha of Rahu.

Moon Dasa-Jupiter Bhukti: It is a favourable period. In the major period of the Moon and in the sub period of Jupiter, there will be gain of wealth, sudden acquisition of robes, jewels and conveyances. There will be promotion in job, meeting with holy persons, or pilgrimage, child birth, purchase of gold ornaments, marriage. When Jupiter is in 6, 8, 12 house, the results will not be so favourable, but loss of prestige or position, mental tension. Acquisition of a kingdom, auspicious celebrations at home, gains of clothes and ornaments, recognition from the king beneficence of the Isht Lord (Isht Devata), gains of wealth, land, conveyances, success in all ventures by the beneficence of the king will be the beneficial effects in the Antar Dasha of Guru in the Dasha of Chandra, if Guru is placed in a Kendra, or in a Trikona to Lagna, or, if Guru is in his own, or in his exaltation Rashi. Destruction of preceptor (and father etc.) and children, loss of position, mental agony, quarrels, destruction of a house, conveyances and agricultural land will be the evil effects in his Antar Dasha, if Guru is in Ari, Randhra, or Vyaya, if Guru is combust, in his debilitation Rashi, or be associated with malefic. Gains of cattle, grains, clothes and happiness from brothers, acquisition of property, valour, patience, oblations, celebrations, like marriage, gain of a kingdom will be the favourable effects, if Guru is in 3rd, or in the 11th from the Lord of the Dasha. Effects, like unpalatable food, journeys to places away from the homeland, will be derived, if Guru is weak and is placed in the 6th, the 8th, or the 12th from Chandra. There will be good effects at the commencement of the Antar Dasha and distress at its end. There will be premature death, if Guru is Dhan, or Yuvati

Lord. Remedial measures for obtaining relief from the above evil effects are recitation of Shiva Sahasranam Japa and giving gold in charity.

Moon Dasa-Saturn Bhukti: There will be separation from his/her own men, fear from diseases and great grief. There will be loss in wealth, mental anxiety, ill health due to bile and misunderstanding with relative and officer. When Saturn is in 3, 6, 11th house from Moon, more favourable results can be seen. There will be marriage or child birth, new venture in the case of business men, new dealings in iron or oil business and gain. Ill health with fatigue will also be the result. The effects like birth of a son, friendship, gain of wealth and property, profits in business with the help of Sudra, increase in agricultural production, gains from son, riches and glory by the beneficence of the king, will be experienced in the Antar Dasha of Sani in the Dasha of Chandra, if Sani is in a Kendra, or in a Trikona from Lagna, or, if Sani is in his own Rashi, in his own Navamsa, in his exaltation Rashi, if Sani receives a Drishti from, or is associated with benefic, or, if Sani is in Labh with strength. The effects like visits to holy places, bathing in holy rivers etc., the creation of troubles by many people and distress from enemies, will be derived in the Antar Dasha of Sani, if Sani is in Ari, Randhra, Vyaya, or Dhan, or, if Sani is in his debilitation Rashi. Effects, like enjoyments and gains of wealth some times, while opposition, or quarrels with wife and children at other times, will be realized, if Sani is in a Kendra, or in a Trikona from the Lord of that Dasha, or is endowed with strength. If Sani is in Dhan, Yuvati, or Randhra, there will be physical distress. The remedial measures to be adopted for obtaining relief from the evil effects are Mrityunjaya Japa, giving in charity a black cow, or female buffalo.

Moon Dasa-Mercury Bhukti: He/She will gain positions with formal seal (or insignia), elephants, horses and wealth and will enjoy incomparable happiness. There will be gain of wealth through maternal side, favourable news in disputes, good results in mental activities, Royal favour, and happiness through females. If Mercury is ill placed, there will be quite contrary results with fear, ill health and mental anxiety. The effects like acquisition of wealth, recognition by the king, gain of clothes etc., discussions on Shashtra gain of knowledge from society with learned and holy people, enjoyments, birth of children, satisfaction, profits in business, acquisition of conveyance and ornaments will be experienced in the Antar Dasha of Buddha in the Dasha of Chandra, if Buddha is in a Kendra, or in a Trikona, if Buddha is in his own Rashi, in his own Navamsa, or in his exaltation Rashi, endowed with strength. The effects like marriage, oblations (Yagya), charities, performance of religious rites, close relations with the king, social contacts with men of learning, acquisition of pearls, corals, Mani (jewels), conveyances, clothes, ornaments, good health, affections, enjoyments, drinking of Soma Rasa and other tasty syrups will be derived in the Antar Dasha of Buddha, if he is in a Kendra, or in a Trikona, or in the 11th, or in the 2nd from the Lord of the Dasha. Pains in the body, loss in agricultural ventures, imprisonment, distress to wife and children will be the inauspicious effects, if Buddha be in the 6th, the 8th, or the 12th from the Lord of the Dasha, or be in his debilitation Rashi. If Buddha is the Lord of Dhan, or Yuvati, there will be fear of fever. The remedial measures to be adopted for obtaining relief from the evil effects are recitation of Vishnu Sahasranam and giving a goat in charity.

Moon Dasa-Ketu Bhukti: This is an unfavourable period. There will be ill health in family circle, eye troubles, mental worry, displeasure with higher authorities or loss in business. There will be pilgrimage and blessing of saints likely when Ketu is well placed. The effects like gain of wealth, enjoyment, happiness to wife and children, religious inclination, will arise in the Antar Dasha of Ketu in the Dasha of Chandra, if Ketu is in a Kendra, in a Trikona, or Sahaj and is endowed with strength. There will be some loss of wealth at the commencement of the Antar Dasha. Later all will be well. Gain of wealth, cattle will be the effects, if Ketu is in a Kendra, in the 9th, the 5th, or the 11th from the Lord of the Dasha and is equipped with strength. There will be loss of wealth at the end of the Antar Dasha. There will be obstacles in ventures, due to interference by enemies and quarrels, if Ketu be in the 8th, or the 12th from the Lord of the Dasha, or receives a Drishti from, or is associated with malefic. If Ketu is in Dhan, or in Yuvati, there will be danger of affliction of the body with diseases. Mrityunjaya Japa will give relief in all the evil effects and will ensure gain of wealth and property with the beneficence of Lord Shiva.

Moon Dasa-Venus Bhukti: This is a favourable period. There will be acquisition of conveyances run in waters, robes, ornaments and many wives. There will be birth of child or marriage, royal favour good clothing, sudden wealth, help from wife side people, gain in agriculture. When Venus is badly placed, there will be bad name, evil company, and mental worry. The effects like acquisition of a kingdom, gaining of clothes, ornaments, cattle, conveyances etc., happiness to wife and children, construction of a new house, availability of sweet preparations every day, use of perfumes, affairs with beautiful women, sound health will be experienced in the Antar Dasha of Sukra in the Dasha of Chandra, if Sukra is in a Kendra, in a Trikona, in Labh, Bandhu, or Dharma, or in his exaltation Rashi, or in his own Rashi. Physical soundness, good reputation, acquisition of more land and houses, will result, if Sukra is yuti with the Lord of the Dasha. There will be loss of landed property, children, wife and cattle and opposition from government, if Sukra is in his debilitation Rashi, combust, or receives a Drishti from, or is associated with malefic. If Sukra is in Dhan in his exaltation Rashi, or in his own Rashi, or is there, associated with the Lord of Labh, there will be acquisition of an underground hidden treasure, gain of land, enjoyment, birth of a son. Advancement of good fortune, fulfilment of ambitions with the beneficence of the king, devotion to deities and Brahmins, gain of jewels, like pearls will result, if Sukra is yuti with Dharma, or Labh Lord. Acquisition of more house property and agricultural land and gain of wealth and enjoyment will be the good effects, if Sukra is in a Kendra, or in a Trikona from the Lord of the Dasha. Deportation to foreign lands, sorrows, death and danger from thieves and snakes will be the results, if Sukra is in the 6th, the 8th, or the 12th from the Lord of the Dasha. There will be danger of premature death, if Sukra be the Lord of the 2nd, or 7th. The remedial measures are Rudra Japa and giving in charity a white cow and silver.

Moon Dasa-Sun Bhukti: This is a mixed period. He/She will suffer from tuberculosis, but will be valorous, will earn wealth through kings and will enjoy all kinds of monetary gains. There will be happiness and general prosperity, ill health to parents, general debility, new position or appointment in case of unemployed. Recovery of a lost kingdom and wealth, happiness in the family, acquisition of villages and land with the kind assistance of one's friends and the king, birth of a son, beneficence of Goddess Lakshmi, will be the beneficial results in the Antar Dasha of Surya in the Dasha of Chandra, if Surya is in his exaltation Rashi, in his own Rashi, in a Kendra, or in Putra, or in Dharma, or in Labh, or in Dhan, or in Sahaj. At the end of the Antar Dasha there is the likelihood of attacks of fever and lethargy. Danger from the government, thieves and snakes, affliction with fever and troubles in foreign journey are the likely results, if Surya is in the 8th, or 12th from the Lord of the Dasha. If Surya is the Lord of Dhan, or Yuvati, there will be sufferings from fever in his Antar Dasha. Worship of Lord Shiva is the remedial measure to obtain relief from the above evil effects.

13.3.3 Predictions by Antar Dasha in the Dasha of Mars:

Mars Dasa-Mars Bhukti: This is an unfavourable period. There will be misunderstandings between brothers and sisters and superiors in office, quarrel, loss, ill health due to excess heat in the body. When Mars is exalted or in own house or in favourable position, there will be gain through landed property or acquisition of house, favourable results in family litigation, purchase of ornaments and other luxury items. The effects like gains of wealth by the beneficence of the king, beneficence of Goddess Lakshmi, recovery of a lost kingdom and of wealth, birth of a son, will arise in the Antar Dasha of Mangal in his own Dasha, if he is in a Kendra, in Putra, in Dharma, in Labh, in Sahaj, or in Dhan, or be associated with the Lord of Lagna. Fulfilment of ambitions by the beneficence of the king and acquisition of a house, land, cow, buffalo will be the effects, if Mangal is in his exaltation, in his own Rashi, or in his own Navamsa and is endowed with strength. Urinary troubles, wounds, danger from snakes and the king will be the results, if Mangal is in Randhra, or Vyaya, or is associated with, or receives a Drishti from malefic. There will be mental agony and body pains, if Mangal is the Lord of Dhan, or Yuvati. Lord Shiva will give relief by restoring health and providing gains of wealth and happiness, if the person concerned performs Rudra Japa and gives a red-coloured bull in charity.

Mars Dasa-Rahu Bhukti: This is an unfavourable period. There will be Royal displeasure, mind going towards evil doings, changing the living place, punishments, loss of cattle, wife, long journey, bad name, fall and injury. The effects like recognition from government, gain of house, land etc., happiness from son, extraordinary profits in business, bathing in holy rivers, like Ganges and foreign journeys, will be the auspicious effects in the Antar Dasha of Rahu in the Dasha of Mangal, if Rahu is in his Moolatrikona, in his exaltation Rashi, in a Kendra, in Labh, Putra, or Dharma and is associated with benefic. Danger from snakes, wounds, destruction of cattle, and danger from animals, diseases, due to imbalance of bile and wind, imprisonment will be the results, if Rahu is in Randhra, or Vyaya, or receives a Drishti from, or is associated with malefic. There will be loss of wealth, if Rahu is in Dhan and great danger of premature death, if he is in Yuvati. The remedial measure is Naga Puja, offering food to Brahmins and Mrityunjaya Japa to be obtained relief from the above evil effects. They will help in the prolongation of longevity.

Mars Dasa-Jupiter Bhukti: This is a favourable period. In Mars Dasa and in Jupiter's sub period there will be good conduct, virtues, and meritorious acts. Favourable Jupiter indicates that there will be pilgrimage, birth of child, promotion, devotion to God. Unfavourable Jupiter indicates loss of brothers, loss of money, failure in undertakings, mental worry. The effects like good reputation and renown, honours by government, increase in wealth and grains, happiness at home, gain of property, happiness from wife and children will be realized in the Antar Dasha of Guru in the Dasha of Mangal, if Guru is in Dharma, or Putra, in a Kendra, or in Labh, or in Dhan, or, if Guru is in his exalted, or own Navamsa. Acquisition of a house, land, well-being, gain of property, sound health, good reputation, gains of cattle, success in business, happiness to wife and children, reverence from government, gain of wealth will be beneficial effects, if Guru is in a Kendra, in a Trikona, or in the 11th from the Lord of the Dasha, or, if Guru is associated with the Lord of Dharma, Karma, or Bandhu, or Lagna, or, if Guru is in a benefic Navamsa. Danger from thieves, snakes, wrath of the king, bilious diseases, oppression by goblins (Prot), loss of servants and co-born, will be evil effects, if Guru is in Ari, Randhra, or in Vyaya, or, if Guru is in his debilitation Rashi, or, if Guru is associated with, or receives a Drishti from malefic, or, if Guru is otherwise weak. There will be suffering from fever, or danger of premature death, if Guru is the Lord of Dhan. The remedial measure to be adopted to combat the above evil effects is recitation of Shiva Sahasranam.

Mars Dasa-Saturn Bhukti: This is an unfavourable period. There will be evils after evils and loss of wealth and of near and dear. There will be loss of wealth, danger in operation, quarrels, fight, court troubles, displeasure in office and loss of position and mental worry, hard period and scarcity of food. The effects like recognition from the king, increase in reputation, gain of wealth and grains, happiness from children and grandchildren, increase in the number of cows will be experienced in the Antar Dasha of Sani in the Dasha of Mangal, if Sani is in a Kendra, in a Trikona, in his Moolatrikona, in his exalted, or his own Navamsa, or, if Sani is associated with the Lord of Lagna, or, if Sani is associated with benefic. Results will generally fructify on Saturdays in the month of Sani. Danger from Yavana kings (foreign dignitaries), loss of wealth, imprisonment, possibility of affliction with diseases, loss in agricultural production will result, if Sani is in his debilitation Rashi, or in an enemy Rashi, or, if Sani is in Randhra, or in Vyaya. The effects like great danger, loss of life, wrath of king, mental agony, danger from thieves and fire, punishment by the king, loss of co-born, dissensions amongst members of the family, loss of cattle, fear of death, distress to wife and children, imprisonment will be felt, if Sani is Dhan, or Yuvati Lord and is associated with malefic. There will be journeys to foreign lands, loss of reputation, violent actions, loss from sale of agricultural lands, loss of position, agony, defeat in battle, urinary troubles, if Sani is in a Kendra, in the 11th, or in the 5th from the Lord of the Dasha. Effects, like death, danger from the king and thieves, rheumatism, pains, danger from the enemy and members of the family, will be experienced, if Sani is in the 8th or the 12th from the Lord of the Dasha and is associated with malefic. There will be relief from the evil effects by the beneficence of Lord Shiva, if Mrityunjaya is performed in the prescribed manner.

Mars Dasa-Mercury Bhukti: He/She have fear from thieves and enemies, loss of horses and elephants and burning sensation. There will be mind diversion towards holy activities, favourable results in education, travel, marriage, gain in business, and respect in society. When mercury is ill placed, the period will be of mental illness, robbery, money through illegal way, bad name. The effects like association with pious and holy persons, performance of Ajaya Japa, charities, observance of religious rites, gain of reputation, inclination towards diplomacy, availability of sweetish preparations, acquisition of conveyances, clothes and cattle, conferment of authority in the king's retinue, success in agricultural projects, will be experienced in the Antar Dasha of Buddha in the Dasha of Mangal, if Buddha is in a Kendra, or in a Trikona from Lagna. Diseases of heart, imprisonment, loss of kinsmen, distress to wife and children, destruction of wealth and cattle will result, if Buddha is in his debilitation Rashi, if Buddha is combust, or, if Buddha is in Ari, Randhra, or Vyaya. There will be journeys to foreign lands, increase in the number of enemies, affliction with much kind of ailments, antagonism with the king, quarrels with kinsmen, if Buddha be associated with the Lord of the Dasha. Fulfilment of all ambitions, gain of wealth and grains, recognition by the king, acquisition of a kingdom, gain of clothes and ornaments, attachment to many kind of musical instruments, attainment of the position of a Commander of an Army, discussions on Shashtra and Puranas, gain of riches to wife and children and beneficence of Goddess Lakshmi will be the very auspicious results, if Buddha is in a Kendra, or Trikona from the Lord of the Dasha, or, if Buddha is in his exaltation Rashi. The effects like defamation, sinful thinking, harsh speech, danger from thieves, fire and the king, quarrels without reason, fear of attacks by thieves and dacoits (armed robber bands) during travels, will be derived, if Buddha is in the 6th, the 8th, or the

12th from Mangal, or is associated with malefic. There will be a possibility of critical illness in the Antar Dasha of Buddha, if he is Dhan, or Yuvati Lord. Remedial measures to obtain relief from these evil effects are recitation of Vishnu Sahasranam and giving a horse in charity.

Mars Dasa-Ketu Bhukti: This is an unfavourable period. There will be sufferings on account of misunderstandings with brothers or relatives or in family, disease due to infection or fire. Beneficence of the king, gain of wealth, little gains of land at the commencement of the Dasha and substantial later, birth of a son, conferment of authority by government, gain of cattle will be the results in the Antar Dasha of Ketu in the Dasha of Mangal, if Ketu is in a Kendra, in a Trikona, in Sahaj, or Labh, or, if Ketu is associated with, or receives a Drishti from benefic. Birth of a son, increase in reputation, beneficence of Goddess Lakshmi, gains of wealth from employees, attainment of the position of a Commander of an Army, friendship with the king, performance of oblations, gains of clothes and ornaments will be the beneficial effects, if Ketu is a Yoga Karaka and is endowed with strength. (Ketu assumes the role of a Yoga Karaka, if he is yuti with a Yoga Karaka Graha (Lord of a Kendra and a Trikona)). Effects, like quarrels, tooth trouble, distress from thieves and tigers, fever, dysentery, leprosy and distress to wife and children will be experienced, if Ketu is in the 6th, the 8th, or the 12th from the Lord of the Dasha. If Ketu is in Dhan, or in Yuvati, there will be diseases, disgrace, agony and loss of wealth.

Mars Dasa-Venus Bhukti: He/She will have fear of war, diseases, evils and loss of wealth. There will be happiness in family, success in matrimonial alliances, religious travels, purchase of property through wife, victory over enemies. When Venus is badly, it will give anxiety and mental depression. The effects like acquisition of a kingdom, great enjoyment and comfort of luxuries, gain of elephants, horses, clothes, will be derived in the Antar Dasha of Sukra in the Dasha of Mangal, if Sukra is in a Kendra to Lagna, if Sukra is in his exaltation, or in his own Rashi, or, if Sukra is Lagna, Putra, or Dharma Lord. If Sukra is related to Lagna Lord, there will be happiness to wife and children, opulence and glory and increased good fortune. Gain of property, celebrations on the birth of a son, gain of wealth from the employer, acquisition of a house, land, villages etc. by the beneficence of the sovereign, will be the results, if Sukra is in the 5th, the 9th, the 11th, or the 2nd from the Lord of the Dasha. In the last part of the Dasha there will be functions of songs and dances and bathing in holy water. If Sukra is connected with, or related to the Lord of Karma, there will be construction of wells, reservoirs etc. and performance of religious, charitable and pious deeds. There will be sorrows, physical distress, loss of wealth, danger from thieves and the king, dissensions in the family, distress to wife and children and destruction of cattle, if Sukra be in the 6th, the 8th, or the 12th from the Lord of the Dasha, or be associated with malefic. If Sukra be the Lord of the 2nd, or the 7th, there will be pains in the body in his Antar Dasha. For regaining good health the remedial measure to be adopted is giving a cow or female buffalo in charity.

Mars Dasa-Sun Bhukti: This is a favourable period. He/She will be fierce, valorous, be honoured by the king, be successful in battle and will acquire wealth in various ways. There will be contact with holy persons or saints; mind occupied with religious thoughts, long travel, and health of father improves. When Sun is badly placed, there will be diseases due to excess heat and mixed events. The effects like acquisition of conveyances, gain of reputation, birth of a son, growth of wealth, amicable atmosphere in the family, sound health, potency, recognition by the king, extraordinary profits in business and audience with the king will be experienced in the Antar Dasha of Surya in the Dasha of Mangal, if Surya is in his exaltation, in his own Rashi, or, if Surya is in a Kendra, in a Trikona, or in Labh along with Karma Lord and with Labh Lord. Distress to the body, agony, and failure in ventures, possibilities of suffering from troubles in the forehead, fever, and dysentery will be the effects, if Surya is in the 6th, the 8th, or the 12th from the Lord of the Dasha, or, if Surya is associated with malefic. There will be attacks of fever, danger from snakes and poison and distress to son, if Surya be the Lord of the 2nd, or the 7th. The remedial measure to gain good health and wealth is to perform worship of Surya in the prescribed manner.

Mars Dasa-Moon Bhukti: This is a better period. He/She will procure wealth in various ways, be happy, will have many friends and plenty of corals. There will be profit, good health to children, purchase of ornaments, repairs in the living place. When Mars is the lord of 5 or 9 or 10 or placed in such position with good benefic aspect, the Dasa will give good understandings between brothers and professional betterment and gain. Acquisition of more kingdom, gain of perfumes, clothes, construction of reservoirs, shelters for cows, celebrations of auspicious functions, like marriage, happiness to wife and children, good relations with parents, acquisition of property by the beneficence of the sovereign, success in the desired projects will be the effects in the Antar Dasha of Chandra in the Dasha of Mangal, if Chandra is in her exaltation Rashi, or in her own Rashi, or in a Kendra, or in

Dharma, or in Bandhu, or in Karma, or in Lagna along with the Lords of those Bhava. The good effects will be realized in full, if Chandra is waxing. Waning Chandra will reduce the impact of the effects to some extent. The effects, like death, distress to wife and children, loss of lands, wealth and cattle and danger of a war will be experienced, if Chandra is in his debilitation Rashi, or, if Chandra is in his enemy Rashi, or, if Chandra is in Ari, in Randhra, or in Vyaya from Lagna, or from the Lord of the Dasha. There will be the possibility of premature death, distress to the body and mental agony, if Chandra is Dhan, or Yuvati Lord. The remedial measures to be adopted to obtain relief from the above evil effects are recitation of Mantras of the Goddess Durga and the Goddess Lakshmi.

13.3.4 Predictions by the Antar Dasha in the Dasha of Rahu:

Rahu Dasa-Rahu Bhukti: This is an unfavourable period. There will be mental anxiety, ill health to wife or other members of family, transfer in the place of work, bad name, poisonous bites, court troubles, leaving home and wandering. The effects like acquisition of a kingdom, enthusiasm, cordial relations with the king, happiness from wife and children and increase in property, will be derived in the Antar Dasha of Rahu in the Dasha of Rahu, if Rahu is in Karka, Vrischika, Kanya, or Dhanu or is in Sahaj, Ari, Karma, or Labh, or is yuti with a Yoga Karaka Graha in his exaltation Rashi. There will be danger from thieves, distress from wounds, antagonism with government officials, destruction of kinsmen, and distress to wife and children, if Rahu is in Randhra, or Vyaya, or be associated with malefic. If Rahu is Dhan, or Yuvati Lord, or is in Dhan, or Yuvati, there will be distress and diseases. To obtain relief from the above evil effects Rahu should be worshipped (by recitation of his Mantras) and by giving in charity things, connected with, or ruled by Rahu.

Rahu Dasa-Jupiter Bhukti: This is a better period. There will be promotion in job, child birth or marriage, favourable atmosphere in office, good health, pilgrimage to holy places, litigations or court troubles are also likely. The effects like gain of position, patience, destruction of foes, enjoyment, cordial relations with the king, regular increase in wealth and property, like the growth of Chandra of the bright half of the month (Shukla Paksha), gain of conveyance and cows, audience with the king by performing journey to the West, or South-East, success in the desired ventures, return to one's homeland, doing good for Brahmins, visit to holy places, gain of a village, devotion to deities and Brahmins, happiness from wife, children and grand children, availability of sweetish preparations daily will be derived in the Antar Dasha of Guru in the Dasha of Rahu, if Guru is in his exaltation, in his own Rashi, in his own Navamsa, or in his exalted Navamsa, or, if Guru is in a Kendra, or in a Trikona with reference to Lagna. Loss of wealth, obstacles in work, defamation, and distress to wife and children, heart disease, entrustment of governmental authority will result, if Guru is in his debilitation Rashi, is combust, is in Ari, Randhra, or Vyaya, is in an enemy Rashi, or is associated with malefic. There will be gains of land, good food, gains of cattle, inclinations towards charitable and religious work, if Guru is in a Kendra, in a Trikona, the 11th, the 2nd, or the 3rd from the Lord of the Dasha and is endowed with strength. Loss of wealth and distress to body will result, if Guru is in the 6th, the 8th, or the 12th from the Lord of the Dasha, or, if Guru is associated with malefic. There will be danger of premature death, if Guru is Dhan, or Yuvati Lord. The person will get relief from the above evil effects and enjoy good health by the beneficence of the Lord Shiva, if he worships his idol, made of gold.

Rahu Dasa-Saturn Bhukti: It is, generally, a very unfavourable period. There will be extreme difference of opinion between husband and wife, if 7th Bhava or Venus is bad. There will be, even, divorce or separation, diseases due to pains in joints, excess wind and leaving to remote place. The effects like pleasure of the king for devotion in his service, auspicious functions, like celebration of marriage at home, construction of a garden, reservoir, gain of wealth and cattle from well-to-do persons, belonging to the Sudra class, loss of wealth caused by the king during journey to the West, reduction in income, due to lethargy, return to homeland, will be derived in the Antar Dasha of Sani in the Dasha of Rahu, if Sani is in a Kendra, in a Trikona, in his exaltation, in his own Rashi, in his Moolatrikona, in Sahaj, or in Labh. Danger from menials, the king and enemies, distress to wife and children, distress to kinsmen, disputes with the coparceners, disputes in dealings with others, but sudden gain of ornaments, will result, if Sani is in his debilitation Rashi, in his enemy's Rashi, or in Randhra, or Vyaya. There will be heart disease, defamation, quarrels, danger from enemies, foreign journeys, affliction with Gulma, unpalatable food and sorrows, if Sani is in the 6th, the 8th, or the 12th from the Lord of the Dasha. Premature death is likely, if Sani is Dhan, or Yuvati Lord. Remedial measure to obtain relief from the above evil effects and to regain good health is giving a black cow or a she-buffalo in charity.

Rahu Dasa-Mercury Bhukti: It is, generally, a very favourable period. There will be marriage, promotion in the job, expansion of business, new circle of friends. In the second part, there will be birth of child, vehicles, enjoyments in life, evil ways in enjoyments and illegal methods of earning. Auspicious effects, like Raja Yoga, well being in the family, profits and gain of wealth in business, comforts of conveyances, marriage and other auspicious functions, increase in the number of cattle, gain of perfumes, comforts of bed, women, will be derived in the Antar Dasha of Buddha in the Dasha of Rahu, if Buddha is in his exaltation Rashi, in a Kendra, or in Putra and, if Buddha is endowed with strength. Good results, like Raja Yoga, beneficence of the king and gain of wealth and reputation, will be realized particularly on Wednesday in the month of Buddha. Sound health, Isht Siddhi, attending discourse on Puranas and ancient history, marriage, offering of oblations, charities, religious inclination and sympathetic attitude towards others will result, if Buddha is in a Kendra, in the 11th, 3rd, 9th, or 10th from the Lord of the Dasha. There will be opprobrium (Ninda) of deities and Brahmins by the native, loss of fortune, speaking lies, unwise actions, fear from snakes, thieves and the government, quarrels, distress to wife and children, if Buddha is in Ari, Randhra, or Vyaya, or, if Buddha receives a Drishti from Sani. If Buddha is Dhan, or Yuvati Lord, there will be fear of premature death. Remedial measure to obtain relief from the above evil effects is recitation of Vishnu Sahasranam.

Rahu Dasa-Ketu Bhukti: This is again a period of strain with disease, ill health due to some poison; wife will become enemy, displeasure with superiors in office, loss of wealth and blame. During the Antar Dasha of Ketu in the Dasha of Rahu there will be journeys to foreign countries, danger from the king, rheumatic fever etc. and loss of cattle. If Ketu is yuti with Randhra Lord, there will be distress to the body and mental tension. Enjoyment, gain of wealth, recognition by the king, acquisition of gold will be the results, if Ketu is associated with, or receives a Drishti from benefic. There will be Isht Siddhi, if Ketu is related to the Lord of Lagna. If he is associated with the Lord of Lagna, there will definitely be gain of wealth. There will also definitely be increase in the number of cattle, if Ketu is in a Kendra, or in a Trikona. The effects like danger from thieves and snakes, distress from wounds, separation from parents, antagonistic relations with kinsmen, mental agony will be derived, if Ketu is without strength in Randhra, or Vyaya. If Ketu is Dhan, or Yuvati Lord, there will be distress to the body. The remedial measure to obtain relief from the above evil effects is giving a goat in charity.

Rahu Dasa-Venus Bhukti: This will be a better period than others. There will be purchase of vehicles; wife would be a source of happiness, marriage if unmarried, child birth, benefits like promotion, other favour in the office, gain in agricultural holdings and general happiness. Some troubles from enemies and ill health are also likely. The effects like gains of wealth through Brahmins, increase in the number of cattle, celebrations for the birth of a son, well-being, recognition from government, acquisition of a kingdom, attainment of a high position in government, great enjoyment and comforts will be experienced in the Antar Dasha of Sukra in the Dasha of Rahu, if Sukra is with strength in a Kendra, in a Trikona, or in Labh. Construction of a new house, availability of sweet preparations, happiness from wife and children, association with friends, giving of grains in charity, beneficence of the king, gain of conveyances and clothes, extraordinary profits in business, celebration of Upasayan ceremony of wearing the sacred thread (Janou) will be the auspicious results, if Sukra be in his exaltation, in his own Rashi, in is exalted, or in his own Navamsa. There will be diseases, quarrels, separation from one's son, or father, distress to kinsmen, disputes with coparceners, danger of death to oneself, or to one's employer, unhappiness to wife and children, pain in the stomach etc., if Sukra is in Ari, Randhra, or Vyaya, in his debilitation, or in an enemy's Rashi, or, if Sukra is associated with Sani, Mangal, or Rahu. Enjoyments from perfumes, bed, music, gain of a desired object, fulfilment of desires will be the results, if Sukra is in a Kendra, in a Trikona, in the 11th, or in the 10th from the Lord of the Dasha. Effects, like danger from the wrath of Brahmins, snakes and the king, possibility of affliction with diseases, like stoppage of urine, diabetes, pollution of blood, anaemia, availability of only coarse food, nervous disorder, imprisonment, loss of wealth, as a result of penalties, or fines, imposed by government, will be derived, if Sukra is associated with malefic in the 6th, 8th, or 12th from the Lord of the Dasha. There will be distress to wife and children, danger of premature death to oneself, if Sukra is Dhan, or Yuvati Lord. Remedial measures to obtain relief from the above evil effects are worship of Goddess Durga and Goddess Lakshmi.

Rahu Dasa-Sun Bhukti: There will be transfer in job or transfer in place of working, disease due to excess heat, changing the place of living, educational achievements and charity, mental worry and uneasiness will also prevail. The effects like cordial relations with the king, increase in wealth and grains, some popularity/respect, some possibility of becoming head of a village, will be experienced in the

Antar Dasha of Surya in the Dasha of Rahu, if Surya is in his exaltation, in his own Rashi, in Labh, in a Kendra, or in a Trikona, or in his exalted, or own Navamsa. There will be good reputation and encouragement and assistance by government, journeys to foreign countries, acquisition of the sovereignty of the country, gains of elephants, horses, clothes, ornaments, fulfilment of ambitions, happiness to children, if Surya is associated with, or receives a Drishti from Lagna, Dhan, or Karma Lord. Fevers, dysentery, other diseases, quarrels, antagonism with the king, travels, danger from foes, thieves, fire will be the results, if Surya is in his debilitation Rashi, or, if Surya is in the 6th, 8th, or 12th from the Lord of the Dasha. Well-being in every way and recognition from kings in foreign countries will be the results, if Surya is in a Kendra, in a Trikona, in the 3rd, or in the 11th from the Lord of the Dasha. There will be danger of critical illness, if Surya is Dhan, or Yuvati Lord. Worship of Surya is the remedial measure, recommended to obtain relief from the above evil effects.

Rahu Dasa-Moon Bhukti: There will be condition of health changing, some kind of loss through wife, enjoyments in life, travels, sea travel also likely, financial improvements, gain in lands, and death of relatives. The effects like acquisition of a kingdom, respect from the king, gains of wealth, sound health, gains of garments and ornaments, happiness from children, comforts of conveyances, increase in house and landed property, will be derived in the Antar Dasha of Chandra in the Dasha of Rahu, if Chandra is in his exaltation, in his own Rashi, in a Kendra, Trikona, or in Labh, or, if Chandra is in a friendly Rashi, receiving a Drishti from benefic. Beneficence of the Goddess Lakshmi, all-round success, increase in wealth and grains, good reputation and worship of deities will be the results, if Chandra is in the 5th, 9th, in a Kendra, or in the 11th from the Lord of the Dasha. There will be the creation of disturbances at home and in the agricultural activities by evil spirits, leopards and other wild animals, danger from thieves during journeys and stomach disorders, if Chandra is bereft of strength in the 6th, 8th, or 12th from the Lord of the Dasha. There will be the possibility of premature death, if Chandra is Dhan, or Vyaya Lord. The remedial measure to obtain relief from the above evil effects is to give in charity a white cow, or a female buffalo.

Rahu Dasa-Mars Bhukti: This is a period of test and indicates displeasure with officers, failure in court cases, loss through brothers or cousins, bad habits, severe mental agony and decrease of mental power. The effects like the recovery of a lost kingdom and recovery of lost wealth, property at home and increase in agricultural production, gain of wealth, blessings by the household deity (Isht Dev), happiness from children, enjoyment of good food, will be derived in the Antar Dasha of Mangal in the Dasha of Rahu, if Mangal is in Labh, Putra, or Dharma, or, if Mangal is in a Kendra, if Mangal receives a Drishti from benefic, or, if Mangal is in his exaltation, or in his own Rashi. There will be acquisition of red-coloured garments, journeys, audience with the king, well-being of children and employer, attainment of the position of a Commander of the Army, enthusiasm and gain of wealth through kinsmen, if Mangal is in a Kendra, in the 5th, 9th, 3rd, or in the 11th from the Lord of the Dasha. Distress to wife, children and co-born, loss of position, antagonistic relations with children, wife and other close relations, danger from thieves, wounds and pain in the body will result, if Mangal is in the 6th, 8th, or 12th from the Lord of the Dasha, receiving a Drishti from malefic. There will be lethargy and danger of death, if Mangal is Dhan, or Yuvati Lord. The remedial measure to obtain relief from the above evil effects is giving a cow or a bull in charity.

13.3.5 Predictions by the Antar Dasha in the Dasha of Jupiter:

Jupiter Dasa-Guru Bhukti: There will be favourable news in office and promotion, good health, success in activities, pilgrimage, loss of wealth, failures, un-attachment in family and children. The effects like sovereignty over many kings, very well endowed with riches, revered by the king, gains of cattle, clothes, ornaments, conveyances, construction of a new house and a decent mansion, opulence and glory, dawn of fortune, success in ventures, meetings with Brahmins and the king, extraordinary profits from the employer and happiness to wife and children, will be experienced in the Antar Dasha of Guru in his own Dasha, if Guru is in his exaltation Rashi, in his own Rashi, in a Kendra, or Trikona. Association with the menials, great distress, slander by coparceners, wrath of the employer, danger of premature death, separation from wife and children and loss of wealth and grains will be the results, if Guru is in his debilitation Rashi, in his debilitated Navamsa, or in Ari, Randhra, or Vyaya. There will be pains in the body, if Guru is the Lord of Yuvati (or of Dhan). The remedial measure to obtain relief from the above evil effects and to get fulfilment of ambitions is recitation of Rudr Japa and Shiva Sahasranam.

Jupiter Dasa-Saturn Bhukti: He/She will be attached to prostitutes and intoxicants and will be insulted in gambling. He will be endowed with buffaloes and asses and be devoid of virtues. There will be misunderstanding in family and relations, failure in business, debts, court troubles, mental uneasiness, funeral ceremonies for others, evil ways and habits, and pain in foot or joints. The effects like acquisition of a kingdom, gain of clothes, ornaments, wealth, grains, conveyances, cattle and position, happiness from son and friends, gains specially of a blue-coloured horse, journey to the West, audience with the king and receipt of wealth from him, will be derived in the Antar Dasha of Sani in the Dasha of Guru, if Sani is in his exaltation, in his own Rashi, in a Kendra, or Trikona endowed with strength. Loss of wealth, affliction with fever, mental agony, and infliction of wounds to wife and children, inauspicious events at home, loss of cattle and employment, antagonism with kinsmen will be results, if Sani is in Ari, Randhra, or Vyaya, if Sani is combust, or, if Sani is in an enemy's Rashi. There will be gain of land, house, son and cattle, acquisition of riches and property through the enemy, if Sani is in Kendra, Trikona, the 11th, or in the 2nd from the Lord of the Dasha. Effects, like loss of wealth, antagonistic relations with kinsmen, obstacles in industrial ventures, pains in the body, danger from the members of the family will be realized, if Sani is in the 6th, 8th, or 12th from the Lord of the Dasha, or, if Sani is associated with a malefic. There will be fear of premature death, if Sani is Dhan, or Yuvati Lord. The remedial measures to obtain relief from these evil effects and to enjoy sound health are recitation of Vishnu Sahasranam and giving in charity a black cow or a female buffalo.

Jupiter Dasa-Mercury Bhukti: There will be destruction of diseases, acquisition of friends and respect for parents in the sub period of Mercury in the major period of Jupiter. There will be improvement in finance, auspicious ceremonies at home, expansion of business, favours from superiors, mental activities in artistic lines and birth of a good child. The effects like gains of wealth, bodily felicity, acquisition of a kingdom, gain of conveyances, clothes and cattle, will be derived in the Antar Dasha of Buddha in the Dasha of Guru, if Buddha is in his exaltation, in his own Rashi, or in Kendra, in Trikona, or, if Buddha is associated with the Lord of the Dasha. There will be increase in the number of enemies, loss of enjoyment and comforts, loss in business, affliction with fever and dysentery, if Buddha receives a Drishti from Mangal. Gains of wealth in his own country, happiness from parents and acquisition of conveyances by the beneficence of the king will result, if Buddha is in a Kendra, in the 5th, or 9th from the Lord of the Dasha, or, if Buddha is in his exaltation Rashi. There will be loss of wealth, journeys to foreign countries, danger from thieves while travelling, wounds, burning sensations, eye troubles, wanderings in foreign lands, if Buddha is in the 6th, 8th, or 12th from the Lord of the Dasha, or, if Buddha is associated with a malefic without receiving a Drishti from a benefic. Distress without reason, anger, loss of cattle, loss in business, fear of premature death will be the results, if Buddha be associated with a malefic, or malefic in Ari, in Randhra, or in Vyaya. There will be enjoyment, gains of wealth, conveyances and clothes at the commencement of the Antar Dasha, even if Buddha is associated with a malefic, but receives a Drishti from a benefic. At the end of the Dasha, however, there will be loss of wealth and bodily distress. Premature death may be expected, if Buddha is Dhan, or Yuvati Lord. The most effective and beneficial remedial measure for prolongation of longevity and to obtain relief from other evil effects is recitation of Vishnu Sahasranam.

Jupiter Dasa-Ketu Bhukti: There will be sacrificing for the sake of others, change of living place, separation from relations, pilgrimage to holy places, loss in wealth and illness due to poison. Moderate enjoyment, moderate gain of wealth, coarse food, or food, given by others, food, given at the time of death ceremonies and acquisition of wealth through undesirable means will be the results, in the Antar Dasha of Ketu in the Dasha of Guru, if Ketu is associated with, or receives a Drishti from a benefic. The effects like loss of wealth by the wrath of the king, imprisonment, diseases, loss of physical strength, antagonism with father and brother and mental agony, will be experienced, if Ketu be in the 6th, 8th, or 12th from the Lord of the Dasha, or be associated with malefic. Acquisition of a palanquin, elephants etc., beneficence of the king, success in the desired spheres, profits in business, increase in the number of cattle, gain of wealth, clothes from a Yavana king (Muslim dignitary) will be the auspicious effects, if Ketu is in the 5th, 9th, 4th, or 10th from the Lord of the Dasha. There will be physical distress, if Ketu is Dhan or Yuvati Lord (or, if Ketu is in Dhan, or in Yuvati. The remedial measure to obtain relief from the above evil effects is performance of Mrityunjaya Japa in the prescribed manner.

Jupiter Dasa-Venus Bhukti: He/She will have fear from enemies, will face destruction and grief and will live through Brahmins. There will be employment if unemployed, success or promotion in job or business, happiness at home, improvement in children's life, purchase of jewels especially in diamond, auspicious ceremonies at home, bad name and difficulties for women folk. The effects like acquisition of conveyances, like palanquin, elephants, gain of wealth by the beneficence of the king, enjoyment, gain of blue and red articles, extraordinary income from journeys to the East, well-being in the family, happiness from parents, devotion to deities, construction of reservoirs, charities, will be derived in the Antar Dasha of Sukra, if Sukra is in a Kendra, Trikona, or in Labh, or, if Sukra is in his own Rashi and receives a Drishti from a benefic, or from benefic. Evil effects, like quarrels, antagonism with kinsmen, distress to wife and children, will be felt, if Sukra is in the 6th, 8th, or 12th from the Lord of the Dasha, or Lagna, or, if Sukra is in his debilitation Rashi. There will be quarrels, danger from the king, antagonism with the wife, disputes with the father-in-law and with brothers, loss of wealth, if Sukra is associated with Sani, or Rahu, or with both. There will be gain of wealth, happiness from wife, meeting with the king, increase in the number of children, conveyances and cattle, enjoyment of music, society with men of learning, availability of sweetish preparations, giving help and assistance to kinsmen, if Sukra is in a Kendra, Trikona, or in the 2nd from the Lord of the Dasha. Loss of wealth, fear of premature death, antagonism with wife will be experienced, if Sukra is Dhan, or Yuvati Lord. The remedial measure to obtain relief from these evil effects is giving a tawny-coloured cow or a female buffalo in charity.

Jupiter Dasa-Sun Bhukti: He/She will be free from enemies and diseases and will be honoured by the king. He will either be endowed with valour and happiness. There will be increase in financial standard, favour from superiors, and improvement in health, more pious and holy activities. Gain of wealth, reverence, happiness and acquisition of conveyances, clothes, ornaments, birth of children, cordial relations with the king, success in ventures will be the auspicious results in the Antar Dasha of Surya in the Dasha of Guru, if Surya is in his exaltation, in his own Rashi, in a Kendra, Trikona, or in Sahaj, Labh, or Dhan and be endowed with strength. The effects like nervous disorder, fever, laziness, or reluctance in the performance of good deeds, indulgence in sins, antagonistic attitude towards all, separation from kinsmen and distress without reasons, will be experienced, if Surya is in Ari, Randhra, or Vyaya, or, if Surya is in the 6th, 8th, or 12th from the Lord of the Dasha. There will be physical distress, if Surya is Dhan, or Yuvati Lord. The remedial measure to obtain relief from the above evil effects and to enjoy good health is recitation of Adhitya Hridaya Path.

Jupiter Dasa-Moon Bhukti: He/She will marry a thousand women, will conquer diseases and enemies and will prosper in all fronts akin to a king. There will be happiness from Children, marriage, birth of child, purchase or acquisition of property, home comforts, name and fame and benefit in mental activities like writing. The effects like reverence from the king, opulence and glory, happiness from wife and children, availability of good food, gain of reputation by performance of good deeds, increase in the number of children and grandchildren, comforts by the beneficence of the king, religious and charitable inclinations, will be derived in the Antar Dasha of Chandra in the Dasha of Guru, if Chandra is in a Kendra, Trikona, or in Labh, or, if Chandra is in her exaltation, or in her own Rashi and, if Chandra is full and strong and in an auspicious Bhava from the Lord of the Dasha. There will be loss of wealth and kinsmen, wanderings in foreign lands, danger from the king, thieves, quarrels with co-partners, separation from a maternal uncle, distress to mother, if Chandra is weak, or is associated with malefic, or, if Chandra is in Ari, Randhra, or Vyaya, or, if Chandra is in the 6th, 8th, or 12th from the Lord of the Dasha. Physical distress will be experienced, if Chandra is Dhan, or Yuvati Lord. The remedial measure to obtain relief is Durga Saptashati Path.

Jupiter Dasa-Mars Bhukti: He/She e will be rude, will win over enemies, be valorous, be famous in war and will enjoy all kinds of happiness. There will be pilgrimage to holy temples, new ventures and wealth, loss due to robbery, difficulties in job and displeasure with boss or authorities. The effects like the celebration of functions, such as marriage, gain of land, or villages, growth of strength and valour and success in all ventures, will be derived in the Antar Dasha of Mangal in the Dasha of Guru, if Mangal is in his exaltation, in his own Rashi, or in his exalted, or own Navamsa. There will be gain of wealth and grains, availability of good sweetish preparations, pleasure of the king, happiness from wife and children and other auspicious effects, if Mangal is in a Kendra, Trikona, in Labh, or Dhan and is associated with, or receives a Drishti from benefic. Loss of wealth and house, eye trouble and other inauspicious effects will be the results, if Mangal is in the 8th, or 12th from the Lord of the Dasha, or, if Mangal is in his debilitation Rashi, associated with, or receiving a Drishti from malefic. The effects will be particularly adverse at the commencement of the Antar Dasha. There will be some mitigation of evil effects later. There will be physical distress and mental agony, if Mangal is the Lord of Dhan, or Yuvati. The remedial measure to obtain relief from the above evil effects and to get gains of wealth and property is to give a bull in charity.

Jupiter Dasa-Rahu Bhukti: There will be loss and financial strain, some income from people in low standard, disease, anxiety to wife /spouse and mental tension. The effects like attachment to Yogi, gain of wealth and grains during the first five months, sovereignty over a village, or country, meeting with a foreign king, well-being in the family, journeys to distant lands, bathing in holy places, will be derived in the Antar Dasha of Rahu in the Dasha of Guru, if Rahu is in his exaltation, in his own Rashi, in his Moolatrikona, or, if Rahu is in a Kendra, or Trikona, or, if Rahu receives a Drishti from the Lord of a Kendra, or, if Rahu is associated with, or receives a Drishti from a benefic. Danger from thieves, snakes, the king, wounds, troubles in domestic affairs, antagonism with co-born and coparceners, bad dreams, quarrels without reason, danger from diseases will result, if Rahu is associated with a malefic, if Rahu is in the 8th, or 12th from the Lord of the Dasha. There will be physical distress, if Rahu is in Dhan, or in Yuvati. The remedial measures to obtain relief from the above evil effects are Mrityunjaya Japa and giving a goat in charity.

13.3.6 Predictions by the Antar Dasha in the Dasha of Saturn:

Saturn Dasa-Saturn Bhukti: There will be ill health, mental tension, worry from sons, wife and relatives and some loss is also indicated. The effects like acquisition of a kingdom, happiness from wife and children, acquisition of conveyances, like elephants, gain of clothes, attainment of the position of a Commander of the Army by the beneficence of the king, acquisition of cattle, villages and land, will be derived in the Antar Dasha of Sani in the Dasha of Sani, if Sani is in his own, in his exaltation Rashi, or in deep exaltation, or, if Sani is in a Kendra, or Trikona, or, if Sani is a Raja Yoga Karaka. Fear or danger from the king, getting inflicted with injuries with some weapon, bleeding gums, and dysentery will be the evil effects at the commencement of the Dasha, if Sani is in Randhra, or Vyaya, or, if Sani is associated with malefic in his debilitation Rashi. There will be danger from thieves, going away from the homeland, mental agony in the middle portion of the Dasha. The last part of the Dasha will yield beneficial results. There will be danger of premature death, if Sani is Dhan, or Yuvati Lord. Lord Shiva will afford protection and render relief, if Mrityunjaya Japa is performed in the prescribed manner.

Saturn Dasa-Mercury Bhukti: He/She will enjoy fortunes, happiness, success and honour and will financially gain. There will be expansion of education and knowledge, financial improvement, marriage, birth of child, favourable news in place of work and holy ceremonies at home. The effects like reverence from the people, good reputation, gain of wealth, comforts of conveyances, inclination towards performance of religious sacrifices (Yagya), Raja Yoga, bodily felicity, enthusiasm, well-being in the family, pilgrimage to holy places, performance of religious rites, listening to Purana, charities, availability of sweetish preparations, will be derived in the Antar Dasha of Buddha in the Dasha of Sani, if Buddha is in a Kendra, or Trikona. Acquisition of a kingdom, gain of wealth, headship of a village will be the effects at the commencement of the Dasha, if Buddha is in Ari, Randhra, or Vyaya from Lagna, or from the Lord of the Dasha, or, if Buddha is associated with Surya, Mangal and Rahu. Affliction with diseases, failure in all ventures, anxiety and feeling of danger will be experienced in the middle portion and in the last part of the Dasha. There will be physical distress, if Buddha is Dhan, or Yuvati Lord. The remedial measures to obtain relief from the above evil effects and to regain enjoyment in life are recitation of Vishnu Sahasranam and giving grains in charity.

Saturn Dasa-Ketu Bhukti: There will be ill health due to swelling in joints, especially knee joints, loss of money, quarrel with son, fear of poison and trouble through women. Evil effects, like loss of position, dangers, poverty, distress, foreign journeys, will be derived in the Antar Dasha of Ketu in the Dasha of Sani, even if Ketu is in his exaltation, in his own, in a benefic Rashi, or in a Kendra, or Trikona, or, if Ketu is associated with, or receives a Drishti from benefic. If Ketu is related to the Lagna Lord, there will be gain of wealth and enjoyment and bathing in holy places and visit to a sacred shrine at the commencement of the Antar Dasha. Gain of physical strength and courage, religious thoughts, audience with the king (high dignitaries of government, like president, prime minister, governor, ministers) and all kinds of enjoyments will be experienced, if Ketu is in a Kendra, in a Trikona, in the 3rd, or 11th from the Lord of the Dasha. Fear of premature death, coarse food, cold fever, dysentery, wounds, danger from thieves, separation from wife and children, will be the results, if Ketu is in Randhra, or Vyaya from Lagna, or from the Lord of the Dasha. There will be physical distress, if Ketu is in Dhan, or Yuvati. Remedial measure to obtain relief from the above evil effects and to regain enjoyments of life by the beneficence of Ketu is giving a goat in charity.

Saturn Dasa-Venus Bhukti: He/She will have increased number of friends, freedom from grief, increase of fame, happiness from wife, wealth and success. This is a brighter period. There will be promotion in job, favourable news in the place of work, happiness in family, success in undertaking, coming of wife's property and victory in disputes. The effects like marriage, birth of a son, gain of wealth, sound health, well-being in the family, acquisition of a kingdom, enjoyments by the beneficence of the king, honours, gain of clothes, ornaments, conveyance and other desired objects, will be derived in the Antar Dasha of Sukra in the Dasha of Sani, if Sukra is in a Kendra, Trikona, or in Labh, associated with, or receiving a Drishti from benefic. If during the period of Antar Dasha of Sukra Guru is favourable in transit, there will be dawn of fortune and growth of property. If Sani is favourable in transit, there will be Raja Yoga effects, or the accomplishment of Yogi Rites (Yogi Triya Siddhi). Distress to wife, loss of position, mental agony, quarrels with close relations will be the results, if Sukra is in his debilitation Rashi, if Sukra is combust, or, if Sukra is in Ari, Randhra, or Vyaya. Fulfilment of ambitions by the beneficence of the king, charities, performance of religious rites, creation of interest in the study of Shashtra, composition of poems, interest in Vedanta, listening to Purana, happiness from wife and children will be experienced, if Sukra is in Dharma, Labh, or Kendra from the Lord of the Dasha. There will be eye trouble, fevers, loss of good conduct, dental problems, heart disease, pain in arms, danger from drowning, or falling from a tree, antagonism towards relations with the officials of government and brothers, if Sukra is in the 6th, 8th, or 12th from the Lord of the Dasha. There will be physical distress, if Sukra is Dhan, or Yuvati Lord. The remedial measures to obtain relief from the above evil effects and to regain enjoyment and good health is by the beneficence of Goddess Durga and the performance of Durga Saptashati Path and giving a cow, or a female buffalo in charity.

Saturn Dasa-Sun Bhukti: There will be destruction of wealth, wife and children and abundant fear from enemies. There will be diseases due to poison of blood, theft, affliction in eyes, wife and children badly affected and mental suffering. The effects like good relations with one's employer, well-being in the family, happiness from children, gain of conveyances and cattle, will be derived in the Antar Dasha of Surya in the Dasha of Sani, if Surya is in his exaltation, in his own Rashi, or, if Surya is associated with Dharma Lord, or, if Surya is in a Kendra, or Trikona, associated with, or receiving a Drishti from benefic. There will be heart disease, defamation, loss of position, mental agony, separation from close relatives, obstacles in industrial ventures, fevers, fears, loss of kinsmen, loss of articles, dear to the person, if Surya is in Randhra, or Vyaya, or, if Surya is in the 8th, or 12th from the Lord of the Dasha. There will be physical distress, if Surya is Dhan, or Yuvati Lord. The worship of Surya is the remedial measure to obtain relief from the above evil effects.

Saturn Dasa-Moon Bhukti: He/She will lose his wife, or she will be kidnapped. He/She will incur separation from relatives, or enmity with them. There will be loss of property and money, debts, changing of house due to dispute, enmity among relations, and death of some important family member. The effects like gains of conveyance, garments, ornaments, improvement of fortune and enjoyments, taking care of brothers, happiness in both maternal and paternal homes, increase in cattle wealth, will be derived in the Antar Dasha of Chandra in the Dasha of Sani, if Chandra is full, in her exaltation, or in her own Rashi, or in a Kendra, or Trikona, or in the 11th from the Dasha Lord, or, if Chandra receives a Drishti from benefic. There will be great distress, wrath, separation from parents, ill health of children, losses in business, irregular meals, administration of medicines, if Chandra is waning, if Chandra is associated with, or receives Drishti from malefic, or, if Chandra is in his debilitation Rashi, or, if Chandra is in malefic Navamsa, or, if Chandra is in the Rashi of a malefic Graha. There will, however, be good effects and some gain of wealth at the commencement of the Antar Dasha. Enjoyment of conveyances and garments, happiness from kinsmen, happiness from parents, wife, and employer will be the results, if Chandra is in a Kendra, Trikona, or in the 11th from the Lord of the Dasha. The effects like sleepiness, lethargy, loss of position, loss of enjoyments, increase in the number of enemies, antagonism with kinsmen, will be experienced, if Chandra is weak and is in the 6th, 8th, or 12th from the Lord of the Dasha. There will be lethargy and physical distress, if Chandra is Dhan, or Yuvati Lord. The remedial measures to obtain relief from the above evil effects and prolongation of longevity are Havan and giving jiggery, Ghee, rice, mixed with curd, a cow, or a female buffalo in charity.

Saturn Dasa-Mars Bhukti: He/She will leave the country (in adverse circumstances) and will incur diseases and grief in many matters. There will be bad name, wandering, frequent transfer in job, serious illness and loss by theft. The effects like enjoyments, gain of wealth, reverence from the king, gain of conveyances, clothes and ornaments, attainment of the position of a Commander of the Army, increase in agricultural and cattle wealth, construction of a new house, happiness to kinsmen, will be derived from the very commencement of the Antar Dasha of Mangal in the Dasha of Sani, if Mangal is in his exaltation, in his own Rashi, or, if Mangal is associated with Lagna Lord, or with the Dasha Lord. There will be loss of wealth, danger of wounds, danger from thieves, snakes, weapons, gout and other similar diseases, distress to father and brothers, quarrels with co-partners, loss of kinsmen, coarse food, going away to foreign lands, unnecessary expenditure, if Mangal is in his debilitation Rashi, or combust, or in Randhra, or Vyaya and associated with, or receiving a Drishti from malefic. Great distress, dependence on others and fear of premature death, may be expected, if Mangal is in Dhan, or, if Mangal is Yuvati, or Randhra Lord. The remedial measures to obtain relief from the above evil effects are performance of Havan and giving a bull in charity.

Saturn Dasa-Rahu Bhukti: There will be increase of troubles, diseases in limb, insect bites and misery in every walk. The effects like quarrels, mental agony, physical distress, agony, antagonism with the sons, danger from diseases, unnecessary expenditure, discord with close relations, danger from the government, foreign journeys, loss of house and agricultural lands, will be derived in the Antar Dasha of Rahu in the Dasha of Sani, if Rahu not be in his house of exaltation, or any other auspicious position. Enjoyment, gains of wealth, increase in agricultural production, devotion to deities and Brahmins, pilgrimage to holy places, increase in cattle wealth, well-being in the family will be the results at the commencement of the Antar Dasha, if Rahu is associated with Lagna Lord, or a Yoga Karaka Graha, or, if Rahu is in his exaltation, or in his own Rashi, or, if Rahu is in a Kendra, or Labh from Lagna, or from the Lord of the Dasha. There will be cordiality with the king and happiness from friends in the middle portion of the Antar Dasha. There will be acquisition of elephants, opulence and glory, cordial relations with the king, gains of valuable clothes, if Rahu is in Mesh, Kanya, Karka, Vrishabh, Meena, or Dhanu. There will be physical distress, if Rahu is associated with Dhan, or Yuvati Lord. The remedial measures to obtain relief from the above evil effects are Mrityunjaya Japa and giving a goat in charity.

Saturn Dasa-Jupiter Bhukti: He/She will gain a high position and will acquire villages and happiness from his wife. This is, comparatively, better period. There will be purchase of ornaments, physical comforts, and success in expected matters, new friends and new position. The effects like success all-round, well-being in the family, gain of conveyances, ornaments and clothes by the beneficence of the king, reverence, devotion to deities and the preceptor, association with men of learning, happiness from wife and children, will be derived in the Antar Dasha of Guru in the Dasha of Sani, if Guru is in a Kendra, or in a Trikona, or, if Guru is associated with Lagna Lord, or, if Guru is in his own, or in his exaltation Rashi. Results, like death of the near relations, loss of wealth, antagonism with the government officials, failure in projects, journeys to foreign lands, affliction with diseases, like leprosy, will be experienced, if Guru is in his debilitation Rashi, or, if Guru is associated with malefic, or, if Guru is in Ari, Sahaj, or Vyaya.

There will be opulence and glory, happiness to wife, gains through the king, comforts of good food and clothes, religious-mindedness, name and fame in the country, interest in Vedas and Vedanta, performance of religious sacrifices, giving grains in charity, if Guru is in the 5th, 9th, 11th, 2nd, or Kendra from the Lord of the Dasha. Antagonism with kinsmen, mental agony, quarrels, loss of position, losses in ventures, loss of wealth, as a result of imposition of fines, or penalties by government, imprisonment distress to wife and son will be the results, if Guru is weak and is in the 6th, 8th, or 12th from the Lord of the Dasha. There will be physical distress, agony, death, or death of any member of the family, if Guru is Dhan, or Yuvati Lord. Remedial measures to obtain relief from the above evil effects are recitation of Shiva Sahasranam and giving gold in charity.

13.3.7 Predictions by the Antar Dasha in the Dasha of Mercury:

Mercury Dasa-Mercury Bhukti: There will be purchase of home or shifting to more comfortable place, money and help from relatives, learning of astrology or similar subjects and improvement in conditions of life. The gain of jewels, like pearls, learning, increase in happiness and performance of pious deeds, success in the educational sphere, acquisition of name and fame, meeting with new kings, gain of wealth and happiness from wife, children and parents will be the effects in the Antar Dasha of Buddha in his own Dasha, if Buddha is placed in his exaltation Rashi, or is otherwise well placed. There will be loss of wealth and cattle, antagonism with kinsmen, diseases, like stomach pains, piety in discharging duties, as a government official, if Buddha is in his debilitation Rashi, or, if Buddha is in Ari, Randhra, or Vyaya, or, if Buddha is associated with malefic. Distress to wife, death of members of the family, affliction with diseases, like rheumatism and stomach pains will result, if Buddha is Dhan, or Yuvati Lord. Remedial measure to obtain relief from the above evil effects is recitation of Vishnu Sahasranam.

Mercury Dasa-Ketu Bhukti: There will be disease due to excess of bile in the body, unnecessary travels, loss of wealth, mental agony, improvement in education, and success in artistic lines. The effects like physical fitness, little gain of wealth, affectionate relations with kinsmen, increase in cattle wealth, income from industries, success in the educational sphere, acquisition of name and fame, honours, audience with the king and joining a banquet with him, comforts of clothes, will be experienced, if Ketu is associated with benefic in a

Kendra, or Trikona, or, if Ketu is yuti with Lagna Lord, or with a Yoga Karaka. The same will be the results, if Ketu is in a Kendra, or in the 11th from the Lord of the Dasha. Fall from a conveyance, distress to son, danger from the king, indulgence in sinful deeds, danger from scorpions, quarrels with the menials, sorrow, diseases and association with menials will be the results, if Ketu is yuti with malefic in the 8th, or 12th from the Lord of the Dasha. There will be physical distress, if Ketu is Dhan, or Yuvati Lord. The remedial measure to obtain relief is giving a goat in charity.

Mercury Dasa-Venus Bhukti: He/She will honour elders, Gods and guests and will acquire robes and ornaments. There will be religious ceremonies at home; purchase of jewels, marriage or birth of child, family happiness, prosperity to relatives, purchase of landed properties also likely, illegal connections and drinking liquor habit develops. The effects like inclination to perform religious rites, fulfilment of all ambitions through the help of the king and friends, gains of agricultural lands and happiness will be derived in the Antar Dasha of Sukra in the Dasha of Buddha, if Sukra is in a Kendra, in Labh, in Putra, or in Dharma. There will be acquisition of a kingdom, gain of wealth and property, construction of a reservoir, readiness to give charities and to perform religious rites, extraordinary gain of wealth and gains in business, if Sukra is in a Kendra, in the 5th, 9th, or 11th from the Lord of the Dasha. Heart disease, defamation, fevers, dysentery, separation from kinsmen, physical distress and agony will result, if Sukra is weak in the 6th, 8th, or 12th from the Lord, or the Dasha. There will be fear of premature death, if Sukra is Dhan, or Yuvati Lord. The remedial measure to obtain relief from the above evil effects is to recite Mantras of Goddess Durga.

Mercury Dasa-Sun Bhukti: There will be sudden income of wealth, gold, horses, coral and elephants. There will be Royal honour, appointment, vehicles, political career, disease in stomach, fire accidents, sickness to wife and acquisition of some wealth. Effects, like dawn of fortune by the beneficence of the king, happiness from friends, will be derived in the Antar Dasha of Surya in the Dasha of Buddha, if Surya is in his own, or in his exaltation Rashi, or in a Kendra, or Trikona, or in Dhan, or Labh, or in his exalted, or own Navamsa. There will be acquisition of land, if Surya receives a Drishti from Mangal and comforts of good food and clothes, if such a Surya receives a Drishti from Lagna Lord. Fear, or danger from thieves, fire and weapons, bilious troubles, headaches, mental agony and separation from friends will be the results, if Surya is in Ari, Randhra, or Vyaya from Lagna, or from the Lord of the Dasha and, if Surya is weak and associated with Sani, Mangal and Rahu. There will be fear of premature death, if Surya is Dhan, or Yuvati Lord. Worship of Surya is the remedial measure to obtain relief from the above evil effects.

Mercury Dasa-Moon Bhukti: He/She will be afflicted by itch, leprosy, and tuberculosis, loss of limbs, fear from elephants and destruction of conveyances. There will be health troubles, disputes through women, gain through women, jewels and general ill behaviour of relations. The Yoga becomes very strong for beneficial effects, if in the Antar Dasha of Chandra in the Dasha of Buddha. Chandra is in a Kendra, or Trikona from Lagna, or, if Chandra is in her exaltation, or in her own Rashi, associated with, or receiving a Drishti from Guru, or, if Chandra is a Yoga Karaka herself. Then there will be marriage, birth of a son and gain of clothes and ornaments. In the circumstances, mentioned above, there will also be construction of a new house, availability of sweetish preparations, enjoyment of music, study of Shashtra, journey to the South, gains of clothes from beyond the seas, gain of gems, like pearls. There will be physical distress, if Chandra is in her debilitation, or in an enemy's Rashi. If Chandra is in a Kendra, Trikona, in the 3rd, or 11th from the Lord of the Dasha, there will be at the commencement of the Antar Dasha visits to sacred shrines, patience, enthusiasm and gains of wealth from foreign countries. Danger from the king, fire and thieves, defamation, or disgrace and loss of wealth on account of wife, destruction of agricultural lands and cattle will be the results, if Chandra is weak and is in the 6th, 8th, or 12th from the Lord of the Dasha. There will be physical distress, if Chandra is Dhan, or Yuvati Lord. There will be relief, prolongation of longevity and restoration of comforts by the beneficence of Goddess Durga, if the Mantras of the Goddess are recited in the prescribed manner and clothes are given in charity.

Mercury Dasa-Mars Bhukti: There will be diseases of the head, gout, many kinds of miseries. There will be some benefits from superiors, disease due to insect bite, neighbours becoming enemies, visiting house of ill fame, punishments by superiors and mental anxiety. Effects, like well-being and enjoyments in the family by the beneficence of the king, increase in property, recovery of a lost kingdom, birth of a son, satisfaction, acquisition of cattle, conveyances and agricultural lands, happiness from wife, will be derived in the Antar Dasha of Mangal in the Dasha of Buddha, if Mangal is in his exaltation, in his own Rashi, in a Kendra, or Trikona, or, if Mangal is associated with Lagna Lord. Physical distress, mental agony, obstacles in industrial ventures, loss of wealth, gout, distress from wounds and danger from weapons and fever will be the results, if Mangal be associated with, or receives a Drishti from malefic in Randhra, or in Vyaya. There will be gain of wealth, physical felicity, birth of a son, good reputation, affectionate relations with kinsmen, if Mangal receives a Drishti from benefic in a Kendra, Trikona, or in the 11th from the Lord of the Dasha. If Mangal be associated with malefic in the 8th, or 12th from the Lord of the Dasha, there will be distress, danger from kinsmen, wrath of the king and fire, antagonism with the son, loss of position at the commencement of the Antar Dasha, enjoyments and gains of wealth in the middle portion of the Antar Dasha, danger from the king and loss of position at the end of the Antar Dasha. There will be fear of premature death, if Mangal is Dhan, or Yuvati Lord. The remedial measures to be adopted to obtain relief from the above evil effects are Mrityunjaya Japa and giving a cow in charity.

Mercury Dasa-Rahu Bhukti: There will be evil and bad connections with women, change in position, failure in cases, money from friends, disease due to indigestion and the like and acquisition of Divine Knowledge. The effects like reverence from the king, good reputation, gain of wealth, visits to sacred shrines, performance of religious sacrifices and oblations, recognition, gain of clothes, are derived in the Antar Dasha of Rahu in the Dasha of Buddha, if Rahu is in a Kendra, or Trikona, or, if Rahu is in Mesh, Kumbh, Kanya, or Vrishabh. There will be some evil effects at the commencement of the Antar Dasha, but all will be well later. There will be an opportunity to have conversation, or a meeting with the king, if Rahu is in Sahaj, Randhra, Karma, or Labh. In this position, if Rahu be associated with a benefic, there will be a visit to a new king. Pressure of hard work, as a government functionary, loss of position, fears, imprisonment, diseases, agony to self and kinsmen, heart disease, loss of reputation and wealth, will be the results, if Rahu is associated with a malefic, or malefic in the 8th, or 12th from the Lord of the Dasha. There will be fear of premature death, if Rahu is in Dhan, or in Ari. The remedial measures to obtain relief from the above evil effects are recitation of Mantras of Goddess Durga and Goddess Lakshmi in the prescribed manner and giving a tawny-coloured cow or female buffalo in charity.

Mercury Dasa-Jupiter Bhukti: He/She will be free from enemies, diseases and sins, be virtuous and becomes a royal adviser. There will be some benefits from superiors, birth of son, closeness of relations, wife becoming more attached, ill feeling with relations, quarrel, ill health, loss of wealth, misunderstanding with father or son and the like. The effects like physical felicity, gain of wealth, beneficence of the king, celebration of auspicious functions, like marriage etc., at home, availability of sweetish preparations, increase in cattle wealth, attending discourses on Purana, devotion to deities and the preceptor, interest in religion, charities, worship of Lord Shiva, will be derived in the Antar Dasha of Guru in the Dasha of Buddha, if Guru is in a Kendra, Trikona, or in Labh, or, if Guru is in his exaltation, or in his own Rashi. Discord with king and kinsmen, danger from thieves, death of parents, disgrace, punishment from government, loss of wealth, danger from snakes and poison, fever, losses in agricultural production, loss of lands will be the results, if Guru is in his debilitation Rashi, is combust, or is in Ari, Randhra, or in Vyaya, or, if Guru is associated with, or receives a Drishti from Sani and Mangal. There will be happiness from kinsmen and from one's son, enthusiasm, increase in wealth and name and fame, giving grains in charity, if Guru is in a Kendra, Trikona, or in the 11th from the Lord of the Dasha and, if Guru is endowed with strength. Agony, anxiety, danger from diseases, antagonism with wife and kinsmen, wrath of the king, quarrels, loss of wealth, danger from Brahmins will be the results, if Guru is weak and, if Guru is in the 6th, 8th, or 12th from the Lord of the Dasha. There will be physical distress, if Guru is Dhan, or Yuvati Lord, or, if Guru is in Dhan, or Yuvati. The remedial measures to obtain relief from the above evil effects are recitation of Shiva Sahasranam and giving a cow and gold in charity.

Mercury Dasa-Saturn Bhukti: He/She will seek sexual gratification, like a eunuch or, like a bull and be deprived of virtues, wealth, pleasures and issues. There will be diseases, debts, scandal, and gain of money from illegal sources, failure in land holdings or failure in iron or oil business and good actions like building temples or charities or pilgrimage. The effects like well-being in the family, acquisition of a kingdom, enthusiasm, increase in cattle wealth, gain of a position, visits to sacred shrines, will be derived in the Antar Dasha of Sani in the Dasha of Buddha, if Sani is in his exaltation, his in his own Rashi, or in a Kendra, or Trikona, or in Labh. Danger from enemies, distress to wife and children, loss of thinking power, loss of kinsmen, loss in ventures, mental agony, journeys to foreign lands and bad dreams will be the results, if Sani is in the 8th, or 12th from the Lord of the Dasha. There will be fear of premature death, if Sani is Dhan, or Yuvati Lord. The remedial measures to obtain relief from the above evil effects and to regain sound health are performance of Mrityunjaya Japa and giving a black cow and female buffalo in charity.

13.3.8 Predictions by the Antar Dasha in the Dasha of Ketu:

Ketu Dasa-Ketu Bhukti: There will be mental worry due to son or wife, loss of money, poison, fear, a general set back and check in life. The effects like happiness from wife and children, recognition from the king, but mental agony, gain of land; village will be derived in the Antar Dasha of Ketu in his own Dasha, if Ketu is in a Kendra, or Trikona, or, if Ketu is related to Dharma, Karma, or Bandhu Lord. Heart disease, defamation, destruction of wealth and cattle, distress to wife and children, instability of mind will we be the results, if Ketu is in his debilitation Rashi and, if Ketu is in Randhra or Vyaya along with a combust Graha. There will be danger from diseases, great distress and separation from kinsmen, if Ketu is related to Dhan, or Yuvati Lord, or, if Ketu is in Dhan, or Yuvati. The remedial measures to obtain relief from the above evil effects are performance of Durga Saptashati Japa and Mrityunjaya Japa.

Ketu Dasa-Venus Bhukti: There will be success in undertaking, birth of child, ill health to children and fever or dysentery. The effects like beneficence from the king, good fortune, gain of clothes, recovery of lost kingdom, comforts of conveyances, visits to sacred shrines and gain of lands and villages by the beneficence of the king, will be derived in the Antar Dasha of Sukra in the Dasha of Ketu, if Sukra is in his exaltation, in his own Rashi, or, if Sukra is associated with Karma's Lord in a Kendra, or Trikona and there will be dawn of fortune, if in such position he is associated with Dharma Lord also. Sound health, well-being in the family and gains of good food and conveyances will be the results, if Sukra is in a Kendra, Trikona, or in the 3rd, or 11th from the Lord of the Dasha. There will be quarrels without any cause, loss of wealth, distress to cattle, if Sukra is in the 6th, 8th, or 12th from the Lord of the Dasha. If Sukra is in his debilitation Rashi, or, if Sukra is associated with a debilitated Graha, or, if Sukra is in Ari, or Randhra, there will be quarrels with kinsmen, headaches, eye troubles, heart disease, defamation, loss of wealth and distress to cattle and wife. Physical distress and mental agony will be caused, if Sukra is Dhan, or Yuvati Lord. The remedial measures to obtain relief from the above evil effects are performance of Durga Path and giving a tawny-coloured cow, or female buffalo in charity.

Ketu Dasa-Sun Bhukti: There will be check in business, expansion of knowledge, uneasiness, travel, health of wife giving anxiety and worry. The effects, like gains of wealth, beneficence of the king, performance of pious deeds and fulfilment of all ambitions, will be derived in the Antar Dasha of Surya in the Dasha Ketu, if Surya is in his exaltation, in his own Rashi, or, if Surya is associated with, or receives a Drishti from a benefic in a Kendra, Trikona, or in Labh. Danger from the king, separation from parents, journeys to foreign lands, distress from thieves, snakes and poison, punishment by government, antagonism with the friends, sorrows, danger from fever will be the results, if Surya is associated with a malefic, or malefic in Randhra, or in Vyaya. There will be physical fitness, gain of wealth, or the birth of a son, success in performance of pious deeds, headship of a small village, if Surya is in a Kendra, Trikona, in the 2nd, or 11th from the Lord of the Dasha. Obstacles in availability of food, fears and loss of wealth and cattle will be the results, if Surya is associated with evil Graha in the 8th, or 12th from the Lord of the Dasha. There will be distress at the commencement of the Antar Dasha with some mitigation at its end. There will be fear of premature death, if Surya is Dhan, or Yuvati Lord. The remedial measure to obtain relief from the above evil effects and to regain comforts by the beneficence of Surya is to give a cow and gold in charity.

Ketu Dasa-Moon Bhukti: There will be financial improvements, loss or mental uneasiness, disease through water or cold and troubles from children. The effects like recognition from the king, enthusiasm, well-being, enjoyments, acquisition of a house, lands, abnormal gains of food, clothes, conveyances, cattle, success in business, construction of reservoirs and happiness to wife and children, will be derived in the Antar Dasha of Chandra in the Dasha of Ketu, if Chandra is in her exaltation, in her own Rashi, in a Kendra, Trikona, in Labh, or in Dhan. The beneficial results will be realized fully, if Chandra is waxing. Unhappiness and mental agony, obstacles in ventures, separation from parents, losses in business, and destruction of cattle will be caused, if Chandra is in her debilitation Rashi, or in Ari, Randhra, or Vyaya. There will be the acquisition of a cow, or cows, land, agricultural lands, meeting kinsmen and the achievement of success through them, increase in cows milk and curd, if Chandra is in a Kendra, Trikona, or in the 11th from the Lord of the Dasha and, if Chandra is endowed with strength. There will be auspicious results at the commencement of the Antar Dasha, cordial relations with the king in the middle portion of the Antar Dasha and danger from the king, foreign journey, or journeys to distant places at its end. Loss of wealth, anxiety, enmity with kinsmen and distress to brother, will be the results, if Chandra is in the 6th, 8th, or 12th from the Lord of the

Dasha. If Chandra is Dhan, Yuvati, or Randhra Lord, there will be fear of premature death. The remedial measures to obtain relief from the above effects are recitation of Mantras of Chandra and giving in charity things, connected with Chandra.

Ketu Dasa-Mars Bhukti: There will be a general anxiety about children, quarrels in family, increase of enemies, punishments, death, and operation in the body. The effects like acquisition of land, village, increase in wealth and cattle, laying out of a new garden, gain of wealth by the beneficence of the king, will be derived in the Antar Dasha of Mangal in the Dasha of Ketu, if Mangal is in his exaltation, in his own Rashi, if Mangal is associated with, or, receives a Drishti from benefic. If Mangal is related to Dharma, or Karma Lord, there will definitely be gain of land and enjoyment. There will be recognition from the king, great popularity and reputation and happiness from children and friends, if Mangal is in a Kendra, Trikona, or in the 3rd, or 11th from the Lord of the Dasha. There will be fear of death/disaster during a foreign journey, diabetes, unnecessary troubles, danger from thieves and the king and quarrels, if Mangal is in the 8th, 12th, or 2nd from the Lord of the Dasha. In the above circumstances amidst evil effects there will be some auspicious effects also. High fever, danger from poison, distress to wife, mental agony and fear of premature death will be the results, if Mangal is Dhan, or Yuvati Lord. By the beneficence of Mangal there will be enjoyment and gain of property, if, as a remedial measure, a bull is given in charity.

Ketu Dasa-Rahu Bhukti: There will be Royal or government punishments, blood poison, loss of wealth or property, loss in business and visiting prostitute homes for pleasure. The effects like increase of wealth and gain of wealth, grains, cattle, lands, village from a Yavan king, will be derived in the Antar Dasha of Rahu in the Dasha of Ketu, if Rahu is in his exaltation, his own, in a friends Rashi, or in a Kendra, or Trikona, or in Labh, or Sahaj, or Dhan. There will be some trouble at the commencement of the Dasha, but all will be well later. Frequent urination, weakness in the body, cold fever, danger from thieves, intermittent fever, opprobrium, quarrels, diabetes, and pain in stomach will be the results, if Rahu is associated with a malefic in Randhra, or in Vyaya. There will be distress and danger, if Rahu is in Dhan, or in Yuvati. The remedial measure to obtain relief from the above evil effects is Durga Saptashati Path.

Ketu Dasa-Jupiter Bhukti: There will be contact with persons of high status, happiness through wife, marriage, and increase in holdings and profits in business. The effects like increase in wealth and grains, beneficence of the king, enthusiasm, gain of conveyances, celebration, like birth of a son at home, performance of pious deeds, Yagya, conquest of the enemy and enjoyments, will be derived in the Antar Dasha of Guru in the Dasha of Ketu, if Guru is in his exaltation, in his own Rashi, or is associated with Lagna, Dharma, or Karma Lord in a Kendra, or Trikona. Danger from thieves, snakes and wounds, destruction of wealth, separation from wife and children, physical distress will be the results, if Guru is in his debilitation Rashi, or in Ari, Randhra, or Vyaya. Though some good effects may be felt at the commencement of the Antar Dasha, there will be only adverse results later. There will be gains of many varieties of garments, ornaments by the beneficence of the king, foreign journeys, taking care of kinsmen, availability of decent food, if Guru is associated with a benefic in a Kendra, Trikona, in the 3rd, or 11th from the Lord of the Dasha. Fear of premature death will be caused, if Guru is Dhan, or Yuvati Lord. The remedial measures to obtain relief from the above evil effects are Mrityunjaya Japa and recitation of Shiva Sahasranama.

Ketu Dasa-Saturn Bhukti: There will be prison life conditions, loss of money in many ways, strained feelings with relations, exile to far off places and change of house. The effects like distress to oneself and one's kinsmen, agony, increase in cattle wealth, loss of wealth, as a result of imposition of fines by government, resignation from the existing post, journeys to foreign lands and danger of thieves during travelling, will be derived in the Antar Dasha of Sani in the Dasha of Ketu, if Sani is deprived of strength and dignity. There will be loss of wealth and lethargy, if Sani is in Randhra, or Vyaya. Success in all ventures, happiness from the employer, comforts during journeys, increase in happiness and property in ones own village, audience with the king will be the results, if Sani is in a Trikona in Meena, in Tula, in his own Rashi, or, if Sani is in an auspicious Navamsa, or is associated with a benefic in a Kendra, Trikona, or in Sahaj. (According to Brihat Jatak, Sani in Tula, Meena, Dhanu, Makara and Kumbh in Lagna gives Raja Yoga). There will be physical distress, agony, obstacles in ventures, lethargy, defamation, death of parents, if Sani is associated with a malefic, in the 6th, 8th, or 12th from the Lord of the Dasha. Fear of premature death may be expected, if Sani is Dhan, or Yuvati Lord. The remedial measures to obtain relief from the above evil effects are performance of Havan with sesame seeds (Til) and giving a black cow, or female buffalo in charity.

Ketu Dasa-Mercury Bhukti: There will be money from mental pursuits, children giving worry and anxiety, failure of ideas or plans and fear from relations. The effects like acquisition of a kingdom, enjoyments, charities, gain of wealth and land, birth of a son, celebration of religious functions and functions, like marriage suddenly, well-being in the family, gain of clothes, ornaments, will be derived in the Antar Dasha of Buddha in the Dasha of Ketu, if Buddha is in a Kendra, or Trikona, or, if Buddha is in his exaltation, or in his own Rashi. There will be association with men of learning, dawn of fortune and listening to religious discourses, if Buddha is associated with Dharma, or Karma Lord. Antagonism with government officials, residing in other people's houses, and destruction of wealth, clothes, conveyances and cattle will be the results, if Buddha is associated with Sani, Mangal, or Rahu in Ari, Randhra, or Vyaya. There will be some beneficial effects at the commencement of the Dasha, still better results in the middle, but inauspicious at the end. There will be good health, happiness from one's son, opulence and glory, availability of good food and clothes and abnormal profits in business, if Buddha is in a Kendra, Trikona, or in the 11th from the Lord of the Dasha. Distress, unhappiness and troubles to wife and children and danger from the king may be expected at the commencement of the Antar Dasha, if Buddha is weak in the 6th, 8th, or 12th from the Lord of the Dasha. There will, however, be visits to sacred places in the middle of the Dasha. Fear of premature death will be caused, if Buddha is Dhan, or Yuvati Lord. The remedial measure to obtain relief from the above evil effects is recitation of Vishnu Sahasranam.

21.4.9 Predictions by the Antar Dasha in the Dasha of Venus:

Venus Dasa-Sukra Bhukti: There will be a general easiness will prevail in life conditions increase in finance, fame and birth of male child. The effects like gain of wealth, cattle through Brahmins, celebrations in connection with the birth of a son, well-being, recognition from the king, acquisition of a kingdom, will be derived in the Antar Dasha of Sukra in his own Dasha, if Sukra is in a Kendra, Trikona, or in Labh and, if Sukra is endowed with strength. Construction of a new house, availability of sweet preparations, happiness to wife and children, companionship with a friend, giving grains in charity, beneficence of the king, gain of clothes, conveyances and ornaments, success in business, increase in the number of cattle, gain of garments by performing journeys in the western direction will be the results, if Sukra is in his exaltation, in his own Rashi, or, if Sukra is in his exalted, or own Navamsa. There will be acquisition of a kingdom, enthusiasm, beneficence of the king, well-being in the family, increase in the number of wives, children and wealth, if Sukra is associated with, or receives a Drishti from a benefic and is in a friendly Navamsa, in Sahaj, Ari, or Labh. Danger from thieves, antagonistic relations with government officials, destruction of friends and kinsmen, distress to wife and children may be expected, if Sukra is associated with, or receives a Drishti from a malefic in Ari, Randhra, or Vyaya. There will be fear of death, if Sukra is Dhan, or Yuvati Lord. Remedial measures to obtain relief from the above evil effects are Durga Path and giving a cow in charity.

Venus Dasa-Sun Bhukti: There will be disease, of the throat, stomach and eyes and punishments from the king. This is not so much a favourable period. There will be general anxiety, troubles in family, quarrel, damage to property and wealth. There will be a period of agony, wrath of the king, quarrels with the coparceners in the Antar Dasha of Surya in the Dasha of Sukra, if Surya is in any Rashi, other than his exaltation, or debilitation Rashi. (This verse does not appear to be correctly worded, because Surya does produce good effects in a position, other than exaltation, or debilitation). The effects like acquisition of a kingdom and wealth, happiness from wife and children, happiness from employer, meeting with friends, happiness from parents, marriage, name and fame, betterment of fortune, birth of a son, will be experienced, if Surya is in his exaltation, in his own Rashi, in a Kendra, Trikona, in Dhan, or Labh, or in Kendra, Trikona, in the 2nd, or 11th from the Lord of he Dasha. Distress, agony, distress to members of the family, harsh language, distress to father, loss of kinsmen, wrath of the king, danger at home, many diseases, destruction of agricultural production will be the results, if Surya is in Ari, Randhra, or Vyaya, or, if Surya is in his debilitation, or in an enemy's Rashi. There will be evil influence of the Graha, if Surya is Dhan, or Yuvati Lord. Worship of Surya is the remedial measure to obtain relief from the above evil effects.

Venus Dasa-Moon Bhukti: He/She will suffer from diseases of nails, teeth and head apart from jaundice. There will be gain through women, expansion of mental activity, vehicles, success in undertakings, intense devotion to God, nervous troubles due to excess sexual pleasure and troubles through women. The effects like gain of wealth, conveyances, clothes by the beneficence of the king, happiness in the family, great opulence and glory, devotion to deities and Brahmins, will be derived in the Antar Dasha of Chandra in the Dasha of Sukra, if Chandra is in her exaltation, or in her own Rashi, or is associated with the Lord of Dharma, benefic, or with Karma's Lord, or, if Chandra is in a Kendra, Trikona, or Labh. In the above circumstances there will also be association with musicians and men of learning and receiving of decorations, gain of cows, buffaloes and other cattle, abnormal profits in business, dining with brothers. Loss of wealth, fears, physical distress, agony, wrath of the king, journeys to foreign lands, or pilgrimage, distress to wife and children and separation from kinsmen will be the results, if Chandra is in her debilitation Rashi, is combust, or is in Ari, Randhra, or Vyaya, or, if Chandra is in the 6th, 8th, or 12th from the Lord of the Dasha. There will be sovereignty over a province, or village by the beneficence of the king, clothes, construction of a reservoir, increase in wealth, if Chandra is in a Kendra, or Trikona, or in the 3rd, or 11th from the Lord of the Dasha. There will be physical fitness at the commencement of the Antar Dasha and physical distress in its last portion.

Venus Dasa-Mars Bhukti: He/She will suffer from diseases related to blood and bile, will gain lands, royal patronage and enthusiasm. There will be increase of family holdings, marriage, and gain of wealth through women, materialistic outlook, and diseases of eye or bile. The effects like acquisition of kingdom, property, clothes, ornaments, land and desired objects, will be derived in the Antar Dasha of Mangal in the Dasha of Sukra, if Mangal is in a Kendra, or Trikona, or in Labh, or, if Mangal is in his exaltation Rashi, or is in one of his own Rashi, or is associated with the Lagna, Dharma, or Karma Lord. There will be fever from cold, diseases (like fever) to parents, loss of position, quarrels, antagonism with the king and government officials, extravagant expenditure, if Mangal is in Ari, Randhra, or Vyaya, or, if Mangal is in the 6th, 8th, or 12th from the Lord of the Dasha. Physical distress, losses in profession, loss of village, land will be the results, if Mangal is the Dhan, or Yuvati Lord.

Venus Dasa-Rahu Bhukti: There will be change of place of living, getting property by lottery or race or by unexpected ways, silent prayers, name and fame increasing. The effects like great enjoyment, gain of wealth, visits of friends, successful journeys, gain of cattle and land, will be derived in the Antar Dasha of Rahu in the Dasha of Sukra, if Rahu is in a Kendra, or Trikona, or in Labh, or, if Rahu is in his exaltation, or in his own Rashi, or is associated with, or receives a Drishti from benefic. Enjoyments, destruction of enemy, enthusiasm and beneficence of the king will be the results, if Rahu is in Sahaj, or Ari, or Karma, or Labh. Good effects will be experienced up to 5 months from the commencement of the Antar Dasha, but at the end of the Dasha there will be danger from fevers and indigestion. In the above circumstances, except for obstacles in ventures and journeys and worries, there will be all enjoyment, like those of a king. Journeys to foreign lands will bring success and the person will return safely to his homeland. There will also be blessings from Brahmins and auspicious results consequent to visits to holy places. There will be inauspicious effects on oneself and one's parents and antagonism with people, if Rahu be associated with a malefic in the 8th, or 12th from the Lord of the Dasha. Physical distress will be caused, if Rahu is Dhan, or Yuvati Lord. The remedial measure to obtain relief is Mrityunjaya Japa.

Venus Dasa-Jupiter Bhukti: There will be growth of virtues, good qualities and wealth and acquisition of kingdom. There will be help from persons in rank, patronage from wife and children, Royal honour and wealth. When Jupiter is in 6, 8, 12 houses there would be unnecessary travel, failures and general disgust in life. The effects like recovery of the lost kingdom, acquisition of desired grains, clothes and property, reverence from one's friend and the king and gain of wealth, recognition from the king, good reputation, gain of conveyances, association with an employer and with men of learning, industriousness in the study of Shashtra, birth of a son, satisfaction, visits of close friends, happiness to parents and son, will be derived in the Antar Dasha of Guru in the Dasha of Sukra, if Guru is in his exaltation, in his own Rashi, or in a Kendra, or Trikona to Lagna, or to the Lord of the Dasha. There will be danger from the king and from thieves, distress to oneself and to kinsmen, quarrels, mental agony, loss of position, going away to foreign lands and danger of many kinds of diseases, if Guru is in the 6th, 8th, or 12th from the Lord of the Dasha and be associated with a malefic. There will be physical distress, if Guru is Dhan, or Yuvati Lord. The remedial measure to obtain relief from the above evil effects is Mrityunjaya Japa.

Venus Dasa-Saturn Bhukti: He/She will be associated with old women, will head over cities, men etc. and will destroy enemies. There will be disease due to evil habits, bad company, loss of health and money. The effects like great enjoyments, visits of friends and kinsmen, recognition from the king, birth of a daughter, visits to holy places and sacred shrines, conferment of authority by the king, will be derived in the Antar Dasha of Sani in the Dasha of Sukra, if Sani is in his exaltation, in his own Rashi, in a Kendra, Trikona, or in his own Navamsa. There will be lethargy and more expenditure than income, if Sani is in his debilitation Rashi. Many kinds of distresses and troubles at the commencement of the Antar Dasha, like stress to parents, wife and children, going away to foreign lands, losses in profession, destruction of cattle, will be the results, if Sani is in Randhra, or Labh, or Vyaya, or, if Sani is in the 8th, 11th, or 12th from the Lord of the Dasha. There will be physical distress, if Sani is Dhan, or Yuvati Lord. The remedial measures to obtain relief from the above evil effects are Havan with sesame seeds (Til), Mrityunjaya Japa, Durga Saptashati Path (by oneself, or through a Brahmin).

Venus Dasa-Mercury Bhukti: He/She will gain through women and will enjoy happiness and fulfilment of desires. There will be marriage, success in court affairs, and increase in financial standards, children giving mental satisfaction and ailments in the body. The effects like dawn of fortune, birth of a son, gain of wealth through judgement of court, listening to stories from the Purana, association with persons, competent in poetry, visits of close friends, happiness from employer, availability of sweetish preparations, will be derived in the Antar Dasha of Buddha in the Dasha of Sukra, if Buddha is in a Kendra, or Trikona, or in Labh (from Lagna, or from the Lord of the Dasha), or is in his exaltation, or in his own Rashi. If Buddha is in the 6th, 8th, or 12th from the Lord of Dasha, or, if Buddha is weak, or is associated with a malefic, there will be agony, loss of cattle, residence in other people's houses and losses in business. There will be some good effects at the commencement, moderate in the middle portion and distress from fever at the end of the Antar Dasha. There will be physical distress, if Buddha is Dhan, or Yuvati Lord. The remedial measure to obtain relief from the above evil effects is the recitation of Vishnu Sahasranam.

Venus Dasa-Ketu Bhukti: There will be pilgrimage, worship, visiting saints, good education to children, danger from animals, weakness of body, anxiety but the end will be in happiness. Auspicious effects, like availability of sweetish preparations, abnormal gains in profession and increase in cattle wealth, will be derived from the very commencement of the Antar Dasha of Ketu in the Dasha of Sukra, if Ketu is in his exaltation, or in his own Rashi, or is related to a Yoga Karaka Graha, or, if Ketu is possessed of positional strength. (It is not laid down anywhere, in which Bhava Ketu does get positional strength). In the above circumstances there will be definite victory in war at the end of the Antar Dasha. Moderate results will be experienced in the middle portion of the Antar Dasha and sometimes there will also be the feeling of distress. There will be danger from snakes, thieves and wounds, loss of power of thinking, headache, agony, quarrels without any cause, or reason, diabetes, excessive expenditure, antagonism with wife and children, going away to foreign land, loss in ventures, if Ketu is in the 8th, or 12th from the Lord of the Dasha, or, if Ketu is associated with a malefic. There will be physical distress, if Ketu is Dhan, or Yuvati Lord. The remedial measures to obtain relief from the above effects are Mrityunjaya Japa and giving a goat in charity. Remedial measures for appeasing Sukra will also prove beneficial.

13.4 Predictions by Pratyantar Dasha

13.4.(1) Pratyantar Dasha in the Antar Dasha of Surya: Surya-Surya:

There will be argument with other persons, loss of wealth, distress to wife and headache. The above are general effects. Such inauspicious effects will not be produced, if Surya is in Trikona, if Surya is the Lord of an auspicious Bhava, or is in an auspicious Bhava and in a benefic Varga. All other Pratyantar effects should be judged in this manner.
Surya- Chandra: Excitement, quarrels, loss of wealth, mental agony. **Surya- Mangal**: Danger from the king and from weapons, imprisonment and distress from enemies and fire. **Surya- Rahu**: Disorder of phlegm, danger from weapons, loss of wealth, destruction of a kingdom and mental agony. **Surya- Guru**: Victory, increase in wealth, gains of gold, garments, conveyances etc. **Surya- Sani**: Loss of wealth, distress to cattle, excitement, diseases etc. **Surya- Buddha**: Affectionate relations with kinsmen, availability of good food, gains of wealth, religious-mindedness, reverence from the king. **Surya- Ketu**: Danger to life, loss of wealth, danger from the king, trouble with enemies. **Surya- Sukra**: Moderate effects or some gains of wealth may be expected.

13.4.(2) Pratyantar Dasha in the Antar Dasha of Chandra: Chandra-Chandra:

There will be acquisition of land, wealth and property, reverence from the king and availability of sweetish preparations. Chandra- Mangal: Wisdom and discretion, reverence from the people, increase in wealth, enjoyments to kinsmen, but there will be danger from an enemy. Chandra- Rahu: Well-being, gain of wealth from the king and danger of death, if Rahu is yuti with a malefic. Chandra- Guru: Enjoyments, increase in dignity and glory, gain of knowledge through the preceptor, acquisition of a kingdom and acquisition of gems etc. Chandra- Sani: Bilious troubles, loss of wealth and name and fame. Chandra- Buddha: Birth of a son, acquisition of a horse and other conveyances, success in education, progress, gain of white garments and grains. Chandra- Ketu: Quarrels with Brahmins, fear of premature death, loss of happiness and distress all-round. Chandra- Sukra: Gain of wealth, enjoyments, birth of a daughter, availability of sweet preparations and cordial relations with all. **Chandra- Surya:** Gain of happiness, grains and garments, victories everywhere.

13.4.(3) Pratyantar Dasha in the Antar Dasha of Mangal: Mangal-

Mangal: There is danger from enemies, quarrels and fear of premature death on account of blood diseases. **Mangal- Rahu**: Destruction of wealth & kingdom (fall of government), unpalatable food and quarrels with the enemy. **Mangal- Guru**: Loss of intelligence, distress, sorrows to children, fear of premature death, negligence, quarrels and no fulfilment of any ambition. **Mangal- Sani**: Destruction of the employer, distress, loss of wealth, danger from enemies, anxiety, quarrels and sorrows. **Mangal- Buddha**: Loss of intelligence, loss of wealth, fevers and loss of grains, garments and friends. **Mangal- Ketu**: Distress from diseases, lethargy, premature death, danger from the king and weapons. **Mangal- Sukra**: Distress from bad man, sorrows, danger from the king and from weapons, dysentery and vomiting. **Mangal- Surya**: Increase in landed property and wealth, satisfaction, visits of friends, happiness all-round. **Mangal- Chandra**: Gains of white garments etc. from the southern direction, success in all ventures.

13.4.(4) Pratyantar Dasha in the Antar Dasha of Rahu: Rahu-Rahu:

There will be imprisonment, disease, danger of injuries from weapons. **Rahu- Guru**: Reverence everywhere, acquisition of conveyances, like elephants, gain of wealth. **Rahu- Sani**: Rigorous imprisonment, loss of enjoyments, danger from enemies, affliction with rheumatism. **Rahu- Buddha**: Gain in all ventures, abnormal gain through wife. **Rahu- Ketu**: Loss of intelligence, danger from enemies, obstacles, loss of wealth, quarrels, excitement. **Rahu- Sukra**: Danger from a woman Yogi, danger from the king, loss of conveyances, availability of unpalatable food, loss of a wife, sorrow in the family. **Rahu- Surya**: Danger from enemies, fevers, distress to children, fear of premature death, negligence. **Rahu- Chandra**: Excitement, quarrels, worries, loss of reputation, fear, distress to father. **Rahu- Mangal**: Septic boil in the anus, distress, due to a bite and pollution of blood, loss of wealth, excitement.

13.4.(5) Pratyantar Dasha in the Antar Dasha of Guru: Guru-Guru:

There will be acquisition of gold, increase in wealth. **Guru- Sani**: Increase in lands, conveyances and grains. **Guru- Buddha**: Success in the educational sphere, acquisition of clothes and gems, like pearls, visits of friends. **Guru- Ketu**: Danger from water and thieves. **Guru- Sukra**: Several kinds of learning, gain of gold, clothes, ornaments, well-being and satisfaction. **Guru- Surya**: Gain from the king, friends and parents, reverence everywhere. **Guru- Chandra**: No distress, gain of wealth and conveyances, success in ventures. **Guru- Mangal**: Danger from weapons, pain in anus, burning in the stomach, indigestion, distress from enemies. **Guru- Rahu**: Antagonism with menials and loss of wealth and distress through them.

(6) Pratyantar Dasha the Antar Dasha of Sani: Sani-Sani: There will be physical distress, quarrels, danger from menials. **Sani- Buddha**: Loss of intelligence, quarrels, dangers, anxiety about availability of food, loss of wealth, danger from enemy. **Sani- Ketu**: Imprisonment in the camp of the enemy, loss of lustre, hunger, anxiety and agony. **Sani- Sukra**: Fulfilment of ambitions, well-being in the family, success in ventures and gains there from. **Sani- Surya**: Conferment of authority by the king, quarrels in the family, fevers. **Sani- Chandra**: Development of intelligence, inauguration of big a venture, loss of lustre, extravagant expenditure, association with many women. **Sani- Mangal**: Loss of valour, distress to son, danger from fire and enemy, distress from bile and wind. **Sani- Rahu**: Loss of wealth, clothes, land, going away to foreign lands, fear of death. **Sani- Guru**: Inability to prevent losses, problems caused by women, quarrels, excitement.

13.4.(7) Pratyantar Dasha in the Antar Dasha of Buddha: Buddha –

Buddha: There will be gain of intelligence, education, wealth, clothes. **Buddha- Ketu**: Course food, stomach troubles, eye troubles, distress from bilious and blood disorders. **Buddha- Sukra**: Gains from a northern direction, loss of cattle, acquisition of authority from government. **Buddha- Surya**: Loss of splendour and distress through diseases, distress in the heart. **Buddha- Chandra**: Marriage, gain of wealth and property, birth of a daughter, enjoyments all-round. **Buddha- Mangal**: Religious-mindedness, increase in wealth, danger from fire and enemies, gain of red clothes, injury from a weapon. **Buddha- Rahu**: Quarrels, danger from wife, or some other woman, danger from the king. **Buddha- Guru**: Acquisition of a kingdom, conferment of authority by the king, reverence from the king, education, intelligence. **Buddha- Sani**: Bilious and windy troubles, injuries to the body, loss of wealth.

13.4.(8) Pratyantar Dasha in the Antar Dasha of Ketu: Ketu-Ketu:

There will be sudden disaster, going away to foreign lands, and loss of wealth. **Ketu- Sukra**: Loss of wealth through a non-Hindu king, eye troubles, headache, loss of cattle. **Ketu- Surya**: Antagonism with friends, premature death, defeat, exchange of arguments. **Ketu- Chandra**: Loss of grains, physical distress, misunderstanding, dysentery. **Ketu- Mangal**: Injury from weapons, distress from fire, danger from menials and enemies. **Ketu- Rahu**: Danger from women and enemies, distress, caused by menials. **Ketu- Guru** Loss of friends, wealth and garments, opprobrium in the house, troubles from everywhere. **Ketu- Sani**: Death of cattle and friends, physical distress, very meagre gain of wealth. **Ketu- Buddha**: Loss of understanding, excitement, failure in education, dangers, failure in all ventures.

13.4.(9) Pratyantar Dasha of Sukra in the Antar Dasha of Sukra:

Sukra- Sukra: There will be gains of white clothes, conveyances, gems, like pearls, association with beautiful damsel. **Sukra- Surya**: Rheumatic fever, headache, danger from the king and enemies and meagre gain of wealth. **Sukra- Chandra**: Birth of a daughter, gain of clothes from the king, acquisition of authority. **Sukra- Mangal**: Blood and bile troubles, quarrels, many kinds of distresses. **Sukra- Rahu**: Quarrels with wife, danger, distress from the king and enemies. **Sukra- Guru**: Acquisition of kingdom, wealth, garments, gems, ornaments and conveyance, like elephants. **Sukra- Sani**: Acquisition of donkey, camel, goat, iron, grains, sesame seeds, physical pains. **Sukra- Buddha**: Gains of wealth, knowledge, authority from the king, gain of money, distributed by others. **Sukra- Ketu**: Premature death, going away from homeland, gains of wealth at times.

Vishottary Dasa Timing of Events: In a particular planet's operative Dasha time, the planet gives its Dasha results as per its Natural signification, lordship, placements, aspects and strength of planet and not the functional status of the planet. The Divisional Charts helps to know specific area for further guidance. Vimshotari Dasa can be of immense help to predict the Timing of Marriage, Divorce, Birth of children, Gain of wealth & finance, Education, Family happiness, Birth of Brothers/Sisters, Higher Education, Travels, Acquisition of Lands or Properties, Happiness, purchase of Vehicles, Anxieties, Depression, Offences, Legality, Diseases, Loans, Liabilities, Religious activity, Charities, Foreign Travel or Settlement, Change of Professions and Business, Bed comforts, Expenditure, Investments and Gain in Speculation etc. The Timing of the above events will come to pass in the Dasa, Antar Dasa and Pratyantar Dasa of the concerned planets or the planets involved to affect the events. The planets in Seershodaya Signs, such as 3, 5, 6, 7, 8, 11, will give their effects in the beginning of the Dasa; those in Ubhayodaya, such as 12, will give in the middle of the Dasa and those in Prishtodaya Signs, such as 1, 2, 4, 9, and 10, as well as those in fall (Debilitation) will reveal their effects in the concluding parts of their Dasa. The prediction of the timing of the important events can be done with the help of following given clues:

Dasha	Antar Dasha	Events being controlled
1st & 2nd Dasha	1st, 3rd & 10th lord	Childhood, Early Childhood, Education, Prosperity or down fall of parents, Birth of brothers and sisters.
2nd & 3rd Dasha	4th, 5th & 7th lord	Education, Marriage, Prosperity and rise or down fall or fame or defame to father, entry into service or business, completion of education and Marriage.
2nd, 3rd & 4th Dasha	4th, 7th & 10th lord	Marriage, Father or mother death, Prosperity and success or down fall in profession, marriage, first and second issues, own health, Father's death.
2nd, 3rd & 4th Dasha	1st, 4th & 5th lord	Child birth, Increase or decrease of happiness, help to close relatives, financial prosperity.
3rd & 4th Dasha	10th & 6th lord	Occupation, profession, Increase or decrease of happiness, help to close relatives, financial prosperity.

14
Planetary Combinations (Yoga)

General:
Let us understand the meaning of 'Yoga' first. When a person is born with a combination of predetermined planets, these are called 'Yoga'. Yoga is planetary combinations that lead to a certain result. Yoga means union of the planets. Many gets confuse whether Yoga has positive or negative influence. Yoga may be a favorable combination of planets or the unfavourable combinations of planets. The unfavourable combinations are called 'Arishtas'. It depends on how much powerful the yoga is on a person. For some people have very good Yoga, but their planets are placed in either of 3rd, 6th, 8th or 12th house, the positives of the yoga gets reduced. Note that positive effects may reduce but does not get vanished completely since the required combination of planet is present. As per ancient Astrological books, planet combines in a favourable yoga, it results is enhanced. How and when Yoga will produce its effects in his/her life is generally determined through the Dasa of the planets. If a planet producing yoga is involved in much good, strength it will give its beneficial results at the time he/she enters the Dasa of that planet. There are auspicious yoga for wealth, success, and happiness as well as inauspicious yoga for downfall, poverty and ill health. Persons with yoga are not always rich and famous because it depends on how certain yoga is placed in the horoscope. We shall have to examine the strength of the planets connected to the yoga.

Principles for Qualifying the Strong Yogas

Finding Yogas in a person's chart is an essential part of Vedic horoscope analysis. Positive Yoga show a person has his greatest gifts and achievements; while negative Yoga indicates the areas a person has most challenge. There is a simple rule that the blessings of a positive yoga will occur for the house themes of the house where the yoga-forming planets reside. Example: If a person has a powerful yoga in the 4th house, they may own wonderful homes, or drive great cars, or come from a revered lineage, or have a deep understanding of psychology, or have an amazing mother, or any other 4th house theme might be enhanced, depending upon the nature of the planets involved. Yogas vary in their impact upon a person's life. It is important to determine how effectively the yoga is going to operate. This is a process that qualifies yoga.

Analysis for qualifying Yoga: After finding yoga in a person's chart, we need to assess to what degree each yoga will manifest by examining additional conditions. There are many different factors to evaluate the Yoga. A good yoga is formed by strong, exalted, own sign or, Moolatrikona sign, and having association with benefic planet, but not combust or debilitated and having no association with malefic planets. Then, it becomes more auspicious; otherwise if a bad yoga is formed by a weak or corrupt planet, it becomes more harmful and giving chalanges, bad results.

Relationship of Yoga forming Planet to the Specific Ascendant of the Chart: Vedic astrologers understand that planets perform differently for each Ascendant. In fact, this is the basis of the concept of functional or temporal benefic and malefic. E.g. Jupiter does not function well for Libra and Taurus (Venus) Ascendants because of the houses it rules for each Lagna. For Taurus rising, Jupiter rules the 8th and 11th houses; for Libra rising, Jupiter rules the 3rd and 6th houses. Positive yoga formed by Jupiter for these Ascendants have less potency and more chalanges. This illustrates how important the Ascendant's influence is upon the effectiveness of yoga. The following table shows the most productive and least productive planets for each Ascendant.

Table: Best and Worst Planets for each Ascendant
(Yogakaraka planet is shown in bold.)

Ascendant	Benefic Planets	Malefic Planets	Most Malefic	Neutral Planets
Aries	Mars, Sun, **Jupiter.**	Venus, Saturn	Mercury	Moon
Taurus	Venus, Sun, Mars, Mercury, **Saturn**	Moon	Jupiter	--
Gemini	Mercury, Venus, **Saturn**	Sun, Jupiter	Mars	Moon
Cancer	Moon, **Mars,**	Mercury,	Saturn	Sun

Sign				
	Jupiter	Venus		
Leo	Sun, **Mars**, Jupiter	Moon, Mercury, Venus	Saturn	--
Virgo	Mercury, **Venus,** Saturn	Sun, Moon, Jupiter	Mars	--
Libra	Venus, **Mercury, Saturn Rahu**	Sun, Moon	Jupiter	Mars
Scorpio	Mars, Sun, **Moon**, Jupiter	Venus	Mercury	Saturn
Sagittarius	Jupiter, **Sun, Ketu,** Mars	Moon, Mercury, Saturn	Venus,	--
Capricorn	Saturn, **Mercury,** Venus	Moon, Mars, Jupiter	Sun	--
Aquarius	Saturn, Sun, Mars, **Venus**	Moon, Mercury	Jupiter	--
Pisces	Jupiter, Moon, **Mars**	Sun, Mercury, Saturn	Venus	--

(i) A planet having challenging ruler ships for a specific Ascendant noticeably weakens its ability to uplift a life through the yoga it forms and magnifies the challenges in a person's life through its negative effects. It does not, however, nullify positive effects completely. One must always take all of the relevant factors into account.

A. **The positive Yoga:** Much yoga, by definition, requires at least one of the yoga-forming planets to be in its own sign or exalted. Nevertheless, for any positive yoga to have a notably uplifting effect upon a person's life, at least one of the yoga-forming planets should be strong. If none of the yoga-forming planets is strong, then the yoga may still be beneficial to a lesser degree. Planets gain strength by being in their own sign, or exalted (or even in own nakshatra), by direction (dig bala), by brightness (Bright Moon or retrograde planet), in the same sign as in the Navamsha (Vargottama), or in Parivartana Yoga (mutual exchange, I.e. in each other's signs) with another planet. The more types of strength a yoga-forming planet have, the more powerful the yoga.

B. **The negative Yoga:** A negative yoga to be seriously problematic in a person's life, if the yoga-forming planets are weak and association with malefic planet. Weakness can be by the sign of debilitation, by dimness (combustion or Dark Moon), or by proximity to another malefic or planetary war.

Checklist for Qualifying Yogas: In order to use the principles in this article easily, I decided to create the following checklist for assessing positive yoga.

Major considerations: 1. Yoga-forming planets shall be strong. If not, then the yoga is not likely to give rise to much success unless the yoga-forming planets are participating in more yoga.

2. Yoga shall be recuring from the Chandra or Surya Lagnas. If so, then the yoga is more powerful.

3. Yoga-forming planets shall receive aspects from benefic. If not so, the magnitude of the yoga's effects will be notably diminished, and additional negative issues might arise.

4. Yoga-forming planets shall have some relationship with the First House or the ruler of the First House. If not, the abilities and gifts represented by the yoga will be hard to access.

5. The Ascendant and lord of the Ascendant shall be predominately strong. The yoga good effects will be easily expressed, if both are strong and influenced by benefic, but it is not, if both are weak and influenced by malefic.

6. The person shall be running or going to run the Dasa of one of the yoga-forming planets. If so, then the yoga is likely to reach full fruition. If not, then if the person has already run the Dasa of one of the yoga-forming planets, then the yoga may not reach full fruition.

7. Either or both of the yoga-forming planets have good house ruler ships and placement. If not, then the yoga's effect will be significantly diminished.

8. Yoga has taken place in houses 6, 8, or 12. If so, then the themes of that house will be enhanced (e.g. medical skill for the 6th, or occult gifts for the 8th), but the yoga may not give the notable results in conventional ways.

9. The yoga-forming planets shall be close to each other by orb. If so, then the yoga may be a bit more powerful.

15.1 Planetary Combinations (Yoga)

Adhi Yoga: There must be at least two, but preferably more, benefic, in any combination, in the sixth, seventh and eighth houses calculated from the Moon. These planets are in the opposing houses or opposition from the Moon. For example, Jupiter in the 6th house from the Moon, and Venus and Mercury in the 8th from the Moon, or all the 3 are in one of these houses, or Venus in the 8th, Mercury in the 7th and Jupiter in the 6th houses from the Moon. It indicates great leadership. It makes an individual polite, trustworthy, affluent and capable of defeating his adversaries.

Akhand Samrajya Yoga: A planetary combination producing a long life of affluence formed by Leo and Scorpio, as Ascendant, making Jupiter rule either the 5th or the 11th house in a natal chart.

Amala yoga: A benefic in the 10th house from the Moon or the ascendant is Amala Yoga. Amala Yoga will confer fame, and will make him/her honoured by the king, enjoy abundant pleasures, charitable, fond of relatives, helpful to others, pious and virtuous.

Amar Yoga: An auspicious planetary combination formed in two ways, (1) all cardinal houses occupied by all malefic, or by all benefices. Here, the native owns landed property and real estate, and in the latter case, he becomes rich and affluent; (2) Sun in Aries or in Leo occupies the Ascendant, or any other cardinal or trine house while Moon is in exaltation or in its own sign, i.e., in Cancer or Taurus, and Jupiter and Venus occupy the 8th or 12th house in the natal chart. This yoga nullifies all evils in the horoscope.

Amarak Yoga: A planetary combination formed by the lord of the 7th house placed in the 9th house and the lord of the 9th in the 7th house while both these planets are strong. It bestows to the native long arms, big eyes, knowledge of law and religious scriptures. His wife is faithful to him and he leads a pure and moral life.

Anapha yoga: Planets, excluding the Sun, Rahu or Ketu, in the house 12th from the Moon, creates Anapha Yoga. Mars in this position makes the person powerful, self-controlled and a leader of persons engaged in undesirable activities. Mercury makes him proficient in oratory, and absorbing conversations, and skilled in social arts. Jupiter makes the native a serious-minded, righteous person spending money on charity. Venus makes the person a womanizer yet respected by persons in authority. Saturn leads to disenchantment, and the nodes, to perversity. The Moon under the yoga bestows well-formed organs, good manners and self-respect. The natural benefic planet gives name, fame, wealth and success while natural malefic gives trouble. This effect increases if there are more benefic in the twelfth house but create more concern, if there are more malefic. If the planets Sun, Rahu and Ketu are with other planets in 12th from the Moon, they will drastically weaken or cancel this yoga. This is because when the Sun is close to the Moon, the Moon is weak or dark. The full Moon is considered very auspicious because it reflects more light. The nodes of the Moon (Rahu and Ketu) when situated close to the Moon will definitely change the Moon's effects. Rahu and Ketu will distort the qualities of the planets they conjoin.

Ara Sauri Yoga: Planetary combination between Saturn and Mars. It produces serious afflictionsto the native.

Ardha Chandra Yoga: If the seven Graha occupy continuously seven Bhava, commencing from a Bhava, which is not angular to the Lagna, the Yoga produced is known, as Ardha Chandra Yoga. He/She born in Ardha Chandra Yoga will lead an Army, will possess a splendorous body, be dear to king, be strong and endowed with gems, gold and ornaments.

Aristha Yoga: Planetary combinations producing misfortune. These combinations nullify auspicious results and produce hardships. Some of these combinations are: (i) Malefic associated with the 6th, 8th and 12th houses or their lords; (ii) Malefic aspects on a weak Moon (iii) Sun, Mars, and Saturn in the 5th house (iv) Mars, Saturn, or Sun in the 8th house (v) malefic aspect on weak Ascendant lord, Sun or Moon (vi) Sun, Mars, Rahu and Saturn in Ascendant (vii) exchange of signs between Jupiter and Mars (viii) Mars and Saturn in the 2nd house while Rahu occupies the 3rd house (ix) Rahu in the 4th, and Moon in the 6th or 8th house x) Mars in the 7th, Venus in the 8th and Sun in the 9th house (xi) Malefic in the 7th and 12th houses (xii) Jupiter, Sun, Rahu and Mars occupy signs of malefic Planets while Venus is in the 7th house (xiii) lord of Ascendant associated with a malefic or flanked by two malefic, and a malefic positioned in the 7th house (xiv) Saturn in the 8th house, Moon in Ascendant, or Venus and Moon in the 6th or 8th House (xv) Moon and Mercury in the 6th or 8th house. This planetary combination produces misfortune for the nativeand also nullifies the auspicious results of other yoga and produce hardships.

Ashta Lakshmi Yoga: When Rahu is in 6th and Jupiter in Kendra this is formed. It gives wealthy condition and happiness is seen.

Asubha Mala Yoga: When all the benefices are placed in the sixth, eighth and twelfth houses, the benefice become powerless and then, it indicates misfortune and a difficult life.

Avatara Yoga: A planetary combination formed by (i) the Ascendant occupying a cardinal Sign, and Venus and Jupiter in 1st, 4th, 7th or 10th houses, and (ii) Saturn in exaltation. The combination bestows spiritual blessings and he becomes deep student of religious and esoteric literature with psychic susceptibilities. The combination also bestows high status in society, renown for meritorious deeds and pilgrimages to religious and historical places.

Bajra Yoga: A planetary combination under which all benefic are located in the 1st and 7th houses in a natal chart. It makes the individual good-natured and lucky during the whole of his life.

Bhairi Yoga: If Vyaya, Thanu, Dhan and Yuvati Bhava are occupied, as Dharma Lord is strong, he/she obtains Bhairi Yoga. Again another kind of Bhairi Yoga is formed, if Sukra, Guru and Lagna Lord are in a Kendra, while Dharma Lord is strong. He/She will be endowed with wealth, wife and sons. He/She will be a king, be famous, virtuous and endowed with good behaviour, happiness and pleasures.

Bhaskara Yoga: Bhaskar is one of the several names of Sun. This yoga is formed when Mercury is placed in the 2nd house from the Sun, Moon is placed in the 11th house from the Mercury and Jupiter is placed either in the 5th or the 9th house from the Moon. It is a very rare yoga. A native with this yoga in his or her Kundali is as great as is this yoga. This auspicious yoga provides immense wealth, happiness and luxury to its natives. They are interested in arts and have a very caring nature. A person born under this combination is courageous, powerful, learned and has deep knowledge of religious scriptures, mathematics and classical music.

Bhavya Yoga: A planetary combination formed by Moon in the 10th house, the navamsa lord of Moon in exaltation, and the lord of the 9th house associated with the lord of the 2nd house. This combination makes a person rich, respected and learned. He may be renowned as a botanist and a collector of artefacts.

Bheri Yoga: The planetary combination which is formed in 3 ways: (i) all planets occupy the Ascendant, in the 2nd, 7th, and 10th houses (ii) Venus and the lord of Ascendant are placed in a cardinal house from the Ascendant, and the lord of the 9th house is strong (iii) Venus and the lord of Ascendant and Jupiter are in mutual angles and the lord of the 9th house is strong. All these combinations make the individual learned in scientific subjects, practical in mundane affairs, and well provided with wealth and luxuries of life.

Bhujang Yoga: Malefic placed in 3 Kendra will produce malefic results.

Bhupa Yoga: A planetary combination formed by the lord of the 5th or 9th house from the sign where the Navamsa lord of Rahu is posited occupying its own sign and expected by Mars. The combination makes the individual born under it victorious in warfare and bestows on him high military status.

Budha-Aditya Yoga (Sun-Mercury conjunction): The conjunction of Sun and Mercury constitutes a famous, Budha-Aditya Yoga, an asset of the horoscope. It is synonymous to Bhadra Yoga, (one of Pancha-Mahapurusha yoga). The legitimacy of this yoga is when it gets extra strength from the sign of own, exaltation or friends such as Aries, Gemini, Cancer, Leo, Virgo, and Sagittarius. Then, it uplifts the life of the native with certainty. If the yoga germinates in sign of debilitation such as Libra or Pisces or it is ruined by the association or aspect of malefic such as Mars, Saturn, Rahu or Ketu, its natural strength is pitiful. Budhaditya yoga with the Sun and Mercury placed in the 10th house, the first, seventh, fifth or eleventh house is very benefic. This is a very common yoga as Mercury can never be more than one Sign away from the Sun. The sign and house the yoga occurs in, is important and will indicate skills in these areas. Mercury and the Sun combination will make the Mercury brilliant in effects in areas that planet rules. This applies to any planet not only Mercury. It indicates an intellectual.

Chamar Yoga: If Lagna Lord is exalted in a Kendra and receives a Drishti from Guru, Chamar Yoga is formed. This Yoga also occurs, if two benefic are in Lagna, or Dharma, or Karma, or Yuvati Bhava. He/She will be a king, or honoured by the king, long lived, scholarly, eloquent and versed in all arts. Such an individual is generally born in a royal family.

Chandra Mangal yoga: When Moon and Mars are together, or there is natural aspect between them. If Mars is conjunct, or opposed the Moon, Chandra Mangal Yoga is formed. When Mars is in 7th or in combination with Moon, this yoga is formed. This gives great financial prosperity, richness, probably made by unscrupulous means and great business sense. In a man's chart it can bring gain through women. It can also cause problems with the mother. Because of this yoga man is passionate. This indicates more landed properties, reputation, and wealth for the native. Restlessness and worry is also there.

Chandra Yoga: A planetary combination constituted by an exalted planet in ascendant expected by Mars, while the lord of the 9th house is placed in the 3rd house. It makes the individual administer, an adviser, or the commander of an army. The individual is courageous and lives for more than six decades.

Chandrika Yoga: A planetary combination in which the Ascendant is occupied by the lord of the sign in which the 9th lord is also placed while Mars is posited in the 5th house. Persons born under it are powerful, attain a high social status in life but do not have male issues.

Chapa Yoga: It is a planetary combination in which all planets occupy the 10th to the 4th houses. Persons born under it are expert thieves, and are socially despised, will be liar, will protect secrets, be a thief, be fond of wandering, forests, be devoid of luck and be happy in the middle of the life. Chapa Yoga is also formed if the Sun is in Aquarius, Mars in Aries, and Jupiter in its own sign, which makes the individual a globetrotter.

Charussagara yoga: All planets placed in all the Kendra houses indicate good reputation and good fortune. It is better when it is formed by benefice. Planets placed there are extremely effective. When planets are occupying all the Kendra houses, their energy will be powerfully felt, which in most cases, will be a good thing, especially when it concerns natural or temporary benefic. When there is Srik yoga, the natural benefices will be very influential in life, which of course, is very positive.

Chatra Yoga: If all the Graha occupy the seven Bhava from Yuvati, Chatra Yoga occurs. Here again the Graha should occupy seven continuous Bhava. He/She born in Chatra Yoga will help his/her own men, be kind, dear to many kings, very intelligent, happy at the beginning and end of his/her life and be long-lived.

Chaturmukh Yoga: A planetary combination formed by Jupiter in a cardinal house from the sign occupied by the lord of the 9th house, Venus in a cardinal house in respect of the sign occupied by the lord of the 11th house, and the Ascendant lord and the lord of the 10th house themselves placed in the cardinal houses. This combination makes the individual erudite and successful in his undertakings and much respected. He is well provided with material possessions and lives a very long life.

Chatursagar Yoga: An auspicious planetary combination formed by all planets, benefic as well as malefic, occupying cardinal houses. It bestows wealth, affluence, and high status in life. Such a person becomes famous even after death.

Chhatra Yoga: A planetary combination formed by all planets situated in the first seven houses of the chart. It makes the individual very happy from the beginning of his life till the very end.

Daamini Yoga: He/She born in Daamini Yoga will be helpful to others, will have righteously earned wealth, be very affluent, famous, will have many sons and gems, be courageous and red-lettered.

Dainya yoga: This yoga occurs when one of the planets participating in the yoga is lord of the sixth, eighth or twelfth house. These are

Dustasthana houses that generally have an unfavourable influence. In most cases, the lord of the sixth, eighth or twelfth house will be strengthened by the benefice influence of the other planet. However, the other planet involved in this yoga is always damaged.

Danda Yoga: An auspicious planetary combination formed in several ways. If Venus aspects Jupiter placed in the 3rd house, while the lord of the 3rd is in exaltation, Danda Yoga is formed. Alternatively, if all the 7 Graha in the 4 Bhava, commencing from Karma form Danda Yoga and he/she will lose sons and wife, will be indigent, unkind, away from his men and will serve mean people. Or, it it is formed, when all planets are placed only in Gemini, Cancer, Virgo, Sagittarius and Pisces signs. Danda Yoga makes a person respected, very rich, an able administrator, and a pious person. An inauspicious combination under the name is formed when all planets occupy only the 10th, 11th, and 12th houses. It makes the individual depraved, dependent on others for livelihood, and discarded by his kith and kin.

Daridra Yoga: Planetary combination producing indigence and personal infirmities. They are: (i) Jupiter as lord of the 8th house or the 1st house exceeds the strength of the lord of the 9th house, and the lord of the 11th house is neither placed in a cardinal house nor is combust (ii) Debilitated and combust Jupiter, Mars, Saturn, or Mercury occupies the 11th, 6th, 12th, 8th or the 5th Bhava(iii) Saturn in 9th house aspect by malefic planets while Mercury is associated with the Sun and occupies the Ascendant and has Pisces Navamsa (iv) Jupiter, Mercury, Venus, Saturn, and Mars occupy in any order 8th, 6th, 12th, 5th, and 10th Bhava, and the lord of the 12th house, weakened by Sun's aspect, has greater strength than the Ascendant lord (v) Depressed Venus, Jupiter, Moon, and Mars occupy any four of the 1st, 10st, 11th, 6th, 7th , and 8th Bhava s (vi) Venus in Ascendant in its debilitation sign, while Jupiter, Mars, and Moon are also in debilitation (vii) The Ascendant is in a cardinal sign, while the rising Navamsa is aspect by Saturn and depressed Jupiter (viii) if Jupiter is in the 6th or the 8th Bhava in a sign not belonging to itself (ix) Ascendant in a fixed sign, malefic in cardinal and trine houses in strength, and angles devoid of benefices (x) Night time birth, Ascendant in a cardinal sign, weak benefics occupy angles and trine, and malefic not in cardinal houses. Persons born in Daridra Yoga suffer deprivations of different intensities and meet unlucky and trying conditions of life. They earn by foul means. Their social life is dishonourable. They meet unexpected failures in life.

Data Yoga: A planetary combination formed by Jupiter in Ascendant, Venus in 4th house, Mercury in the 7th, and Mars in the 10th house. It makes an individual very affluent and generous.

Devendra Yoga: A planetary combination formed by Ascendant placed in a fixed sign, Ascendant lord in 11th, the lord of 11th in Ascendant, and the lords of 2nd and 10th houses in mutual exchange. This combination is powerful in making the individual extremely beautiful, loved by pretty women, owner of vast wealth and villas. He attains a very high social status.

Dhana Yoga: Bhava houses 2, 6, 10 are primary and 7 and 11 are secondary for Dhana yoga. 2nd House in a horoscope signifies self generated wealth, 6th through loans, and 10th through employment. If the 2nd is strong money comes by inheritance and investment. If 6th is stronger than 2 and 10 money comes as interest on lending. If 12th becomes very strong the native borrows but never repays fully there by always in debt. If 10th is stronger than 2 and 6 the native can make money at all times through various means. The Dhana karakas or significators of wealth are Sun and Jupiter. So, both Jupiter and Sun should be well placed to be wealthy. Another important rule is that "Sun's arch rival Saturn should not be in the 2nd house by position, aspect or association with 2nd lord from the Lagna or Moon. This is a must condition for Dhana yoga. Other conditions for wealth are as below: 1) if the lord of the ascendant is in 10th House the native will be richer than his parents. 2) The same will apply if 10th lord is in the ascendant, 3) If Jupiter is placed in 9 or 11th and Sun in the 5th the native becomes rich, 4) Mutual exchange of Houses of 2nd and the 9th lord (except Saturn) makes a person rich. 5) Sun in 6th or 11th the person becomes rich, particularly when Sun is in Rahu star and Rahu in Sun's Star. 6) 2nd lord in 8th a person becomes rich by self-efforts. 7) If Moon and Jupiter or Moon and Venus be in 5th the person becomes rich. 8) If Mercury happens to be in Aries or Cancer the person becomes rich. 9) If 7th house has Mars or Saturn and 11th house has any planet other than Ketu the person will earn huge wealth by doing business. If Ketu is in the 11th he will earn through foreign agency. 10) If the 7th house has either Mars or Saturn and the 11100th house has Saturn or mars or Rahu the person earns wealth by sports, Gambling, Commissions, rent lawyer's fees etc.

Dharma Yoga: A planetary combination constituted by the occupancy of Jupiter and Venus along with the lord of the 2nd in the 9th house. It makes the person very pious, fond of warfare, chivalrous, and the commander of an army. He also becomes rich and charitable.

Durudhura Yoga: This is the planetary combinations constituted by planets, other then the Sun, Rahu and Ketu, are placed in both sides of Moon, the twelfth and the second house from Moon or are situated on either side of Moon. They generally produce affluence such as wealth, comforts in life, and high social status. Sun and the nodes must not be involved in this combination. If Moon is flanked by Mars and Mercury the combination makes the individual cruel, greedy, fond of old women, and a liar. Mars and Jupiter in the position make the individual renowned, clever, and rich and a defender of others from adversaries. Venus and Mars make the individual fond of warfare, rigorous physical exertion, and courageous deeds. With Saturn and Mars, the individual becomes an expert in sexual art, accumulates much money, indulges in a fast life, and is surrounded by enemies. The combination of Jupiter and Mercury bestows religiosity, knowledge of scriptures, all round affluence and renown. Mercury and Venus in the situation make the individual beautiful, attractive, affluent, courageous, and the recipient of high official status. Moon in between Saturn and Mercury enables the individual to travel to different countries in pursuit of wealth. Jupiter and Venus make the individual patient, intelligent, balanced, and ethical, he acquires jewels, renown, and good administrative position. When Venus and Saturn form the combination, they enable the individual to acquire an aged wife from a respectable family, they make the individual skilled in many trades, loved by women, and respected by government officials. Saturn and Jupiter flanking Moon create many difficulties in the personal life of the individual; he is often surrounded by scandals, difficulties, and litigation, though from these he emerges unscathed. Durudhara Yoga makes the individual endowed with much physical comfort, wealth, loyal helpers, and sincere followers, but towards the end in the individual's life there arises a strong impulse for renouncing the worldly possessions. Benefic give good result and make rich and high morals and malefic cause trouble. Benefic and malefic both combined give neutral results. Planets on both sides surround the Moon. This totally supports the Moon. He/She is happy and feels supported in life. He/She will receive as well as give back to the world.

Foreign Travel Yoga: According to Astrology, Planetary Yoga (planetary combinations) for travel abroad is studied from 7th, 9th or 12th houses of the horoscope. Rahu is the significator of foreign travel. The Sun, Mercury and Saturn are chief significators of travel abroad. In the main period or sub period of planets owning or occupying these houses there may be foreign tours. Even the transit of Saturn, Rahu and Jupiter from these houses facilitates foreign tours.

Gada Yoga: A planetary combination formed in two ways: Moon posited in 2nd house along with Jupiter and Venus, or the lord of 9th house aspecting them and (ii) all planets, excluding the nodes are posited in adjoining cardinal houses. Gada Yoga makes the individual engaged in philanthropic and religious activities but fierce in appearance and free from any enemy. He earns much money. He is also happily married, be skilful in Shashtra and songs and endowed with wealth, gold and precious stones.

Gaja Yoga: A planetary combination in which the lord of the 7th house from Ascendant, which would be 9th from the 11th house, is in the 11th house along with Moon, and the lord of the 11th house aspects them. A person born under this combination is always happy, rich, religious, and lives in luxurious style.

Gajakesari Yoga: An auspicious planetary combination formed by certain relationships between Moon and Jupiter. It postulates that Jupiter must be in an angle from Moon or the Ascendant or that benefic such as Venus, Jupiter, and Mercury without being debilitated or combust, aspect the Moon. An alternative condition is that Jupiter in a quadrant from the Ascendant or Moon is in association with or expected by benefic which are neither combust nor posited in the 6th house. The combination makes the individual bright, affluent, and intelligent, accomplished and favoured by the government, is good for leadership, authority and financial prosperity. This yoga is both protective from evil consequences of other maleficent as well as productive of auspicious results.

Go Yoga: The planetary combination formed by exaltation of the lord of the Ascendant, and strong Jupiter placed in its Moolatrikona in association with the lord of the 2nd house. It makes the individual hail from an elite family and bestows upon him happiness, attractive appearance, and high social standing.

Guru Chandala Yoga: A planetary combination relating Jupiter and Rahu. When these two planets are associated together in a house, it produces inauspicious results. It makes the individual depraved and inclined to indulge in socially and morally unethical activities.

Hal/Hala Yoga: It arises when all the planets are located in a group of triangular houses other than the Ascendant. According to another version; all planets occupying the 5th and 9th houses also give rise to Hala Yoga. Persons born under this combination are engaged in agricultural activities in an important way. He/She born in Hal Yoga will eat a lot, will be very poor, will be miserable, agitated, given up by friends and relatives. He/She will be a servant.

Harihara Brahma Yoga: It refers to 3 sets of planetary combinations: (i) Benefic placed in the 2nd, 8th, and 12th houses from the sign in which the lord of the 2nd house is placed (ii) Jupiter, Moon, and Mercury posited in the 4th, 9th, and 8th houses from the sign where the lord of the 7th house is situated (iii) the Sun, Venus and Mars in the 4th, 10th, or 11 house from the Ascendant lord. These combinations make the individual truthful, effective speaker, victorious, well-versed in religious scriptures, and philanthropic.

He will be a Jyotish, virtuous, strong, beautiful, famous, learned and pious.

Ichchita Mrityu Yoga: The planetary combination formed by Mars in a cardinal house and Rahu in the 7th. It leads the individual to self-destruction.

Ikkbala Yoga: A term used in annual forecasting. It is formed by all the planets placed in Panphara houses. It produces many desired events during the year.

Induvara Yoga: It is an inauspicious planetary combination. The seven planets (excluding Rahu and Ketu) in a progressed horoscope based on solar ingress principle, situated in Apoklima houses produce obstacles and thereby nullify the fructification of an auspicious combination in the kundali that may otherwise be present.

Kahal Yoga: This combination is constituted in several ways: the lords of 9th and 4th houses should be in mutual angles and the lord of the Ascendant should be strong. The combination is also formed if the lord of the 4th occupies its exaltation or own sign and is aspect by or is in conjunction with the 10th lord. Persons born under the combination are courageous, virile, commanding a well-equipped army, and ruling over an extensive area, yet they are intellectually foolish, ignorant. Kalanidhi Yoga: A planetary combination formed by Jupiter in 2nd or 5th house, Mercury and Venus expecting or combining with it. An individual born under it is honoured by many heads of state as a great Artist Salman Khan. He becomes affluent, accomplished, and healthy and occupies a very high position in life. A person born under it is prosperous, respectful, benevolent, and God-fearing, happy, charitable and regal in demeanour.

Kahala Yoga: This occurs when one of the planets of this yoga is lord of the third house. The other planet may not be lord of the sixth, eighth or twelfth house (because we then have Dainya yoga). The third house is the house of energy and strength. Kahala yoga would mean that the person involved would invest a lot of energy and vigour in improving his circumstances. Example: If Taurus is on the ascendant, the Moon is lord of the third house and Saturn lord of the ninth and tenth houses. Supposing the Moon is in Capricorn and Saturn in Cancer. Then we have Kahala yoga. He/She will put a lot of energy (third house) into his/her career (10th) and into broadening his/her horizons (9th house).

Kalanidhi Yoga: If Guru is placed in Dhan or Putra Bhava and receives a Drishti from Buddha and Sukra or associated with them, Kalanidhi Yoga is caused. An individual born under it is honoured by many heads of state. He becomes affluent, accomplished, and healthy and occupies a very high position in life. He will be virtuous, honoured by the kings, bereft of diseases, be happy, wealthy and learned.

Kalasarpa Yoga: When all planets are between Rahu and Ketu this yoga is formed. Even though favourable planets are placed between them or in own house this has force. This is a bad yoga. Generally first half of life will be in struggles.

Kalpa Drum Yoga: Lagna Lord, the dispositor of Lagna Lord (a), the dispositor of the Graph "a" (b), the Navamsa dispositor of the Graha "b". If all these are disposed in Kendra and in Kona from Lagna, or are exalted, Kalpa Drum Yoga exists. One with Kalpa Drum Yoga will be endowed with all kinds of wealth, be a king, pious, strong, fond of war and merciful.

Kamal Yoga: If all the Graha are in the 4 Kendra, Kamal Yoga is produced or it is formed by all planets situated in 1st, 4th, 7th and 10th houses. One born in Kamal Yoga will be rich and virtuous, be long lived, very famous and pure. He/She will perform hundreds of auspicious acts and he/she will be a king. This makes the individual born under it renowned, happy and accomplished in many arts?

Kamboola Yoga: A planetary combination produced by Ithasala relationship between the lord of the Ascendant and the lord of the 10th house, especially when one of them is associated with Moon. The Kamboola Yoga is of 3 kind, Shrestha (the best), Madhyama (ordinary) and Adhama (the worst) depending upon the strength of the planets concerned.

Kanduka Yoga: A planetary combination formed by the lord of 10th house placed in 9th house, the lord of 2nd in Ascendant, and the 2nd and 10th houses posited by benefic. An individual born under this combination is charitable but very materialistic in his approach to life. He seeks enjoyment of all kinds of physical comforts and a luxurious life.

Karagar Yoga: A planetary combination formed by one, two or three malefic planets un-aspect by any benefic and posited simultaneously in 2-12, 3-11, or 4-10 houses. It produces the possibility of imprisonment, or detention. Similar results also occur if malefic occupy 12th and 9th houses.

Kedar Yoga: A planetary combination formed by all planets occupying four houses in a natal chart. It makes the individual ever ready to wage a righteous war, undertake a righteous mission, follow traditional religious practices and be humble, patient, philanthropic, interested in agriculture and respected in his society.

Kema Druma Yoga: This is formed when there are no planets on both sides of Moon or the Moon has no connection to any planets excluding the Sun, Rahu and Ketu. He/She is very poor and feels isolated, alienated, with no support. He/She are a dark mind, lacking intelligence, feeling completely disconnected. He/She will feel lonely and lead a rather difficult life. It makes the individual devoid of any education and intelligence. He suffers from penury and meets many difficulties in life.

Kesari Yoga: When Jupiter and Moon are mutually Kendra this yoga is formed. More so when Jupiter is in Kendra, 7th to Moon (bright period Moon) this yoga is formed. It gives wealth, name and fame as a big man.

Khadg Yoga: If there is an exchange of Rashi between the Lords of Dhan and Dharma Bhava, as Lagna Lord is in a Kendra, or in a Kona, he/she will be endowed with wealth, fortunes and happiness, is learned and is intelligent, mighty, grateful and skilful.

Koot Yoga: If all the Graha occupy the seven Bhava from Bandhu, Koot Yoga is formed. Here again the Graha should occupy seven continuous Bhava. One born in Koot Yoga will be a liar, will head a jail, be poor, crafty, and cruel and will live in hills and fortresses.

Kshema Yoga: A planetary combination formed by the lord of the Ascendant and the lords of 8th, 9th and 10th houses occupying their own signs. It makes the individual support his family members and other relations. He becomes personally rich, happy and lives for long.

Kurma Yoga: A planetary combination produced by benefices in 5th, 6th, and 7th houses either in exaltation, own signs, or those of friendly planets, or in the Navamsa of friendly planets. Alternatively, it is formed if the benefic are in Ascendant, 3rd, and 11th houses occupying exaltation, own signs, or their Moola Trikona positions. Persons born under this yoga become leaders, very renowned, charitable, helpful, and they lead a very happy life and he/she will be a king, be courageous, virtuous, famous, helpful, and happy. He/She will be a leader of men.

Kusuma Yoga: The planetary combination formed by Saturn occupying the 10th house, Venus placed in a cardinal house with fixed signs and a weak Moon in a trine house. Alternatively, Jupiter should be in Ascendant, Moon in the 7th and Sun occupying 8th position from Moon (that is, in the 2nd house in this situation). He/She will be a king, or equal to him/her. Persons born with this combination belong to an aristocratic family, they attain high status in the society, and possess charitable disposition and enjoy unblemished glory.

Kuta Yoga: A planetary combination formed by all planets consecutively placed from 4th to 10th houses. It makes the individual dwell in forests or mountainous regions; they are very cruel in temperament.

Lagna Adhi Yoga (Lagnadhi Yoga): If benefic exclusively occupy the 6th, 7th and 8th house, without the company of malefic, it confers minister ship, leadership over Army, justice ship, kingship and lordship over many women, long life, freedom from diseases and miseries, possession of virtues and happiness. If benefic are in Yuvati and Randhra Bhava and is devoid of Yuti with and/or Drishti from malefic, it makes him/her a great person, learned in Shashtra and happy.

Lakshmi Yoga: If Dharma Lord is in a Kendra identical with his Moolatrikona Rashi, or own Rashi, or in exaltation, while Lagna Lord is endowed with strength, Lakshmi Yoga occurs. He/She will be charming, virtuous, and kingly in status, endowed with many sons and abundant wealth. He/She will be famous with high moral merits. Persons born under it are graceful, religious, wealthy, accomplished, famous, and enjoy high status in the society. Their offspring's are very bright.

Maal Yoga: If 3 Kendra are occupied by benefic, Maal Yoga is produced. This Yoga produces benefic results. He/She will be ever happy, endowed with conveyances, robes, food and pleasures, be splendorous and endowed with many females.

Madan Yoga: A planetary combination constituted by the lord of the 10th house posited in Ascendant along with Venus and the lord of 11th house occupying the 11th house itself. The combination makes the individual born under it very attractive and highly placed in political circles. He begins prospering at an early age of twenty years.

Maha Bhagya Yoga: For a man, if the ascendant, the Sun and the Moon are placed in uneven signs and he is born in the daytime, which indicate very fortunate. Uneven signs are, by the way, positive;

masculine Signs (the air and fire signs). Daytime is also masculine. For a woman, the ascendant, the Sun and the Moon are placed in even signs, and she is born during the night, which indicates very fortunate. Even signs are negative; feminine signs (the earth and water signs). The night is feminine.

Maha Pataka Yoga: A planetary combination formed by Moon associated with Rahu and aspect by Jupiter conjunct with a malefic. It leads the individual, even if highly intellectual and well placed in society, to indulge in mean behaviour and acts.

Maha Yoga: When Parivartana Yoga occurs between the lords of the first, second, fourth, fifth, seventh, ninth, tenth or eleventh houses, we have Maha yoga. The lords of the third, sixth, eighth or twelfth houses may not be involved, because then we would have either a Dainya yoga or a Kahala yoga. This yoga enriches the planets involved. Another way in which planets can be connected is through Parivartana Yoga. This is called mutual reception in western astrology. The planets receive connection through being in each other's sign of ruler ship.

Mala Yoga: A planetary combination formed by the lords of the 2nd, 7th, 9th and 11th houses posited in their respective signs. It bestows high administrative status on the individual and makes him a minister, a royal treasurer, or a leader of the people. His fortune brightens after the age of 33 years.

Marud Yoga: A planetary combination formed by Jupiter in a trine house from Venus, Moon in 5th from Jupiter, and Sun in a cardinal house in respect with Moon. The combination makes the person very rich and a successful businessman.

Mathanda Yoga: If a benefic, such as Jupiter, Venus, Mercury or Moon or at least two stays in lagna or in Kendra or in Kona; or stays in Arudha lagna, or in Arudha kendra or in aarudha kona, the main period and sub-periods of that benefic give the best beneficial results (first grade, above 75%). His wife has all the best benefits due to him and leads happy married life and he has meteoric rise in life and gained huge income and houses, lands and properties.. This person has enjoyed all the above-said benefits. He is a rich business tycoon.

Matsya Yoga: A planetary combination formed by malefic and benefic both occupying the 5th house, malefic in Ascendant and in 9th house, and a malefic in either the 4th or the 8th house from the Ascendant. A person born under this combination is compassionate, religious, intelligent and renowned.

Mridang Yoga: If Lagna Lord is strong and others occupy Kendra, Kona, own Bhava, or exaltation Rashi, Mridang Yoga is formed. He/She concerned will be a king, or equal to a king and be happy.

Mriga Yoga: A planetary combination formed by the placement of Navamsa lord of the 8th house in an auspicious sign along with some auspicious planet, and the lord of the 9th house in exaltation. It makes the individual respected, rich, immensely charitable, and powerful in personality.

Mukuta Yoga: A planetary combination formed by Jupiter in the 9th house from the sign occupied by the lord of the 9th house, a benefic posited in the 9th house from Jupiter and Saturn in the 10th house from the Ascendant. A person born in this combination possesses farms and forests, becomes a leader of tribal people, and is erudite. He is learned, yet cruel by temperament. His prosperity begins quite early in life.

Musala Yoga: A planetary combination formed by all planets placed in fixed signs or alternatively, Rahu in the 10th house, the lord of the 10th house in exaltation and expected by Saturn. It makes the person born under it very rich with immovable assets. He becomes an advisor to the government or a powerful commercial organization or he occupies a high status in administration. He will be endowed with honour, wisdom, wealth, dear to king, famous, have many sons and be firm in disposition.

Nabhi Yoga: A planetary combination formed by Jupiter in the 9th house, lord of the 9th house in the 11th from Jupiter, i.e., 7th from Ascendant, and Moon associated with Jupiter. The combination bestows on the individual born under it auspicious events in life, especially after the age of 25 years. He also receives many honours from the State and accumulates huge wealth.

Naga Yoga: A planetary combination formed by the lord of the Navamsa sign of the 10th house lord occupying the 10th house along with the Ascendant lord. An individual born under it receives his education especially after the age of 16 years. He finally receives state honours and riches. By temperament he is polite.

Nagendra Yoga: A planetary combination formed by the placement of the lord of 9th house in 3rd house inspected by Jupiter. It makes the individual physically well proportioned, good-natured and learned. His prosperity increases after the age of 6 years.

Naktya Yoga: A planetary combination used in Tajaka system. It relates to the relationship between planets with different motions in close association. If the lord of the Ascendant and the lord of the house whose result is being studied do not have mutual aspect but there is a fast-moving planet in between them, then the fast-moving planet in between them transfers the benefic influence of the anterior planet to the forward one.

Nala Yoga: A planetary combination formed the exaltation of the lord of the Navamsa sign in which the lord of the 9th house is placed and is in association with the Ascendant lord. It makes the individual powerful after 7 years of age. He receives many state honours and is interested in the scriptures.

Nala Yoga: All the Graha in Dual Rashi cause Nala Yoga. He/She born in Nala Yoga will have uneven physique, be interested in accumulating money, very skilful, helpful to relatives and charming.

Nalika Yoga: A planetary combination formed by the placement of the lord of the 5th house in the 9th house while the lord of the 11th house occupies the 2nd house along with Moon. The combination makes the individual very creative and an excellent orator.

Nanda Yoga: A planetary combination formed by two planets in each of the two signs and one planet in each of the three signs. It bestows affluence and long life.

Nasir Yoga: A planetary combination formed by the Ascendant lord and Jupiter placed in the 4th house, Moon associated with the lord of the 7th house, and the Ascendant aspect by a benefic. An individual born under this combination is very charitable, rich, well-proportioned yet stocky in constitution. He gains repute after the age of 33 years.

Nau Yoga: A planetary combination formed by all planets occupying consecutively the first seven houses without any gap. It makes the individual earn his livelihood from professions connected with navigation, fishing, import-export, and international commerce.

Nauka Yoga: If all the Graha occupy the seven Bhava from Lagna, Nauka Yoga occurs. Here again the Graha should occupy seven continuous Bhava. He/She born in Nauka Yoga will derive his/her livelihood through water, be wealthy, famous, wicked, wretched, dirty and miserly.

Neechabhangha Yoga: The planetary combination for the cancellation of adverse effects of a debilitated planet. The cancellation enables the person to attain the status of a king. The combination is formed in several ways, such as (i) a planet at birth in its depression has the lord of that sign, or that of its exaltation sign in a cardinal house either with respect to ascendant or Moon sign (ii) the lord of the Navamsa occupied by the depressed planet at birth posited in a cardinal house, or in a trine house with respect to the Ascendant while the Ascendant lord itself is in a Navamsa owned by a movable sign.

Nripa Yoga: A planetary combination formed by the lord of Navamsa sign of Ascendant associated with the lord of Moon sign and the lord of the 10th house aspecting it. An individual born under this combination occupies a very high status in society and is much renowned. The Yoga fructifies early in life.

Obhayachari Yoga: A planetary combination formed by planets, other than Moon, situated on both sides of Sun. It makes the individual well proportioned, handsome, skilled and effective in many undertakings, full of enthusiasm, tolerant, and balanced in approach even to complicated problems. Such a person is affluent like a king, enjoys good health and possesses all good things in life.

PADMA YOGA: A planetary combination formed by the lords of the 9th house from Ascendant and from Moon situated together in the 7th house from Venus. Individuals born with this stellar configuration are very happy, live in luxury and are engaged in auspicious activities. After the age of fifteen years, they are granted favours by the state and elders.

Padma Yoga: A planetary combination formed by the lords of the 9th house from Ascendant and from Moon situated together in the 7th house from Venus. Individuals born with this stellar configuration are very happy, live in luxury and are engaged in auspicious activities. After the age of fifteen years, they are granted favours by the state and elders.

Pakshin Yoga: A planetary combination produced by all planets in the 4th and 10th houses. It makes the individual a bearer of messages; he may even be an ambassador. He would be quarrelsome and always travelling.

Panch Mahapurush Yoga: Planetary combinations which indicate maturity of the soul. These are formed if any of the five luminaries, namely, Saturn, Jupiter, Mars, Mercury and Venus, possessed of strength occupies its own or exaltation, Kendra or Trikona. These planets produce five kinds of illustrious personages, and the combinations are known as Sasa, Ruchaka, Bhadra, Hamsa, and Malavya yogas formed by Saturn, Mars, Mercury, Jupiter, and Venus, respectively. These combinations induce the individual to liberate himself from involuntary actions and to direct one's conscious efforts.

Panch Mahapurush (Hamsa/Hansa) Yoga: Hamsa Yoga: This Yoga is formed when Jupiter is in the Kendra house of Janma Kundali and is placed in Sagittarius, Pisces or Cancer sign. Jupiter is exalted in Cancer sign while it is in its own sign in Sagittarius or Pisces sign. This Yoga is also known as Hansak or Hamsa or Hansa Yoga. Hansa Yoga is considered very auspicious since Jupiter is a natural Karak planet for wealth. It is believed that lord Vishnu resides in the Kendra house of Kundali. Therefore if a strong Jupiter is in the Kendra, it gives auspicious results throughout your life. It makes the person fortunate, well built and having the voice of a swan. He gets a beautiful wife and possesses all comfort. He is religiously inclined and favourably disposed towards spiritual studies. The combination is said to bestow a life of more than 82 years. You may be beautiful, fortunate and will have a good character because of taking birth in Hans Yoga. You will be fond of good food if you are born in Hans Yoga. Sweet food will fascinate you the most. You will be appreciated by good and influential people due to the auspicious effects of this Yoga. Your work will be noticed by important people. Jupiter is the Karaka planet of intelligence and therefore, you would be good in decision making. You may even be influential in the society. A child born in Hansa Yoga usually has several birthmarks on his arms and legs. These marks might resemble the image of a shell, lotus or fish. He/She is religious, very fortunate. Hamsa yoga could be seen in the horoscope of a learned person or a clergyman.

Panch Mahapurush (Malavya) Yoga: Malavya Yoga is formed when Venus is in the Kendra house of a Janma Kundali and is exalted in Pisces sign or is placed in its own signs, Taurus or Libra. You will remain healthy and fit due to the auspicious effect of this Yoga. The formation of Malavya Yoga involves planet Venus. Shukracharya was believed to be intelligent and a great scholar. Therefore, a person born in this Yoga is intelligent, patient, calm and content. You will be famous and own good vehicles and will be the head of a cultural organization, gives him a life-span of 70 years. The individual dies at a sacred place, practicing yoga and penance and possesses a graceful appearance with the lustre of Moon. He lives happily up to a ripe old age. This Yoga will also make you fond of good food. You will always be interested in trying new delicacies and dishes. Your wife and children will be the reason of your happiness if you are born in Malavya Yoga. Venus is the Karaka planet of material desires in Vedic astrology. It can be associated with lust, greed and fame. Therefore if you are born in this Yoga, you will be inclined towards material desires and may want to show off your assets and wealth.

Panch Mahapurusha (Bhadra) Yoga: Bhadra Panch Mahapurush Yoga is formed when Mercury is in the Kendra house of your Janma Kundali and is placed in Gemini or Virgo sign. Mercury is considered to be a prince of all planets. A prince is always meant to be strong and therefore, you will always be healthy and fit because of this Yoga. Mercury is considered to be a Karaka planet of intelligence. With the presence of Bhadra Yoga, you may be very intelligent and well-mannered. Bhadra Yoga makes you polite and decent due to the effect of planet Mercury which is a prince among all planets. Due to this, you are usually liked by everyone around you. You might also become intelligent, famous, rich and influential in the society due to the effect of this Yoga. You will also inspire others and will always be fortunate in life. In ancient times, it was believed that a person born in Bhadra Yoga lives his life like a king. He could also be a member of the king's court. Hence, a person born in Bhadra Yoga during modern times might get a high post in a government office.

Panch Mahapurusha (Ruchaka) Yoga: This Yoga is formed when Mars is placed in its own or exalted house in Kendra houses (1, 4, 7, and 10) from Lagna or Moon. This is formed when Mars is in Kendra house of your Kundali in his own house or house of exaltation, such as, placed in Aries, Scorpio or Capricorn sign. A person born in this Yoga has a big face. Mars is considered a strong and energetic planet. It is considered a commander of all the planets. Therefore, a person born in this Yoga is strong and courageous since birth. You may also be manipulative, convincing and filled with pride. You will be talented and will work in the field of your interest. Due to a number of different talents, you will be liked by most of the people around you. A person born in this Yoga gives a lot of importance to freedom and independence. Therefore, you will want to set up an independent business. You may even achieve success on doing so. You will be wealthy as well as famous. A person with Ruchaka yoga has a strong physique like that of a warrior. He is known for his valour and intelligence. He is disciplined, traditional, and a conqueror who always abides by customs and laws. He will get good health, wealth, a higher position in defence or police. He has strong physique, well versed in ancient love, conforms to tradition and customs, and becomes famous. Such an individual also becomes wealthy, lives for long, and leads a group of men or an army.

Panch Maha Purusha (Sasa) Yoga: This is one of the Panch Maha Purusha Yogas. This yoga is formed when Saturn is placed in its own or exalted house in Kendra houses (1, 4, 7, and 10) from Lagna or Moon. A person born in this yog is healthy, wealthy, and famous, has many servants and can be a great leader or ruler. It makes the individual command many retinues. His libidinous proclivity is unbridled. He commands over a region. He is psychologically at a point where a radical transformation in his attitude is imminent; disenchantment with sex life could lead him towards spirituality. He may turn out to be a desireless philanthropist, sensuous, occultist, leader of non-traditional and anti-social elements, fearless and capable of performing arduous deeds. But the mentioned yoga gets weak under the following conditions: If sun joins any of the 5 planets; if weak Moon joins the planet making Raj Yoga; if any debilitated planet joins any of the 5 planets; if a planet is positioned with any of the 5 planets in inimical or harmful sign being owner of 6th, 8th and 12th house, and then Panch Mahapurush Yoga won't produce good effect.

Papakartri Yoga: Malefic planets flanking any house or a planet. It destroys the auspicious nature of the same and imparts malefic influence. The house or the planet thus afflicted does not prosper.

Parijata Yoga: A planetary combination related with the position of the Ascendant lord. If the lord of the sign where the Ascendant lord is situated, or if the lord of the navamsa where the lord of the sign in which the Ascendant lord is situated is placed in a cardinal or trine house, Parijata Yoga is formed. It makes the person born under it a sovereign, destined to be happy during the middle or the later part of his life. Such a person is respected by other kings. He is fond of wars, possesses immense wealth, is mindful of his duties towards the state, and is compassionate in disposition.

Parivartana Yoga: This means exchange of mutual houses. Any two lords may exchange their houses. This confers very good results. If the lords of 5 and 10 exchange or 7 and 10 exchange or 5 and 9 exchange very good results can be seen. Even when the lord of 10 is in 9 or 9 th in 10th this yoga is formed partially. Even in cases where the lord of 10th has aspect of the lord of 5 or 9 this yoga can form to an extent.

Parivrajya Yoga: Some important ascetic Yoga are as follows: (1) Four or more planets in strength occupying a single house with Raja Yoga present in the horoscope (2) The lord of Moon sign with no aspect on itself, aspects Saturn or Saturn aspects the lord of the sign occupied by Moon which is also weak (3) Moon occupies Drekkana of Saturn and is expected by it. Such an individual renounces the world and mundane relationships (4) Moon occupies the navamsa of Saturn or Mars, and is expected by Saturn. Such a person is disenchanted with mundane existence and leads the life of a recluse (5) Jupiter, Moon and the Ascendant expected by Saturn, and Jupiter occupying the 9th house in the horoscope make a person born in Raja Yoga a holy and illustrious founder of a system of philosophy (6) Saturn unaccepted by a planet occupies the 9th House and there is Raja Yoga in the horoscope. The combination will make the individual enter a, holy order and become a lord of men.

Parvat Yoga: Planetary combinations of this name are of two kinds. First, benefic in a cardinal house from Ascendant and 6th and 8th houses either posited by benefic or vacant. Second, the Ascendant lord and the 12th house lord both in cardinal houses from each other, and expected by friendly planets. Persons born under these combinations are very fortunate, fond of learning different subjects, charitable and considerate. They become political or social leaders. They, however, have a great attraction for women.

Preshya Yoga: It is formed by, (1) when Sun is in the 10th house, Moon is in the 7th, Saturn in the 4th, and Mars in the 3rd and the Ascendant are in a cardinal sign while Jupiter is placed in the 2nd house. Persons born under this combination during night will be a servant of another person (2) If Venus occupies the 9th house, Moon the 7th house, Mars the 8th, and Jupiter owns the 2nd house or the Ascendant while the Ascendant is in a fixed sign. The person born in this combination lives always in servitude (3) When a person is born during night time and has the lord of the movable rising sign in Sandhi and a malefic planet occupies a cardinal house (4) Jupiter attains Iravathamsa and occupies a Sandhi, and Moon is not situated in a cardinal house but possesses Uttam-Varga and Venus is in the rising sign and birth is in the night time during the dark half of a lunar month. The person is born as a menial (5) if at the time of birth of a person, Mars, Jupiter, and Sun occupy, respectively, the Sandhis of 6th, 4th, and 10th Bhavas, or (6) If Moon while occupying the Amsa (q.v.) of a malefic planet is in a Benefic sign, or (7) When Jupiter is in Capricorn occupying the 6th, 8th, or the 12th Bhava, and Moon is in the 4th Bhava from the rising sign, -the individual born will have to work at the biddings of others.

Raj Rajeshwar Yoga: This yoga is formed when Sun is placed in Jupiter's native sign Pisces and Moon is placed in Cancer. It is the most powerful Rajyoga in astrology. A native of this yoga is endowed with power and self-esteem akin to Sun and a charming and pleasant nature like Moon. This yoga makes a person very creative and intelligent. He earns name and fame in the society. Natives of this yoga benefit from government jobs and earn good money. This yoga makes its natives very wealthy and prosperous. A native of this yoga reaches the top most level in his profession and is rewarded. He has a personality that attracts people from all sections of the society. The natives of Raj Rajeshwar yoga lead an amazing family life, enjoying the sweetest fruits of life.

Budhaditya Yoga: Mercury is the closest planet to Sun. It is usually placed in combination with Sun. The auspicious placement of Mercury and its combination with Sun in a house negates all the unfavourable effects associated with the house. This yoga is also a Rajyoga. A native who has this yoga in his or her Kundali is courageous and energetic like Sun. As Mercury is the significator of intelligence, the native acquires knowledge in different subjects. These people are honoured in the society. They lead a happy and flourishing life.

Raja (Chakra) Yoga: A planetary combination formed by Rahu in the 10th, lord of the 10th house in Ascendant, and Ascendant lord in the 9th house. It makes the individual the administrator of a region. He commands an army and is much respected. Alternatively, all planets in odd houses beginning with Ascendant also produce Chakra Yoga. It bestows high social status to the individual and he will be an emperor, at whose feet will be the prostrating kings, heads, adoring gem studded diadems.

Raja (Dwaja) Yoga: A planetary combination formed by all the malefic placed in the 8th house and all benefices in the Ascendant. Under this combination, a leader is born in this combination.

Raja (Ekwali) Yoga: The planetary combination in which all planets occupy different houses in a sequential manner. It makes the individual an emperor.

Raja Pad Yoga or Dhan Yoga: Some important Raja Yoga is listed below, such as, (1) when more than three planets are in exaltation sign or own sign or placed in Kendra, this yoga is formed; (2) When Lord of 2nd, 9th, 5th, 11th or one of these occupies Kendra from Lagna or Moon, this Yoga is formed; (3) When three or four planets having Digbala, this Yoga is formed; (4) When Kendra lord and Trikona lord make relationship between them, like, 9th lord is in 1st, 4th, 7th and 10th or 5th lord in is 1st, 4th 7th and 10th or 1st lord is in 9th or 5th or 4th lord is in 9th and 5th or 7th lord is in 9th and 5th or 10th lord is in 9th and 5th or Exchange of house between Kendra and Trikona. All Raja yoga and Dhan yoga give name, fame, wealth, all kind of comforts, wealth, and royal status, and authority to the native. Other important Raja Yoga is listed below: (1) Mutual relationship between (a) Karakamsa (q.v.) and ascendant; (b) Atma Karaka and Putra Karaka; (c) Signs occupied in Navamsa by Atma Karaka and Putra Karaka planets; and (d) between Ascendant lord and the lord of the 5th house. Benefices or malefic aspects on these relationships significantly affect the result. (2) Ascendant, 2nd, and 4th houses associated with benefices and the 3rd house occupied by a malefic. (3) The 2nd house occupied by any of the planets, viz., Moon, Jupiter, Venus, or a strong Mercury occupying their own signs. (4) Debilitated planets in 6th, 8th and 3rd houses, while the ascendant lord occupies its own or its exaltation sign in ascendant. (5) Lord of the 10th house while occupying its own or its exaltation sign aspects the ascendant. (6) All the benefices occupy cardinal houses. (7) Debilitated lords of the 6th, 8th, and 12th houses aspect the ascendant. (8) Any relationship between the lords of the 5th and 9th houses. (9) The association of the lords of the 5th, 10th, and 4th houses and ascendant with the lord of the 9th house. (10) Lord of the 5th house in association with the lord of the 9th house or with the ascendant lord in the 1st, 4th, or the 10th house. (11) Venus associated with Jupiter in the 9th house if it happens to be the sign of Sagittarius or Pisces, or with the lord of the 5th house. (12) Moon in the 3rd or 11th house and Venus placed in the 7th house from it. (13) The lords of a Kendra and a Trikona house are in conjunction, mutually aspect or in reception. (14) With an Aries ascendant, the Moon is lord of the fourth house and the Sun is lord of the fifth house. (15) The Sun and the Moon is in conjunction in a horoscope.

Raja Pada Yoga: It is very auspicious combination formed by Moon and Ascendant in Vargottama Navamsa, and four or more planets expecting them. It makes the individual head of a state or its equivalent.

Raja Rajya Yoga: Following Planetary combinations are Rajarajya Yoga, which produce affluence, wealth, and royal status: (1) Mutual relationship between Karakamsa and Ascendant. (2) Ascendant, 2nd, and 4th houses associated with benefic and the 3rd house occupied by a malefic. (3) The 2nd house occupied by any of the planets, Moon, Jupiter, Venus, or strong Mercury occupying their own signs. (4) Debilitated planets in 6th, 8th and 3rd houses, while the Ascendant lord occupies its own or its exaltation sign in Ascendant. (5) Lord of the 10th house while occupying its own or its exaltation sign aspects the Ascendant. (6) All the benefic occupy cardinal houses. (7) Debilitated lords of the 6th, 8th, and 12th houses aspect the Ascendant. (8) Any relationship between the lords of the 5th and 9th houses. (9) The association of the lords of the 5th, 10th, and 4th houses and Ascendant with the lord of the 9th house. (10) Lord of the 5th house in association with the lord of the 9th house or with the Ascendant lord in the 1st, 4th, or the 10th house. (11) Venus associated with Jupiter in the 9th house if it happens to be the sign of Sagittarius or Pisces, or with the lord of the 5th house. (12) Moon in the 3rd or 11th house and Venus placed in the 7th house from it. Many other benefic combinations such as Gaja Kesari Yoga, Pancha Maha Purusha Yoga and Lakshmi Yoga are also important Rajya Yoga.

Raja Rajyapad Yoga: Raja Yoga is calculated from the Karakans Lagna and the natal Lagna. On the one hand the pair of Atma Karaka and 5th Karaka should be considered, other hand the natal Lagna Lord and 5th Lord should be taken into consideration. The effects, due to such association, will be full, or a half, or a quarter, according to their strengths. Maha Raja Yoga: If Lagnas Lord and 5th Lord exchange their Rasi, or, if Atma Karaka and 5th Karaka (Char) are in Lagna, or in 5th House, or in the exaltation Rasi, or in own Rasi, or in own Navamsa, receiving an Aspect from a benefic, Maha Raja Yoga is produced. The native so born will be famous and happy. If Lagnas Lord and Atma Karaka are in 1st, 5th, or 7th House, conjunct with, or receiving an Aspect from a benefic, a Raja Yoga is formed. If there be benefic in the 2nd, the 4th and the 5th counted either from Lagnas Lord, or from Atma Karaka Rasi, one will become a king. Similarly malefic in the 3rd and 6th from Lagnas Lord, or from Atma Karaka Rasi will make one a king. One will be related to royal circles, if Venus is the Karakans, or in the 5th there from, or in Lagna, or in Arudh Lagna, receiving an Aspect from, or conjunct with Jupiter, or Moon. Even, if a single Planet gives an Aspect to the natal Lagna, or Hora Lagna, or Ghatika Lagna, the native will become a king. If the Shad Vargas of Lagna is occupied, or receives an Aspect from one and the same Planet, a Raja Yoga is doubtlessly formed. Accordingly, if the Aspect is full, half, or one fourth, results will be in order full, medium and negligible. If the 3 Lagnas (natal, Hora and Ghatika) are occupied by Planets in exaltation, or in own Rasi, or, if the natal Lagna, the Drekana Lagna and the Navamsa Lagna have exalted Planets, Raja Yoga is formed. If Moon and a benefic are in the Arudh Lang, as Jupiter is in the 2nd from the natal Lagna and both these places are receiving Aspects from Planets in exaltation, or Planets in own Rasi, there will be a Raja Yoga. If Lagna, 2nd and 4th House are occupied by benefic, while a malefic is in 3rd House, one will become a king, or equal to a king. The native will be wealthy, if one among Moon, Jupiter, Venus and Mercury is exalted in 2nd House. If 6th, 8th and 3rd House are occupied by debilitated Planets, as Lagnas Lord is exalted, or is in own House and gives a Aspect to Lagna, there is a Raja Yoga. Again a Raja Yoga is formed, if 6th, 8th and 12th Lords are in fall, or in inimical Rasi, or in combustion, as Lagnas Lord, placed in his own Rasi, or in its exaltation Rasi, gives a Aspect to Lagna. If 10th Lord, placed in his own House, or in its exaltation Rasi, gives an Aspect to Lagna, a Raja Yoga is formed. Similar is the case, if benefic are in Kendra. If the Atma Karaka Planet is in a benefic Rasi/Navamsa, the native will be wealthy. If there are benefic in

Kendra from Karakans Lagna, he will become a king. If the Arudh Lagna and Dar Pad are in mutual Kendra, or in mutual 3rd/11th Houses, or in mutual Kona, the native will doubtlessly become a king. If two or all of House, Hora, and Ghatika Lagnas are receiving an Aspect from exalted Planets, a Raja Yoga is formed. If House, Hora and Ghatika Lagnas, their Drekana and Navamsa, or the said Lagnas and their Navamsa, or the said Lagnas and their Drekana receive an Aspect from a Planet, a Raja Yoga is formed. If Arudh Pad is occupied by an exalted Planet, particularly Moon in exaltation, or by Jupiter and/or Venus (with, or without exaltation), while there is no Argala by a malefic, the native will become a king. If the Arudh Pad is a benefic Rasi, containing Moon, while Jupiter is in 2nd House, the same effect will prevail. Even, if one among 6th, 8th and 12th Lords, being in debilitation, gives an Aspect to Lagna, there will be a Raja Yoga. The native will become a king, if a Planet, ruling 4th, 10th, 2nd, or 11th, gives an Aspect to Lagna, while Venus gives an Aspect to the 11th from Arudh Lagna, as Arudh Lagna is occupied by a benefic. The same effect will be obtained, if a debilitated Planet gives an Aspect to Lagna and is placed in 6th, or 8th House. Again similar result will prevail, if a debilitated Planet, placed in 3rd, or 11th House, gives an Aspect to Lagna. 9th Lord is akin to a minister and more especially 5th Lord. If these two Planets mutually give an Aspect, the native will obtain a kingdom. Even, if these two are conjunct in any House, or, if they happen to be placed in mutually 7th places, one born of royal scion will become a king. The native will attain a kingdom, if 4th Lord is in 10th House and 10th Lord is in 4th House and, if these Planets give an Aspect to 5th and 9th Lords. If the Lords of 5th, 10th, 4th and Lagna are conjunct in 9th House, one will become a ruler with fame, spreading over the four directions. If the Lord of 4th or of 10th House joins either the 5th Lord, or 9th Lord, the native will obtain a kingdom. If 5th Lord is in Lagna, 4th, or 10th House, conjunct with 9th Lord, or Lagna Lord, the native will become a king. Should Jupiter be in his own Rasi identical with 9th House and conjunct with either Venus, or 5th Lord, the native will obtain royal status. Two and a half Ghati from mid-day or from mid-night is auspicious time. A birth during such an auspicious time will cause one to be a king, or equal to him. If Moon and Venus is mutually in 3rd and 11th House and receiving Aspects from each other, while they are placed elsewhere, a Raja Yoga is obtained. If Moon, endowed with strength, is Vargaottama and receives an Aspect from four or more Planets, the native will become a king. One will become a king, if Lagna in Uttamamsa receives an Aspect from four or more Planets, out of which Moon should not be one. If one, or two, or three Planets are in exaltation, one of a royal scion will become a king, while another will be equal to a king, or be wealthy. If four or five Planets occupy their exaltation Rasi, or Moolatrikona Rasi, even a person of base birth will become king. 46. If six Planets are exalted, the native will become emperor and will enjoy various kinds of royal paraphernalia. Even, if one among Jupiter, Venus and Mercury is in exaltation, while a benefic is in a Kendra, the native will become a king, or be equal to him. If all benefic are relegated to Kendra, while malefic are in 3rd, 6th and 11th House, the native, though may be of mean descent, will ascend the throne.

Raja Yoga: It is also formed by, if one among Chandra, Guru, Sukra and Buddha is exalted in Dhan Bhava. If Ari, Randhra and Sahaj Bhava are occupied by debilitated Graha, as Lagna Lord is exalted, or is in own Bhava and gives a Drishti to Lagna, there is a Raja Yoga. Again a Raja Yoga is formed, if Ari, Randhra and Vyaya Lords are in fall, or in inimical Rashi, or in combustion, as Lagna Lord, placed in his own Rashi, or in its exaltation Rashi, gives a Drishti to Lagna. If Karma Lord, placed in his own Bhava, or in its exaltation Rashi, gives a Drishti to Lagna, a Raja Yoga is formed. Similar is the case, if benefic are in Kendra. If the Atma Karaka Graha is in a benefic Rash/Navamsa, he/she will be wealthy. If there are benefic in Kendra from Karakans Lagna, he/she will become a king. If the Arudha Lagna and Dar Pad are in mutual Kendra, or in mutual Sahaj/Labh Bhava, or in mutual Kona, he/she will doubtlessly become a king. If two or all of Bhava, Hora, and Ghati Lagna are receiving a Drishti from exalted Graha, a Raja Yoga is formed. All Raja Pad Yoga gives name, fame, wealth, all kind of comforts and authority and makes him/her head of a state or its equivalent or a king, or equal to a king. He/She will be wealthy,

Raja/Rajya Yoga: Planetary combinations which produce affluence, wealth, and royal status like Gaja Kesari Yoga, Pancha Maha Purusha Yoga and Lakshmi Yoga. Some important Rajya Yogas are listed below: (1) Mutual relationship between Karakamsa and Ascendant. (2) Ascendant, 2nd, and 4th houses associated with benefic and the 3rd house occupied by a malefic. (3) The 2nd house occupied by any of the planets, Moon, Jupiter, Venus, or strong Mercury occupying their own signs. (4) Debilitated planets in 6th, 8th and 3rd houses, while the Ascendant lord occupies its own or its exaltation sign in Ascendant. (5) Lord of the 10th house while occupying its own or its exaltation sign aspects the Ascendant. (6) All the benefic occupy cardinal houses. (7) Debilitated lords of the 6th,

8th, and 12th houses aspect the Ascendant. (8) Any relationship between the lords of the 5th and 9th houses. (9) The association of the lords of the 5th, 10th, and 4th houses and Ascendant with the lord of the 9th house. (10) Lord of the 5th house in association with the lord of the 9th house or with the Ascendant lord in the 1st, 4th, or the 10th house. (11) Venus associated with Jupiter in the 9th house if it happens to be the sign of Sagittarius or Pisces, or with the lord of the 5th house. (12) Moon is in the 3rd or 11th house, and Venus placed in the 7th house from it. Many other benefic combinations such as Gaja Kesari Yoga, Pancha Maha Purusha Yoga and Lakshmi Yoga are also important Raja/Rajya Yoga.

RajaYoga: This is a shubha ('auspicious') yoga that gives success and a grand rise in career or business, and a greater degree of financial prosperity particularly during the dasha of the planets that give rise to Raja yogas. Raja yoga-causing planets during the course of their respective dashas confer their most auspicious results if they happen to own the lagna-bhava (the Ascendant) or the Putra-bhava (the 5th house) or the Bhagyasthana (the 9th house); the person remains healthy, wealthy, happy and successful enjoying yoga and Raja yoga results in case the lagna, the 3rd, the 6th, the 8th, the 9th and the 12th houses counted from the lagna are also not occupied by any planet, and the Kendra's (quadrants) are occupied only by benefic planets. Raja yoga is formed owing to a lord of a Kendra (the lord of the 1st, the 4th, the 7th or the 10th counted from the Lagna or Natal Moon and a lord of a Trikona (1st, 5th or the 9th house) establish mutual relationship. It becomes more pronounced if the lord of another Trikona joins them or if their dispositors, preferably, the lord of the Ascendant, is in its exaltation in a Kendra or a Trikona. For any yoga or Raja yoga to produce more effective results the yoga-causing planets possessing requisite six kinds of strength (Shadbala), must form an immediate relationship with the Lagna ('Ascendant'), which is possible by any one of them occupying or aspecting the lagna or by directly associating with the lord of the lagna but without any one of them being afflicted by either natural or functional malefic, or by a lord of a trikasthana (6th, the 8th or the 12th house). However, these results get adversely modified by the presence of other Ashubha ('inauspicious') Arista yogas. Mihira states that the results of powerless planets are enjoyed in dreams and thoughts only. Moreover, if any planet occupying a particular sign as part of a yoga formation happens to be aspect by the lord of that sign and both occupy auspicious houses then alone Raja yoga is formed. If the lord of the 9th or the lord of the 10th respectively own the 8th and the 11th, their association will not give rise to an effective yoga or Raja yoga or if they do not conjoin either in the 9th or 10th house. According to Parasara, the most powerful Raja yoga arises when the strong lord of the lagna is in the 5th house and the strong lord of the 5th house occupies the lagna-Kendra or if the Atmakaraka ('the planet most advanced in the sign') and the Putrakaraka (chara karaka) are jointly or severally in the lagna or in the 5th house or occupy their exaltation or own sign or Navamsa in aspect to a benefic planet and adds that one will be a king if benefic planets occupy the Kendra from the Karakamsha ('the Navamsa occupied by the Atmakaraka') or if the Arudha lagna and the Darapada are in mutual Kendra or trikonas or in the 3rd and the 11th from each other or if the lord of the 10th house placed in its own or exaltation sign aspects the lagna or if the lagna is aspect by the debilitated lord of the 6th, the 8th or the 12th house. If the dispositors of Gulika (Mandi) is in a Kendra or a Trikona vested with requisite strength in own or exaltation or friendly sign then one possesses a pleasing personality, is popular and famous and enjoys the benefits of Raja yoga, he becomes a powerful ruler.

Rajju Yoga: All the Graha in Movable Rashi cause Rajju Yoga. He/She born in Rajju Yoga will be fond of wandering, be charming, will earn and settle in foreign countries. He/She will be cruel and mischievous.

Rasatala Yoga: It is a planetary combination constituted by the lord of the 12th house in exaltation, and Venus posited in the 12th house and aspect by the lord of the 4th house. Persons born under it attain the status of head of state. They may find wealth buried under the earth.

Ravi Yoga: A combination of planets formed by Sun in the 10th house and the lord of the 10th house in 3rd house along with Saturn. It makes the individual born with this combination a Scientist who attains a powerful status in administration. He eats very little, is much occupied with his studies, and is greatly respected.

Rekha Yoga: It is a planetary combination leading to poverty. It arises when a weak lord of Ascendant is aspect by the lord of the 8th house, and Jupiter is combust by Sun. Alternatively, if the lord of Navamsa occupied by the lord of the 4th house is obscured by Sun while Sun it is aspect by the lord of the 12th house.

Royal Association Yoga: 'Karakendra' is the dispositor of Atma Karaka. Similarly 'Amatyesa' means the dispositor of Amatya Karaka. Royal Association Yoga is formed, (1) If Karma Lord is yuti with, or receives a Drishti from the dispositor of Amatya Karaka, or even, if

Karma Lord is yuti with, or receives a Drishti from Amatya Karaka himself, he/she will be a chief in the king's court. (2) If Karma and Labh Bhava are devoid of malefic occupation and devoid of Drishti from a malefic, while Labh Bhava receives a Drishti from its own Lord, he/she will be a chief in the king's court. (3) If Amatya Karaka and the dispositor of Atma Karaka be together, he/she will be endowed with great intelligence and will be a king's minister. (4) If Atma Karaka is strong and is with a benefic, or Amatya Karaka is in its own Bhava, or in exaltation, he/she will surely become a king's minister. (5) There is no doubt in his/her becoming a king's minister and famous, if Atma Karaka is in Thanu, or Putra, or Dharma Bhava. (6) If Atma Karaka or Amatya Karaka is placed in a Kendra, or in a Kona, he/she will beget royal mercy, royal patronage and happiness thereof. (7) If malefic be in the 3rd and the 6th from Atma Karaka, or from Arudha Lagna, or in Sahaj and Ari Bhava, he/she will become Army chief. (8) If Atma Karaka is in a Kendra, or in a Kona, or in exaltation, or in its own Bhava and gives a Drishti to Dharma Lord, he/she will be a king's minister. (9) If the Lord of the Rashi, where Chandra is placed becomes Atma Karaka and, if this Lord is placed in Thanu Bhava along with a benefic, he/she will become a king's minister at his advanced age. (10) If the Atma Karaka be in Putra, Yuvati, Karma, or Dharma Bhava and happen to be with a benefic, he/she will earn wealth through royal patronage. (11) If the Arudha of Dharma Bhava happens to be itself the Janma Lagna, or, if Atma Karaka is placed in Dharma Bhava, he/she will be associated with royal circles. (12) He/She will gain through royal association, if Labh Bhava is occupied by its own Lord and is devoid of a Drishti from a malefic. The Atma Karaka should at the same time be yuti with a benefic. (13) An exchange of Rashi between Karma Lord and Lagna Lord will make him/her associated with the king in a great manner. (14) If Sukra and Chandra are in the 4th from Karakans Lagna, he/she will be endowed with royal insignia. (15) If Lagna Lord or the Atma Karaka be yuti with Putra Lord in a Kendra, or in a Kona, he/she will be a king, or minister.

Sakata Yoga: It is an inauspicious planetary combination described variously in classical texts. It is formed by all the planets in the 1st and 7th houses which make the individual accept low professions. He/She born in Sakata Yoga will be afflicted by diseases, will have diseased or ugly nails, be foolish, will live by pulling carts, be poor and devoid of friends and relatives. It also occurs when Jupiter occupies the 6th or 8th position from Moon posited in a house other than the cardinal houses in relation with the Ascendant. It brings poverty even to those born in a royal family. Such a person is troubled throughout his life and is disliked by the head of the state. Moon in the 12th, 8th or 6th house from Jupiter causes Sakata Yoga unless Moon is situated in a cardinal house. A person with this combination loses his wealth or position in life, but regains them. Sakata Yoga produces cyclic fluctuation in fortune, just like the wheel of a chariot, rotating on its axis.

Samrajya Yoga: A planetary combination formed by the lord of the Navamsa sign of the lord of the 9th house, along with Jupiter in the 2nd house. It makes the individual a top-ranking administrative officer living in luxury.

Samudra Yoga: Samudra Yoga is produced by all planets in even houses such as the 2nd, 4th, 6th, etc., while odd houses, such as the 1st, 3rd, etc., are vacant. He/She will have many precious stones and abundant wealth, be endowed with pleasures, dear to people, will have firm wealth and be well disposed. It makes the individual a top-ranking administrative officer living in luxury. It bestows much renown on the individual and he is provided with all conveniences of life.

Sankhya Yoga: There are 7 kind of Sankhya Yoga, such as: **i) Gola Yoga:** If all Graha are in one Rashi, Gola Yoga is formed. He/She born in Gola Yoga will be strong, be devoid of wealth, learning and intelligence, and be dirty, sorrowful and miserable. **ii) Yuga Yoga:** If all Graha are in 2 Rashi, Yuga Yoga is formed. One born in Yuga Yoga will be heretic, be devoid of wealth, be discarded by others and be devoid of sons, mother and virtues. **iii) Sool Yoga:** If all Graha are in 3 Rashi, Sool Yoga occurs. He/She born in Sool Yoga will be sharp, indolent, bereft of wealth, be tortuous, prohibited, valiant and famous through war. **iv) Kedara Yoga:** If all Graha are in 4 Rashi, Kedara Yoga occurs. He/She born in Kedara Yoga will be useful to many, be an agriculturist, and be truthful, happy, fickle-minded and wealthy. **v) Pasha Yoga:** If all Graha are in 5 Rashi, Pasha Yoga is formed. He/She born in Pasha Yoga will be liable to be imprisoned, be skilful in work, and be deceiving in disposition, will talk much, be bereft of good qualities and will have many servants. **vi) Daam Yoga:** If all Graha are in 6 Rashi, Daam Yoga occurs. **vii) Veena Yoga:** If all Graha are in 7 Rashi, Veena Yoga is produced. He/She will be fond of songs, dance and musical instruments, be skilful, happy, and wealthy and be a leader of men.

Saraswati yoga: When Jupiter, Venus and Mercury are placed in Kendra houses, in a friendly sign, or even better, in the second house, which indicate writer, learned person, speaker, and intellectual. When they are in a favourable position, Jupiter, Venus and Mercury can provide wisdom. In Gandhiji horoscope, Jupiter, Venus and Mercury are placed in Kendra houses. It is true that Jupiter and Mercury are placed in, for them, neutral signs, but as all the named planets are either in conjunction with each other, or aspect benefices, the Saraswati yoga here is extremely powerful.

Sarpa Yoga: Only malefic are placed in the Kendra houses, which indicate misfortune, a difficult life.

Shakata Yoga: This is when Jupiter is in the 6th, 8th, or 12th house from the Moon. Shakata means "cart", or wheel. His/Her the fortunes will raise and fall. It means fluctuating fortunes. If there is either the Adhi or Basmati Yoga, then the Shakata Yoga is cancelled. Or, When the Moon is placed in the sixth, eighth or twelfth house, calculated from Jupiter, which indicate fluctuating fortunes. This is more or less the opposite of the Gaja Kesari yoga. When the Moon is placed in a Dustasthana house calculated from Jupiter, we have a difficult relationship between these two planets.

Shakti Yoga: A planetary combination formed by all planets situated in the 7th, 8th, 9th, and 10th houses. It makes the individual lazy and devoid of wealth and happiness but he acquires great skill in arguing for criminal litigants. Or, If all the 7 Graha are in the 4 Bhava, commencing from Yuvati, Shakti Yoga occurs. He/She born in Shakti Yoga will be bereft of wealth, be unsuccessful, miserable, mean, lazy, long lived, interested and skilful in war, firm and auspicious.

Shankh Yoga: If Lagna Lord is strong, while the Lords of Putra and Ari Bhava is in mutual Kendra, then Shankh Yoga is produced. Alternatively, if Lagna Lord along with Karma Lord is in a Movable Rashi, while Dharma Lord is strong, Shankh Yoga is obtained. He/She born with Shankh Yoga will be endowed with wealth, spouse and sons. He/She will be kindly disposed, propitious, intelligent, meritorious and long-lived.

Shankha Yoga: A planetary combination formed in 2 ways,(i) the lords of the 5th and 6th houses in cardinal houses from one another, while the Ascendant is strong, and (ii) the lords of the ascendant and the 10th house placed in movable signs while the lord of the 9th house is strong. These combinations make the individual born under them, well versed in scriptures, a man of principles and ethics, and engaged in laudable activities. Such individuals have a long life.

Shar Yoga: A planetary combination formed by the placement of all planets in the 4th, 5th, 6th and 7th houses. It makes the individual born under it cruel and related with prisons. If all the 7 Graha are in the 4 Bhava, commencing from Bandhu, Shar Yoga occurs. He/She born in Shar Yoga will make arrows, be head of a prison, will earn through animals, will eat meat, will indulge in torture and mean handiworks.

Sharda Yoga: A planetary combination formed in two ways, (i) the lord of the 10th house posited in the 5th house, Mercury placed in a cardinal house, and Sun either in its own sign or in a very strong position, and (ii) Jupiter situated in a trine house from Moon, and Mars in a trine house from Mercury. Under these combinations, an individual becomes well behaved, dutiful, and God-fearing and is honoured by the state and will obtain wealth, spouse and sons, be happy, scholarly, dear to the king, pious and virtuous.

Shashya Panch Mahapurush Yoga: If Saturn is in its own sign or in exaltation and in a Kendra house or Shashya Yoga is formed when Saturn is in the Kendra house of a Janma Kundali and is placed in Capricorn, Aquarius or Libra sign. You may achieve success in your life with the auspicious effects of Shash Yoga. You will be strong, rich, happy and influential. You might get attracted to people of other gender with the effect of this Yoga. Men can get attracted to women and vice versa and these relationships may result in marriage. The internal development of an individual can be faster in Shash Yoga as it is formed with the effect of Saturn. You are likely to get powers due to which you may get involved in the company of corrupt people. You should be cautious while exercising your authority and should not use them for illegal purpose. He/She is powerful, strict, position of authority. Shasya yoga is fitting for the horoscope of a politician or a director.

Shrinath Yoga: If Yuvati Lord is in Karma Bhava, while Karma Lord is exalted and yuti with 9th Lord, Shrinath Yoga takes place. He/She will be equal to Lord Devendra.

Shringatak & Hal Yoga: All Graha in Lagna, Putra and Dharma Bhava cause Shringatak Yoga, while all Graha in Dhan, Ari and Karma Bhava, or in Sahaj, Yuvati and Labh Bhava, or in Bandhu, Randhra and Vyaya Bhava cause Hal Yoga. He/She born in Shringatak Yoga will be fond of quarrels and battles, be happy, dear to king, endowed with an auspicious wife, be rich and will hate women.

Srik Yoga: This occurs when only natural benefice placed in the Kendra houses indicate comfort, good luck and abundance.

Subha Kartari Yoga: If the enclosure of a planet or a house is by benefice, we call it Subha Kartari yoga. **Example:** If Venus is in Taurus, Jupiter (a benefice) in Aries, and the Moon (a benefice) in Gemini, then Venus will function exceptionally well. This is Subha

Kartari Yoga. A house can also be enclosed without any planet in that house.

Sunapha Yoga: Any planet excluding the Sun, Rahu and Ketu in the house second from the Moon create Sunapha Yoga. This indicates richness and financial benefits; acquisition and the planet itself will describe the manner in which things will be acquired. For example, Mars in the 2nd house from the Moon is Sunapha Yoga indicating an aggressive manner in acquiring money, or possessions. This Yoga is about receiving. The effects of this are stronger if benefice are situated in the second house calculated from the Moon. Should there be one or more malefic in this second house from the Moon, the effect will be limited, or there will be no effect at all.

The 10th house is to be examined regarding native's professional and financial opportunities as it rules livelihood. If the 10th house and its lord and the significators (Sun, Jupiter, Saturn and Mercury) are strong, the native occupy a respectable occupation which helps in improving the financial status. There is a large number of Yoga which has been described in the various Astrological texts. Even the Panch maha Purush Yoga bestow special characteristic.

The 2nd house itself, lord and Jupiter represent accumulated wealth. The association of the 1st, its lord and the sun with the second and its lord increases the native's wealth through his own efforts. If the 9th, its lord and the significators Sun and Jupiter are related to them, wealth is inherited. According to the Bhavarth Ratnakar, the native is rendered poor if the 2nd lord occupy the 12th and 12th lord is placed in the Lagna aspect by a Maraka or a malefic planet.

The 9th house represents fortunes and, therefore, if the 9th house, its lord, Sun and Jupiter are strong and associated with good houses, planets, luck favours the native with financial prosperity. When the 9th lord is placed in the 1st house of Lagna lord is in 9th house, the native is a self made person.

Trimurthi Yoga: There are 3 Yoga in Trimurthi Yoga. Counted from Dhan Lord, if benefic occupy the 2nd, 12th and 8th, Hari Yoga is formed. If the 4th, 9th and 8th with reference to the Rashi, occupied by Yuvati Lord, are occupied by benefic, Hara Yoga is obtained. Brahma Yoga is generated, if, counted from Lagna Lord, benefic are in the 4th, 10th and 11th Rashi. He/She, born in anyone of the above Yoga, will be happy, learned and endowed with wealth and sons.

Ubhayachari yoga: When there is a planet, other than the Moon, Rahu and Ketu, in both the second and the twelfth house from the Sun, it forms Ubhayachari yoga. This yoga empowers the planets that surround the Sun. It is like the king is supported. This yoga gives prominence, power and wealth. When the planets that surround the Sun are benefices it is far more beneficial whereas malefic give effects in a more stressful manner. This brings good fortune in many areas, but again only if it involves benefice. Malefic will create trouble. When a person has both a Veshi and Voshi yoga, he actually has Ubhayachari yoga. It would be logical to expect yoga when there are no planets in the second or the twelfth house calculated from the Sun. This is not the case. The Sun is in itself; a powerful planet and it can function very well alone, without the support of other planets. A benefice in the vicinity is always appreciated, but it is not necessary. Several of the aforementioned yoga will appear in everyone's horoscope. Because they appear so often, they are yoga that can only lead to sizeable results, when they are related to other indications.

Vajra Yoga: Vajra Yoga is caused by all benefic in Lagna and Yuvati Bhava, or all malefic in Bandhu and Karma Bhava. He/She will be happy in the beginning and at the end of life, be valorous, charming, and devoid of desires and fortunes and be inimical.

Vapi Yoga: If all of them happen to be in all the Apoklima, or in all the Panapharas, Vapi Yoga occurs.

Vasi Yoga: Vasi yoga is formed when the 12th house from the Sun is occupied by any planet other than Moon. This yoga provides auspicious or inauspicious results depending upon the planet placed in the 12th house. The placement of benefic planet in this house makes its natives intelligent and virtuous. A native is skilled and wise and develops the art of communication under the auspicious Vasi yoga. It blesses a happy and successful family life to its natives. On the other side, a malefic planet in this house results in weak memory and selfish nature. A native may not have a sympathetic attitude towards other people and may have to live away from his or her family and thus face several difficulties in life.

Vasumati Yoga: This is formed when two out of three benefic, Venus, Jupiter and Mercury, provided Mercury is not associated with any malefic, occupy the Upachaya houses (3, 6, 10, or 11) from the Moon or Ascendant. This yoga gives wealth, and life becomes very prosperous improving with age. Any benefic in Upachaya houses will produce the effect of wealth. Two benefic will be good and one will be moderately good. From the Moon it is definitely more powerful since the Moon represents growth. It produces possibilities for the individual to become a billionaire. Example: Kennedy had Jupiter and Venus in the tenth house calculated from the Moon.

Veshi Yoga: The Veshi yoga is formed when there is any planet except Moon in the 2nd house from the Sun. The result of this yoga depends upon the placement of auspicious or inauspicious planet in the house. A benefic planet in this house makes the personality of a native very beautiful and attractive. She becomes a good orator and an efficient leader. They are active social workers. Inauspicious Veshi yoga is made up of malefic planets. A person who has this malefic yoga may face many difficulties, especially in his or her business or professional life. A native's financial condition deteriorates and he may face rejection in his professional life. In this case Venus 2nd from the Sun (evening star) this person would be more careful and deliberate in matters concerning love. Persons born under Vesi Yoga are truthful, lazy, unbiased and rich. This yoga gives riches and status, at least when this yoga is formed by benefice. If it is formed by malefic it can lead to all kinds of problems.

Vibhasu ayoga: A planetary combination formed by Mars either exalted or placed in the 10th house, exalted Sun in the 2nd house, and Moon along with Jupiter in the 9th house. It enables the individual born under the combination to get an adorable wife and lead a happy personal life. He will be rich and will occupy a high status in life.

Vibhavasu Yoga: A planetary combination formed by Mars either exalted or placed in the 10th house, exalted Sun in the 2nd house, and Moon along with Jupiter in the 9th house. It enables the individual born under the combination to get an adorable wife and lead a happy personal life. He will be rich and will occupy a high status in life.

Vidyuta Yoga: A planetary combination formed by the lord of 11th house in exaltation along with Venus occupying a cardinal house in relation with the house occupied by the Ascendant lord. It makes the individual charitable, affluent, and enables him to occupy a high status in life.

Vihag Yoga: If all planets confine to Bandhu and Karma Bhava, then Vihag Yoga occurs. He/She born in Vihag Yoga will be fond of roaming, be a messenger, will live by sexual dealings, and be shameless and interested in quarrels.

Vipareeta Raja (Harsha) Yoga (Maha Saint Yoga): If the lord of 6th or 8th or 12th is in 6th house, it forms and causes Harsha Yoga. If the lord of 6th or 8th or 12th is in 6th Bhava in the enemy house, it forms and causes a strong Harsha Yoga or Vipareeta (Viparita) Raja Yoga. The native is fortunate, happy, invincible, physically strong and sturdy, wealthy, famous and afraid of vices and sins. This is the natal chart of the great sage, and spiritual giant like Sri Ramanjacharya, or Shri Stya Sai Baba. Saturn is the lord of 8th house (Aquaius); he is in 6th house (enemy). This causes a powerful Harsha yoga. So he enjoyed all the above benefits all his life. The lord of 6th house (Sagittariu) is Jupiter; he is in 12th house (Vyaya; loss and expenses). This causes a strong Vimala yoga, so he enjoyed its benefits.

Vipareeta Raja (Sarala) Yoga (Big Celebrity Yoga): The lord of 6th, or 8th or 12th is in 8th house causes Sarla yoga. Benefits: This yoga makes the person long-lived, fearless, happy, invincible, learned, a terror to his enemies, wealthy, famous, Celebrity and prosperous. Example : This is the natal chart of the Swami Vidyaranya, who was mainly instrumental to start the famous Vijayanagara Kingdom at Hampi which stood as bulwark for Hindu religion, culture, social and political revival against the invading Muslim and Bahamani Sultans in south India for about two and half centuries. Moon is waxing; he is the lord of 8th house (Cancer); He is in his own house in 8th house, forming strong Sarala yoga. He enjoyed all the benefits of this yoga in life.

Vipareeta Raja (Vimala) Yoga (Empire Yoga): The lord of 6th, or 8th or 12th is in 12th house causes Vimala yoga. Benefits: This yoga renders the person frugal, happy, independent, with noble qualities and wealthy. Example: This is the natal chart of the Queen Victoria, Empress of the huge British Empire (famed as where the Sun never sets). Venus is the lord of 6th house (Libra); he is in 12th house (Vyaya; loss and expenses). This caused Vimala yoga. She enjoyed its benefits all her life. This Vipareetha Raja yoga blessed her to accede to the British throne unexpectedly. Lady Victoria became heiress-apparent of the British crown on the death of her uncle in 1837; she became Queen of Great Britain. She was crowned the next year. So she gained the position of the Empress by the death of his childless uncle, (unfortunate event). This is an excellent illustration of Vipareetha Raja yoga.

Vishnu Yoga: A combination formed by the lords of the 9th and 10th houses as well as the lord of the Navamsa sign lord of the 9th posited in the 2nd house. It makes the individual favoured in an important way by the state. By nature, he is patient, is erudite, skilled in debating and is an engaging conversationalist. He becomes rich and lives for long.

Voshi yoga: If benefic planets, other then the Moon, Rahu or Ketu, are placed in the twelfth house calculated from the Sun, he/she is spiritually developed and intelligence. A malefic in the twelfth house will give many problems.

Wealth and Prosperity Yoga: The 2nd and 11th houses are mainly responsible for finance. Lagna, 9th and 5th houses also play important roles. Even 10th house, Jupiter and Moon are important for finance. The person is a pauper ever since his birth and lives by alms if the lords of the 2nd and 11th houses are placed in 6, 8 or 12 along with malefic. The association or affliction of 6th, 8th or 12 houses destroy the ambitious effects of wealth or dampen the financial prospects. The lord of 2nd with lord of 9th occupying the 11th with Moon and Jupiter will make a native a millionaire. Lord of Lagna in a decent position is itself a great asset which sustains a person throughout his life. The best place for Lagna Lord to occupy is the Lagna itself. Even strong Sun at a benefic place from the lord of Ascendant is a great asset.

Wealth Yoga: Lord of 10th house occupying 5th house will prove a great asset for a native bestowing abundant wealth which will never leave him. Laxmi Yoga: If lord of Lagna is powerful and lord of 9th house occupies own or exaltation position in Kendra or Trikona. One who has strong Sun in Horoscope can never is mean, vicious, criminal or untrustworthy. Therefore, he can not adopt wrong means for livelihood. If lord of 2nd and 11th be placed or related with 6th/8th/12th while Mars is in 11th and Rahu is in 2nd, native will loose his wealth on account of royal punishment. If Lagna, the lord and Sun are weak or afflicted and also have the association with Rahu or a malefic, the very foundation of Horoscope is deeply shaken. The native suffers setbacks in his financial matters owing to wrong decisions.

Yav Yoga: If all benefic in Bandhu and Karma Bhava, or all malefic in Lagna and Yuvati Bhava, Yav Yoga is generated. He/She will observe fasts, religious rules, auspicious acts, and will obtain much happiness, wealth, sons in his/her mid-life. He/She will be charitable and firm. It makes the individual courageous and his mid-span.

Yugma Yoga: A planetary combination formed by the lord of the 4th in 9th along with some benefic and Jupiter aspecting it. He/She receives valuable gifts from the state and from persons in authority, and leads a happy and affluent life.

YUP YOGA: A planetary combination formed by all planets in Ascendant, 2nd, 3rd, and 4th houses.

Yup Yoga: A planetary combination formed by all planets in Ascendant, 2nd, 3rd, and 4th houses. He/She will have spiritual knowledge and will be interested in sacrificial rites. He/She will be endowed with a wife, be strong, interested in fasts and other religious observations and be distinguished. It makes the individual religious, charitable, generous, and he performs many important rites.

Miscellenius other Yoga

11) Mars in 4th Sun in 5th Or Jupiter in 11th or 5th, earning will be through ancestral properties, crops, or building, which will increase. 12) Saturn in 4th identical with Libra, Capricorn or Aquarius the native will earn through Numbers (accountants, statisticians, mathematician etc) 13) If Mars, Jupiter and Moon join Cancer for a cancer native he becomes very rich through employment resource and divine grace. 14) If Rahu, Saturn, or Mars or Sun in the 11th the native becomes rich gradually. 15) If quadrant or trinal houses are occupied by Jupiter, Venus, Moon, Mercury or houses 3, 6, 11 by Sun, Rahu, Mars, Saturn the native will become exceedingly rich during Dasa of Rahu/mercury/Saturn /Venus. 16) If all Kendra are occupied by planets OR All trines are occupied by benefic OR All "Upachayas" (3,6, 11) occupied by Malefic The native becomes exceedingly rich. 7) Benefic planets or 10th lord in Taurus or Libra and Venus or lord of 7th in 10th the native will become rich by marriage or through wife's earnings. 18) Jupiter in Cancer, Sagittarius, or Pisces, and 5th lord in 10th the person becomes rich through son/daughter. 18) Mercury + Jupiter + Venus in any house, earning by religious means (purohit, Pundit, Astrologer, Preacher, Head of religious institutions etc). 19) Mercury +Venus + Saturn in any House, the person will earn by business leadership 20). Jupiter in 10/11 and Sun/Mercury in 4/5 or vice versa, native becomes rich by Good administrative manipulations 21) Lords of 6,8,12 join 6, 8, or 12 or 11th the native becomes suddenly rich. 11th house is the house of gains and rules income. The 11th and its lord along with Jupiter must be strong to generate wealth. According to the Sarvartha Chintamani, if a benefic occupy 11th house, the native gets wealth through honest and novel means. If the malefic is in the 11th house, the native resorts to unfair and unscrupulous methods of earning. The source of gains and incomes can be known through the nature of the planet from the Bhavas related to the 11th, its lord and its significators. For example 7th lord and its Karaka Venus associated with 11th may help the native to gain through the spouse.

16

Jaimini Astrology

16.1 Predictions by Jaimini Astrology:

Jaimini Karaka: Jaimini was the student of the great sage Parasara, who wrote the Vedic astrology, known as the Brihat Parasara Hora Shashtra. It was the sage Jaimini who fully developed these basic principles outlined by Parasara in his Sutra, which is called "Jaimini Sutra" or "Jaimini astrology". It is basically a subset of the astrology by sage Parasara. Sage Jaimini emphasises on the use of Variable Karakas or Chara Karakas along with the Fixed Karakas. Jaimini makes use of an entirely different set of Karaka than regular Parasara astrology Karaka. In Parasara astrology the Karakas of each house are fixed, which represents certain signification (Karkatva), in the chart. This is not the case in Jaimini astrology. Jaimini astrology determines the Karaka of the event or house based on the degrees of the planet in any sign attained by each of the 7 Planets in a Natal Chart of the individual.

Jaimini Variable karaka: Variable karakas are 1 Atmakarka 2. Amatyakarka. 3. Bharatrikarka 4. Matrukarka. 5. Pitru Karka 6. Gnati Karka 7. Dara Karka.

In Prashari Astrology the Sun, apart from being karka for father, is Atmakarka for each and every horoscope but according to Jaimini the Sun may be Atmakaraka in one horoscope and may be Daiv karaka in the other horoscope or any other Karaka in the third horoscope. Hence each and every Karaka is changeable from horoscope to horoscope. So it is called Variable Karaka. With the following sutras the sage defines the Variable Karaka starting with Atmakarka. Atmakarka, i.e. the 1st Karaka or "soul-indicator" is a very important planet. In Parasara astrology, the Sun always indicates the soul, but in Jaimini astrology, any planet attains Atma Karkatva position depending on his highest degree in a particular Sign of the Natal Chart. This is a very flexible system, compared to the Parasara

Karaka. It is this variable Karaka system that constitutes the first major difference between Jaimini astrology and Parasara astrology.

Fixed Karaka: Moon is a fixed karaka for mother, 4th House, but as per Parashari rules it becomes Atmakarka. Mars is a fixed karaka for brothers, 5th house, but it becomes Amatya karka. Jupiter is fixed Karaka for Putra (son), 5th house, but it becomes Bharatri karka, as per jaimini system. The Seven Bhava in Horoscope is the Karaka as given here. 1st Bhava is Atmakaraka; 2nd Bhava is Amatyakaraka; 3rd Bhava is Bharatrukaraka; 4th Bhava is Matru and Putra Karaka, 5th Bhava is Pitrukaraka; 6th Bhava is Gnatikaraka and 7th Bhava is Darakaraka (Spousekaraka)

1. Atmakarka: The Planet of the highest degree in the Natal chart represents Atmakarka, i.e. the 1st Karaka. Atma Karaka is the most important and has a prime say on the native, just as the king is the most famous among the men of his country and is the head of all affairs and is entitled to arrest and release men and the minister cannot go against the king, the other Karaka, viz. Putra Karka, Amatya Karka etc. cannot predominate over Atma Karka in the affairs of the native. If the Atma Karka is adverse, other Karaka cannot give their benefic effects. Similarly, if Atma Karka is favourable, other Karaka cannot predominate with their malefic influences. This indicates for the 1st house matters. It is considered very favourable, if the Atmakarka is in a Kendra, conjunct, square, or opposed to the Amatyakaraka. When the Atmakarka is in exaltation or in beneficial Rasi or conjunctions to benefic; though the person is imprisoned he/she will be liberated, will live in holy places and will have Moksha or Final Emancipation and ardently coveted by all devotees and the yogis. But when it is in Neecha Rasi or with evil conjunctions and aspects, he/she will be imprisoned, will suffer from chains and other tortures, and will not have Moksha. But if the debilitated Atmakarka has beneficial aspects or conjunctions, he/she will be liberated. The idea seems to be the securing of final salvation. Here Bandhana and Moksha may be interpreted as malefic and benefic results.

Here the sage is strict on choosing the seven karakas from seven planets i.e. from Sun to Saturn. Since the Sutra does not state about Rahu & Ketu, they cannot be considered under any circumstances as any Karaka. Atmakarka is the supreme lord of the planets. If a benefic planet becomes Atmakarka and it is completely free from malefic influences and well placed in the horoscope, it is capable of final annunciation of soul or the Moksha. In case the Atmakarka is a benefic planet but it is under the influence of malefic, the native performs both good and evil deeds. Next to Amatyakarka it is

Bharatrikarka. Next to Bharatrikarka it is Matrikarka. Next to Matrikarka it is Putrakarka. Next to Putrakarka it is Gnatikarka. Next to Gnatikaraka it is Darakarka.

Example 1: DOB: 4.9.1986, TOB: 09.35 AM IST, POB: Delhi, Longitude: 77E13, Latitude: 28N40. Longitude of All the planets:
Sun 12 S 19°44'; Moon 10 S 26°27'; Mars 8 S 22°33'; Mercury 12 S 00°25'; Jupiter 10 S 22°16'; Venus 1 S 15°18'; Saturn(R) 7 S 14°26'; Rahu 12 S 06°18' & Ketu 6 S 06°18'.

Since, the planet with highest longitudes, irrespective of sign, is known as Atmakarka, Planet with highest longitudes is the Moon 10 S 26°27'. Here the longitude of Moon is 26°27". Hence the Moon becomes Atmakarka.

Next comes Mars with Longitudes 8 S 22°33', i. e. 22°33', it becomes Amatyakarka and so on as per table below:

Planet	Longitude	Jamini/Char Karaka
Moon	10 S 26^0 27'	Atmakaraka
Mars	8 S 22^0 33'	Amatyakaraka
Jupiter	10 S 22^0 16'	Bharatrukaraka
Sun	12 S 19^0 44'	Matrukaraka
Venus	1 S 15^0 18'	Putrakaraka
Saturn (R)	7 S 14^0 26'	Gnatikaraka
Mercury	12 S 00^0 25'	Darakaraka

When two planets attain identical number of degrees, in such a case the planet with natural strength qualifies to become that karaka while the second becomes the Next karaka.

Example 2: Female Native, DOB: 16.7.1975, TOB: 17.05 IST, POB: Moga (Punjab), Longitude: 75E°10', Latitude: 30N°48'. Sun 2 S 29°48'; Moon 06 S 08°19'; Mars 0 S 17°12'; Mercury 2 S 13°04'; Jupiter 11 S 29°48'; Venus 4 S 11°12'; Saturn 2 S 29°05'.

In this case both the Sun and Jupiter have attained identical number of degrees and minutes. The Sun is placed in sign Gemini, a neutral sign but in a bad i.e. 8th house while Jupiter is placed in its own sign

Pisces and in eternal house. Thus Jupiter is stronger than the Sun, hence qualifies to become Atmakarka.

Planet	Longitude	Jamini/Char Karaka
Jupiter	11 S 29^0 48'	Atmakaraka
Sun	2 S 29^0 48'	Amatyakaraka
Saturn (R)	2 S 29^0 05'	Bharatrukaraka
Mars	0 S 17^0 12'	Matrukaraka
Mercury	2 S 13^0 04'	Putrakaraka
Venus	4 S 11^0 12'	Gnatikaraka
Moon	6 S 08^0 19'	Darakaraka

2. Amatyakaraka: Planet of highest degree in next to Atmakarka in the Natal chart represents Amatyakaraka, i.e. the 2nd Karaka. This indicates ups and downs in his/her career and represent the 10th house factors. When the Amatya or Mantrikaraka is powerful and well combined and aspect, he/she will become a great Minister or Councillor. But when it is ill combined and badly aspect and debilitated he/she becomes an evil Councillor or an adviser who brings disgrace on himself/herself and also on those to whom he/she offers his counsel.

3. Bhratrukaraka: Planet of highest degree in any Sign next to Amatyakaraka in the Natal chart becomes Bhratrukaraka or gets lordship over brothers, i.e. the 3rd Karaka. This acts on siblings and represents 3rd house affairs like the 3rd house lord or Mars. If the Bhratrukaraka is debilitated, joins evil planets and has malicious aspects then there will be ruination to brothers. He/She will have no brothers or, if he/she gets them, they will die or become wretched, poor and disgraceful. If, on the other hand, the Bhratrukaraka is exalted, well combined and well aspect there will be plenty of brothers and prosperity and success will attend on them.

4. Matrukaraka: Planet of highest degree in any sign next to Bhratrukaraka in the Natal chart becomes lord of the mother or Matrukaraka (MP), i.e. the 4th Karaka. This acts like the 4th house lord or Moon, indicating the mother or other 4th house affairs. It helps in prediction of the children by the Saptamsa chart.

5. Putrakaraka: Planet of highest degree in any Sign next to Matrukaraka in the Natal chart becomes the lord of the children or Putrakaraka, i.e. the 5th Karaka. This acts like the 5th house lord or Jupiter, indicating children or the 5th house affairs. It also helps in prediction of the children by the Saptamsa chart.

6. Gnatikaraka: Planet of highest degree in any Sign next to Putrakaraka in the Natal chart represents Gnatikaraka, i.e. the 6th Karaka. This acts like the 6th house lord, indicating diseases or other 6th house affairs.

7. Darakaraka: Planet of highest degree in any sign next to Gnatikaraka in the Natal chart represents Darakaraka, i.e. the 7th Karaka. This acts like the 7th house lord or Venus, indicating the relationship. Darakaraka can be investigated to help understand events concerning the marriage or relationship partner like Venus or Jupiter, and the Lord of the Seventh House in Para Sara.

Parasara Graha Fixed Karkatva: As per Para Sara, each Graha has the constant Karkatva. The stronger among Surya and Sukra indicates the father, while the stronger among Chandra and Mangal indicates the mother. Mangal denotes sister, brother-in-law, younger brother and mother. Buddha rules maternal relative, while Guru indicates paternal grand father. Husband and sons are, respectively, are denoted by Sukra and Sani. These constant significances are derivable from the Bhava, counted from the said constant Karaka. Such as the 9th from Surya denotes father, the 4th from Chandra mother, the 3rd from Mangal brothers, the 6th from Buddha maternal uncle, the 5th from Guru Sons, the 7th from Sukra wife and the 8th from Sani death.

Parasara Yoga Karaka: Graha becomes Yoga Karaka, if they are in mutual angles identical with own Rashi, exaltation Rashi, or friendly Rashi. In Karma Bhava, a Graha will be significantly so. Graha simply (not being in friendly, own, or exaltation Rashi) in Lagna, Bandhu and Yuvati Bhava do not become such Yoga Karaka. Even, if they be placed in other Bhava, but with such dignities, as mentioned, shall become Yoga Karakas. With such Graha even a person of mean birth will become a king and be affluent. One born of royal scion then will surely become a king. Thus the effects are declared, considering the number of such Graha and the order the native belongs to.

Jaimini Sign Aspects: The second big difference in Jaimini astrology and Parasara Astrology is the way the "aspects" are determined. In Parasara astrology, every planet casts an aspect on the seventh house from it, and three planets, such as Jupiter, Mars, and Saturn cast special aspects too. However, in Jaimini astrology, it is the signs that cast aspects. It is a fairly simple system as given below:

Each Movable sign aspects all fixed signs except for the sign adjacent to it. Each Fixed sign aspects all Movable signs except for the sign adjacent to it. Each Dual Sign aspects all other Dual signs except for the sign adjacent to it. In the language of Western astrology, we would describe it in the following way: The Movable (Cardinal) and Fixed signs each cast sextile, an in conjunct, and a trine. The Dual (Mutable) signs all cast squares and oppositions. For easy reference, the following list gives the aspect cast by each sign of the zodiac:

Aries: It casts aspect on Leo, Scorpio, and Aquarius. **Taurus:** It casts aspect on Cancer, Libra, and Capricorn. **Gemini:** It casts aspect on Virgo, Sagittarius, and Pisces. **Cancer:** It casts aspect on Scorpio, Aquarius, and Taurus. **Leo:** It casts aspect on Libra, Capricorn, and Aries. **Virgo:** It casts aspect on Sagittarius, Pisces, and Gemini. **Libra:** It casts aspect on Aquarius, Taurus, and Leo. **Scorpio:** It casts aspect on Capricorn, Aries, and Cancer. **Sagittarius:** casts aspect on Pisces, Gemini, and Virgo. **Capricorn:** It casts aspect on Taurus, Leo, and Scorpio. **Aquarius:** It casts aspect on Aries, Cancer, and Libra. **Pisces:** It casts aspect on Gemini, Virgo, and Sagittarius.

Diagram of Rashi Drishti: Lord Brahma narrated the diagram of Drishti, so that Drishti are easily understood by a mere sight of the diagram. Draw a square, or a circle marking the 8 directions (4 corners and 4 quarters thereof). Mark the zodiacal Rashi, such as Mesh and Vrishabh in East, Mithuna in the North-East, Karka and Simha in the North, Kanya in the North-West, Tula and Vrischika in the West, Dhanu in the South-West, Makara and Kumbh in the South and Meena in the South-East.

Predictions by Aspect Effects: The followings are the effects of the aspects made up in the Rashi:

1. Cardinal Sign Aspect: If the aspect is made up in Cardinal Sign, he/she tends to be critical, impersonal, and in many ways dominate. He/She has a lot of energies and he is constantly looking for ways to spend it. The cardinal cross produces a lot of impatience in his nature and he often makes hasty decisions that he may regret later.

2. Fixed Sign Aspect: If the aspect is made up in Fixed Sign, his/her life can be hard for him/her to deal with abstractions, values and ideals. He/She may need to be more flexible in dealing with people; less determined and rigid. This aspect implies endurance and resistance in the face of adversity, but it can produce inflexibility, which inhibits progress.

3. Mutable Sign Aspect: If the aspect is made up in Mutable Sign, he/she tends to get emerged in criticism, ideals, and values. He/She should be aware of the importance for him/her to show his/her individuality and avoid merely following the leader and trend of the times. The easiest course for him/her may be to try to avoid or ignore difficulties, but such action does nothing to resolve them.

Jaimini Chara Dasha: The third way that Jaimini astrology differs from Parasara astrology is in the Dasa periods it uses. In Parasara astrology the most used Dasa system is the Vimshotari Dasa system. In Jaimini astrology, however, it is the Chara Dasa system that is most popular. In the Chara Dasa, rather than planets determining the Dasa sequence, Signs are the defining factor. Therefore, Chara Dasa is a Sign-based Dasa system. For example, in Chara Dasa we say that the individual is running Sagittarius Period. If there are planets located in the any of the Dual signs, he will be activated during this period. Furthermore, we can rotate the Lagna of the chart to Sagittarius and examine it from there. That is a good way to understand the character of the Chara Dasa period.

The Chara Dasa system can be used in isolation or in combination with the Vimshottary Dasa system to get excellent predictive results.

Example: If a Jyotish see one factor in the chart, it has a little significance. But when he sees that the factor is repeated three or four times, then it becomes highly significant. Therefore, it is of benefit to the astrologer to look at the same chart from as many angles as possible. Look at it from the birth Ascendant, look at it from the Moon Ascendant (Moon-Sign), look at it from the Sun Ascendant (Sun-Sign) and see that which patterns repeat them. You can be sure that the repeated patterns are powerful and will show great significance in the life of the individual. That is why it is so useful to combine Parasara techniques with Jaimini techniques. Look at the chart through the eyes of sage Parasara and then look at the same chart through the eyes of sage Jaimini. Find out whether it tells the same story. Then it can be sure that this factor is correct and 100% applicable to the native. If it tells a somewhat different story, then combine both factors. If it tells a completely different story, look deeper, we will find something we are missing. This is this secret to powerful, accurate, astonishingly good astrology.

Prediction of Longevity by Jaimini Astrology:
The most powerful and unique feature of Jaimini astrology is its ability to determine the lifespan of an individual.

The enable us to predict the span of life allotted to an individual such short, middle, or long. Mainly, the Lagna lord and the 8th lord are

taken into consideration and checked for their positions in the natal chart. However, the longevity is predicted by the following three other methods:
1) Find out Lord of Ascendant and Lord of 8th House.
2) Find out Ascendant and Moon.
3) Find out Ascendant and Hora Ascendant

If the result found by all the three steps is same in 2 or 3 outcomes, then it is correct. If all three outcomes are different, differ to the Ascendant + Hora.

Term of Life	if signs are:	or signs are:	Lifespan
Alpayu (short life)	C, D	S, S	< 32
Madhyayu (middle life)	C, S	D, D	< 64
Purnayu (long life)	C, C	D, S	> 64

Where C = Chara (Moveable, Cardinal) Signs such as Aries, Cancer, Libra, Capricorn.
S = Sthira (Fixed) Signs such as Taurus, Leo, Scorpio, Aquarius.
D = Dwiswabhava (Common, Mutable) Signs such as Gemini, Virgo, Sagittarius, Pisces.

Exception: The Ascendant and Moon method is used when the Moon is in the Lagna or 7th. The best two among the above three are taken into consideration to predict about the life span, i.e. longevity of a person as per thumb-rule as below:

Long Life: Both of them (Lagna lord and 8th lord) should be in moving signs or one should be in a fixed sign and other should be in a dual sign.

Medium Life: Both of them (Lagna lord and 8th lord) should be in a dual sign or one should be in a fixed and another should be in a moving sign.

Short Life: Both of them (Lagna lord and 8th lord) should be in a fixed sign or one should be in a dual sign and another should be in a moving sign.

Nature of Death: The nature of the death will be in accordance with nature of the lord of the 22nd Decanate or the sign containing it. In other opinion, if the 8th house is not occupied or aspect by any planet, the 22nd Decanate from the rising Decanate will indicate the cause of death of the native. The cause of the death for those born in different Decanate is as follow: **Aries:** 1st Decanate: It will be by serpent, poison, water, bilious disorder. 2nd Decanate: It will be by water, insects, ice. 3rd Decanate: It will be by drowning in water, ocean, and tank. **Taurus:** 1st Decanate: It will be by camel, horse, and ass. 2nd Decanate: It will be by bilious disorder, fire, rheumatic complaints, robbery. 3rd Decanate: It will be by falls from vehicle, ocean, and battle. **Gemini:** 1st Decanate: It will be by bronchial-disorder. 2nd Decanate: It will be by poison, quadrupeds, fevers. 3rd Decanate: It will be by rheumatism, four-legged animal, falls from a high place or in forest. **Cancer:** 1st Decanate: It will be by inauspicious dream, trouble from crocodile or watery-animal or enemy. 2nd Decanate: It will be by lightening, fire, poison, assault. 3rd Decanate: It will be by venereal diseases overheat. **Leo:** 1st Decanate: It will be by drowning, poison, diseases on feet. 2nd Decanate: It will be by oedema, in forest. 3rd Decanate: It will be by fall, assault, injury from instrument or weapons. **Virgo:** 1st Decanate: It will be by disease on header rheutism. 2nd Decanate: Poisonous creatures or animals, death in neglected place such as forest. 3rd Decanate: It will be by lightening, quadrupeds, elixir prepared by women. **Libra:** 1st Decanate: It will be by women, quadrupeds, fall from a high places. 2nd Decanate: It will be by stomach disorders. 3rd Decanate: It will be by serpent, ferocious creatures and underwater fish. **Scorpio:** 1st Decanate: It will be by implements, poison, drink prepare by women. 2nd Decanate: It will be by diseases in loins. 3rd Decanate: It will be by disorders in bones or legs stones. **Sagittarius:** 1st Decanate: It will be by venereal diseases, rheutism, and blood-pressure. 2nd Decanate: It will be by poison, rheumatism. 3rd Decanate: It will be by drowning in water, stomach complaints. **Capricorn:** 1st Decanate: It will be by serpent, underwater creatures, and ferocious animals. 2nd Decanate: It will be by fire, implements, and sprits (evil), fever. 3rd Decanate: It will be by all diseases caused by women. **Aquarius:** 1st Decanate: It will be by forest, water, bilious disorder. 2nd Decanate: It will be by venereal diseases, mental diseases, and women. 3rd Decanate: It will be by quadrupeds, diseases in face. **Pisces:** 1st Decanate: It will be by venereal diseases, stomach complaints, and women. 2nd Decanate: It will be by water, conveyance travelling in attar. 3rd Decanate: It will be by unsightly diseases due to poison, germs or bacteria.

Prediction of Profession by Jaimini Astrology:

Karakamsa Chart: According to Jaimini Astrology, the planet with highest longitude becomes the Atmakaraka. The sign in which the Atmakaraka is situated in the Navamsa becomes the Karakamsha. A planet in Karakamsha becomes the key planet. Here are some of the astrological combinations of planets in Karakamsa, which shall give a glimpse of the profession of the Native: **Sun:** Sun is responsible to give more power, ability to rule and command and, will provide a

Government employment, or a high profile business executive with status and dignity. (a) If Sun occupies the Karakamsha lagna, the native will be fond of public work, and will take interest in political activities. (b) If Sun and Venus aspect Karakamsha, the native will be employed by Royal people or Top Class politicians. (c) Sun occupying 10th house in Karakamsha chart and aspect by Jupiter, the native will be a successful Trader of Cattle. (d) If Sun occupies 5th house in Karakamsha chart the native will become a Philosopher or a Musician. (e) If Sun is in conjunction with Rahu and occupies Karakamsha, and Sun being located in benefic Vargas, the native will become a Doctor treating poisonous afflictions. **Moon:** Moon is responsible for fluctuations in occupation or business, but encourages the native for business in aqua products, Marine produces, Liquid chemicals, Restaurant; Dairy products etc. (a) if full moon and Venus joins Atmakaraka in Navmansha money will be earned by education. (b) If Moon is conjunct with Jupiter in Karakamsha or in 5th from Karakamsha, the native becomes an author and earns his livelihood by authoring books. (c) If Moon is alone in Karakamsha or in 5th from Karakamsha, the native become a musician. (d) Moon in Karakamsha aspect by Venus makes the native an Alchemist. (e) Instead of Venus if aspect by mercury he will be a Doctor. **Mars:** Mars denotes employment in the field where courage is required and a need to be skilful, brave and also makes the native Dentists, Surgeons, and Engineers. (a) If Mars occupies the Karkamsha, the native earns his livelihood by metallurgy. (b) If Mars occupies the Karakamsha with benefic aspects, the native will be a Judge. (c) If mars occupies 5th place from Karakamsha under the aspect of Saturn, the native will become an Engineer/Mechanic. **Mercury**: Mercury is solely responsible for the field of literature, arts and occupations requiring study, skill and intelligence. (a) If mercury occupies the karakamsha the native becomes Merchant and well versed in social and political matters. (b) If mercury occupies 5th place from Karakamsha, the native will become a Vedic Scholar. **Jupiter:** Usually Jupiter gives the profession of physicians, lawyers, ministers, bankers, authors, journalists, philosophers etc. (a) When Jupiter occupies the karakamsha the native becomes a philosopher, religious head and will have good knowledge of Vedanta. (b) If Jupiter occupies 5th place from karakamsha, he will be a scholar in Vedas and Upnishads. **Venus:** Venus is responsible to gain financial assistance through marital sources, jewellery pleasure and luxury items. (a) If Venus joins Atmakaraka in Navmansha, the native will become a big politician. (b) If Venus joins karakamsha or the 5th from karakamsha, the native will become a great poet. **Saturn:** Saturn is responsible to the professions, such as builders, miners, land surveyors, agriculturists, labour contractors, municipal officials and Government service. (a) If Saturn occupies the Karakamsha the native will adopt the profession of his forefathers. (b) If Saturn alone occupies 4th or 5th place from karakamsha, the native will be skilful in archery and will adopt the same skill for livelihood. (c) Saturn occupying karakamsha under the aspect of Mars the native will become a builder. **Rahu:** Rahu will give results like Saturn but some of the combination indicates different professions also. (a) If Rahu joins atmakaraka in navmansha, the native earns his bread as a thief or decoit. (b) Rahu having close association with Mandi and occupying 4th place from Karakamsha the native will be specialised Doctor to treat snake bites and poisonous problems. (c) Rahu in 5th from Karakamsha makes the native to be a good Engineer/Mechanic. **Ketu:** Ketu will give results partially like Jupiter and Mars. (a) If Ketu occupies 4th place from karakamsha the native will become a clock maker (Repairer). (b) Ketu occupying karakamsha makes the native a mathematician. (c) If occupied in 5th house from karakamsha under benefic aspects the native will be mathematician. (d) Ketu occupying karakamsha under malefic aspects makes the native a thief. (e) Ketu occupying 5th place from karakamsha under the aspect of Jupiter makes the native an astrologer.

Planet and House Strengths by Jaimini Astrology:

The Planet attains the strength due to its position from Atmakaraka (**AK**) and due to "Amsa" **like** AK, AmK, BK, MK, PK, GK and DK position gained by Jaimini Astrology and it is given in tabulated form. This is the simplified, a straightforward and balanced approach to determining Graha and Bhava Balas, using guidelines given in the Jaimini sutram. Rules for Determining Strengths of Planets and Houses Graha (planet) Balas are as below.

Planets are strong based on the following three criteria:

1. Mulatrikona Bala: The strengths based on Exaltation Sign, Mulatrikona Sign, Own Sign, Friendly Sign, Neutral Sign, Enemy Sign and Debilitation Sign position.

2. Amsa Bala: The strengths based on the planet's position as Atmakaraka, Amatyakaraka, etc., or in other words highest degree position of the planet to downward or to lowest degree Bala.

3. Kendra Bala: The strength based on the Kendra (angular) relationship of the planet with the Atmakaraka of Jaimini Astrology.

Houses are strong based on the following three criteria:

1. Chara Bala: The natural strength of signs, such as, Dual signs is stronger than fixed signs; and fixed signs are stronger than cardinal signs.

2. Sthira Bala: The strength of signs due to occupation of planets. The more planets occupy a sign; the Sign/House receives more strength.

3. Drishti Bala: The strength due to aspect of the sign lord, Jupiter and Mercury.

Planet Position	Unit of Strength	Sun	Moo	Mar	Mer	Jup	Ven	Sat	Rah	Ket
Exaltation	60									
Mulatrikona	45									
Own Sign	30									
Friend Sign	22.50									
Neutral Sign	15									
Enemy Sign	07.50									
Debilitation	03.75									
1-4-7-10toAK	60									
2-5-8-11toAK	30									
3-6-9-2 to AK	15									
AK	60									
AmK	45									
BK	30									
MK	22.50									
PK	15									
GK	07.50									
DK	03.75									
Total	472.50									

17

Previous Birth Curse

Births on Amavasya (last day of the Krishna Paksha), on Chaturdasi (14th Tithi) in Krishna Paksha (dark half of the month), in Bhadra Karan, in the Nakshatra of the brother, in the Nakshatra of father or mother, at the time of entry of Surya in a new Sign, the time of Pata, at the time of solar and lunar eclipses, at the time of Vyati Pata, in Gandanta of all the three kinds, in Yamaghantak, Tithikshaya, in Dagdha Yoga, is considered inauspicious. The birth of a son after three daughters or the birth of a daughter after three sons and the birth of a freak are also inauspicious.

1) Birth on Amavasya: The person, born on Amavasya is always poverty stricken.

2) Birth on Krishna Chaturdasi: The person, born Krishna Chaturdasi is always curse. Divide the span of Chaturdasi in 6 parts. The birth in the first part is auspicious. Second part causes destruction, or death of father. Third part causes death of the mother. Fourth part takes away the maternal uncle. Fifth part destroys the entire family. Sixth part causes loss of wealth, or destruction (death) of the native.

3) Birth in Bhadra Karan: The birth in Bhadra Karan, Tithi Kshaya, Vyatipata, Paridha, Vajra and Yamaghants are always inauspicious birth.

4) Birth in Brother or Parents' Nakshatra: If the birth takes place in the Nakshatra of the brother or the parents, death of the brother or the father or mother takes place without doubt or they have to undergo death-like suffering.

5) Birth on Sankranti: The child born at the Sankranti (entry of Surya in a new Sign), is poor and unhappy, but he becomes well-to-do and happy by remedial measures undertaken.

6) Birth in Eclipses: A child born at the time of solar, or lunar eclipse, suffers from ailments, distress and poverty and faces danger of death.

7) Birth in Gandanta: The birth in Gandanta is inauspicious for the child. Gandanta is of three kinds, namely of Tithi, Nakshatra and Lagna. Birth, travelling and performance of auspicious functions, like marriage during Gandanta are likely to cause death of the person concerned. The last 2 Ghatikas of Purna Tithi (5th, 10th, 15th) and the first 2 Ghatikas of Nanda Tithi (1st, 6th, 11th) are known, as Tithi Gandanta. Similarly the last two Ghatikas of Revati and first two Ghatikas of Ashvini, the last two Ghatikas of Ashlesha and first two Ghatikas of Magha and the last two Ghatikas of Jyeshtha and first two Ghatikas of Moola are known, as Nakshatra Gandanta. The last half Ghatika of Meena and first half Ghatika of Mesh, the last half Ghatika of Karka and first half Ghatika of Simha, the last half Ghatika of Vrischika and first half Ghatika of Dhanu are known, as Lagna Gandanta. Amongst these Gandanta the last 6 Ghatikas of Jyeshtha and first 8 Ghatikas of Moola are known, as Abhukta Moola. The remedial measures will ensure long life, good health and prosperity for the child.

8) Birth in Abhukta Moola: The ruling deity of Jyeshtha is Indra and the ruling deity of Moola is Rakshasa. As both the deities are inimical to each other, this Gandanta is considered, as the most evil. A child born during the period of Abhukta Moola should either be abandoned, or the father should not see the face of the child for 8 years.

9) Birth in Jyeshtha Gandanta: A girl, born in Jyeshtha Nakshatra destroys or causes the death of the elder brother of her husband; and a girl, born in fourth quarter of Visakha Nakshatra, destroys her husband's younger brother. A child, born in the 2nd, 3rd, or 4th quarter of Ashlesha Nakshatra destroys his/her mother-in-law and child, born in 1st, 2nd, or 3rd quarter of Moola Nakshatra, becomes the destroyer of his/her father-in-law. Therefore suitable measures, as may be possible within one's means, should be taken at the time of the marriage of such boys and girls. There will be no evil effect, if the husband has no elder brothers.

10) Birth after Three Daughter/Sons: The birth of a daughter after the birth of three sons or the birth of a son after the birth of three daughters is ominous for both the maternal and paternal families of such children. By the performance of the above remedial rites the evil effects are wiped out and the child and his parents etc. enjoy happiness.

Remedial Measures: The remedial measures rites should be performed for relief from the evil effects of birth in and on following "Inauspicious Birth" by Previous Birth Curse as per the advice of a proficient Jyotish so that he can enjoy all the happiness.

www.ingramcontent.com/pod-product-compliance
Lightning Source LLC
Chambersburg PA
CBHW080615190526
45169CB00009B/3187